Optimal Control and Stochastic Estimation: Theory and Applications Volume 2

MICHAEL J. GRIMBLE
and
MICHAEL A. JOHNSON
Industrial Control Unit
University of Strathclyde
Glasgow, Scotland, UK

A Wiley–Interscience Publication

JOHN WILEY & SONS
Chichester · New York · Brisbane · Toronto · Singapore

Library of Congress Cataloging-in-Publication Data:

Grimble, Michael J.
Optimal control and stochastic estimation.

'A Wiley–Interscience publication.'
Includes bibliographies.
1. Control theory, 2. Mathematical optimization.
3. Stochastic processes. 4. Estimation theory.
I. Johnson, Michael A. II. Title.
QA402.3.G7 1987 629.8'312 86-22374
ISBN 0 471 90593 3 (v. 1)
ISBN 0 471 91265 4 (v. 2)

British Library Cataloguing in Publication Data:

Grimble, Michael J.
Optimal control and stochastic estimation: theory and applications.
1. Control theory 2. Mathematical optimization
I. Title II. Johnson, Michael A.
629.8'312 QA402.3
ISBN 0 471 90593 3 V.1
ISBN 0 471 91265 4 V.2

Typeset by Mathematical Composition Setters Ltd, Salisbury
Printed in Great Britain by St Edmundsbury Press Ltd, Bury St. Edmunds

Dedication

To my wife Wendy, and my children
Andrew and Claire

Michael J. Grimble

To my children Eleanor and Laurence,
and my mother and father

Michael A. Johnson

Contents

Foreword

The development and application of linear optimal control theory are the central themes of this book. As is well known, the LQP/LQG theory has provided a paradigm for control system design whose potential applications stretch far and wide, and the authors include new and unusual applications of the theory to challenging industrial design problems. Also it is clear that many of the theoretical developments they describe are strongly motivated by their experiences with serious applications, and by the demands of 'real' problems.

I first met the authors in the early 'seventies when they joined my group to undertake postgraduate research into the advanced problems of automation in the metal industries. Our aim was the application of rigorous intellectual approaches to the applied research problems of modelling and control, and the use whenever sensible of advanced control-theoretical ideas and techniques. In those early days as on many subsequent occasions, we were both surprised and delighted by the theoretical and conceptual challenges which emerged naturally during the pursuit of solutions to the 'real' and apparently banal problems of industry which at first sight appeared little connected with 'pure' research.

My work with Mike Grimble and the late and very gifted Martin Fuller was concerned with obtaining solutions to the roll gap equations for tinplate rolling. Here, 'limiting' reductions are in prospect, the physical phenomena encountered are somewhat surprising and the process is not straightforward to control. The solution procedures we developed utilized function space concepts and iterative techniques developed originally for the solution of two-point boundary value problems arising in control optimizations. We succeeded in computing what are, as far as I know, the only published solutions to these equations and obtained considerable insight into tinplate phenomena.

Optimization also played a key role in the work I undertook with Mike Johnson. We set ourselves the very considerable task of exploiting the properties of optimal controllers to isolate 'good' multivariable control 'structures' for a tandem mill. We successfully produced new-bound formulae for linear optimal control systems and plausibly managed to isolate the main structures for the control of gauge and tension in a tandem mill.

Since those early days these researches have matured and resulted in the design of a whole range of advanced mill computer control schemes which have proved both incisive and effective in industrial terms. This much I perhaps expected. What I did not, at the time, anticipate was the Odyssey on which my former associates had been inadvertently launched as they pursued and developed their initial interests in optimal control theory and its application. Apart from this book, these now long-standing interests have led them to establish the Industrial Control Unit at the University of Strathclyde, Glasgow, Scotland.

The IEEE Special Issue on *LQG/LQP and Kalman–Bucy Filtering*, published in the 'seventies, represented a milestone in technical publication in this area, containing over three hundred pages of new technical material and of the order of a thousand key references. It serves to reflect how the subject had even then attracted wide interest and acquired a rich theoretical structure. This initial interest and activity seems to have continued unabated over the last decade as widely differing concepts and results have emerged. In seeking to make this material fall into place and to establish a fresh conceptual perspective in the whole area, the authors have undertaken a task which is both daunting and worthwhile and perhaps overdue! As is widely accepted, publications in this area are generally 'heavy' on theory and 'light' on significant applications outside the aerospace industry. By the inclusion of non-trivial application studies in refreshingly new areas the authors have further enhanced the book. I wish them well.

GREYHAM F. BRYANT

Professor of Control Engineering
Imperial College, London

Authors' Preface

Our interest in optimal control and stochastic estimation theory and its applications developed from our experience on the Industrial Automation Group at Imperial College, London and our industrial experience with GEC Electrical Projects Ltd., Rugby, and the London Research Station of the British Gas Corporation.

The Industrial Automation Group at Imperial College provided a stimulating research environment which we have endeavoured to emulate at the Industrial Control Unit of the University of Strathclyde. Professor Bryant's group at Imperial College was mainly concerned with Steel Industry Applications. Optimal control and optimization theory found their way into many areas of modelling and control not usually found in textbook theory. The first motivation for the present text was therefore to record some of the many useful areas of optimal control which are of value in applications.

The authors greatly admired the 1972 text of Kwakernaak and Sivan (*Linear Optimal Control Systems*) in producing a very accessible but comprehensive collection of work on linear optimal control and filtering systems. However, in many of the areas considered by Kwakernaak and Sivan, advances have now been made which are not covered in any follow-on text. This was the second motivation for this book and it is hoped that this text fills this gap and extends the work of these authors, so that the combined volumes present a modern and comprehensive overview of the field.

Chapter 1 includes a review of basic systems theoretic results needed in the following chapters. It also introduces the Hilbert space-gradient approach to the solution of optimal control problems which is an underlying theme of much of the analysis.

Very few recent texts include the Wiener–Hopf optimization theory which provides a useful transfer-function approach to solving linear optimal control and estimation problems. This topic is introduced in the Chapter 2 and the simplicity of the basic ideas are illustrated. However, recent theoretical analyses in the *s*- or *z*-domains have been based upon a polynomial systems approach. Hence, in Chapter 11, these two different methods are brought together and shown to be parallel approaches to the same problem. Indeed the insight gained by looking at the two techniques enables the connection between Diophantine equations and partial fraction expansions, for example, to be explained. Chapter 3 introduces our approach to the multivariable root-loci control problem. That is, an analysis is presented for describing the locations of and properties of the closed-loop poles of an optimal system as the control weighting becomes vanishingly small. The approach taken in this chapter has not been described in previous text books.

The exponents of non-optimal multivariable design methods often point to a weakness of optimal control as being the problem of selecting the cost-weightings matrices. There are, of course, some problems such as ship-steering control systems design which give rise to natural cost functions to be minimized. However, if optimal control is only to be used as a design procedure, where optimality is not important, then other methods must be chosen for selecting these matrices. In Chapter 4, a collection of different techniques is described which, for example, enable the eigenstructure in the optimal system to be predetermined.

Some of the Q and R optimal cost-weighting selection procedures are based upon the asymptotic properties of optimal systems. Chapter 5 includes a discussion of the maximal accuracy which can be achieved with optimal regulators as the control weighting tends to zero. This topic gives some insight into possible control difficulties which may be present in a system. For example, the presence of non-minimum phase zeros in a plant will mean the error component of the cost function cannot be driven to zero, however large the control signals become. The maximal accuracy results are used later in the chapter to enable the best set of input actuators to be selected. In chemical plants and large systems there is often a difficulty of making a choice between many different possible combinations of input actuators or output measurements. The maximal accuracy results throw some light on the best set of input actuators to employ.

The subject of system structure assessment is considered further in Chapter 6. An optimal control problem can be solved for each possible system structure and the relative merits of each design can be judged based upon the minimum costs which can be achieved and the number of inputs and outputs employed. In practice, because of the large number of possible combinations of inputs and outputs which may be tried, it would be computationally very expensive to solve a large number of optimal control problems and hence the subject dealt with in this chapter is that of establishing easily computable bounds on the optimal cost. Given such bounds, it is then possible to make a rapid assessment of different system structures based upon the estimated cost values. The middle section of this chapter concentrates on the subject of fixed structure optimal control problems. There has been a lot of interest over the last few years from the academic community in the use of constant output feedback optimal control laws and the many results in this area are surveyed and the properties discussed.

The Kalman filter has been one of the most successful developments of the modern control theory. There are numerous applications in many industries including the aerospace and marine industries. The Kalman filter and the time domain aspects of filtering theory are considered in Chapter 8.

The first part of the chapter concentrates on a solution of the continuous-time filtering problem. Most modern books on filtering theory use the Wiener process rather than the older white-noise models. From a practical point of view, the white-noise system representation is to be preferred since it simplifies the notation and analysis. However, since many books include chapters of this type the former approach is taken in Chapter 8. The latter parts of the chapter discuss the discrete-time Kalman filtering problem. A great strength of the Kalman filter is that it is so appropriate for discrete-time implementation. The numerical problems in implementing the algorithm are considered and prediction and smoothing problems are discussed.

The frequency-domain properties of Kalman filters are considered in Chapter 9. These properties provide insight into the behaviour of the filter and also show the relationship of the filter to more classical frequency-domain-based techniques. The chapter also includes an introduction to finite-time filtering, prediction and smoothing problems. Finite impulse response filters find wide application in signal processing but they are not so common in the control field. The approach taken in this chapter is novel and has not been published before.

The main control system application for the Kalman filter is as part of a state estimate feedback control loop as described in Chapter 10. The separation principle of stochastic optimal control theory is introduced in this chapter using the Hilbert space techniques once again. As in the filtering chapters, both continuous-time and discrete-time problems are considered. The digital version of the LQG controller is used extensively in applications and is probably the only controller for multivariable systems which adequately caters for the different characteristics of the noise sources and disturbances. It is interesting that the multivariable frequency-domain design methods developed by MacFarlane, Rosenbrock and Mayne seem particularly suited to deterministic systems design but the need to include so many design objectives in stochastic applications leads naturally to the use of an LQG design framework.

It would have been difficult to imagine a few years ago that any new methods of representing systems would have a very large impact on control design techniques and analysis. However, the polynomial systems approach pioneered by Kučera has resulted in a much greater insight into the frequency-domain properties of optimal systems and has also influenced applications areas such as multivariable self-tuning control systems design. There are now basically three approaches to the s- or z-domain solution of optimal control problems. The first approach considered in Chapter 2 reflects the Wiener transfer-function based solution of optimal control and filtering problems. The second is the fractional system description used by workers such as Desoer and Vidyasagar which is not considered in this text. The third method is the polynomial systems approach which provides much insight and also has the tremendous advantage of simplifying the computational algorithms which are needed. Hence there are several objectives for Chapter 11:

1. To present what is becoming the standard approach to estimation z-domain optimal control and filtering problem solutions: the polynomial systems approach.
2. To show the relationship between the Wiener and polynomial systems methods.

The idea of a stabilizing controller parameterization is also introduced in this chapter. This avoids difficulties with unstable pole-zero cancellations which can occur in the older Wiener–Hopf approaches.

The area of adaptive and self-tuning control is discussed in Chapter 12. Although the self-tuning controller was first developed in the early 1970's, it is only recently that it has now found its way into real applications in any substantial numbers. The number of adaptive controllers on the market is increasing rapidly and many different types of devices are being produced for different areas of application and duties. The majority of adaptive controllers are based on either optimal control concepts or on PID controller structures. Chapter 12 on self-tuning control algorithms therefore illustrates an excellent applications area for optimal control. The chapter also details many of the different control laws which may be employed for this type of design. Both implicit and explicit self-tuning schemes are discussed and the practical problems of implementation are considered.

It is hoped that the applications Chapters 7 and 13 will illustrate the power of the optimal control techniques in several industrial applications. The design of deterministic optimal multivariable systems is considered in Chapter 7 and stochastic systems are considered in Chapter 13. Most of the material in Chapter 7 is concerned with the design of a shape-control system for a cold-rolling mill. The control of shape or flatness in mills involves the measurement of a large number of variables typically over 30 and the control of 8 to 10 actuators. This chapter is based on experience gained on the design on the shape-control system for the Sendzimir mills of the British Steel Corporation at Sheffield. In fact, the actual control law which was developed by BSC and Unit engineers and which was implemented, was not based on optimal control theory since it was necessary to have particularly simple control structure. However, the details of the modelling and control problem presented in

Chapter 7 are based on the study. It is often the case that optimal and non-optimal designs are completed for particular projects and the design which is most appropriate is chosen for implementation.

By way of contrast, the optimal control approach has particular advantages in stochastic systems as illustrated in the ship-positioning design procedure described in Chapter 13. In this case, optimal and non-optimal designs were completed for the control problem but the optimal control approach was selected for implementation by the GEC Elcctrical Projects Co. Ltd. The stochastic optimal control solution to the ship-positioning control problem was first introduced by Jens Balchens' Group at Trondheim. It is now the accepted standard solution to the problem and the Kalman filter design which was used has demonstrated its flexibility by being used in various roles without requiring major changes to the scheme. Chapter 13 also includes a description of the application of the Kalman filter to soaking-pit temperature estimation problems. There are many industrial problems which fall into the category of estimating temperatures in an environment where direct measurement is difficult or impossible. The chapter illustrates one of the key requirements of a Kalman filter, that of producing a good model on which to base the filter. The art of good Kalman filter design lies in the expertise of the systems modeller.

It is hoped that this text will be of value to engineers in industry by providing an indication of the power and scope of the optimal control and estimation approach. It should also be clear that there are many areas which are suitable for further research. Several of the topics discussed are open research questions which should be of value to postgraduate research engineers. The work of Kwakernaak and Sivan has been widely adopted for undergraduate teaching and we hope the book will complement this volume and will be of value to students.

There are many individuals we should thank for contributions to the book. In particular we are indebted to Professor Mark Davis (Imperial College), Professor Martins de Carvalho (Instituto de Engeharia de Sistemas e Computadores, Portugal), Professor Adrian Roberts (Queens University, Belfast) and Dr. William Leithead (Industrial Control Unit) for their detailed comments on Chapters 8, 10 and 11. Many of the research engineers of the Industrial Control Unit have provided simulation results and we should like to thank our former colleagues: Dr. Patrick Fung (Spar Aero-Space, Canada), Dr. Tom Moir (Paisley College of Technology, Scotland), Dr. Munie Gunawardene (Robert Gordons Institute of Technology, Aberdeen, Scotland), Dr. John Ringwood (NIHE Dublin), Dr. John Fotakis (Public Petroleum Corporation, Greece), Dr. Ken Dutton (formerly British Steel Corporation, Sheffield) and Dr. Malcolm Daniels (University of Dayton, Illinois, USA).

Similarly, there are many individuals whose lectures and talks have influenced our thinking on the subject of Control Engineering and we should particularly like to note the work of Professors Åström, MacFarlane, Rosenbrock, and Professor Greyham Bryant. Indeed, it was our time spent at Imperial College, London which awakened and stimulated our interest in this subject. There are many members of the present Industrial Control Unit who are introducing new ideas and who will no doubt be mentioned in future publications. We highly value this interaction with the members of our group.

Most of the diagrams were drawn by Wendy Grimble and the manuscript was typed by Ann Taylor and Sheena Dinwoodie; their unstinting and enthusiastic help is gratefully acknowleged. We must record also our thanks to Ian McIntosh (John Wiley and Sons) for his encouragement and advice.

One final point before leaving you to your studies. Students often believe that their lecturers or professors know all about their subjects and have infinite wisdom in the topic of their choice. This is, of course, far from the truth but we hope it will not be too evident from the text that you are about to enjoy. We wish you good reading.

MICHAEL J. GRIMBLE AND MICHAEL A. JOHNSON

Spring 1986

CHAPTER 8

Time Domain Analysis of Filtering and Smoothing Problems

8.1. INTRODUCTION

The Kalman filter [1], [2] has been one of the most useful developments to emerge from the growth of modern control theory which began in the early 1960s. It has a proven record of success in application over a wide range of disciplines (Gelb [3]). The filter is based upon a plant or signal model which is in state equation form. In control engineering applications, the most important role for the Kalman filter is as a state estimator. Some would argue that 'Kalman estimator' is a more appropriate title for the device. In the context of the stochastic Linear Quadratic Gaussian (LQG) optimal control problem, a separation principle arises which retains the deterministic optimal control gain but replaces the feedback of the system state by the feedback of an estimate of system state. (See Kwakernaak and Sivan [4] and Chapter 10 of this text.) This estimated state vector is generated by a Kalman filter in the LQG problem.

Historically, the Wiener filter (Wiener [5]) was derived before the Kalman filter. This earlier optimal filter was defined as the linear filter which minimized the estimation error covariance and was derived using a frequency-domain analysis. Unfortunately, the transfer-function form for the Wiener filter was difficult to compute and implement. Consequently, the Kalman filter which has a very convenient recursive algorithm, ideal for digital computations, is the most frequently used. The Kalman filter is also applicable to a wider class of estimation problems, accommodating time-varying process models and non-stationary noise sources. However, for problems where both filters apply, the Wiener and Kalman filters have identical transfer characteristics. The Wiener filter arises in the steady-state or infinite-time-filtering problem ($t_0 = -\infty$) where the system is assumed time-invariant and the noise sources are stationary. The continuous-time Kalman filter is sometimes termed the Kalman–Bucy filter to acknowledge the seminal contributions of the joint originators (Kalman and Bucy [6]).

The discrete Kalman filter has an additional advantage that it can be extended in a rather obvious way to solve non-linear filtering problems. At each sample instant, the discrete filter provides an estimate of the system state and this can be used to calculate the current operating point on any non-linear characteristics within the system. This enables a first-order linear approximation to be used to represent the system at this operating point. These linearized equations for the system can then be employed, as in the Kalman filter, to calculate the filter gain matrix. Since the Kalman filter incorporates a state-space representation (that is, a dynamic model) of the system, any non-linearities within the system must also be included in the extended filter model. This approximation to an optimal filter for a non-linear system is appropriately called an *extended Kalman filter* (Jazwinski [7]).

The first successful applications of the Kalman filter for control purposes were in the aerospace industries where exact system equations could be readily derived. The growth of localized computing power is slowly enabling Kalman filtering techniques to be employed in the manufacturing and process industries. Some applications of this type are detailed in Chapter 13. For further details of applications, reference may also be made to the recent special issue of the *IEEE Transactions on Automatic Control* (Vol. AC-28, No 3, March 1983) which was wholly devoted to applications of Kalman filtering.

It is a problem to derive the Kalman filter in a simple yet rigorous way. Most authors employ readily-accepted engineering intuition and avoid obscure results in measure theory and functional analysis. This chapter also follows such a route, delving only when necessary into the stochastic background to the problem whilst aiming for the Kalman filter and the Wiener–Hopf equations. In subsequent chapters, some of the possible extensions to the Kalman filtering theory are examined but the extended Kalman filter *per se* is not discussed further; a text such as Gelb [3] is recommended for an introduction to this topic.

The first part of the chapter is concerned with the background probability and Hilbert space theory needed to derive the continuous-time Kalman filter. The filter is defined in the major theorem of the chapter. This is followed by a brief introduction to discrete-time estimation problems and various forms of the discrete-time Kalman filter are described. Smoothing and prediction problems are also introduced and finally numerical problems in estimation are considered.

8.2. THE MATHEMATICAL INSTRUMENTS OF FILTERING THEORY

Some twenty years after the appearance of the classic paper on estimation theory by Kalman and Bucy [6], there now exists an extensive literature for the derivation of what is usually called the Kalman filter. The tutorial paper by Rhodes [56] is, for example, an excellent simple introduction to the subject. To introduce and derive the Kalman filter theorem poses major difficulties in deciding the level of mathematical sophistication to be used. One approach is to use the rigorous mathematical approach exemplified by Curtain and Pritchard [8]. This arrives at a Kalman filter but imparts rather less understanding of the engineering details en route. A much more leisurely introduction to the mathematical nuances is given in Davis [9], but even in this text the reader is directed to other sources for some of the mathematical details, particularly whenever measure theory threatens to intrude too far into the analysis. This gives some indication of the difficulty in determining just the right level of complexity. In this respect, the original source paper due to Kalman and Bucy still remains a salutory read!

The analysis begins with a brief overview of the main concepts and definitions for stochastic systems. (A more detailed introduction can be found in Davis [9] or Maybeck [14].)

These concepts and results are eventually drawn together in Section 7.3 to derive the Kalman filter algorithm and the Wiener–Hopf equation. (The latter being the necessary and sufficient condition for the Wiener filter.)

8.2.1. Probabilistic concepts

Probability spaces

The *probability space* designated by the triple (Ω, β, P) is essential to the mathematical development for random phenomena. The *sample space* Ω is a non-empty set of sample points or simple events $\omega(\omega \in \Omega)$ containing all possible outcomes of the random experiment of interest. The *event set* β is a collection or class of subsets of Ω each associated with a particular experimental outcome. (This set forms a σ-algebra, that is, the set is closed under all denumerable intersections or unions.) Given a particular event A, a finite number $P(A)$ can be assigned to give the probability of the event occurring, thus

$$0 \leq P(A) \leq 1$$

The set function $P(A)$ is called the *probability measure* of A or simply the probability of A and it satisfies $P(\Omega) = 1$.

Real scalar random variable

A scalar random variable $X(\cdot)$ is a real valued point function which assigns a real scalar value to each point $\omega \in \Omega$, such that every set $B \subset \Omega$ of the form $B = \{\omega: X(\omega) \leq \xi, \xi \in \mathbb{R}\}$ is an element of $\beta(B \in \beta)$.

Distribution function

Consider a random experiment with random variable X then the probability of the event $\{\omega: X(\omega) \leq x\}$ is specified by a real number depending upon the function

$$F_X(x) = P(X \leq x)$$

which is called the *distribution function* of X. The distribution function always exists and satisfies:

$$F_X(-\infty) = 0 \quad \text{and} \quad F_X(\infty) = 1.$$

Continuous and discrete random variables

The random variable X is called a continuous random variable if its distribution function is absolutely continuous. The random variable is called discrete when the distribution function assumes the form of a staircase with a finite or countably infinite number of jumps. Thus, a discrete random variable may assume only discrete point values, or values on a countable set.

Probability density function

The probability density function for a continuous random variable X, is defined as:

$$f_X(x) = \frac{dF_X(x)}{dx}$$

and has the following properties:

$$f_X(x) \geq 0, \qquad \int_a^b f_X(x)\, dx = F_X(b) - F_X(a), \qquad \int_{-\infty}^{\infty} f_X(x)\, dx = 1.$$

Alternatively, the probability density function (pdf) may be defined via the relationship:

$$P(\omega: X(\omega) \leq x) = F_X(x) = \int_{-\infty}^{x} f_X(\alpha)\, d\alpha.$$

The probability density function encapsulates the random properties of a particular variable and is the quantity of most interest in Bayesian estimation theory. (In the Bayesian approach, the optimal parameter estimates are based on the evolution or propagation of the conditional probability density function.)

Expected value

The *expected value* or *expectation* of a discrete random variable gives the statistical average and follows readily as:

$$E\{X\} \triangleq \sum_i X(\omega_i)P(\omega_i)$$

or if the probability density function $f_X(x)$ exists the expected value of a continuous random variable is:

$$E\{X\} \triangleq \int_{-\infty}^{\infty} \xi f_X(\xi)\, d\xi. \tag{8.1}$$

The expected value is also termed the first moment or mean of X and is a *first-order statistic* (since only the first moment is involved).

Higher-order moments may also be defined. Let X denote a continuous random variable with density function $f_X(\xi)$, and let $Y = \phi(X)$. Then the expectation of Y is given as:

$$E\{Y\} = \int_{-\infty}^{\infty} \phi(\xi) f_X(\xi)\, d\xi.$$

The *second moment* of X an be obtained directly as:

$$E\{X^2\} = \int_{-\infty}^{\infty} \xi^2 f_X(\xi)\, d\xi \tag{8.2}$$

and is is a *second-order statistic*. Similarly, moments may be generated about any real number α. The nth moment about α is defined as $E\{(X-\alpha)^n\}$. The nth *central moment* is of particular interest where α is taken equal to the mean value $m_X \triangleq E\{X\}$. The second central moment:

$$\sigma^2 = E\{(X - m_X)^2\} \tag{8.3}$$

is called the *variance of the distribution* and it measures the spread or dispersion of the random variable about its mean value m_X.

Random vectors

The concept of a random vector arises when considering an n tuple of random variables, namely, a random vector $\boldsymbol{x}$ comprises a n-tuple of the random variables $x_i \triangleq X_i$, $i = 1, 2, \ldots, n$. The *mean* or *first moment* of $\boldsymbol{x}$ can be written as:

$$\mathbf{m}_x = E\{\boldsymbol{x}\} = \int_{-\infty}^{\infty} \boldsymbol{\xi} f_x(\boldsymbol{\xi})\, d\boldsymbol{\xi} \tag{8.4}$$

where $\mathbf{m}_x \in \mathbb{R}^n$ and $f_x(\boldsymbol{\xi})$ is the probability density function for the random vector $\boldsymbol{x}$ (the integral here acts on each element of the vector $\boldsymbol{\xi}$ in (8.4)). The *second moment* of $\boldsymbol{x}$ becomes:

$$\boldsymbol{\Phi}_{xx} = E\{\boldsymbol{x}\boldsymbol{x}^T\} = \int_{-\infty}^{\infty} \boldsymbol{\xi}\boldsymbol{\xi}^T f_x(\boldsymbol{\xi})\, d\boldsymbol{\xi}, \tag{8.5}$$

where $\boldsymbol{\Phi}_{xx} \in \mathbb{R}^{n \times n}$ (the integral here acts on each element of the matrix in (8.5)). The *second central moment* of $\boldsymbol{x}$ may be written as:

$$\mathbf{R}_{xx} = E\{\boldsymbol{x} - \mathbf{m}_x)(\boldsymbol{x} - \mathbf{m}_x)^T\} = \int_{-\infty}^{\infty} (\boldsymbol{\xi} - \mathbf{m}_x)(\boldsymbol{\xi} - \mathbf{m}_x)^T f_x(\boldsymbol{\xi})\, d\boldsymbol{\xi}, \tag{8.6}$$

$$= E\{\boldsymbol{x}\boldsymbol{x}^T\} - \mathbf{m}_x\mathbf{m}_x^T$$

where $\mathbf{R}_{xx} \in \mathbb{R}^{n \times n}$. The *crosscorrelation matrix* becomes:

$$\boldsymbol{\Phi}_{xy} = E\{\boldsymbol{x}\boldsymbol{y}^T\} = \int_{-\infty}^{\infty}\int_{-\infty}^{\infty} \boldsymbol{\xi}\boldsymbol{\zeta}^T f_{xy}(\boldsymbol{\xi}, \boldsymbol{\zeta})\, d\boldsymbol{\xi}\, d\boldsymbol{\zeta}, \tag{8.7}$$

where $f_{xy}(\boldsymbol{\xi}, \boldsymbol{\zeta})$ is the joint probability density function of the two random vectors $\boldsymbol{x}$ and $\boldsymbol{y}$. The *crosscovariance* matrix follows similarly as:

$$\mathbf{R}_{xy} = E\{(\boldsymbol{x} - \mathbf{m}_x)(\boldsymbol{y} - \mathbf{m}_y)^T\}$$

$$= \int_{-\infty}^{\infty}\int_{-\infty}^{\infty} (\boldsymbol{\xi} - \mathbf{m}_x)(\boldsymbol{\zeta} - \mathbf{m}_y)^T f_{xy}(\boldsymbol{\xi}, \boldsymbol{\zeta})\, d\boldsymbol{\xi}\, d\boldsymbol{\zeta}. \tag{8.8}$$

Gaussian or normal random vectors

The *Gaussian* or *normal* random vector is particularly important in the modelling of physical systems. The n-dimensional vector $\boldsymbol{x}$ is said to be a Gaussian random vector if it can be described by a probability density function of the form:

$$f_x(\boldsymbol{\xi}) = \frac{1}{(2\pi)^{n/2}(\det(\mathbf{P}))^{1/2}} \exp(-\tfrac{1}{2}[\boldsymbol{\xi} - \mathbf{m}]^T\mathbf{P}^{-1}[\boldsymbol{\xi} - \mathbf{m}]). \tag{8.9}$$

The vector $\mathbf{m}$ denotes the mean value of $\boldsymbol{x}$ and the matrix $\mathbf{P} > \mathbf{O}$ denotes the covariance matrix of $\boldsymbol{x}$. Davis [9] gives a more general definition of normality in terms of characteristic functions which is valid for singular covariance functions.

Correlation and independence

Using the definition of cross-correlation (8.7), two random vectors $\boldsymbol{x}$ and $\boldsymbol{y}$ are termed *uncorrelated* if,

$$E\{xy^T\} = E\{x\}E\{y^T\}. \tag{8.10}$$

That is, x and y are uncorrelated if the second-order moment $\boldsymbol{\Phi}_{xy}$ can be expressed as the product of the first-order moments $\mathbf{m}_x$ and $\mathbf{m}_y$. Clearly, if one of the random vectors is zero mean, then being uncorrelated implies

$$\boldsymbol{\Phi}_{xy} = E\{xy^T\} = \mathbf{O} \text{ or } \mathbf{R}_{xy} = \operatorname{cov}[x, y] = \mathbf{O}.$$

The *independence* of two or more random vectors is a condition where the joint probability density function can be written as a product of the respective marginal probability density functions. Hence, x and y independent implies x and y uncorrelated but not vice versa.

This last result can easily be demonstrated using a scalar example. Let X be normally distributed with zero mean and variance σ^2, ($X \sim N(0, \sigma^2)$) and let $Y = X^2$. Clearly X and Y are *not independent*. However, the probability density function for the scalar Gaussian random variable X is:

$$f_X(\xi) = \frac{1}{((2\pi)^{1/2}\sigma)} \exp\left(-\frac{1}{2\sigma^2}\,\xi^2\right).$$

Thus, the correlation function becomes:

$$\begin{aligned} E\{XY\} &= E\{X^3\} \\ &= \int_{-\infty}^{\infty} \xi^3 f_X(\xi)\, d\xi, \end{aligned}$$

Since $\xi^3 f_X(\xi)$ is an odd function then the above integral is zero and X and Y are *uncorrelated.*

8.2.2. A Hilbert space of random variables

A simple introduction to a Hilbert space of random variables is given in Luenberger [10]. One of the distinguishing features of a Hilbert space is the existence of an inner product defined on the space.

Inner-product and metric

A random variable is said to be square integrable if $E\{X^2\} < \infty$ (this implies $E\{X\} < \infty$ also) and this condition yields the basic set from which to fashion a Hilbert space of random variables. In particular the inner product is defined as:

$$\langle X, Y\rangle_H = E\{XY\} \tag{8.11}$$

and the metric or distance function is given by:

$$d(X, Y) \triangleq \| X - Y \| = [\langle X - Y, X - Y\rangle_H]^{1/2}$$

$$= [E\{X - Y)^2\}]^{1/2}.$$

A random variable X which satisfies $E\{X^2\} < \infty$ is termed second order.

Convergence of a sequence of random variables

The first step in the development of a calculus for stochastic processes is to define convergence of a sequence of random variables. Two types of convergence are discussed below.

(a) Convergence in mean square. A sequence of random variables $\{X_n\}$ converges in mean square to a random variable X if

$$\lim_{n\to\infty} \| X_n - X \| = 0. \tag{8.12}$$

This convergence is also called convergence in quadratic mean or second-order convergence and is written as:

$$\operatorname*{l.i.m.}_{n\to\infty} X_n = X$$

where l.i.m. denotes limit in the mean.

(b) Convergence almost surely or convergence with probability one. A sequence $\{X_n\}$ of random variables is said to converge *almost surely* (a.s.) or *with probability one* (w.p.1) if

$$P\left\{\lim_{n\to\infty} X_n = X\right\} = 1. \tag{8.13}$$

This type of convergence is often written as:

$$X_n \underset{\rightarrow}{\text{a.s.}} X$$

and is also referred to as convergence almost everywhere (Thomas [60]).

The following examples illustrate that neither of the above concepts implies the other. Let $\Omega \triangleq [0,1]$, $\beta \triangleq$ Borel sets of $[0,1]$ (namely 'almost' all the subsets of $[0,1]$) and P be the Borel measure (the 'length' of the subset). Thus, the random variable $\omega \in \Omega$ is uniformly distributed on $[0,1]$. Consider now a function defined via:

$$f_i(x) = \begin{cases} i, & x \in [0, 1/i] \\ 0, & x \in [1/i, 1] \end{cases}$$

and a sequence of random variables:

$$X_i(\omega) \triangleq f_i(\omega).$$

This sequence of random variables, as shown in Fig. 8.1 converges a.s. to zero since $f_i(\omega) = 0$ whenever $i > 1/\omega$. However, convergence in the mean square sense fails because $E\{X_i^2\} = i$ for all i. The sequence shown in Fig. 8.2 has the alternative properties since it converges to $X(\omega) = 0$ for all ω in the mean square sense but convergence a.s. fails since there exist no $\omega \in [0,1]$ such that $X_i(\omega) \to 0$, as $i \to \infty$.

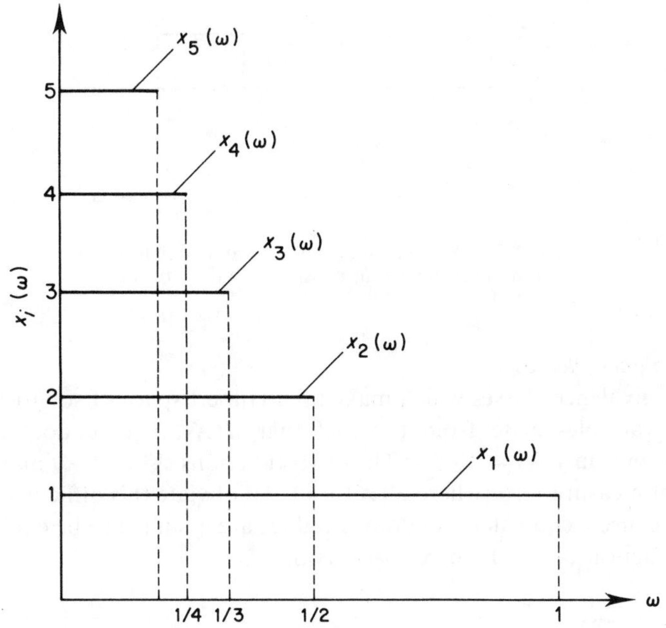

Fig. 8.1. Sequence of random variables converging a.s. but not in mean square to $X(\omega) = 0$, for all $\omega \in [0,1]$

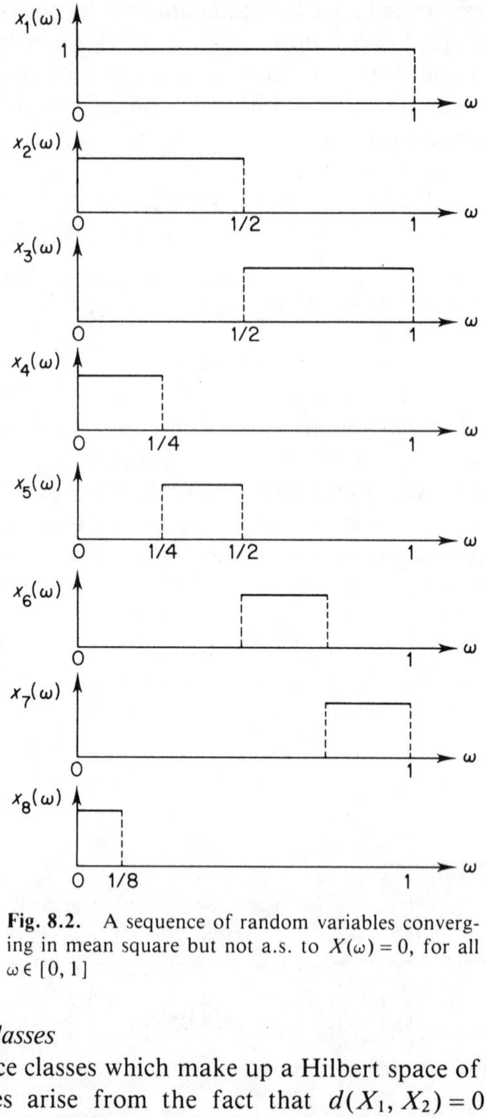

Fig. 8.2. A sequence of random variables converging in mean square but not a.s. to $X(\omega) = 0$, for all $\omega \in [0, 1]$

Equivalence classes

The equivalence classes which make up a Hilbert space of square integrable random variables arise from the fact that $d(X_1, X_2) = 0$ does not imply $X_1 = X_2$, but only $X_1 = X_2$ a.s. Thus two functions X_1 and X_2 may differ on a set $\tilde{X}$ of measure zero (where $P\{\tilde{X}\} = 0$). To obviate this difficulty, such sets of almost surely equivalent random variables are assumed to be equivalent for the satisfaction of the Hilbert space axioms.

Completeness

The final step in the definition of the Hilbert space is to demonstrate that the space that has been constructed is complete (Luenberger [10]). The completeness of this space (preHilbert) with respect to the metric $d(X, Y)$ has been verified by Taylor [59].

Geometric concepts

The geometric structure of Hilbert spaces (of which orthogonality is just one concept) played an important clarifying role in the original derivation of the Kalman–Bucy filter (Kalman and Bucy [6]). The following brief survey opens with the idea of a closed subspace.

DEFINITION 8.1. *Subspace.* Let $H_s \subset H$, then H_s is a subspace of H if given any $\boldsymbol{h}_1, \boldsymbol{h}_2 \in H_s$, every linear combination $\alpha \boldsymbol{h}_1 + \beta \boldsymbol{h}_2 \in H_s \subset H$ for all $\alpha, \beta \in \mathbb{R}$. □

DEFINITION 8.2. *Closed subspace.* Subspace $H_s \subset H$ is a closed subspace if every Cauchy sequence $\{\boldsymbol{h}_n\}$ in H_s is such that $\boldsymbol{h} = \lim_{n \to \infty} \boldsymbol{h}_n \in H_s \subset H$. □

If $H_s \subset H$ is a closed subspace it may be treated as a Hilbert space in its own right.

Orthogonality

A concept of importance in the geometry of Hilbert spaces is that of orthogonality. If $\boldsymbol{h}_1$ and $\boldsymbol{h}_2$ are vectors from a real Hilbert space H, then $\boldsymbol{h}_1$ and $\boldsymbol{h}_2$ are orthogonal if and only if $\langle \boldsymbol{h}_1, \boldsymbol{h}_2 \rangle = 0$ where $\langle \cdot , \cdot \rangle$ inner product defined on H.

Two elements $\boldsymbol{x}, \boldsymbol{y}$ from a Hilbert space of random variables are orthogonal if and only if:

$$\langle \boldsymbol{x}, \boldsymbol{y} \rangle_{H_n} = E\{\langle \boldsymbol{x}, \boldsymbol{y} \rangle_{E_n}\} = 0.$$

Hence, if the Hilbert space comprises zero-mean random variables, orthogonality is equivalent to the elements being uncorrelated.

Combining orthogonality and the notion of a closed subspace the following geometric ideas result. An element $\boldsymbol{h} \in H$ is orthogonal to the subspace $H_s \subset H$ (written $\boldsymbol{h} \perp H_s$) if and only if $\langle \boldsymbol{h}_i, \boldsymbol{h} \rangle = 0$ for all $\boldsymbol{h}_i \in H_s$. If all such elements are collected together, the orthogonal complement to the closed subspace $H_s \subset H$ may be defined as:

$$H_s^{\perp} \triangleq \{\boldsymbol{h} \in H \text{ such that } \boldsymbol{h} \perp H_s \subset H\}.$$

The geometric ideas culminate in an orthogonal decomposition of the Hilbert space.

Theorem 8.1. ***Orthogonal decomposition of a Hilbert space.*** *Let $H_s \subset H$ be a subspace of the Hilbert space H, then:*

(*i*) *$H_s^{\perp} \subset H$ is a closed subspace of Hilbert space H.*

(*ii*) *Hilbert space H can be decomposed as two orthogonal subspaces: $H = \bar{H}_s \oplus H_s^{\perp}$ where $\bar{H}_s$ denotes the closed subspace arising from $H_s \subset H$.*

(*iii*) *Any vector $x \in H$ can be given the unique representation: $x = x_1 + x_2$, $x_1 \in \bar{H}_s$ and $x_2 \in H_s^{\perp}$, clearly $\langle x_1, x_2 \rangle = 0$.*

These geometric ideas form the foundation of the orthogonal projection theorem which figures so prominently in linear estimation theory.

Theorem 8.2. *The orthogonal projection theorem. If H is a Hilbert space, and $H_s \subset H$ a closed subspace then any $x \in H$ may be uniquely written as*

$$x = y + z,$$

where $y \in H_s$ and $z \in H_s^{\perp}$. Furthermore

$$\| x - y \| = \min_{v \in H_s} \| x - v \| \tag{8.14}$$

and the minimum of the norm is attained if and only if y is the projection of x on H_s.

The geometric notion given abstract form here is that the shortest distance between a subspace and a point external to the subspace is along the perpendicular from the point to the subspace (see Fig. 8.3).

A simple example which exploits the projection theorem follows. If H_0 denotes the space of zero-mean, square integrable random variables, then $H_0^{\perp}$ is the space of constant random variables, viz.:

$$\{C \in H_0^{\perp} \Leftrightarrow C(\omega) = C \in \mathbb{R} \text{ for all } \omega \in \Omega\}.$$

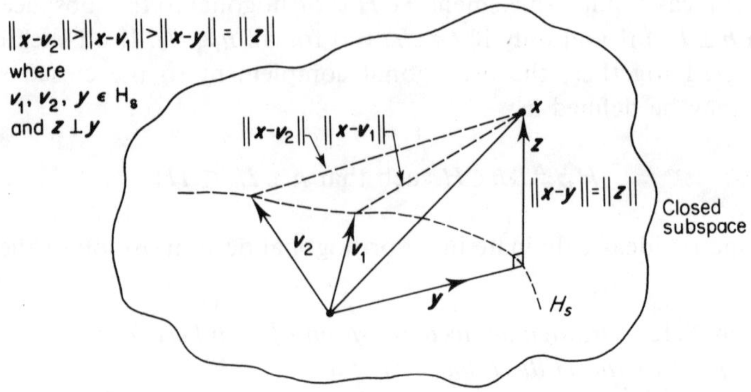

Fig. 8.3. Orthogonal projection theorem

Thus, any $X \in L_2\ (\Omega, \beta, P)$ has the unique decomposition:

$$X = (X - E\{X\}) + E\{X\}$$

where

$$X - E\{X\} \in H_0 \quad \text{and} \quad E\{X\} \perp (X - E\{X\}).$$

A further example of the use of the projection theorem in an estimation problem may be found in Luenberger [10].

8.2.3. Stochastic processes

A random signal or stochastic process occurs when the random variables introduced in the previous section appear as an indexed set. They are of considerable interest to the control engineer since stochastic processes can be used to represent the noise or random disturbances which naturally arise in many industrial or manufacturing processes.

DEFINITION 8.3. *Stochastic process.* Let Ω be a sample space and T be a subset of the real line representing the time-interval of interest. A stochastic process is a real valued function $x(\cdot\,;\cdot)$, defined on $T \times \Omega$, such that for any given $t \in T$, $x(t;\cdot)$ is a random variable. A *vector stochastic process* is an n-tuple of stochastic processes $\boldsymbol{x}(t;\omega) \in \mathbb{R}^n$. □

Note that a stochastic process can theoretically be complex valued but attention here is restricted to physical system models and hence real valued processes. Two parameter indexing ω and t is required to describe both the random nature and the time evolution of a stochastic process.

It follows that $\boldsymbol{x}(\cdot\,;\cdot)$ is a vector stochastic process if for any $t \in T$, and $\xi \in \mathbb{R}^n$, all the sets of the form:

$$A = \{\omega \in \Omega;\ x_i(t;\omega) \le \xi_i;\ i = 1, \ldots, n\}$$

are in β. This encapsulates the randomness of $\boldsymbol{x}(\cdot\,;\cdot)$. If the second argument $\omega \in \Omega$ is fixed, a signal $\boldsymbol{x}(\cdot\,;\omega)$ which is a *sample* from the stochastic process is generated. In the sequel, the argument ω is usually omitted and $\boldsymbol{x}(t;\omega)$ is denoted $\boldsymbol{x}(t)$. A stochastic process is clearly an infinite family or an ensemble of time functions and each time function is one possible *realization* of the process. For simplicity the discussions up to Section 8.2.5, will be concerned with scalar processes.

Statistics of a scalar stochastic process

Assume $x(t)$ and $y(t)$ are scalar processes defined on a *common* probability space. Some useful statistics of scalar stochastic processes are defined as:

(i) Process mean:

$$m_x(t) = E\{x(t)\}. \tag{8.15}$$

(ii) Autocorrelation (or simply correlation) function:

$$\phi_{xx}(t_1, t_2) = E\{x(t_1)x(t_2)\}. \tag{8.16}$$

(iii) Crosscorrelation function:

$$\phi_{xy}(t_1, t_2) = E\{x(t_1)y(t_2)\}. \tag{8.17}$$

(iv) Autocovariance function (or covariance function):

$$\begin{aligned} r_{xx}(t_1, t_2) &= \operatorname{cov}[x(t_1), x(t_2)] \\ &\triangleq E\{(x(t_1) - m_x(t_1))(x(t_2) - m_x(t_2))\}. \end{aligned} \tag{8.18}$$

(v) Cross-covariance function:

$$r_{xy}(t_1, t_2) = \operatorname{cov}[x(t_1), y(t_2)]. \tag{8.19}$$

(vi) Second-order moment function:

$$\phi_{xx}(t, t) = E\{x^2(t)\} \geq 0$$

and the variance function:

$$\begin{aligned} v_x(t) &= r_{xx}(t, t) \\ &= E\{(x(t) - m_x(t))^2\} \geq 0. \end{aligned}$$

The autocorrelation function is sometimes called the second-order joint moment function. If the process is zero mean the correlation and covariance functions are identical.

Second-order process

A stochastic process $\{x(t), t \in T\}$ whose variables are defined on the probability space (Ω, β, P), is known as a second-order process if $E\{x^2(t)\} < \infty$, for each $t \in T$. It has the useful property that both the mean value (8.15) and the covariance function (8.18) exist. Each element of a second order process is by definition a member of the Hilbert space of random variables $H(\Omega, \beta, P)$. In the particular case where T is taken to be the real line, the second-order process may be visualized as tracing out a one parameter family of points in the Hilbert space $H(\Omega, \beta, P)$ when t varies over $T \subset \mathbb{R}$. A zero-mean second-order process occupies a Hilbert subspace defined by:

$$H_0 = \{ x \in H(\Omega, \beta, P); E\{ x\} = 0\}$$

In subsequent sections it may be convenient to use a translation to create a new zero-mean second-order process. For this case, the newly defined process evolves in the H_0 subspace.

DEFINITION 8.4. *Quadratic mean continuous process.* A process is termed quadratic mean continuous if:

$$\lim_{t \to t_1} E\{ (x(t_1) - x(t))^2 \} = 0$$

or in Hilbert space notation if:

$$\lim_{t \to t_1} \| x(t_1) - x(t) \| = 0.$$

□

Proposition 8.1. ***Covariance function of a quadratic mean continuous process.*** *(i) A process $\{x(t), t \in \mathrm{T} \subset \mathbb{R}^+\}$ is quadratic mean continuous at t if and only if $r(\cdot, \cdot)$ is continuous at the diagonal point (t, t).*

(ii) If a process $\{x(t), t \in \mathrm{T} \subset \mathbb{R}^+\}$ is quadratic mean continuous for all $t \in \mathrm{T} \subset \mathbb{R}^+$ then $r(\cdot, \cdot)$ is continuous at every point $(s, t) \in T \times T \subset (\mathbb{R}^+)^2$.

Proof. (a) If $r(\cdot, \cdot)$ is continuous at (t, t) then

$$\begin{aligned} E\{ (x(t_1) - x(t))^2 \} &= E\{ x^2(t_1) - 2x(t_1)x(t) + x^2(t)\} \\ &= E\{ x^2(t_1)\} - 2E\{ x(t_1)x(t)\} + E\{ x^2(t)\} \\ &= r(t_1, t_1) - 2r(t_1, t) + r(t, t) \\ &= r(t_1, t_1) - r(t, t) - 2(r(t_1, t) - r(t, t)) \end{aligned}$$

thus $\lim_{t_1 \to t} E\{ (x(t_1) - x(t))^2 \} = 0$ by continuity of $r(\cdot, \cdot)$ at (t, t).

(b) Let process $\{x(t)\}$ be quadratic mean continuous at (t_1, t_2) then

$$\lim_{t \to t_1} E\{(x(t) - x(t_1))^2\} = \lim_{t \to t_2} E\{ (x(t) - x(t_2))^2 \} = 0.$$

Then

$$r(\tilde{t}_1, \tilde{t}_2) - r(t_1, t_2) = E\{x(\tilde{t}_1)x(\tilde{t}_2)\} - E\{x(t_1)x(t_2)\}$$

$$= E\{x(\tilde{t}_1)(x(\tilde{t}_2) - x(t_2))\} + E\{x(t_2)(x(\tilde{t}_1) - x(t_1))\}$$

hence

$$|r(\tilde{t}_1, \tilde{t}_2) - r(t_1, t_2)| \leq E\{x^2(\tilde{t}_1)\}^{1/2} E\{(x(\tilde{t}_2) - x(t_2))^2\}^{1/2}$$

$$+ E\{x^2(t_2)\}^{1/2} E\{(x(\tilde{t}_1) - x(t_1))^2\}^{1/2}$$

giving

$$\lim_{\substack{\tilde{t}_1 \to t_1 \\ \tilde{t}_2 \to t_2}} |r(\tilde{t}_1, \tilde{t}_2) - r(t_1, t_2)| = 0 \text{ and } r(\cdot\,,\,\cdot) \text{ is continuous at } (t_1, t_2)$$

provided $\{x(t)\}$ is quadratic mean continuous at (t_1, t_2). Properties (a) and (b) prove (i) and ensure the assertions of (ii). □

The continuity of the covariance function does not in any way imply the continuity of the sample functions. The counterexample cited is usually the Poisson process (see Assefi [61] for example).

Properties of stochastic processes

Stochastic processes may be categorized in a number of ways including, for example, the underlying type of probability distribution. A process may exhibit a trend in some or all of its statistics. The absence of any such trends identifies a stationary process discussed later in this section.

A different type of categorization arising from the relationships existing between the process values at different time instants. A process increment is defined below for this purpose.

An *increment* of the random process $x(t)$ on $[t_1, t_2)$ is denoted by $x(t_2) - x(t_1)$. Unless otherwise stated, the following discussion is concerned with zero-mean, finite variance stochastic processes.

Process with orthogonal increments

DEFINITION 8.5. *Process with orthogonal increments.* A process $\{x(t), t \in T\}$ is deemed to have orthogonal increments if,

$$E\{(x(t_1) - x(t_2))^2\} < \infty \quad \text{for all } t_1, t_2 \in T \tag{8.20}$$

and if for any non-overlapping intervals (t_1, t_2), (t_3, t_4)

$$E\{(x(t_1) - x(t_2))(x(t_3) - x(t_4))\} = 0. \tag{8.21}$$

□

In terms of Hilbert space concepts, property (8.21) may be expressed by the orthogonality relationship:

$$(x(t_1) - x(t_2)) \perp (x(t_3) - x(t_4)) \tag{8.22}$$

hence the term *orthogonal* increments. Similarly, a process with uncorrelated increments is defined if the increments $(x(t_1) - x(t_2))$ and $(x(t_3) - x(t_4))$ are uncorrelated random variables for non-overlapping intervals. Clearly, a zero-mean process with orthogonal increments also has uncorrelated increments. The incremental covariance of an orthogonal incremental process is defined as:

$$r(t, t) - r(\tau, \tau) = \text{cov}[x(t), x(t)] - \text{cov}[x(\tau), x(\tau)] \tag{8.23}$$

for $t \geq \tau$.

Equation (8.23) is often written formally as:

$$dr = \text{cov}[dx, dx]. \tag{8.24}$$

(This notation arises from result (iii) of Proposition 8.2 below.)

Given a stochastic process $\{x(t), t \in T\}$ with constant mean $m_x = E\{x(t)\}$, which has orthogonal increments, it is readily demonstrated that this incremental property is independent of translation by an arbitrary random variable. In particular consider the mean preserving translation:

$$\tilde{x}(t) = x(t) - (x(t_0) - m_x)$$

then

(i) $E\{x(t)\} = m_x$,

(ii) $\tilde{x}(t_0) = m_x$

and

(iii) both $x(t)$ and $\tilde{x}(t)$ have orthogonal increments.

This last result may be demonstrated, by direct substitution:

$$E\{(\tilde{x}(t_1) - \tilde{x}(t_2))(\tilde{x}(t_3) - \tilde{x}(t_4))\} = E\{(x(t_1) - x(t_2))(x(t_3) - x(t_4))\}.$$

The translated time series is centred on the process mean at index t_0, thus there is no loss of generality in assuming the original series to satisfy $x(t_0) = m_x$.

The properties of the variance and covariance functions for an orthogonal increments process are summarized in the following:

Proposition 8.2. *Variance and covariance functions.* *Let* $\{x(t), t \in T\}$ *denote a process with orthogonal increments, constant mean value,* m_x *and initial value* $x(t_0) = m_x$. *If the variance function is:*

$$v(t) \triangleq E\{(x(t) - m_x)^2\}, \quad \text{where } m_x \triangleq E\{x(t)\}$$

then

(i) $v(\cdot)$ *is a non-decreasing function of* t.

(ii) Covariance function $r(\cdot, \cdot)$ *satisfies:*

$$\begin{aligned} r(t_1, t_2) &= \operatorname{cov}[x(t_1 \wedge t_2), x(t_1 \wedge t_2)] \\ &= v(t_1 \wedge t_2). \end{aligned} \tag{8.25}$$

(iii) Incremental covariance function, for $t \geq \tau$, *satisfies:*

$$\begin{aligned} r(t, t) - r(\tau, \tau) &= E\{(x(t) - x(\tau))(x(t) - x(\tau))\} \\ &= v(t) - v(\tau) \end{aligned} \tag{8.26}$$

where $t_1 \wedge t_2 \triangleq \min(t_1, t_2)$, *viz. the minimum of* t_1 *and* t_2.

Proof. Let $t_0, t_1, t_2 \in T$ such that $t_0 < t_1 < t_2$ and recall $m_x = E\{x(t)\}$, then

$$x(t_2) - m_x = (x(t_2) - x(t_1)) + (x(t_1) - m_x).$$

Using $x(t_0) = m_x$ (to centre the process) and the orthogonality of the increments obtain:

(i) $$E\{(x(t_2) - m_x)^2\} = E\{(x(t_2) - x(t_1))^2\} + E\{(x(t_1) - m_x)^2\}, \tag{8.27}$$

or

$$v(t_2) - v(t_1) = E\{(x(t_2) - x(t_1))^2\} \geq 0$$

and hence $v(\cdot)$ is a non-decreasing function.

(ii) $$\begin{aligned} r(t_1, t_2) &= E\{(x(t_1) - m_x)(x(t_2) - m_x)\} \\ &= E\{(x(t_1) - m_x)(x(t_2) - x(t_1) + x(t_1) - m_x)\} \\ &= E\{(x(t_1) - m_x)^2\} = v(t_1). \end{aligned}$$

Introduce the notation $t_1 \wedge t_2 \triangleq \min(t_1, t_2)$, the above yields

$$r(t_1, t_2) = r(t_1, t_1) = v(t_1 \wedge t_2).$$

Similarly, reversing the time instants, $t_2 < t_1$ obtains

$$r(t_1, t_2) = r(t_2, t_2) = v(t_1 \wedge t_2). \tag{8.28}$$

(iii) For $\tau \leq t$, $E\{(x(t) - x(\tau))(x(t) - x(\tau))\}$

$$= E\{(x(t) - m_x - (x(\tau) - m_x))(x(t) - m_x - (x(\tau) - m_x))\}$$

$$= r(t, t) - r(t, \tau) - r(\tau, t) + r(\tau, \tau). \tag{8.29}$$

Using $r(t, \tau) = r(\tau, t)$ and the property (ii) yields

$$E\{(x(t) - x(\tau))(x(t) - x(\tau))\} = r(t, t) - r(\tau, \tau)$$

$$= v(t) - v(\tau).$$ □

The covariance condition (ii) can be visualized with the aid of Fig. 8.4, where the time indices satisfy $t_2 > t_1$. The increments are assumed to be orthogonal, hence, if t_1 is fixed, the projection of $x(t_2)$ on $x(t_1)$ remains constant however large $t_2 \in T$ becomes (from condition (ii) Prop. 8.1). Obviously the 'size' of $x(t_2)$, $(E\{x^2(t_2)\})^{1/2}$ cannot be less than that of $x(t_1)$ (using condition (i) Prop. 8.1) since:

$$E\{x^2(t_2)\} = E\{(x(t_2) - x(t_1))^2\} + E\{x^2(t_1)\}.$$

A more specialized incremental process is considered next.

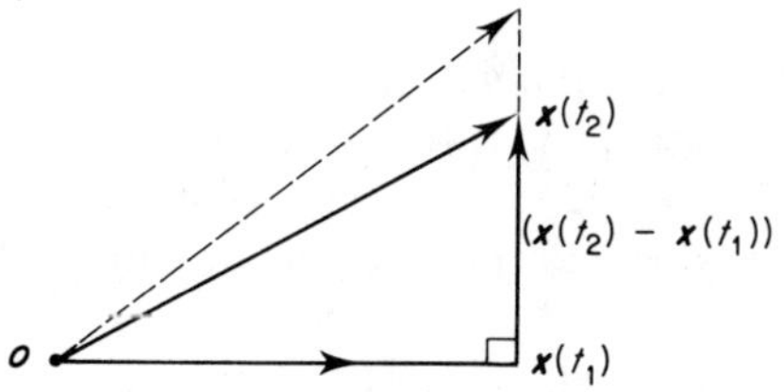

Fig. 8.4. Variance of an orthogonal increments process

Processes with independent increments

A process $\{x(t), t \in T\}$ is said to have independent increments if for any

$t_1 \leq t_2 \leq \cdots \leq t_k$ and $\{t_i \in T; i = 1 \ldots, k\}$, the increments

$$x(t_k) - x(t_{k-1}),\ x(t_{k-1}) - x(t_{k-2}), \ldots,\ x(t_2) - x(t_1),$$

are mutually independent random variables.

An equivalent definition (McGarty [40]) is as follows.

DEFINITION 8.6. *Process with independent increments.* The random process $\{x(t), t \in T\}$ is an *independent increments process* if, for any partition $t_0 \leq t_l \leq \cdots \leq t_n \subseteq T$, the increments $x(t_i) - x(t_j)$, $x(t_k) - x(t_l)$ with $t_j < t_i \leq t_l < t_k$ have the conditional probabilities.

$$P[x(t_i) - x(t_j) \in \lambda \mid x(t_k) - x(t_l) = \xi] = P[(x(t_i) - x(t_j) \in \lambda] \qquad (8.30)$$

for any intervals $\lambda \subset \mathbb{R}$, and a point $\xi \in \mathbb{R}$. □

If a zero-mean process has independent increments and property (8.20) holds the process also has orthogonal increments. Second-order independent increment processes are therefore a subclass of the orthogonal increment processes. Processes with orthogonal increments may be called processes with independent increments in the 'wide sense'. Clearly, more than just the first- and second-order moments must be known to determine whether a particular process has independent increments (unless the process is Gaussian, since a Gaussian orthogonal increments process has independent increments). It is interesting to note that any process with independent increments is both a Markov process and a Martingale. These are stochastic processes often found in more advanced texts on stochastic systems (Doob [12]). Before continuing with the incremental characterization of stochastic processes the concept of stationarity is introduced and discussed.

Stationary processes

A process is termed *stationary* if all of the calculated moments are invariant under a time shift.

DEFINITION 8.7. *Stationary process.* A process $\{x(t), t \in T\}$ is stationary (or strictly stationary) if for all sets of $t_1, t_2, \ldots, t_k \in T$ and any $\tau \in T$ such that $t_i + \tau \in T$, $i = 1, 2, \ldots, k$, the joint distribution of $x(t_1 + \tau)$, $x(t_2 + \tau), \ldots, x(t_k + \tau)$ does not depend on the time shift τ, that is:

$$P(\{\omega : x(t_1 + \tau, \omega) \leq \xi_1, \ldots, x(t_k + \tau, \omega) \leq \xi_k\})$$
$$= P(\{\omega : x(t_1, \omega) \leq \xi_1, \ldots, x(t_k, \omega) \leq \xi_k\}). \qquad (8.31)$$

□

If only the first and second moments of the distributions are invariant under a time shift, the process is *weakly, wide sense* or *second-order* stationary. This idea of restricting stationarity to moments of up to second-order may be extended as follows (see Melsa and Sage, [11]).

DEFINITION 8.8. *Stationarity of order k.* A stochastic process is kth-*order stationary* if the joint density function of order k is independent under a shift of time τ. □

Second-order stationarity is important since it is often a more realistic assumption to make and the assumption can sometimes be tested if the physical signal generating sources are known. Higher-order moments and stationarity are also important in the types of linear system analysis to be described. The explicit criteria to be met for a process to be *wide-sense stationary* are therefore,

(i) $E\{x^2(t)\} < \infty$,
(ii) $E\{x(t)\}$ is constant,
(iii) $E\{(x(t) - m)(x(t + \tau) - m)\}$ is only a function of the time difference τ.

Wide-sense stationarity only involves the first- and second-order moments and does not depend on the probability density function. Clearly, a strictly stationary process is wide-sense stationary if and only if it has finite first-order moments. Wide-sense stationarity does not imply strict stationarity except in the special case of Gaussian processes. The set inclusions are illustrated in Fig. 8.5 (see also Melsa and Sage [11]).

The incremental analysis may now be resumed adding the assumption of stationarity.

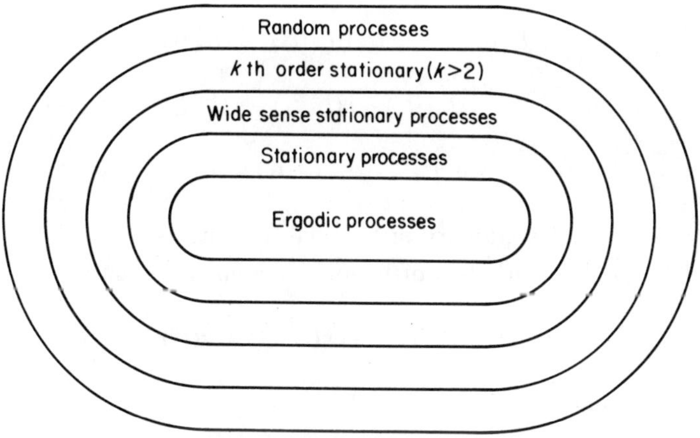

Fig. 8.5. Relationships between stochastic processes

Stationary increments process

A process is said to have stationary increments if it has a constant mean and the variance of the increment $x(t_1) - x(t_2)$ depends only upon the length of the interval $|t_2 - t_1|$, that is,

$$E\{(x(t_1) - x(t_2))^2\} = E\{(x(t_1 + \tau) - x(t_2 + \tau))^2\} \tag{8.32}$$

for all t_1, t_2, τ.

Proposition 8.3. *Variance and covariance functions. A quadratic mean continuous process $\{x(t), t \in T\}$ with constant mean m_x and $x(t_0) = m_x$ has stationary orthogonal increments if and only if its variance and covariance functions have the respective forms:*

$$v(t) = (t - t_0)\sigma^2, \qquad t \geq t_0 \tag{8.33}$$

and

$$r(t_1, t_2) = \min(t_1, t_2)\sigma^2, \qquad t_1, t_2 \geq 0 \tag{8.34}$$

where σ^2 is constant.

Proof. (i) If the covariance function is given by

$$r(t_1, t_2)\min(t_1, t_2)\sigma^2, \text{ so that } r(t_1, t) = \sigma^2 t$$

then by Proposition 8.1 $\{x(t), t \in T\}$ is quadratic mean continuous. For time indices $t_1 \leq t_2 \leq t \leq t_4$

$$\begin{aligned} &E\{(x(t_4) - x(t_3))(x(t_2) - x(t_1))\} \\ &= r(t_4, t_2) - r(t_4, t_1) - r(t_3, t_2) + r(t_3, t_1) \\ &= \sigma^2(t_2 - t_1 - t_2 + t_1) = 0 \end{aligned}$$

and $\{x(t), t \in T\}$ has stationary orthogonal increments.

(ii) If $\{x(t)\}$ has stationary orthogonal increments, equation (8.27) yields:

$$v(t_2) = E\{(x(t_2) - x(t_1))^2\} + v(t_1)$$

but from stationarity

$$v(t_2) = v(t_2 - t_1 + t_0) + v(t_1)$$

or

$$v(t_2 + t_1 - t_0) = v(t_2) + v(t_1),$$

because $v(t_0) = 0$, by the assumption that $x(t_0) = m_x$. The only continuous solution to this equation is given by $v(t) = (t - t_0)\sigma^2$ where $\sigma^2 \in \mathbb{R}$. Also from result (ii) of Proposition 8.2 now obtain:

$$r(t_1, t_2) = \min(t_1, t_2)\sigma^2. \qquad \square$$

Ergodicity

Engineers often assume that a stationary process is also ergodic. This implies that the time averages of the sample functions have limiting values equal to the ensemble mean value, namely

$$\lim_{T \to \infty} \frac{1}{2T} \int_{-T}^{T} x(t, \omega)\ dt = E\{x(t)\}.$$

It is then simple to estimate the mean level of a signal source by measuring the time-average along one sample time response. However, this assumption does not have to be made in the theoretical constructions to follow.

Gaussian process

Gaussian distributed noise sources or disturbance inputs frequently arise in industrial systems. The assumption that the distribution is Gaussian often enables the mathematical analysis involving the noise sources to be simplified.

DEFINITION 8.9. A *Gaussian stochastic process* $\{u(t), t \in T\}$ is a stochastic process where, for the time instants $T = \{t_1, t_2, \ldots, t_j\}$, the random variables $u(t_1), u(t_2), \ldots, u(t_j)$, have a Gaussian joint probability distribution. $\square$

The Gaussian process is therefore a family of Gaussian random variables $\{u(t), t \in T\}$ depending upon a parameter t which is usually time. A sufficient condition for a stationary Gaussian process with covariance matrix $\mathbf{R}(\tau)$ to be ergodic is that $\int_{-\infty}^{\infty} \|\mathbf{R}(\tau)\|\ d\tau < \infty$ (Wong [44]).

The Gaussian process is interesting since it is completely characterized by its mean and covariance matrix (McGarty [40]). It follows that a Gaussian process is stationary if and only if it is wide-sense stationary. The Gaussian process also has the useful property that a linear transformation on such a process produces another Gaussian process. Thus, if driving and measurement noise signals in a *linear* system are Gaussian, all signals within the system are Gaussian.

8.2.4. The Wiener process and the white-noise model

The fundamental independent incremental processes are the Wiener process (Brownian motion) and the Poisson process. Of interest here, is the former, since it may be used to provide a mathematical basis for the white-noise model so frequently used in process control applications.

The Wiener process was originally devised as a model for the effects of molecular collisions on a small particle immersed in a fluid where the requirement for independent increments reflects the assumed independence of successive collisions. The peculiar mathematical properties of a Wiener process originate from this physical model. The sample paths are demonstrably continuous but not differentiable anywhere and are of infinite length when measured over a finite time interval (Davis [9]).

DEFINITION 8.10. *Wiener process or Brownian motion.* The process $\{\omega(t), t \in T\}$ is defined to be a Wiener process if

(i) The process has independent increments,

(ii) The increments are Gaussian random variables, such that for $t_1, t_2 \in T$,

$$E\{\omega(t_1) - \omega(t_2)\} = 0$$

$$E\{(\omega(t_1) - \omega(t_2))^2\} = q \mid t_1 - t_2 \mid, \qquad (q > 0). \tag{8.35}$$

(iii) $\omega(t_0, \tilde{\omega}_i) = 0$ for all $\tilde{\omega}_i \in \Omega$, except possibly on a set of $\tilde{\omega}_i$ with probability zero. □

Note that $\omega(t)$ will denote the random signal from this point in the text and the dependence upon $\tilde{\omega}_i$ need not be explicitly stated, hence no confusion should arise from these different uses of symbol ω.

Thus, a Wiener process is a zero-mean process continuous in quadratic mean, whose increments are stationary, Gaussian and independent. A zero-mean process, with stationary Gaussian independent increments is not necessarily quadratic mean continuous (see, for example, page 79 of Davis [9]). From (8.26) and (8.35) the incremental covariance can be identified as:

$$v(t_1) - v(t_2) = q \mid t_1 - t_2 \mid$$

(where $q = \sigma^2$). This is often written symbolically as:

$$E\{(d\omega(t))^2\} = q\, dt.$$

(See, for example, Melsa and Sage [11].) If $q = 1$, the process is called a standard Wiener process.

An equivalent scalar *time-varying* signal has an incremental covariance:

$$E\{(\omega(t_2)-\omega(t_1))^2\} = \int_{t_1}^{t_2} q(t)\,dt \tag{8.36}$$

for $t_2 \geq t_1$ and $q(t) \geq 0$ for all $t_1, t_2 \in T$, and where the *diffusion function* $q(\cdot)$ is assumed to be at least piecewise continuous. In this case the increments are no longer stationary.

White noise

Electronic engineers use the white-noise analogy to represent a noise signal with a constant power spectral density over the operating frequency range of a system. The time-domain properties of Gaussian white-noise fluctuations $\{\xi(t), t>0\}$ satisfy:

(i) $\xi(t)$ is approximately Gaussian distributed and

(ii) $\xi(t)$ and $\xi(\tau)$ are effectively uncorrelated $|t-\tau|>\delta$ for small δ. A mathematical model for white noise might be constructed from a Gaussian process $\{\zeta(t), t>0\}$ with covariance:

$$r(t,s) = \begin{cases} 1 & s=t \\ 0 & s \neq t. \end{cases}$$

However, the insistence on zero correlation at immediately neighbouring points causes the noise model to have several unacceptable features. The process is not quadratic mean continuous and sample paths are not measurable functions so that terms like $\int_0^1 \zeta(s)\,ds$ have no meaning. The formal mathematical model for white noise is obtained via a normal process $\{\xi(t), t \in T\}$ which has a covariance function $r_\infty(t,\tau) = \delta(t-\tau)$ so that,

(i) $\xi(t)$ is distributed normally,

(ii) $\mathrm{var}(\xi(t)) = \infty$,

(iii) $\mathrm{cov}[\xi(t), \xi(s)] = 0 \qquad t \neq s.$

Fortunately, a connection with the standard Wiener process can be established as follows:

$$\omega(t) = \int_0^t \xi(s)\,ds$$

where for simplicity the initial time t_0 is taken as zero.

Then, the covariance function becomes:

$$
\begin{aligned}
r_\omega(t,s) &= \operatorname{cov}[\omega(t),\omega(s)] = E\{\omega(t)\omega(s)\} \\
&= E\left\{\int_0^t \xi(\tau_1)\, d\tau_1 \int_0^s \xi(\tau_2)\, d\tau_2\right\} \\
&= \int_0^s \int_0^t E\{\xi(\tau_1)\xi(\tau_2)\}\, d\tau_1\, d\tau_2 \\
&= \int_0^s \int_0^t \delta(\tau_1-\tau_2)\, d\tau_1\, d\tau_2 \\
&= \min(t,s). \qquad (8.37)
\end{aligned}
$$

hence, $\omega(t)$ is a standard Wiener process.

This link between the formal representation of white noise and the model for a Wiener process permits the transfer of an engineering model into a mathematically valid formulation. Hence, if white noise appears in integrated form, it may be replaced by a Wiener process equivalent. Heuristically, one can generate Brownian motion by passing white Gaussian noise through an integrator.

It is usual to assume that the noise and disturbance signals in a system are of zero-mean value. The stochastic signals in the following are therefore assumed to be of this type unless otherwise stated.

8.2.5. Stochastic integration, Hilbert spaces and estimation

To define integration with respect to a scalar stochastic signal, the basic limiting process of integration must be considered anew. As in the usual Lebesque (and Riemann) integration theories, the first step is to consider integration with respect to a class of simple functions (step-function approximations) and to follow this with a limit analysis to accommodate more general functions. For a Wiener process whose diffusion function is $q(\cdot)$ (equation (8.36)), the class of functions for which the stochastic integral $I(f)=\int_0^t f(\tau)\, dw(\tau)$ is sought are those satisfying $\int_0^t f^2(\tau)q(\tau)\, d\tau < M < \infty$. Such a class has the property of being square integrable with respect to the measure μ on $[0,\infty)$ where

$$\mu[a,b) = \int_a^b q(\tau)\, d\tau.$$

Clearly, if $q(\cdot)$ is of unit strength, $f \in L_2[0,t]$.

Stochastic integration of a simple function

Partition $(0, t)$ into n intervals with $0 < t_1 < t_2 < \cdots < t_n = t$ and consider the set S of step-functions f_n: $[0, t) \to \mathbb{R}$, such that

$$f_n(s) = \begin{cases} f(0), & s \in [0, t_1) \\ f(t_1), & s \in [t_1, t_2) \\ \vdots & \vdots \\ f(t_{n-1}), & s \in [t_{n-1}, t_n), \qquad 0 \le s < t. \end{cases} \tag{8.38}$$

This is a piecewise constant approximation to the piecewise continuous function of time $f(\cdot)$. The stochastic integral can be constructed as:

$$I(f_n) = \int_0^t g_n(\tau)\ d\omega(\tau) \triangleq \sum_{i=0}^{n-1} f_n(t_i)(\omega(t_{i+1}) - \omega(t_i)). \tag{8.39}$$

This definition must be extended from piecewise constant $f_n(\cdot)$ in the case of piecewise continuous $f(\cdot)$ (see Maybeck [14]) but first some properties of the integral are established.

Integral properties

If $\{\omega(t), t \ge 0\}$ is a scalar orthogonal increments process with diffusion function $q(\tau)$, for all $\tau \in [0, t]$, and the function $f \in L_2(q(\cdot);\ [0, t])$ then the *Wiener integral* of $f(\cdot)$ with respect to the process $\{\omega(t)\}$ is denoted:

$$I(f) = \int_0^t f(\tau)\ d\omega(\tau). \tag{8.40}$$

The restriction of $f(\cdot)$ to the set S of step functions f_n: $[0, t] \to \mathbb{R}$ implies that $I(f_n)$ is just the Stieltjes integral of f_n with respect to the sample path $\omega(t)$, and for this reason $I(f)$is written using the Stieltjes integral notation (Rosenbrock and Storey [15]). The integral satisfies the following algebraic rules (See problem 8.2):

(i) $$E\{I(f_n)\} = \int_0^t f_n(\tau)\ dm(\tau) = 0 \tag{8.41}$$

(ii) $$I(\alpha f_n + \beta g_n) = \alpha I(f_n) + \beta I(g_n) \tag{8.42}$$

(iii) $$E\{I(f_n) I(g_n)\} = \int_0^t f_n(\tau) g_n(\tau) q(\tau)\ d\tau \tag{8.43}$$

for $f_n, g_n \in S$, $m(\tau) = E\{\omega(\tau)\}$ and $\alpha,\ \beta \in \mathbb{R}$.

The first two of these properties follows directly from the definition of the integral (equation 8.39) and the zero mean process. The third result may be demonstrated as follows:

$$E\{I(f_n)I(g_n)\} = E\left\{\left(\sum_i f_n(t_i)(\omega(t_{i+1}) - \omega(t_i))\right)\left(\sum_j g_n(t_j)(\omega(t_{ij+1}) - \omega(t_j))\right)\right\}$$

$$= E\left\{\sum_i f_n(t_i)g_n(t_i)(\omega(t_{i+1}) - \omega(t_i))^2\right\}$$

where the reduction from n^2 to n terms is due to the orthogonality of the increments. The t_i's being the same for f and g represent no loss of generality and the reader may like to prove this fact as an exercise. Hence, from (8.36):

$$E\{I(f_n)I(g_n)\} = \sum_i f_n(t_i)g_n(t_i)E\{(\omega(t_{i+1}) - \omega(t_i))^2\}$$

$$= \sum_i f_n(t_i)g_n(t_i)\int_{t_i}^{t_{i+1}} q(\tau)\, d\tau = \sum_i \int_{t_i}^{t_{i+1}} f_n(t_i)g_n(t_i)q(\tau)d\tau$$

$$= \int_0^t f_n(\tau)g_n(\tau)q(\tau)\, d\tau.$$

Many of the properties of $I(f_n)$ anticipate those of the stochastic integral $I(f)$ which is to be defined subsequently. Two of particular interest are:

(i) $I(f_n)$ is a zero mean random variable,

(ii) $I(f_n)$ is an element of a Hilbert space of random variables.

Wiener integrals

The integral relationships (8.41) to (8.43) are required for more general functions than f_n, in particular for $f \in L_2[0, t]$.

In this case, the time-interval must be partitioned into smaller and smaller steps so that under the appropriate conditions the sequence of random variables $I(f_n)$ converges to some limit as $n \to \infty$. The convergence analysis is described briefly after the following summary of the more general forms of the above integral relationships. Let $f \in L_2[0, t]$ and $\{x(t)\}$ represent an orthogonal increments process which may be non-stationary then (Doob [12], Åström [13]):

(iv) $$E\left\{\int_0^t f(\tau)\, dx(\tau)\right\} = \int_0^t f(\tau)\, dm(\tau), \qquad (8.44)$$

(v) $$\mathrm{var}\left\{\int_0^t f(\tau)\,dx(\tau)\right\} = \int_0^t f^2(\tau)q(\tau)\,d\tau, \tag{8.45}$$

(vi) $$E\left\{\int_0^t g(\tau)\,dx(\tau)\int_0^t f(\tau)\,dx(\tau)\right\} = \int_0^t g(\tau)f(\tau)q(\tau)\,d\tau, \tag{8.46}$$

where $m(\tau) = E\{x(\tau)\}$ and $q(\tau)$ is the diffusion function.

Convergence

The choice of $f \in L_2(q(\cdot),\ [0, t])$ implies the existence of a sequence of simple functions, converging to f such that,

$$\int_0^t (f_n - f)^2 q(s)\,ds \to 0 \quad \text{as } n \to \infty.$$

This basic property of the Lebesque integration of f is used to construct a Cauchy sequence for the random variable $I(f_n)$. It has already been noted that f_m is an element of L_2. Thus,

$$E\{(I(f_n) - I(f_m))^2\} = E\left\{\left(\int_0^t (f_n - f_m)\,d\omega(s)\right)^2\right\}$$

which, using property (8.43) becomes:

$$E\{(I(f_n) - I(f_m))^2\} = \int_0^t (f_n - f_m)^2 q(s)\,ds \to 0$$

as $n, m \to \infty$.

$I(f_m)$ is therefore a Cauchy sequence, in a Hilbert space. Let $I(f)$ denote its limit; then it makes sense to define $I(f) \triangleq \int_0^t f(s)\,d\omega(s)$. where $\{\omega(t), t \in T\}$ is an orthogonal increments process.

A similar definition and properties apply to a Wiener integral defined on an infinite time interval, viz. $I(f) = \int_0^\infty f(s)\,d\omega(s)$; note that the class of time functions $f(\cdot)$ is now restricted to $f \in L_2(q(\cdot);\ [0, \infty))$.

Hilbert subspaces

The evolution of a stationary second-order process $\{x(t), t \in T\}$ can be represented by a family of random variables in a Hilbert space $H(\Omega, \beta, P)$. The family may be characterized as a subspace of H, defined by:

$$H_t^x = \mathscr{L}\{x(t); 0 \le s \le t\}.$$

The spanning operation $\mathscr{L}\{\cdot;\cdot\}$ denotes that elements of H_t^x are taken to be all finite linear combinations $\alpha_1 x(t_1) + \alpha_2 x(t_2) + \cdots + \alpha_n x(t_n)$ where $t_i \leq t$ for all i, together with all q.m. limits of such linear combinations. The Hilbert space H_t^x describes the *totality* of past and present process trajectories. For this reason H_t^x satisfies the trivial inclusion:

$$H_\tau^x \subseteq H_t^x \qquad \tau \leq t.$$

The utility of the Wiener integral has been identified (Davis [9]) as:

Theorem 8.3. ***Hilbert subspaces of Wiener integrals.*** *If $\{x(t), t \in T\}$ is a stationary orthogonal increments process then,*

$$H_t^x \triangleq \left\{\int_0^t f(s)\ dx(s) \text{ where } f \text{ varies over the space } L_2[0,t]\right\}. \tag{8.47}$$

The inclusion result for H_t^x and H_τ^x is formalized as:

Corollary 8.1

$$H_t^x = H_\tau^x \oplus (H_\tau^x)^\perp \qquad (\tau \leq t) \tag{8.48}$$

where $\{x(t), t \in T\}$ is a stationary orthogonal increments process.

Proof.

(a) $H_t^x = \left\{\int_0^t f(s)\ dx(s); f \in L_2[0,t]\right\}$ is a Hilbert space.

(b) $H_\tau^x = \left\{\int_0^\tau g(s)\ dx(s); g \in L_2[0,\tau]\right\}$ is a Hilbert space.

(c) Every $g \in L_2[0,\tau]$ has an extension g_e such that

$$\begin{aligned} g_e &= g(s), \qquad 0 \leq s \leq \tau. \\ &= 0, \qquad \tau < s \leq t \end{aligned}$$

for which $g_e \in L_2[0,t]$.

(d) For all $z(\tau) \in H_\tau^x$

$$\Rightarrow z(\tau) = \int_0^\tau g(s)\ dx(s) = \int_0^t g_e(s)\ dx(s) \in H_t^x$$

thus $H_\tau^x \subseteq H_t^x$.

(e) The result follows by the Hilbert orthogonal subspace decomposition of Theorem 8.1. □

Using the traditional 'white-noise' terminology, the subspace H_t^x which is required in the derivation of the Kalman–Bucy filter (Section 8.3.2) comprises the white noise signals:

$$\eta(t) = f(t)\,\frac{dx}{dt}\,(t).$$

If $f(t)$ is time-varying, the signal $\eta(t)$ is non-stationary.

An important feature of the Wiener integral is that the process $\{x(t), t \in T\}$ can be used to generate a second orthogonal increments process. In particular (Davis [9]):

Theorem 8.4. ***Transformation of a process with orthogonal increments.*** *Let $\{u(t), t \geq 0\}$ be a process with stationary orthogonal increments and $\{p(t), t \geq 0\}$ represent another quadratic mean continuous second-order process ($t_0 = 0$, $T = [0, \infty)$). Then the following statements are equivalent:*

(a) *For each $t \geq 0$, $p(t) \in H_t^u$ and for $\tau \leq t$*

$$p(t) - p(\tau) \perp H_\tau^u.$$

(b) *There exists g: $[0, \infty) \to \mathbb{R}$ such that,*

$$\int_0^t g^2(t)\; dt < \infty$$

for all $t \geq 0$ and

$$p(t) = \int_0^t g(s)\; du(s). \tag{8.49}$$

The main utility of (8.49) is that it permits the identification of the best estimate of a stationary random variable x, given the observations $\{u(s), s \leq t\}$. This is obtained by projecting x onto the subspace spanned by the observations:

$$H_t^u \triangleq \mathscr{L}\{u(s); 0 \leq s \leq t\} \text{ (see Theorem 8.2).}$$

Proposition 8.4. ***Mean square estimator.*** *Consider the random variable $x(t)$ and the orthogonal increments process $\{u(t), t \geq 0\}$. The orthogonal* *projection $\mathscr{P}x(t)$ onto H_t^u is given by:*

$$\hat{x}(t \mid t) = \int_0^t \frac{d}{ds}(E\{x(t)u(s)\})r(s)^{-1}\, du(s) \tag{8.50}$$

where $r(t) \triangleq r(t, t)$ is the variance function for the process $\{u(t)\}$.

Proof.

(1) $\hat{x}(t \mid t) \in H_t^u, \qquad \hat{x}(t \mid t) = \int_0^t \phi(t, \tau)\, du(\tau).$

(2) $x(t) - \hat{x}(t \mid t) \perp H_t^u$ (Orthogonal Projection Theorem 8.1)

$$x(t) - \hat{x}(t \mid t) \perp \int_0^t \psi(\tau)\, du(\tau) \quad \text{for all } \psi \in L_2[0, t].$$

$$\Rightarrow E\left\{(x(t) - \hat{x}(t \mid t)) \int_0^t \psi(\tau)\, du(\tau)\right\} = 0$$

$$E\left\{x(t) \int_0^t \psi(\tau)\, du(\tau)\right\} = E\left\{\int_0^t \phi(t, \tau)\, du(\tau) \int_0^t \psi(\tau_1)\, du(\tau_1)\right\}$$

$$= \int_0^t \phi(t, \tau)\psi(\tau)r(\tau)\, d\tau.$$

This last step follows from integral property (vi) (equation 8.46). Choose $\psi(\tau) = I_{[0,s]}(\tau) =$ indicator or unit-step function on $[0, s]$ for some $s \leq t$.

$$\Rightarrow E\{x(t)(u(s) - u(0))\} = \int_0^s \phi(t, \tau)r(\tau)\, d\tau$$

$$\Rightarrow \phi(t, s) = \left(\frac{d}{ds} E\{x(t)u(s)\}\right) r(s)^{-1}.$$

Clearly, the above steps may be repeated with minor variations in the multivariable case. □

8.2.6. Multivariable stochastic system description

The scalar stochastic processes are collected into an n-tuple to form the vector stochastic processes. The random vector $\boldsymbol{x}(t)$ is defined as follows:

$$x(t) = \begin{pmatrix} x_1(t) \\ x_2(t) \\ \vdots \\ x_n(t) \end{pmatrix}$$

where each $x_i(t) \in H(\Omega, \beta, P)$ for $i = 1, 2, \dots, n$. The Hilbert subspace of vector Wiener integrals is denoted by H_t^ω, as in Theorem 8.3 (equation 8.47) and the space is spanned by the component spaces: $\{H_t^{\omega_1}, H_t^{\omega_2}, \dots, H_t^{\omega_n}\}$.

Statistics of a multivariate stochastic process

Useful statistics of a multivariate stochastic process may be listed as:

(i) The n-vector of process means: $\mathbf{m}_x(t) = E\{x(t)\} \in \mathbb{R}^n$. (8.51)

(ii) Process *autocorrelation* matrix: $\mathbf{\Phi}(t_1, t_2) = E\{x(t_1)x^T(t_2)\} \in \mathbb{R}^{n\times n}$ (8.52)

(iii) Process *autocovariance* matrix:

$$\mathbf{R}(t_1, t_2) = \operatorname{cov}[x(t_1), x(t_2)] \in \mathbb{R}^{n\times n}$$

$$\triangleq E\{(x(t_1) - \mathbf{m}_x(t_1))(x(t_2) - \mathbf{m}_x(t_2))^T\}. \tag{8.53}$$

(iv) Mean square-value or second-order moment matrix is given by $\mathbf{\Phi}(t, t) \geq \mathbf{O}$ and the variance matrix by $\mathbf{R}(t, t) \geq \mathbf{O}$.

The process correlation matrix is also termed the second-order joint moment matrix (Melsa and Sage [11]).

The multivariable stochastic processes are devised using univariate process concepts, thus

(i) An n-vector orthogonal incremental process $\{x(t)\}$ satisfies:

$$(x_i(t_1) - x_i(t_2)) \perp (x_i(t_3) - x_i(t_4)) \quad \text{for each } i,$$

$$(x_i(t_1) - x_i(t_2)) \perp (x_j(t_3) - x_j(t_4)) \quad \text{for } i \neq j;$$

and non-overlapping intervals $(t_1, t_2), (t_3, t_4) \in T \times T$.

(ii) If the process has the constant mean value $\mathbf{m}_x$ the variance matrix $\mathbf{V}(t)$ is defined as:

$$\mathbf{V}(t) = E\{(x(t) - \mathbf{m}_x)(x(t) - \mathbf{m}_x)^T\}. \tag{8.54}$$

(iii) The covariance matrix $\mathbf{R}(t_1, t_2)$ of an orthogonal increments process satisfies:

$$\mathbf{R}(t_1, t_2) = \operatorname{cov}[\boldsymbol{x}(t_1 \wedge t_2), \boldsymbol{x}(t_1 \wedge t_2)]$$

$$= \mathbf{V}(t_1 \wedge t_2). \tag{8.55}$$

(iv) If the increments are additionally assumed to be stationary then the variance matrix:

$$\mathbf{V}(t) = (t - t_0)\mathbf{R}_0 \quad \text{for } \mathbf{R}_0 > \mathbf{O}, \qquad t - t_0 \geq 0 \tag{8.56}$$

and the covariance matrix:

$$\mathbf{R}(t_1, t_2) = (t_1 \wedge t_2)\mathbf{R}_0 \tag{8.57}$$

where $\mathbf{R}_0 \in \mathbb{R}^{n \times n}$.

(v) An n-vector independent increments process $\{\boldsymbol{x}(t), t \in T\}$ satisfies $\boldsymbol{x}(t_2) - \boldsymbol{x}(t_1), \boldsymbol{x}(t_4) - \boldsymbol{x}(t_3), \ldots, \boldsymbol{x}(t_k) - \boldsymbol{x}(t_{k-1})$ are *mutually independent random vectors* for any partition $t_1 < t_2 < \cdots < t_k$ of the time interval T. For a zero-mean second-order process (finite variance) these increments are orthogonal so that such processes form a subset of the orthogonal increment processes. If $\mathbf{R}_0$ above is diagonal then the component processes $x_i(t)$ span orthogonal subspaces.

(vi) An n-vector non-stationary Wiener process $\{\boldsymbol{\omega}(t), t \in T\}$ has independent Gaussian increments with, $E\{\boldsymbol{\omega}(t)\} = \mathbf{o}$

$$E\{(\boldsymbol{\omega}(t_2) - \boldsymbol{\omega}(t_1))(\boldsymbol{\omega}(t_2) - \boldsymbol{\omega}(t_1))^T\} = \int_{t_1}^{t_2} \mathbf{Q}(\tau)\, d\tau \tag{8.58}$$

for $t_2 \leq t_1$. Clearly the variance function satisfies:

$$\mathbf{V}(t) = E\{\boldsymbol{\omega}(t)\boldsymbol{\omega}^T(t)\} = \int_{t_0}^{t} \mathbf{Q}(\tau)\, d\tau$$

and from Proposition 8.1 (amended for the multivariable case) the covariance function:

$$\mathbf{R}(t_1, t_2) = \int_{t_0}^{t_1 \wedge t_2} \mathbf{Q}(\tau)\, d\tau. \tag{8.59}$$

Further details of this process can be found in McGarty [40].

(vii) If the Wiener process is wide sense stationary or stationary, the matrix $\boldsymbol{Q}(\tau) \in \mathbb{R}^{n \times n}$ in (vi) is a constant matrix. The incremental covariance matrix for a stationary Wiener process follows directly from (8.58) as:

$$E\{(\omega(t_1) - \omega(t_2))(\omega(t_1) - \omega(t_2))^T\} = \mathbf{Q}\,|\,t_1 - t_2\,| \tag{8.60}$$

and is written symbolically as:

$$E\{\,d\omega(t)\;d\omega^T(t)\} = \mathbf{Q}\;dt \tag{8.61}$$

(see Åström [13], Doob [12]).

It will now be assumed for convenience that the noise signals in the system are zero mean and that the initial time $t_0 = 0$.

Transformations of multivariable stochastic processes

Linear integral transformations of multivariable stochastic processes are obtained as follows. Let $\{\omega(t); t \geq 0\}$ be a stationary zero-mean orthogonal increments process with covariance matrix $\mathbf{R}(t, \tau) = \operatorname{cov}[\omega(t), \omega(\tau)] = \mathbf{Q}\min(t, \tau)$, where $\mathbf{Q} \in \mathbb{R}^{n \times n}$ is a constant *diagonal* matrix and $t_0 = 0$. The following properties apply.

Proposition 8.5. ***Integral transformations.***

(i) If $\{x(t)\}$ and $\{y(t)\}$ are given by:

$$x(t) = \int_0^t \mathbf{M}(\tau)\;d\omega(\tau), \qquad y(t) = \int_0^t \mathbf{N}(\tau)\;d\omega(\tau)$$

where $\mathbf{M}(\cdot) \in \mathbb{R}^{m \times n}$ and $\mathbf{N}(\cdot) \in \mathbb{R}^{r \times n}$ and the individual elements satisfy conditions of the form $\int_0^t m_{ij}^2(\tau)\;d\tau < \infty$, then:

$$E\{x(t)\} = \mathbf{o},\; E\{y(t)\} = \mathbf{o}$$

and

$$\operatorname{cov}[\mathbf{x}(t), y(\tau)] = E\{x(t)y^T(\tau)\} = \int_0^{t \wedge \tau} \mathbf{M}(u)\mathbf{Q}\mathbf{N}^T(u)\;du. \tag{8.62}$$

(ii) Theorem 8.2 extends to the vector case, viz. $p(t) \in H_t^\omega$ if and only if $p(t) = \int_0^t \mathbf{G}(\tau)\;d\omega(\tau)$, for some matrix,

$$\mathbf{G}(\tau) = \begin{bmatrix} g_{11}(\tau) & \cdots & g_{1n}(\tau) \\ \vdots & g_{22}(\tau) & \vdots \\ \vdots & \ddots & \vdots \\ g_{n1}(\tau) & \cdots & g_{nn}(\tau) \end{bmatrix} \quad \textit{where} \sum_{ij} \int_0^t g_{ij}^2(\tau)\;d\tau < \infty.$$

Proof. The first part of the proposition can be proven by formal incremental methods (see the integral properties established in the previous section). Also,

$$\begin{aligned}\operatorname{cov}[\boldsymbol{x}(t), \boldsymbol{y}(s)] &= E\{\boldsymbol{x}(t), \boldsymbol{y}^T(s)\} \\ &= E\left\{\int_0^t \mathbf{M}(\tau_1)\, d\boldsymbol{\omega}(\tau_1) \int_0^s d\boldsymbol{\omega}^T(\tau_2)\mathbf{N}^T(\tau_2)\right\} \\ &= \int_0^t \int_0^s \mathbf{M}(\tau_1) E\{ d\boldsymbol{\omega}(\tau_1)\, d\boldsymbol{\omega}^T(\tau_2)\}\mathbf{N}^T(\tau_2) \\ &= \int_0^{t \wedge s} \mathbf{M}(\tau_1)\mathbf{Q}\mathbf{N}^T(\tau_1)\, d\tau_1.\end{aligned}$$

The second part of the proposition follows since $H^{\omega_i} \perp H^{\omega_j}$ for $i \neq j$ and p_i can be decomposed uniquely as: $p_i = p_{i1} + p_{i2} + \cdots + p_{in}$. Now $p_{ij} \in H^{\omega_j}$ if and only if $p_{ij}(t) = \int_0^t g_{ij}(s)\, d\omega_j(s)$ for $g_{ij}(s)$ satisfying $\int_0^t g_{ij}^2(s)\, ds < \infty$ (Theorem 8.3). □

Stochastic differential and integral equations:

As previously intimated, the utility of Wiener processes is to bring mathematical rigour to the white-noise-driven linear differential equations so frequently used in stochastic control applications. The usual stochastic model for the state estimation problem is taken as:

$$\dot{\boldsymbol{x}}(t) = \mathbf{A}(t)\boldsymbol{x}(t) + \mathbf{D}(t)\boldsymbol{\xi}(t) \tag{8.63}$$

where $\{\boldsymbol{\xi}(t), t \geq 0\}$ is an m-tuple of white-noise sources and $\mathbf{D}(\cdot) \in \mathbb{R}^{n \times m}$ has piecewise continuous elements. If given in integrated form, this yields:

$$\boldsymbol{x}(t) - \boldsymbol{x}(0) = \int_0^t \mathbf{A}(\tau)\boldsymbol{x}(\tau)\, d\tau + \int_0^t \mathbf{D}(\tau)\boldsymbol{\xi}(\tau)\, d\tau \tag{8.64}$$

or if the transition matrix of the deterministic linear differential equation (Chapter 1) is used, this yields the weighting function form:

$$\boldsymbol{x}(t) = \boldsymbol{\Phi}(t, 0)\boldsymbol{x}_0 + \int_0^t \boldsymbol{\Phi}(t, \tau)\mathbf{D}(\tau)\boldsymbol{\xi}(\tau)\, d\tau. \tag{8.65}$$

The difficulties associated with the properties of the white-noise model for $\{\boldsymbol{\xi}(t)\}$ are therefore obviated by the interpretation of $\{\boldsymbol{\xi}(t)\}$ as the formal derivative of a Wiener process $\{\omega(t), t \geq 0\}$. Thus, the integrated white-noise terms may be replaced by rigorous Wiener integrals, viz.:

$$x(t) - x_0 = \int_0^t \mathbf{A}(\tau)x(\tau)\ d\tau + \int_0^t \mathbf{D}(\tau)\ d\omega(\tau) \tag{8.66}$$

or, in the more usual incremental form:

$$dx(t) = \mathbf{A}(t)x(t)\ dt + \mathbf{D}(t)\ d\omega(t) \tag{8.67}$$

and in the weighting function form:

$$x(t) = \mathbf{\Phi}(t, 0)x_0 + \int_0^t \mathbf{\Phi}(t, \tau)\mathbf{D}(\tau)\ d\omega(\tau). \tag{8.68}$$

It is important to recognize that the above incremental form (8.67) is only a formal representation for the integral equation (8.66). The incremental and weighting function forms are mainly used in the sequel. It is worth noting that the process $\{x(t), t \in T\}$ above is a Gaussian–Markov process (Curtain and Pritchard [8]). If the system parameters are themselves random variables an additional dimension is introduced into the problem (Grimble [62]), the book by Soong [58] is a useful introduction to this generalization.

State response

Equation (8.68) which describes the evolution of the process $x(t)$ deserves further investigation. Assume that $0 \leq t_1 \leq t_2$, then the vector $x(t_2)$ is given as:

$$x(t_2) = \mathbf{\Phi}(t_2, 0)x_0 + \int_0^{t_2} \mathbf{\Phi}(t_2, \tau)\mathbf{D}(\tau)\ d\omega(\tau).$$

The initial state x_0 will be assumed to be zero to simplify the analysis. Thus, since $\{\omega(\tau), 0 \leq \tau \leq t\}$ has stationary orthogonal increments, Theorem 8.3 (suitably extended for the vector case) intimates:

$$x(t_2) = \int_0^{t_2} \mathbf{\Phi}(t_2, \tau)\mathbf{D}(\tau)\ d\omega(\tau)$$

is an orthogonal increments process where,

(i) $x(t_2) \in H_{t_2}^{\omega}$

(ii) $x(t_1) = \int_0^{t_1} \mathbf{\Phi}(t_1, \tau)\mathbf{D}(\tau)\ d\omega(\tau) \in H_{t_1}^{\omega}$

$$\text{(iii)}\quad \mathbf{\Phi}(t_2, t_1)\boldsymbol{x}(t_1) = \int_0^{t_1} \mathbf{\Phi}(t_2, \tau)\mathbf{D}(\tau)\ d\omega(\tau) \in H_{t_1}^{\omega}$$

and

$$\text{(iv)}\quad \boldsymbol{x}(t_2) - \mathbf{\Phi}(t_2, t_1)\boldsymbol{x}(t_1) = \int_{t_1}^{t_2} \mathbf{\Phi}(t_2, \tau)\mathbf{D}(\tau)\ d\omega(\tau) \in (H_{t_1}^{\omega})^{\perp}.$$

Thus, the random variable $\boldsymbol{x}(t_2)$ depends only on $\boldsymbol{x}(t_1)$ and the process $\{\omega(\tau), t_1 \leq \tau \leq t_2\}$ via the orthogonality relationship of (iv) above. Clearly, $\boldsymbol{x}(t_2)$ may be identified as being a 'state vector' having similar properties to that in a deterministic system. The model (8.68) would not be so valuable if $\{\omega(\tau)\}$ were not an orthogonal increments process.

Theorem 8.5. *System transition properties. Let $\{\omega(t),\ t > 0\}$ be an m-vector stationary orthogonal increments process, x_0 an n-vector random variable, orthogonal to H^{ω} with*

$$\operatorname{cov}[\omega(t), \omega(\tau)] = \mathbf{Q}\min(t, \tau),\ \mathbf{m}_0 = E\{\boldsymbol{x}_0\} \quad \textit{and} \quad \mathbf{P}_0 = \operatorname{cov}[\boldsymbol{x}_0, \boldsymbol{x}_0].$$

If the state $\boldsymbol{x}(t)$ is described by the weighting-function equation:

$$\boldsymbol{x}(t) = \mathbf{\Phi}(t, 0)\boldsymbol{x}_0 + \int_0^t \mathbf{\Phi}(t, s)\mathbf{D}(s)\ d\omega(s) \tag{8.69}$$

then the mean $\mathbf{m}(t) \triangleq E\{\boldsymbol{x}(t)\}$ and the covariance $\mathbf{P}(t) = \operatorname{cov}[\boldsymbol{x}(t), \boldsymbol{x}(t)]$ satisfy the equations:

$$\dot{\mathbf{m}} = \mathbf{A}\mathbf{m}, \qquad \mathbf{m}(0) = \mathbf{m}_0$$

$$\dot{\mathbf{P}} = \mathbf{A}\mathbf{P} + \mathbf{P}\mathbf{A}^T + \mathbf{D}\mathbf{Q}\mathbf{D}^T, \qquad \mathbf{P}(0) = \mathbf{P}_0. \tag{8.70}$$

Proof. (i)

$$E\{\boldsymbol{x}(t)\} = E\left\{\mathbf{\Phi}(t, 0)\boldsymbol{x}_0 + \int_0^t \mathbf{\Phi}(t, s)\mathbf{D}(s)\ d\omega(s)\right\}$$

$$= \mathbf{\Phi}(t, 0)E\{\boldsymbol{x}_0\}$$

$$\mathbf{m}(t) = \mathbf{\Phi}(t, 0)\mathbf{m}_0$$

$$\Rightarrow \quad \dot{\mathbf{m}}(t) = \mathbf{A}(t)\mathbf{m}(t), \qquad \mathbf{m}(0) = \mathbf{m}_0.$$

(ii) $x_0 \perp H^{\omega}$, that is the initial condition is uncorrelated with the Wiener process.

$$\mathbf{P}(t) \triangleq \text{cov}[x(t), x(t)] = E\{(x(t) - E\{x(t)\})(x(t) - E\{x(t)\})^T\}$$

$$x(t) = \mathbf{\Phi}(t, 0)x_0 + \int_0^t \mathbf{\Phi}(t, \tau)\mathbf{D}(\tau)\, d\omega(\tau)$$

hence

$$x(t) - E\{x(t)\} = \int_0^t \mathbf{\Phi}(t, \tau)\mathbf{D}(\tau)\, d\omega(\tau) + \mathbf{\Phi}(t, 0)(x_0 - \mathbf{m}_0)$$

$$\mathbf{P}(t) = E\left\{\left(\int_0^t \mathbf{\Phi}(t, \tau)\mathbf{D}(\tau)\, d\omega(\tau)\right)\left(\int_0^t d\omega^T(\tau_1)\mathbf{D}^T(\tau_1)\mathbf{\Phi}^T(t, \tau_1)\right)\right\}$$
$$+ E\{\mathbf{\Phi}(t, 0)(x_0 - \mathbf{m}_0)(x_0 - \mathbf{m}_0)^T\mathbf{\Phi}^T(t, 0)\}. \quad (8.71)$$

From the integral property (vi), (Section 8.2.5):

$$\mathbf{P}(t) = \int_0^t \mathbf{\Phi}(t, \tau)\mathbf{D}(\tau)\mathbf{Q}\mathbf{D}^T(\tau)\mathbf{\Phi}^T(t, \tau)\, d\tau + \mathbf{\Phi}(t, 0)\mathbf{P}_0\mathbf{\Phi}^T(t, 0). \quad (8.72)$$

Leibniz theorem for the differentiation of an integral, namely (Abramowitz and Segun [46]):

$$\frac{d}{dt}\int_{a(t)}^{b(t)} f(t, \tau)\, d\tau = \int_{a(t)}^{b(t)} \frac{\partial}{\partial t} f(t, \tau)\, d\tau + f(t, b)\frac{db}{dt} - f(t, a)\frac{da}{dt}$$

was used along with properties of the transition matrix from Section 2.1, Chapter 1 to obtain (8.70) □

This completes the armoury of probabilistic, stochastic and algebraic results needed before embarking on the derivation of the Kalman–Bucy filter by the Hilbert space approach.

8.3. THE STATE ESTIMATION PROBLEM

The Kalman–Bucy filter (for brevity often referred to as the Kalman filter) is derived in the following two sections. An innovations approach has been adopted so that the interesting properties of innovations may be discussed. These properties are often exploited in adaptive filtering schemes (Moir and Grimble [16]) and are useful for test purposes.

The dependence of the Kalman filter on the matrices of the state equations is very convenient for computer implementation. Another computational advantage is that the Kalman gain and the estimation error covariance matrix $\mathbf{P}(t)$ upon which it depends, can be precomputed and are entirely determined by the plant and noise models. The filtered state estimate is then given as the output of a possibly time-varying but known linear system driven by the measured plant output or observations $\{z(t) \text{ for } t > 0\}$.

8.3.1. The innovations signal process

Assume that the output of a control system or message signal in a communications channel can only be measured in the presence of noise. This measurement or *observation process* is taken to have the incremental form:

$$dz(t) = y(t)\ dt + dv(t) \tag{8.73}$$

where $\{v(t), 0 \leq t\}$ is an orthogonal increments process representing measurement noise $\{y(t), 0 \leq t\}$ is a quadratic mean continuous process representing the system output and $\{z(t), 0 \leq t\}$ represents the observations. Note that since (8.73) is an abbreviated notation for an integral equation, to obtain the quantities which are physically measurable derivatives must be taken as illustrated in Fig. 8.6. Thus, equation (8.73) may be written as:

$$z(t) = \int_0^t y(\tau)\ d\tau + \int_0^t dv(\tau) = \int_0^t y(\tau)\ d\tau + v(t). \tag{8.74}$$

Based on (8.74), it is useful to define some Hilbert subspaces:

$$H_t^z \triangleq \mathscr{L}\{z(\tau),\ 0 \leq \tau \leq t\}, \qquad H_t^y = \mathscr{L}\{y(\tau),\ 0 \leq \tau \leq t\}$$

$$\text{and} \quad H_t^v = \mathscr{L}\{v(\tau), 0 \leq \tau \leq t\}.$$

It is usual to assume that the measurement noise and the system output signal

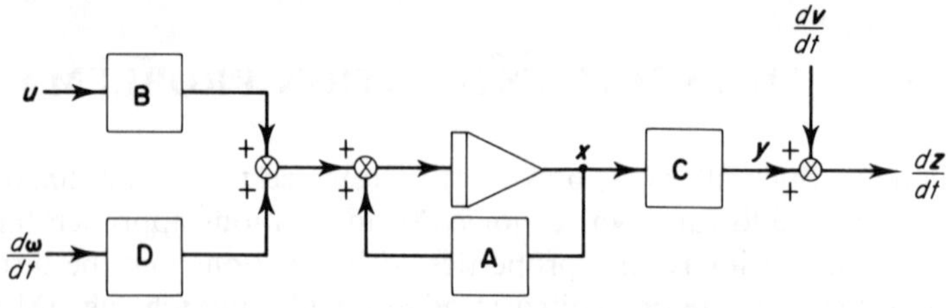

Fig. 8.6. Continuous-time plant or signal process in state equation form

are orthogonal so that $H_t^v \perp H_t^y$ for all $t \geq 0$. This assumption can easily be relaxed, but in many industrial situations the assumption is valid since disturbances and measurement noise are often uncorrelated. The measurement noise may be written as:

$$dv(t) = \mathbf{M}(t)\ dv_1(t) \tag{8.75}$$

where $\{v_1(t), t \geq 0\}$ has *stationary* orthogonal increments, $\mathbf{M}(t) \triangleq [m_{ij}(t)]$, $\text{cov}[v_1(t), v_1(\tau)] = (t \wedge \tau)\mathbf{I}_r$ and matrix $\mathbf{R}(t) \triangleq \mathbf{M}(t)\mathbf{M}^T(t) > \mathbf{O}$ for all $t \geq 0$.

To estimate the process output $y(t) \in H_t^y$ from the process observations up until time instant t, namely subspace H_t^z, the orthogonal projection theorem is utilized. Introduce the projection operator $\mathscr{P}_t^z$: $H_t^y \to H_t^z$, then the optimal estimate of the output obtains as:

$$\hat{y}(t \mid t) = \mathscr{P}_t^z y(t) \quad \text{where } \hat{y}(t \mid t) \in H_t^z.$$

Exploiting the orthogonal principle:

$$(y(t) - \hat{y}(t \mid t)) \perp H_t^z$$

or in terms of inner products:

$$\langle y(t) - \hat{y}(t \mid t),\ z \rangle_{H_r} \triangleq E\{\langle y(t) - \hat{y}(t \mid t), z \rangle_{E_r}\} = 0$$

for any $z \in H_t^z$ and thus the vectors $y(t) - \hat{y}(t \mid t)$ and z are uncorrelated. The projection operator therefore projects $y(t)$ onto the subspace H_t^z to give the best estimate of $y(t)$ (denoted by $\hat{y}(t \mid t)$) where optimality is measured in a mean-square sense (Kalman [1]).

Using this optimal estimate, the innovations process is defined as:

$$\boldsymbol{\nu}(t) = z(t) - \int_0^t \hat{y}(\tau \mid \tau)\ d\tau \in H_t^z. \tag{8.76}$$

It is useful to compare (8.76) with a rearrangement of the observation process (8.73), namely:

$$\int_0^t dv(\tau) = z(t) - \int_0^t y(\tau)\ d\tau. \tag{8.77}$$

Since the plant output $y(t)$ is unknown, it is natural to replace $y(t)$ in equation (8.77) by its optimal estimate $\hat{y}(t \mid t)$ to obtain the so-called innovations signal (8.76). Thus the measurement noise and the innovations process are related in some way. This may be pursued further by defining the output prediction error as:

$$\tilde{y}(t) = y(t) - \hat{y}(t \mid t)$$

and using (8.76) and (8.77) to obtain:

$$\boldsymbol{\nu}(t) = \int_0^t \tilde{y}(\tau)\, d\tau + \int_0^t dv(\tau). \tag{8.78}$$

Among their fascinating properties are two which show the innovation process mean and covariance to be identical to those of the measurement noise process. This is developed in the following theorem which demonstrates that the innovation process has orthogonal increments and spans the same family of subspaces as the observations signal $\{z(t)\}$, namely $H_t^\nu \equiv H_t^z$ for all $t \geq 0$. It is convenient to introduce a scaled and stationary innovation process as:

$$\boldsymbol{\nu}_1(t) = \int_0^t \mathbf{M}^{-1}(\tau)\, d\boldsymbol{\nu}(\tau). \tag{8.79}$$

Theorem 8.6. ***Properties of an innovations process.***

If (*i*) $\{v_1(t)\}$ *has stationary orthogonal increments,*

(*ii*) $\{y(t)\}$ *is a quadratic mean continuous process,*

(*iii*) $H_t^{v_1} \perp H_t^y$, *and* $t \geq 0$,

and

(*iv*) $\mathbf{R}(t) = \mathbf{M}(t)\mathbf{M}^T(t) > \mathbf{O}$ *for all* $t \geq 0$,

then

(a) $\boldsymbol{\nu}(t) - \boldsymbol{\nu}(\tau) \perp H_\tau^z$, (8.80)

(b) $$\operatorname{cov}[\boldsymbol{\nu}(t), \boldsymbol{\nu}(t)] = \int_0^t \mathbf{M}(\tau)\mathbf{M}^T(\tau)\, d\tau, \tag{8.81}$$

for all τ *and* $t > \tau$.

Proof. Choose any $z \in H_\tau^z$ where $0 \leq \tau \leq t$, then

(i) The relationship between subspaces H^z yields

$$H_\tau^z \subseteq H_t^z, \qquad 0 \leq \tau \leq t.$$

(ii) The optimality condition for estimate $\hat{y}(t \mid t)$ gives

$$\langle y(t) - \hat{y}(t \mid t), z\rangle_{H_r} = \langle \tilde{y}(t), z\rangle = 0 \text{ for } z \in H_t^z.$$

But since $H_\tau^z \subseteq H_t^z$ it also follows:

$$\langle y(t) - \hat{y}(t \mid t), z \rangle_{H_r} = \langle \tilde{y}(t), z \rangle = 0 \quad \text{for all } z \in H_\tau^z \text{ and } 0 \le \tau \le t.$$

The final development of this argument is

$$\langle y(s) - \hat{y}(s \mid s), z \rangle_{H_r} = \langle \tilde{y}(s), z \rangle = 0 \quad \text{for all } z \in H_\tau^z$$

$$\text{and for } 0 \le \tau \le s \le t.$$

(iii) Recall

$$z(\tau) = \int_0^\tau y(s)\, ds + \int_0^\tau \mathbf{M}(s)\, dv_1(s) \in H_\tau^z.$$

From Proposition 8.3

$$v(\tau) = \int_0^\tau \mathbf{M}(s)\, dv_1(s) \in H_\tau^{v_1}$$

where $v_1(\cdot)$ is an orthogonal increments process. Thus,

$$\langle z(\tau), v(t) - v(\tau) \rangle = \langle \int_0^\tau y(s)\, ds + v(\tau), v(t) - v(\tau) \rangle = 0$$

since assumption (iii) is $H_t^{v_1} \perp H_t^y$ and $v(\cdot)$ is an orthogonal increments process.

(iv) The innovation increment is

$$\boldsymbol{\nu}(t) - \boldsymbol{\nu}(\tau) = \int_\tau^t \tilde{y}(s)\, ds + \int_\tau^t \mathbf{M}(s)\, dv_1(s) = \int_\tau^t \tilde{y}(s)\, ds + (v(t) - v(\tau)). \tag{8.82}$$

Then for any $z(\tau) \in H_\tau^z$ and using results (ii) and (iii) above gives:

$$\langle z(\tau), \boldsymbol{\nu}(t) - \boldsymbol{\nu}(\tau) \rangle = \langle z(\tau), \int_\tau^t \tilde{y}(s)\, ds + (v(t) - v(\tau)) \rangle = 0 \quad \text{for } 0 \le \tau \le t \tag{8.83}$$

and $\boldsymbol{\nu}(t) - \boldsymbol{\nu}(\tau) \perp H_\tau^z$. (This proves part(a)).

To show part (b) of the theorem, write (8.82) in elemental form, viz.

$$\nu_i(t) - \nu_i(\tau) = \int_\tau^t \tilde{y}_i(s)\, ds + \int_\tau^t \sum_k m_{ik}(s)\, dv_{1k}(s) \tag{8.84}$$

and consider any partition $\tau = t_1 < t_2 < t_2 \cdots < t_n = t$ of the interval $[\tau, t]$ and recall the orthogonal increments property of the innovations to obtain:

$$E\{(\nu_i(t) - \nu_i(\tau))^2\} = \sum_j E\{(\nu_i(t_{j+1}) - \nu_i(t_j))^2\} \tag{8.85}$$

$$= \sum_j E\left\{\left(\int_{t_j}^{t_{j+1}} \tilde{y}_i(\sigma)\, d\sigma + \sum_k \int_{t_j}^{t_{j+1}} m_{ik}(\sigma)\, dv_{1k}(\sigma)\right)^2\right\}. \tag{8.86}$$

Using the Schwarz inequality, as $(t_{j+1} - t_j) \to 0$, three of the terms in the squared expression can be shown (Davis [9]) to converge to zero, hence

$$E\{(\nu_i(t) - \nu_i(\tau))^2\} = \sum_j \sum_k E\left\{\left(\int_{t_j}^{t_{j+1}} m_{ik}(\sigma)\, dv_{1k}(\sigma)\right)^2\right\}$$

$$= \sum_k \int_0^t m_{ik}^2(\sigma)\, d\sigma. \tag{8.87}$$

By similar arguments:

$$E\{(\nu_i(t) - \nu_i(\tau))(\nu_j(t) - \nu_j(\tau))\} = \sum_k \int_0^t m_{ik}(\sigma) m_{jk}(\sigma)\, d\sigma.$$

Equation (8.81) follows by setting $\tau = 0$. □

It is perhaps surprising that $\{\boldsymbol{\nu}(\tau), 0 \le \tau \le t\}$ and $\{z(\tau), 0 \le \tau \le t\}$ generate the same subspaces. To demonstrate this fact, note that by definition

$$\boldsymbol{\nu}(\tau) \in H_\tau^z \subset H_t^z \text{ for all } \tau \le t \text{ thus } H_t^\nu \subset H_t^z.$$

To show the reverse inclusion holds, $z(t)$ must be expressed as a Wiener integral with respect to the innovations process $\boldsymbol{\nu}(t)$. This result is given by Davis ([9]) as:

Theorem 8.7. ***Observation and innovation process subspaces.*** *Suppose the four conditions (i)–(iv) of Theorem* 8.6 *are satisfied, then for each t, $H_t^z \equiv H_t^\nu$.*

In the older white-noise terminology, the innovations process is white and time-varying. The covariance matrix for the innovations and the measurement noise are identical.

8.3.2. The Kalman–Bucy filter

The linear system to be considered is assumed to have a linear stochastic differential equation representation:

$$dx(t) = \mathbf{A}(t)x(t)\ dt + \mathbf{D}(t)\ d\omega(t) + \mathbf{B}(t)u(t)\ dt \tag{8.88}$$

where the state vector $x(t) \in \mathbb{R}^n$. The output equation for the system is:

$$y(t) = \mathbf{C}(t)x(t) \in \mathbb{R}^r$$

and the observation process has the form:

$$dz(t) = \mathbf{C}(t)x(t)\ dt + dv(t). \tag{8.89}$$

The continuous-time process equations are shown in Fig. 8.6. It can be seen that there is a slight notational difficulty in transposing the stochastic differential equation (8.88) to diagrammatic form. The problem arises because equation (8.88) represents an integral equation, thus to obtain Fig. 8.6, plant signals are differentiated and formal white-noise signals result.

The various process assumptions and statistics are now given.

Initial values. Let $t_0 = 0$, and assume that the initial value of the mean $E\{x(0)\} = \mathbf{m}_0 = \mathbf{o} \in \mathbb{R}^n$. The initial state $x_0 = x(0)$ satisfies:

$$\operatorname{cov}[x_0, x_0] = \mathbf{P}_0 \in \mathbb{R}^{n\times n}$$

where $\mathbf{P}_0$ is given *a priori.*

Process model assumptions. The elements of the parameter matrices $\mathbf{A}(t) \in \mathbb{R}^{n\times n}$, $\mathbf{B}(t) \in \mathbb{R}^{n\times m}$, $\mathbf{C}(t) \in \mathbb{R}^{r\times n}$ and $\mathbf{D}(t) \in \mathbb{R}^{n\times q}$ are assumed to be piecewise continuous. The control input $u(t) \in \mathbb{R}^m$ is taken as zero for all $t \geq 0$ in the preliminary derivation. Any deterministic input to the plant must be input to the filter but such inputs do not affect the stochastic solution procedure and it is therefore convenient to neglect them until the last stages of the proof.

Noise statistics. The process noise $\omega(\cdot) \in \mathbb{R}^q$ and the measurement noise $v(\cdot) \in \mathbb{R}^r$ are taken, without loss of generality, to have zero mean, where

$$\operatorname{cov}[\omega(t), \omega(t)] = \mathbf{Q}t, \text{ and } \operatorname{cov}[v(t), v(t)] = \int_0^t \mathbf{R}(\tau)\ d\tau.$$

The constant diagonal matrix $\mathbf{Q} \in \mathbb{R}^{q\times q}$ is positive semidefinite. Noise processes $\omega(\cdot)$, $v(\cdot)$ and the random vector x_0 are assumed to be mutually orthogonal, namely $H^{\omega} \perp H^{v} \perp H^{x_0}$.

Introduce two useful Hilbert subspaces, one comprising state vectors:

$$H_t^x \triangleq \mathscr{L}\{x(\tau), 0 \leq \tau \leq t\}$$

and the other observation vectors:

$$H_t^z \triangleq \mathscr{L}\{z(\tau), 0 \leq \tau \leq t\}.$$

Invoking the orthogonal projection theorem, the optimal estimate of $\boldsymbol{x}(t)$ based on the observation subspace H_t^z is given by:

$$\hat{\boldsymbol{x}}(t \mid t) = \mathscr{P}_t^z \boldsymbol{x}(t) \in H_t^z$$

where $\mathscr{P}_t^z$: $H_t^x \to H_t^z$. The *estimation error* may now be evaluated as:

$$\tilde{\boldsymbol{x}}(t) = \boldsymbol{x}(t) - \hat{\boldsymbol{x}}(t \mid t) \in \mathbb{R}^n. \tag{8.90}$$

Using incremental notation, the innovations process (equation (8.76)) is defined as:

$$d\boldsymbol{\nu}(t) = dz(t) - \hat{y}(t \mid t)\ dt \tag{8.91}$$

but

$$\hat{y}(t \mid t) \equiv \mathbf{C}(t)\hat{x}(t \mid t), \tag{8.92}$$

hence equations (8.89) to (8.92) yield the innovations:

$$d\boldsymbol{\nu}(t) = \mathbf{C}(t)\tilde{x}(t)\ dt + dv(t). \tag{8.93}$$

The four conditions of Theorem 8.6 are satisfied (in particular $H^x \perp H^v$) and $\{\boldsymbol{\nu}(t), 0 \leq \tau \leq t\}$ represents an orthogonal increments process with strength:

$$\operatorname{cov}[\boldsymbol{\nu}(t), \boldsymbol{\nu}(t)] = \int_0^t \mathbf{R}(\tau)\ d\tau \in \mathbb{R}^{r \times r}, \tag{8.94}$$

and $H_t^\nu = H_t^z$. Clearly, the estimate $\hat{\boldsymbol{x}}(t \mid t) \in H_t^z$ and the subspace identification permits an application of Proposition 8.4 to give:

$$\hat{\boldsymbol{x}}(t \mid t) = \int_0^t \mathbf{H}_0(t, \tau)\ d\boldsymbol{\nu}(\tau) \tag{8.95}$$

where the time-varying impulse response of the filter is given by:

$$\mathbf{H}_0(t,\tau)=\frac{d}{d\tau}E\{\boldsymbol{x}(t)\boldsymbol{\nu}^T(\tau)\}\mathbf{R}(\tau)^{-1}. \tag{8.96}$$

Although expressions (8.95) and (8.96) are of little use for filter computation and do not provide a link with the state equations for the system, the optimal estimator problem is solved! These expressions show that the orthogonal projection estimate is obtained as the output of a linear time-varying system (namely a linear filter) driven by a derivative of an innovations process.

The above disadvantages are overcome in the Kalman–Bucy filter.

Theorem 8.8. ***Kalman–Bucy filter.*** *Consider the system* (8.88), (8.89) *and uncorrelated noise sources described above. The optimal estimator which minimises the mean-square estimation error:*

$$\langle \tilde{\boldsymbol{x}}(t),\tilde{\boldsymbol{x}}(t)\rangle \triangleq E\{\tilde{\boldsymbol{x}}^T(t)\tilde{\boldsymbol{x}}(t)\}$$

is given by the following stochastic differential equation:

$$d\hat{\boldsymbol{x}}(t\mid t)=\mathbf{A}(t)\hat{\boldsymbol{x}}(t\mid t)\ dt+\mathbf{K}(t)(d\boldsymbol{z}(t)-\mathbf{C}(t)\hat{\boldsymbol{x}}(t\mid t)\ dt)+\mathbf{B}(t)\boldsymbol{u}(t)\ dt \tag{8.97}$$

where $\hat{\boldsymbol{x}}(t_0\mid t_0)\triangleq E\{\boldsymbol{x}(t_0)\}=\mathbf{m}_0$, *and the Kalman gain matrix:*

$$\mathbf{K}(t)\triangleq \mathbf{P}(t)\mathbf{C}^T(t)\mathbf{R}(t)^{-1},\qquad t\geq t_0. \tag{8.98}$$

The estimation error covariance matrix $\mathbf{P}(t)\triangleq E\{\tilde{\boldsymbol{x}}(t)\tilde{\boldsymbol{x}}^T(t)\}$ *satisfies the matrix Riccati differential equation:*

$$\dot{\mathbf{P}}(t)=\mathbf{A}(t)\mathbf{P}(t)+\mathbf{P}(t)\mathbf{A}^T(t)+\mathbf{D}(t)\mathbf{Q}\mathbf{D}^T(t)-\mathbf{P}(t)\mathbf{C}^T(t)\mathbf{R}(t)^{-1}\mathbf{C}(t)\mathbf{P}(t) \tag{8.99}$$

where

$$\mathbf{P}(t_0)\triangleq \mathbf{P}_0=\operatorname{cov}[\boldsymbol{x}_0,\boldsymbol{x}_0]. \tag{8.100}$$

Proof. The proof of the theorem follows in six stages.

1. Filter dynamics: The filter dynamics are shown to have the form $d\hat{\boldsymbol{x}}(t\mid t)=\mathbf{A}(t)\hat{\boldsymbol{x}}(t\mid t)\ dt+d\boldsymbol{p}(t)$ where $\boldsymbol{p}(t)$ has orthogonal increments with respect to the observations subspace H_t^z.
2. Estimation equation: A weighting-function form for the filter is derived as $\hat{\boldsymbol{x}}(t\mid t)=\int_0^t \mathbf{\Phi}(t,\tau)\mathbf{G}(\tau)\mathbf{M}^{-1}(\tau)\ d\boldsymbol{\nu}(\tau)$ where $\mathbf{\Phi}(t,\tau)$ is the system transition matrix.

3. An expression for $G(t)$: The impulse response matrix for the Kalman–Bucy filter is derived as $\mathbf{\Phi}(t,\tau)\mathbf{P}(\tau)\mathbf{C}^T(\tau)\mathbf{R}(\tau)^{-1}$ where $\mathbf{P}(t) \triangleq E\{\tilde{x}(t)\tilde{x}^T(t)\}$ is the estimation error covariance matrix.
4. Matrix Riccati differential equation: The relationship for the evolution of $\mathbf{P}(t)$.
5. Initial conditions: The modifications for non zero initial state estimates are discussed.
6. Solution of the Riccati equation: The conditions for the solutions of the matrix Riccati equation are presented.

1. *Filter dynamics*

The dynamics of the Kalman–Bucy filter are related to those of the plant and have the form $d\hat{x}(t \mid t) = \mathbf{A}(t)\hat{x}(t \mid t)\, dt + dp(t)$. To demonstrate this result and to obtain an expression for $p(t)$, assume $t_0 = 0$, $\mathbf{m}_0 = E\{x(0)\} = \mathbf{o}$ and let,

$$p(t) \triangleq \hat{x}(t \mid t) - \hat{x}(0 \mid 0) - \int_0^t \mathbf{A}(\tau)\hat{x}(\tau \mid \tau)\, d\tau. \tag{8.101}$$

To show that $(p(t) - p(\tau)) \perp H_\tau^z$ $(t > \tau)$ first obtain:

$$p(t) - p(\tau) = \hat{x}(t \mid t) - \hat{x}(\tau \mid \tau) - \int_\tau^t \mathbf{A}(\tau_1)\hat{x}(\tau_1 \mid \tau_1)\, d\tau_1. \tag{8.102}$$

Now

$$\mathscr{P}_\tau^z\, \hat{x}(t \mid t) = \mathscr{P}_\tau^z \mathscr{P}_t^z x(t) = \mathscr{P}_\tau^z x(t)$$

and also,

$$\mathscr{P}_\tau^z \int_\tau^t \mathbf{A}(\tau_1)\hat{x}(\tau_1 \mid \tau_1)\, d\tau_1 = \int_\tau^t \mathbf{A}(\tau_1)\mathscr{P}_\tau^z \mathscr{P}^z{}_{\tau_1} x(\tau_1)\, d\tau_1 = \mathscr{P}_\tau^z \int_\tau^t \mathbf{A}(\tau_1)x(\tau_1)\, d\tau_1.$$

From these results and (8.102):

$$\mathscr{P}_\tau^z(p(t) - p(\tau)) = \mathscr{P}_\tau^z\left(x(t) - x(\tau) - \int_\tau^t \mathbf{A}(\tau_1)x(\tau_1)\, d\tau_1\right)$$

and from (8.88):

$$x(t) - x(\tau) = \int_\tau^t \mathbf{A}(\tau_1)x(\tau_1)\, d\tau_1 + \int_\tau^t \mathbf{D}(\tau_1)\, d\omega(\tau_1)$$

thus

$$\mathscr{P}_\tau^z(p(t)-p(\tau)) = \mathscr{P}_\tau^z\left(\int_\tau^t \mathbf{D}(\tau_1)\ d\omega(\tau_1)\right) = o$$

and $(p(t)-p(\tau)) \perp H_\tau^z$.

Since $p(t)$ has orthogonal increments with respect to H_t^z and since $H_t^{\nu_1} = H_t^{\nu} = H_t^z$ (recall the definition of ν_1 from equation (8.79)) then from Theorem 8.3 and Proposition 8.3 there exists a matrix $\mathbf{G}(t) = [g_{ij}(t)]$ such that:

$$p(t) = \int_0^t \mathbf{G}(\tau)\ d\nu_1(\tau) \tag{8.103}$$

and

$$\sum_{i,j} \int_0^t g_{ij}^2(\tau)\ d\tau < \infty.$$

2. *Estimation equation*

The estimation equation follows from (8.101) and (8.103) as:

$$\hat{x}(t \mid t) = \int_0^t \mathbf{A}(\tau)\hat{x}(\tau \mid \tau)\ d\tau + \hat{x}(0 \mid 0) + \int_0^t \mathbf{G}(\tau)\ d\nu_1(\tau)$$

or

$$d\hat{x}(t \mid t) = \mathbf{A}(t)\hat{x}(t \mid t)\ dt + \mathbf{G}(t)\ d\nu_1(t) \tag{8.104}$$

and $\hat{x}(0 \mid 0) = E\{x(t_0)\} = \mathbf{o}$. Clearly, from (8.104)

$$\hat{x}(t \mid t) = \int_0^t \mathbf{\Phi}(t,\tau)\mathbf{G}(\tau)\ d\nu_1(\tau)$$

and from (8.79) an alternative expression for the optimal filter becomes:

$$\hat{x}(t \mid t) = \int_0^t \mathbf{\Phi}(t,\tau)\mathbf{G}(\tau)\mathbf{M}^{-1}(\tau)\ d\nu(\tau). \tag{8.105}$$

By comparison with (8.95):

$$\mathbf{H}_0(t,\tau) = \mathbf{\Phi}(t,\tau)\mathbf{G}(\tau)\mathbf{M}^{-1}(\tau) \tag{8.106}$$

and hence $\mathbf{G}(t) = \mathbf{H}_0(t,t)\mathbf{M}(t)$.

3. *Expression for* $\mathbf{H}_0(t, t)$

From (8.75) and (8.93):

$$\boldsymbol{\nu}(\tau) = \int_0^T \mathbf{C}(\tau_1)\tilde{\boldsymbol{x}}(\tau_1)\, d\tau_1 + \int_0^T \mathbf{M}(\tau_1)\, d\boldsymbol{v}_1(\tau_1)$$

and since $\boldsymbol{x}(t) \perp H_\tau^v$ then,

$$E\{\boldsymbol{x}(t)\boldsymbol{\nu}^T(\tau)\} = E\left\{\boldsymbol{x}(t) \int_0^T \tilde{\boldsymbol{x}}^T(\lambda)\mathbf{C}^T(\lambda)\, d\lambda\right\}. \tag{8.107}$$

Also, from (8.68):

$$\boldsymbol{x}(t) = \boldsymbol{\Phi}(t, \lambda)\boldsymbol{x}(\lambda) + \int_\lambda^t \boldsymbol{\Phi}(t, s)\mathbf{D}(s)\, d\boldsymbol{\omega}(s), \qquad t > \lambda$$

and

$$E\{\boldsymbol{x}(t)\tilde{\boldsymbol{x}}^T(\lambda)\} = \boldsymbol{\Phi}(t, \lambda)E\{\boldsymbol{x}(\lambda)\tilde{\boldsymbol{x}}^T(\lambda)\} + \int_\lambda^t \boldsymbol{\Phi}(t, s)\mathbf{D}(s)E\{\, d\boldsymbol{\omega}(s)\tilde{\boldsymbol{x}}^T(\lambda)\}. \tag{8.108}$$

Now

$$\boldsymbol{\omega}(s) \perp H_\lambda^{\omega, v} = \mathscr{L}\{\boldsymbol{\omega}(\tau), \boldsymbol{v}(\tau); \tau \leq \lambda\}, \qquad s > \lambda,$$

and $\tilde{\boldsymbol{x}}(u) \in H_\lambda^{\omega, v}$ and the last term may be neglected in (8.108). Thus,

$$E\{\boldsymbol{x}(t)\tilde{\boldsymbol{x}}^T(\lambda)\} = \boldsymbol{\Phi}(t, \lambda)E\{\boldsymbol{x}(\lambda)\tilde{\boldsymbol{x}}^T(\lambda)\}. \tag{8.109}$$

From (8.90) $\boldsymbol{x}(\lambda) = \tilde{\boldsymbol{x}}(\lambda) + \hat{\boldsymbol{x}}(\lambda \mid \lambda)$ and by definition of the optimal estimate $\tilde{\boldsymbol{x}}(\lambda) \perp H_\lambda^{\xi}$ and $\hat{\boldsymbol{x}}(\lambda \mid \lambda) \in H_\lambda^{\xi}$. Substituting these results in (8.109) obtain:

$$E\{\boldsymbol{x}(t)\tilde{\boldsymbol{x}}^T(\lambda)\} = \boldsymbol{\Phi}(t, \lambda)E\{\tilde{\boldsymbol{x}}(\lambda)\tilde{\boldsymbol{x}}^T(\lambda)\} = \boldsymbol{\Phi}(t, \lambda)\mathbf{P}(\lambda) \tag{8.110}$$

where the *estimation error covariance matrix* is defined as:

$$\mathbf{P}(t) = E\{\tilde{\boldsymbol{x}}(t)\tilde{\boldsymbol{x}}^T(t)\}. \tag{8.111}$$

It now follows from (8.107):

$$E\{\boldsymbol{x}(t)\boldsymbol{\nu}^T(\tau)\} = \int_0^T \boldsymbol{\Phi}(t, \lambda)\mathbf{P}(\lambda)\mathbf{C}^T(\lambda)\, d\lambda$$

and by substituting in (8.96) obtain:

$$\mathbf{H}_0(t,\tau) = \mathbf{\Phi}(t,\tau)\mathbf{P}(\tau)\mathbf{C}^T(\tau)\mathbf{R}(\tau)^{-1}. \tag{8.112}$$

This is the desired *impulse response matrix* for the Kalman–Bucy filter, and from (8.106):

$$\mathbf{G}(t) = \mathbf{P}(t)\mathbf{C}^T(t)\mathbf{R}(t)^{-1}\mathbf{M}(t).$$

4. *Matrix Riccati differential equation*

To derive the matrix Riccati equation from which the estimation error covariance matrix $\mathbf{P}(t)$ can be calculated, first recall from (8.104), (8.79) and (8.93):

$$\begin{aligned} d\hat{x}(t\,|\,t) &= \mathbf{A}(t)\hat{x}(t\,|\,t)\;dt + \mathbf{G}(t)\;d\boldsymbol{\nu}_1(t) \\ &= \mathbf{A}(t)\hat{x}(t\,|\,t)\;dt + \mathbf{G}(t)\mathbf{M}^{-1}(t)(\mathbf{C}(t)\tilde{x}(t)\;dt + dv(t)). \end{aligned} \tag{8.113}$$

Subtracting this equation from (8.88) gives:

$$d\tilde{x}(t) = (\mathbf{A}(t) - \mathbf{G}(t)\mathbf{M}^{-1}(t)\mathbf{C}(t))\tilde{x}(t)\;dt + \mathbf{D}(t)\;d\omega(t) - \mathbf{G}(t)\;dv_1(t). \tag{8.114}$$

Note at this point that if the control signal input is non-zero the estimation error equation (8.114) will be unchanged if an additional term $\mathbf{B}(t)\boldsymbol{u}(t)\;dt$ is added into the estimator (8.113).

Let the transition matrix for $\tilde{x}$ corresponding to the system matrix:

$$\mathbf{A}_c \triangleq \mathbf{A}(t) - \mathbf{G}(t)\mathbf{M}^{-1}(t)\mathbf{C}(t) = \mathbf{A}(t) - \mathbf{P}(t)\mathbf{C}^T(t)\mathbf{R}(t)^{-1}\mathbf{C}(t) \tag{8.115}$$

be denoted by $\mathbf{\Psi}(t,\tau)$. From (8.114), since $\mathbf{G}(t) = \mathbf{P}(t)\mathbf{C}^T(t)\mathbf{R}(t)^{-1}\mathbf{M}(t)$,

$$\begin{aligned} \tilde{x}(t) = \mathbf{\Psi}(t,0)\tilde{x}(0) &+ \int_0^t \mathbf{\Psi}(t,\tau)\mathbf{D}(\tau)\;d\omega(\tau) \\ &- \int_0^t \mathbf{\Psi}(t,\tau)\mathbf{P}(\tau)\mathbf{C}^T(\tau)\mathbf{R}(\tau)^{-1}\mathbf{M}(\tau)\;dv_1(\tau). \end{aligned} \tag{8.116}$$

Since the three terms on the right of this expression are orthogonal, defining $\mathbf{P}_0 = E\{\tilde{x}(0)\tilde{x}^T(0)\}$, obtains:

$$\begin{aligned} \mathbf{P}(t) = \mathbf{\Psi}(t,0)\mathbf{P}_0\mathbf{\Psi}^T(t,0) &+ \int_0^t \mathbf{\Psi}(t,\tau)\mathbf{D}(\tau)\mathbf{Q}\mathbf{D}^T(\tau)\mathbf{\Psi}^T(t,\tau)\;d\tau \\ &+ \int_0^t \mathbf{\Psi}(t,\tau)\mathbf{P}(\tau)\mathbf{C}^T(\tau)\mathbf{R}(\tau)^{-1}\mathbf{C}(\tau)\mathbf{P}(\tau)\mathbf{\Psi}^T(t,\tau)\;d\tau. \end{aligned} \tag{8.117}$$

Differentiating this expression and omitting the time dependence obtain:

$$\dot{\mathbf{P}}(t) = \mathbf{A}_c\mathbf{\Psi}\mathbf{P}_0\mathbf{\Psi}^T + \mathbf{\Psi}\mathbf{P}_0\mathbf{\Psi}^T\mathbf{A}_c^T + \mathbf{D}\mathbf{Q}\mathbf{D}^T$$

$$+ \int_0^t \mathbf{A}_c(t)\mathbf{\Psi}\mathbf{D}\mathbf{Q}\mathbf{D}^T\mathbf{\Psi}^T\, d\tau + \int_0^t \mathbf{\Psi}\mathbf{D}\mathbf{Q}\mathbf{D}^T\mathbf{\Psi}^T\mathbf{A}_c^T(t)\, d\tau + \mathbf{P}\mathbf{C}^T\mathbf{R}^{-1}\mathbf{C}\mathbf{P}$$

$$+ \int_0^t \mathbf{A}_c(t)\mathbf{\Psi}\mathbf{P}\mathbf{C}^T\mathbf{R}^{-1}\mathbf{C}\mathbf{P}\mathbf{\Psi}^T\, d\tau + \int_0^t \mathbf{\Psi}\mathbf{P}\mathbf{C}^T\mathbf{R}^{-1}\mathbf{C}\mathbf{P}\mathbf{\Psi}^T\mathbf{A}^T(t)\, d\tau. \tag{8.118}$$

The integrals were differentiated using Leibniz theorem for the differentiation of an integral (Abramowitz and Segun [46]). Substituting from (8.117) into (8.118) obtain:

$$\dot{\mathbf{P}}(t) = \mathbf{A}_c(t)\mathbf{P}(t) + \mathbf{P}(t)\mathbf{A}_c^T(t) + \mathbf{D}(t)\mathbf{Q}\mathbf{D}^T(t)$$
$$+ \mathbf{P}(t)\mathbf{C}^T(t)\mathbf{R}(t)^{-1}\mathbf{C}(t)\mathbf{P}(t) \tag{8.119}$$

and substituting from (8.115) the desired Riccati equation becomes:

$$\dot{\mathbf{P}}(t) = \mathbf{A}(t)\mathbf{P}(t) + \mathbf{P}(t)\mathbf{A}^T(t) + \mathbf{D}(t)\mathbf{Q}\mathbf{D}^T(t) - \mathbf{P}(t)\mathbf{C}^T(t)\mathbf{R}(t)^{-1}\mathbf{C}(t)\mathbf{P}(t).$$

The initial conditions follows from (8.117) as $\mathbf{P}(0) = \mathbf{P}_0$.

5. *Initial conditions*

If the expected value of the initial state $\mathbf{m}_0 = E\{\mathbf{x}_0\}$ is non-zero the initial estimate $\hat{\mathbf{x}}_0 = \mathbf{m}_0$. This choice of initial condition ensures that the estimation error covariance matrix which depends upon (8.117) and (8.104) is unchanged and the solution of the optimization problem proceeds as before.

6. *Solution of the Riccati equation*

The results of Section 1.5.2 can be used to investigate the solution of the Riccati equation (8.99). The equation has a unique solution which is positive semidefinite and bounded on the interval $[0, t]$ (see Problem 8.3).

This completes the six stages of the proof for Theorem 8.7. □

Filter structure

The Kalman–Bucy filter has the structure shown in Fig. 8.7. Notice that the control signal $\mathbf{u}(t)$ enters the filter in the same manner as it enters the plant, via the input matrix $\mathbf{B}(t)$. When this signal is non-zero it must be included in

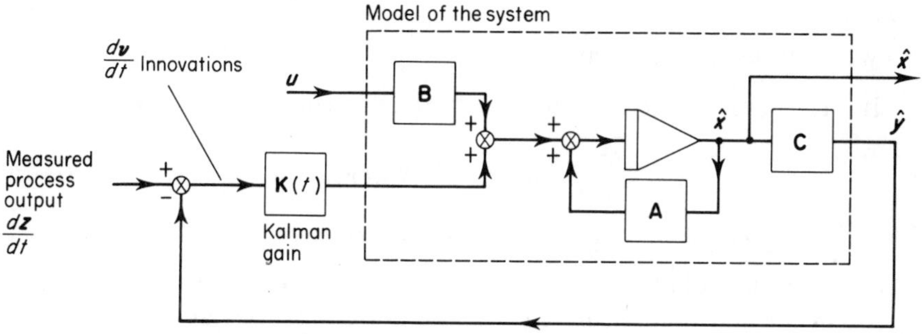

Fig. 8.7. The continuous-time Kalman–Bucy filter

(8.97) if the estimates are to be unbiased. The estimation error equation and covariance expressions are then unchanged, as the note after (8.114) explains. Note that $\boldsymbol{u}(t)$ has been called a control input because this is how the signal arises in later chapters. However, $\boldsymbol{u}(t)$ can represent any known or measurable signal and hence the general rule to follow is that when a plant input is measurable it must be input to the filter. If a disturbance, for example, can be measured then it should be fed into the filter model at the equivalent point of entry into the system equations.

The filter has a feedback structure which may explain why numerous applications have been very successful. Feedback ensures that the filter is relatively robust in the presence of modelling errors and, like the classical feedback control loop, the filter operates well in a wide range of applications.

Kalman filter stability

When considering the 'dual' optimal control problem in Chapter 2, it was noted that the presence of a cross-product term in the cost function changed the usual frequency-domain condition for optimality. That is, the determinant of the return-difference matrix could enter the unit disc in the complex-frequency plane. In estimation problems, it is not unusual to have correlation between the measurement and process noise signals which results in a non-zero cross-covariance matrix term. Thus, it is often the case that the steady-state filter gain and phase margins will be reduced in comparison with the usual infinite gain margin and 60° phase margin cases (Section 2.6).

The stability margins are not, however, so important as in the dual-control problem since the system matrices, within the Kalman filter loop are known (even though they may not match exactly the true plant matrices).

The filter stability margin can be ascertained by calculating the eigenvalues of the matrix $\mathbf{A}_c$ in (8.115). This matrix represents the system matrix for the filter with the feedback loop (via the gain $\mathbf{K}(t)$) closed.

Optimality

Two remarks are pertinent:

(i) The same linear estimator is obtained when

$$J(\tilde{x}(t)) \triangleq E\{\tilde{x}^T(t)\mathbf{Q}_e\tilde{x}(t)\} \tag{8.120}$$

where $\mathbf{Q}_e \in \mathbb{R}^{n\times n}$ satisfies $\mathbf{Q}_e > \mathbf{O}$ and $\mathbf{Q}_e^T = \mathbf{Q}_e$.

(ii) The filter obtained above will also be optimal in the larger class of *linear* and *non-linear* filters if the noise processes are assumed to be Gaussian. When this assumption is not made the optimal filter can only be guaranteed to be the best *linear* estimator.

Time-invariant or steady-state filter

If the system and filter are assumed to be in operation from an initial time $t_0 = -\infty$, the system is time invariant and the noise sources stationary then the steady-state filter is also time invariant. As in the dual-control problem of Section 1.5.2, the system must be assumed to be stabilizable and detectable. This assumption ensures the existence of a constant solution $\mathbf{P}_\infty$ to the appropriate algebraic Riccati equation and the asymptotic stability of the filter. The time-invariant filter is easier to implement than the finite time-interval equivalent since, in this latter case, the gain is time varying. A finite interval filter can, however, be derived (Grimble [57]) which is related to the Kalman filter but which is also time invariant. Filters of this type are described in the next chapter.

The steady-state properties of the optimal filter are summarized below (Kwakernaak and Sivan [4]).

Theorem 8.9. ***Time-invariant Kalman filter.*** *Assume that the system* (8.88) *and* (8.89) *is time invariant and that the noise sources are stationary* ($\mathbf{Q} \geq \mathbf{O}$, $\mathbf{R} > \mathbf{O}$ *are constant matrices*) *and uncorrelated.*

(*a*) *If the system is detectable and stabilizable from the process noise input, then the solution of the Riccati equation* (8.99) *approaches the unique constant value* $\mathbf{P}_\infty$ *as the initial time* $t_0 \to -\infty$ *for every* $\mathbf{P}_0 \geq \mathbf{O}$.

(*b*) *If* $\mathbf{P}_\infty$ *exists it is a positive semidefinite symmetric solution of the algebraic Riccati equation:*

$$\mathbf{O} = \mathbf{A}\mathbf{P}_\infty + \mathbf{P}_\infty\mathbf{A}^T + \mathbf{D}\mathbf{Q}\mathbf{D}^T - \mathbf{P}_\infty\mathbf{C}^T\mathbf{R}^{-1}\mathbf{C}\mathbf{P}_\infty. \tag{8.121}$$

If the system is detectable and stabilizable, then $\mathbf{P}_\infty$ *is the unique positive semidefinite* ($\mathbf{P}_\infty \geq \mathbf{O}$) *solution of the algebraic Riccati equation* (8.121).

(*c*) *If* $\mathbf{Q} > \mathbf{O}$ *and* $\mathbf{P}_\infty$ *exists, it is positive definite if and only if the system is completely controllable.*

(*d*) *The steady-state Kalman filter is asymptotically stable if and only if the system is detectable and stabilizable.*

Proof (Kwakernaak and Sivan [4]). □

8.3.3. Wiener–Hopf equation and the optimal filter

The Kalman–Bucy filter may also be derived using a Hilbert space approach in which the Wiener–Hopf equation provides the necessary and sufficient condition of optimality. Although this derivation does not provide the same insight into the innovations nature of the problem, it does form a valuable link with the gradient optimization approach of previous chapters.

Plant and noise descriptions

The system is given as:

$$dx(t) = \mathbf{A}x(t)\ dt + \mathbf{B}u(t)\ dt + \mathbf{D}\ d\omega(t), \qquad x(0) = x_0 \tag{8.122}$$

$$dz(t) = \mathbf{C}x(t)\ dt + dv(t) \tag{8.123}$$

and the noise and initial condition vectors satisfy:

$$E\{\omega(t)\} = \mathbf{o}, \qquad E\{v(t)\} = \mathbf{o}$$

$$E\{(\omega(t_2) - \omega(t_1))(\omega(t_2) - \omega(t_1))^T\} = \mathbf{Q}\,|\,t_2 - t_1\,|$$

$$E\{(v(t_2) - v(t_1))(v(t_2) - v(t_1))^T\} = \int_{t_1}^{t_2} \mathbf{R}(t)\ dt, \qquad t_2 \geq t_1$$

$$E\{x_0\} = \mathbf{m}_0, \qquad \mathrm{cov}[x_0, x_0] = \mathbf{P}_0$$

$$x_0 \perp H^\omega, \qquad x_0 \perp H^v.$$

where $\mathbf{Q} \geq \mathbf{O}$ and $\mathbf{R}(t) > \mathbf{O}$ for $t \geq 0$.

The system-state response becomes:

$$x(t) = \mathbf{\Phi}(t, 0)x_0 + \int_0^t \mathbf{\Phi}(t, \tau)\mathbf{D}(t)\ d\omega(\tau) \tag{8.124}$$

where the control input is taken as zero. Thus equation (8.124) can be written in the following operator form as:

$$\boldsymbol{x} = \boldsymbol{d}_0 + \boldsymbol{W}_0 \, d\boldsymbol{\omega}. \tag{8.125}$$

(See Chapter 1 for the operator representation of linear systems.) Similarly, the linear filter can be expressed in an integral equation form:

$$\hat{\boldsymbol{x}}(t \mid t) = \boldsymbol{\Psi}(t, 0)\boldsymbol{x}_0 + \int_0^t \mathbf{H}(t, \tau) \; d\boldsymbol{z}(\tau) \tag{8.126}$$

and in operator form as:

$$\hat{\boldsymbol{x}} = \boldsymbol{d} + \boldsymbol{H} \; \mathrm{d}\boldsymbol{z}. \tag{8.127}$$

Using the operator equations (8.125) and (8.127), the linear filtering problem may be stated.

Filtering problem

Construct an estimation error:

$$\tilde{\boldsymbol{x}} = \boldsymbol{x} - \hat{\boldsymbol{x}}$$

and choose the parameters (namely $\mathbf{d}, \mathbf{H}$ and $\boldsymbol{d}, \boldsymbol{H}$) of the linear estimator equation (8.126, 8.127) to minimize a mean-square error cost functional:

$$J = \langle \boldsymbol{x} - \hat{\boldsymbol{x}}, \boldsymbol{x} - \hat{\boldsymbol{x}} \rangle_{H_n} = \langle \tilde{\boldsymbol{x}}, \tilde{\boldsymbol{x}} \rangle_{H_n} \triangleq E\{ \langle \tilde{\boldsymbol{x}}, \tilde{\boldsymbol{x}} \rangle_{E_n} \} \tag{8.128}$$

subject to the linear system constraint (8.125).

If the filter equation form (8.127) is substituted into the cost function (8.128) the following obtains:

$$\begin{aligned} J &= \langle \boldsymbol{x} - \hat{\boldsymbol{x}}, \boldsymbol{x} - \hat{\boldsymbol{x}} \rangle_{H_n} \\ &= \langle \boldsymbol{x} - \boldsymbol{d} - \boldsymbol{H} \; d\boldsymbol{z}, \boldsymbol{x} - \boldsymbol{d} - \boldsymbol{H} \; d\boldsymbol{z} \rangle_{H_n}. \end{aligned} \tag{8.129}$$

Expanding yields:

$$\begin{aligned} J = {} & \langle \boldsymbol{x}, \boldsymbol{x} \rangle_{H_n} - 2\langle \boldsymbol{x}, \boldsymbol{H} \; d\boldsymbol{z} \rangle_{H_n} - 2\langle \boldsymbol{x}, \boldsymbol{d} \rangle_{H_n} \\ & + \langle \boldsymbol{H} \; d\boldsymbol{z}, \boldsymbol{H} \; d\boldsymbol{z} \rangle_{H_n} + 2\langle \boldsymbol{d}, \boldsymbol{H} \; d\boldsymbol{z} \rangle_{H_n} + \langle \boldsymbol{d}, \boldsymbol{d} \rangle_{H_n}. \end{aligned} \tag{8.130}$$

To determine the conditions of optimality, let the optimal filter be $\hat{\boldsymbol{x}} = \hat{\boldsymbol{d}} + \hat{\boldsymbol{H}} \, d\boldsymbol{z}$ and consider perturbations to $\hat{\boldsymbol{H}}$ and $\hat{\boldsymbol{d}}$ of the form $\boldsymbol{H} = \hat{\boldsymbol{H}} + \varepsilon_1 \bar{\boldsymbol{H}}$

and $d = \hat{d} + \varepsilon_2 \bar{d}$ where $\bar{H}$ and $\bar{d}$ are arbitrary non-zero perturbations and $\varepsilon_1, \varepsilon_2 \in \mathbb{R}$. Thus, substituting in the cost functional (8.130) obtains:

$$\begin{aligned} J &= \hat{J} + \delta J(\varepsilon_1, \varepsilon_2) \\ &= \hat{J} - 2\varepsilon_1 \langle x, \bar{H}\, dz \rangle_{H_n} - 2\varepsilon_2 \langle x, \bar{d} \rangle_{H_n} + 2\varepsilon_1 \langle \hat{H}\, dz, \bar{H}\, dz \rangle_{H_n} \\ &\quad + \varepsilon_1^2 \langle \bar{H}\, dz, \bar{H}\, dz \rangle_{H_n} + 2\varepsilon_1 \langle \hat{d}, \bar{H}\, dz \rangle_{H_n} + 2\varepsilon_2 \langle \bar{d}, \hat{H}\, dz \rangle_{H_n} \\ &\quad + 2\varepsilon_1\varepsilon_2 \langle \bar{d}, \bar{H}\, dz \rangle_{H_n} + 2\varepsilon_2 \langle \bar{d}, \hat{d} \rangle_{H_n} + \varepsilon_2^2 \langle \bar{d}, \bar{d} \rangle_{H_n}. \end{aligned} \tag{8.131}$$

The necessary and sufficient conditions of optimality are therefore:

$$\left.\begin{bmatrix} \dfrac{\partial J}{\partial \varepsilon_1} \\[2ex] \dfrac{\partial J}{\partial \varepsilon_2} \end{bmatrix}_{\varepsilon_1 = \varepsilon_2 = 0} = \mathbf{o} \quad \text{and} \quad \begin{array}{l} \dfrac{\partial^2 J}{\partial \varepsilon_1^2} > 0 \\[2ex] \dfrac{\partial^2 J}{\partial \varepsilon_1^2} \dfrac{\partial^2 J}{\partial \varepsilon_2^2} - \left(\dfrac{\partial^2 J}{\partial \varepsilon_1\, \partial \varepsilon_2} \right)^2 > 0 \end{array} \right\}. \tag{8.132}$$

From (8.131),

$$\begin{bmatrix} \dfrac{\partial J}{\partial \varepsilon_1} \\[2ex] \dfrac{\partial J}{\partial \varepsilon_2} \end{bmatrix}_{\varepsilon_1 = \varepsilon_2 = 0} = \begin{bmatrix} -2\langle x, \bar{H}\, dz \rangle_{H_n} + 2\langle \hat{H}\, dz, \bar{H}\, dz \rangle_{H_n} + 2\langle \hat{d}, \bar{H}\, dz \rangle_{H_n} \\[2ex] -2\langle x, \bar{d} \rangle_{H_n} + 2\langle \bar{d}, \hat{H}\, dz \rangle_{H_n} + 2\langle \bar{d}, \hat{d} \rangle_{H_n} \end{bmatrix} = \mathbf{o}$$

$$= \begin{bmatrix} -2\langle x - \hat{x}, \bar{H}\, dz \rangle_{H_n} \\ -2\langle x - \hat{x}, \bar{d} \rangle_{H_n} \end{bmatrix} = \mathbf{o}.$$

This equation yields the two conditions:

$$\langle x - \hat{x}, \bar{H}\, dz \rangle_{H_n} = 0 \quad \text{Wiener–Hopf equation} \tag{8.133}$$

and

$$\langle x - \hat{x}, \bar{d} \rangle_{H_n} = 0 \quad \text{Unbiased estimate condition.} \tag{8.134}$$

The Jacobian conditions become:

$$\frac{\partial^2 J}{\partial \varepsilon_1^2} = 2\langle \bar{H}\, dz, \bar{H}\, dz \rangle_{H_n} > 0 \tag{8.135}$$

and

$$\frac{\partial^2 J}{\partial \varepsilon_1^2}\frac{\partial^2 J}{\partial \varepsilon_2^2}-\left(\frac{\partial^2 J}{\partial \varepsilon_1\,\partial \varepsilon_2}\right)^2=4\{\langle \bar{\boldsymbol{H}}\,dz, \bar{\boldsymbol{H}}\,dz\rangle_{H_n}\cdot\langle \bar{\boldsymbol{d}}, \bar{\boldsymbol{d}}\rangle_{H_n}-(\langle \bar{\boldsymbol{d}}, \bar{\boldsymbol{H}}\,dz\rangle_{H_n})^2\}\rangle 0 \tag{8.136}$$

for arbitrary $\bar{\boldsymbol{H}}$ and $\bar{\boldsymbol{d}}$. (Equation (8.136) is a rearrangement of the Schwartz identity). The next step is to manipulate the Hilbert space stochastic relationships (8.133) and (8.134) into time-domain conditions. In equation (8.134), $\bar{\boldsymbol{d}}$ represents an arbitrary deterministic function defined on $[0, t]$, thus using the inner-product definition (8.128):

$$\langle \boldsymbol{x}-\hat{\boldsymbol{x}}, \bar{\boldsymbol{d}}\rangle_{H_n}=0 \Rightarrow E\{\hat{\boldsymbol{x}}(t \mid t)\}=E\{\boldsymbol{x}(t)\} \tag{8.137}$$

that is $\hat{\boldsymbol{x}}(t \mid t)$ is an unbiased estimator for $\boldsymbol{x}(t)$.

The time-domain Wiener–Hopf condition derives from equation (8.133).

$$\langle \boldsymbol{x}-\hat{\boldsymbol{x}}, \bar{\boldsymbol{H}}\,dz\rangle_{H_n} \triangleq E\left\{\left\langle \boldsymbol{x}(t)-\hat{\boldsymbol{x}}(t \mid t), \int_0^t \bar{\mathbf{H}}(t,\tau)\,dz(\tau)\right\rangle_{E_n}\right\}=0$$

using (8.137) this becomes:

$$\begin{aligned}\langle \boldsymbol{x}-\hat{\boldsymbol{x}}, \bar{\boldsymbol{H}}\,dz\rangle_{H_n}&=E\left\{\left\langle \boldsymbol{x}(t)-E\{\boldsymbol{x}(t)\}-(\hat{\boldsymbol{x}}(t \mid t)-E\{\hat{\boldsymbol{x}}(t \mid t)\}), \int_0^t \bar{\mathbf{H}}(t,\tau)\,dz(\tau)\right\rangle_{E_n}\right\}\\&=E\left\{\left\langle \boldsymbol{x}(t)-E\{\boldsymbol{x}(t)\}, \int_0^t \bar{\mathbf{H}}(t,\tau)\,dz(\tau)\right\rangle_{E_n}\right.\\&\qquad\left.-\left\langle \hat{\boldsymbol{x}}(t \mid t)-E\{\boldsymbol{x}(t \mid t)\}, \int_0^t \bar{\mathbf{H}}(t,\tau)\,dz(\tau)\right\rangle_{E_n}\right\}=0.\end{aligned} \tag{8.138}$$

If the infinitesimal sum origin of the stochastic integral terms in (8.138) are considered, a time-domain condition can be derived. Using the finite-dimensional identity: $\operatorname{trace}\{\mathbf{x}\mathbf{y}^T\}=\langle \mathbf{x}, \mathbf{y}\rangle_{E_n}$, the first term in (8.138) becomes:

$$\begin{aligned}&E\left\{\left\langle \boldsymbol{x}(t)-E\{\boldsymbol{x}(t)\}, \int_0^t \bar{\mathbf{H}}(t,\tau)\,dz(\tau)\right\rangle_{E_n}\right\}\\&\qquad=\operatorname{trace} E\{(\boldsymbol{x}(t)-E\{\boldsymbol{x}(t)\})\int_0^t dz^T(\tau)\bar{\mathbf{H}}^T(t,\tau).\end{aligned} \tag{8.139}$$

Thus,

$$(x(t) - E\{x(t)\}) \int_0^t dz^T(t)\overline{\mathbf{H}}^T(t, \tau) \triangleq \lim_{n\to\infty} \sum_{i=0}^{n-1} (x(t) - E\{x(t)\})$$

$$(z(\tau_{i+1}) - z(\tau_i))^T \overline{\mathbf{H}}^T(t, \tau_i).$$

Also define

$$E\{ (x(t) - E\{x(t)\})(z(\tau_{i+1}) - z(\tau_i))^T\} = \int_{\tau_i}^{\tau_{i+1}} \mathbf{R}_{xz}(t, \tau)\ d\tau \quad (8.140)$$

and in infinitesimal terms:

$$\int_{\tau_i}^{\tau_{i+1}} \mathbf{R}_{xz}(t, \tau)\ d\tau \cong \mathbf{R}_{xz}(t, \tau_i)\ \delta\tau_i. \quad (8.141)$$

Together (8.139) to (8.141) yield:

$$E\left\{\left\langle x(t) - E\{x(t)\},\ \int_0^t \overline{\mathbf{H}}(t, \tau)\ dz(\tau)\right\rangle E_n\right\} = \operatorname{trace}\left\{\lim_{n\to\infty} \sum_{i=0}^{n-1} \mathbf{R}_{xz}(t, \tau_i)\overline{\mathbf{H}}^T(t, \tau_i)\ \delta\tau_i\right\}$$

$$= \operatorname{trace}\left\{\int_0^t \mathbf{R}_{xz}(t, \tau)\overline{\mathbf{H}}^T(t, \tau)\ d\tau\right\}. \quad (8.142)$$

Similarly, for the second term in equation (8.138):

$$E\left\{\left\langle \hat{x}(t \mid t) - E\{\hat{x}(t \mid t)\},\ \int_0^t \overline{\mathbf{H}}(t, \tau)\ dz(\tau)\right\rangle_{E_n}\right\}$$

$$= \operatorname{trace}\ E\left\{(\hat{x}(t \mid \mathrm{t}) - E\langle \hat{x}(t \mid t)\rangle \int_0^t dz^T(\tau)\overline{\mathbf{H}}^T(t, \tau)\right\}. \quad (8.143)$$

From the estimator equation and noting (8.137): $x - E\{\hat{x}\} = \hat{H}\ dz$, hence (8.143) becomes:

$$E\left\{\left\langle \hat{x}(t \mid t) - E\{\hat{x}(t \mid t)\},\quad \int_0^t \overline{\mathbf{H}}(t, \tau)\ dz(\tau)\right\rangle_{E_n}\right\}$$

$$= \operatorname{trace}\ E\left\{\int_0^t \hat{\mathbf{H}}(t, \tilde{\tau}_1)\ dz(\tilde{\tau}_1) \int_0^t dz^T(\tau)\overline{\mathbf{H}}^T(t, \tau)\right\}. \quad (8.144)$$

In infinitesimal terms, the kernel of (8.144) may be written as:

$$\int_0^t \hat{\mathbf{H}}(t, \tilde{\tau}_1)\ dz(\tilde{\tau}_1) \int_0^t dz^T(\tau)\overline{\mathbf{H}}^T(t, \tau) \triangleq \lim_{n\to\infty} \left(\sum_{i_1=0}^{n-1} \hat{\mathbf{H}}(t, \tilde{\tau}_{i_1})(z(\tilde{\tau}_{i_1+1}) - z(\tilde{\tau}_{i_1}))\right)$$

$$\times \left(\sum_{i=0}^{n-1} (z(\tau_{i+1}) - z(\tau_i))^T \overline{\mathbf{H}}^T(t, \tau_i)\right).$$

Define

$$E\{ (z(\tilde{\tau}_{i_1+1}) - z(\tilde{\tau}_{i_1}))(z(\tau_{i+1}) - z(\tau_i))^T \} = \int_{\tilde{\tau}_{i_1}}^{\tilde{\tau}_{i_1+1}} \int_{\tau_i}^{\tau_{i+1}} \mathbf{R}_{zz}(\tilde{\tau}, \tau)\ d\tilde{\tau}\ d\tau$$

and in infinitesimal terms:

$$\int_{\tilde{\tau}_{i_1}}^{\tilde{\tau}_{i_1+1}} \int_{\tau_i}^{\tau_{i+1}} \mathbf{R}_{zz}(\tilde{\tau}, \tau)\ d\tilde{\tau}\ d\tau = \mathbf{R}_{zz}(\tilde{\tau}_{i_1}, \tau_i)\ \delta\tilde{\tau}_{i_1}\ \delta\tau_i.$$

Equation (8.144) becomes therefore,

$$E\left\{\left\langle \hat{x}(t) - E\{\hat{x}(t \mid t)\},\ \int_0^t \overline{\mathbf{H}}(t, \tau)\ dz(\tau)\right\rangle_{E_n}\right\}$$

$$= \operatorname{trace}\left\{\lim_{n\to\infty} \sum_{i=0}^{n-1} \left\{\sum_{i_1=0}^{n-1} \hat{\mathbf{H}}(t, \tilde{\tau}_{i_1})\mathbf{R}_{zz}(\tilde{\tau}_{i_1}, \tau_i)\ \delta\tau_i\right\}\overline{H}(t, \tau_i)\ \delta\tau_i\right\}$$

$$= \operatorname{trace}\left\{\int_0^t \int_0^t \hat{\mathbf{H}}(t, \tilde{\tau})\mathbf{R}_{zz}(\tilde{\tau}, \tau)\ d\tilde{\tau}\ \overline{\mathbf{H}}(t, \tau)\ d\tau\right\}. \tag{8.145}$$

Using the equations (8.142) and (8.145) in the optimality condition (8.138) obtains:

$$\langle x - \tilde{x}, \hat{H}\ dz\rangle_{H_n}$$

$$= \operatorname{trace}\left\{\int_0^t \{\mathbf{R}_{xz}(t, \tau) - \int_0^t \hat{\mathbf{H}}(t, \tilde{\tau})\mathbf{R}_{zz}(\tilde{\tau}, \tau)\ d\tilde{\tau}\}\overline{\mathbf{H}}^T(t, \tau)\ d\tau\right\} = 0. \tag{8.146}$$

The kernel $\overline{\mathbf{H}}(\cdots)$ is chosen arbitrarily, consequently (8.146) yields the deterministic Wiener–Hopf equation for the optimal filter $\hat{\mathbf{H}}(t, :)$ as:

$$\mathbf{R}_{xz}(t, \tau) - \int_0^t \hat{\mathbf{H}}(t, \tilde{\tau})\mathbf{R}_{zz}(\tilde{\tau}, \tau)\ d\tilde{\tau} = \mathbf{O} \tag{8.147}$$

for $0 \leq \tau \leq t$, where

$$E\{(x(t) - E\{x(t)\})(z(\tau_{i+1}) - z(\tau_i))^T\} = \int_{\tau_i}^{\tau_{i+1}} \mathbf{R}_{xz}(t, \tau)\, d\tau \quad (8.148)$$

and

$$E\{z(\tilde{\tau}_{i_1+1}) - z(\tilde{\tau}_{i_1}))(z(\tau_{i+1}) - z(\tau_i))^T\} = \int_{\tau_{i_1}}^{\tau_{i_1+1}} \int_{\tau_i}^{\tau_{i+1}} \mathbf{R}_{zz}(\tilde{\tau}, \tau)\, d\tilde{\tau}\, d\tau \quad (8.149)$$

defines the two covariance functions required to solve (8.147).

The Wiener and Kalman filters may be derived from the Wiener–Hopf equation (8.147) as demonstrated in Meditch [19]. To derive the Wiener filter, equation (8.147) is transformed directly into the complex frequency domain using the two-sided Laplace transform (Kailath [18], Grimble [20]). This type of solution is described in the following chapter when considering the finite-time discrete estimation problem.

8.4. DISCRETE-TIME FILTERING, SMOOTHING AND PREDICTION PROBLEMS

In many practical situations the observation process is monitored at discrete times only. That is, at time t, the following observations are available:

$$z(t_1), z(t_2), \dots, z(t_n),$$

where

$$0 < t_1 < t_2 < \cdots < t_n, \qquad z(t_i) \in \mathbb{R}^r,$$

The system from which such observations are taken can be either continuous time or discrete time but it is usual to treat the estimation problem as being a discrete-time problem. In this case, the plant- or signal-generating process is represented by discrete-state equations.

In the following, the optimal filter for a discrete-time system is described and presented in an algorithmic form. The discrete-time filter is better suited to implementation on a digital computer. This gives an advantage over the continuous-time Kalman filter since the time-varying gain matrix is not readily implemented using analogue circuitry. Consequently, there tend to be fewer applications of the continuous-time Kalman filter.

On the infinite time domain, viz. $t_0 \to -\infty$, the solution of the Kalman filtering problem becomes identical to a Wiener filter and the associated gain matrix becomes constant. However, this too is used far less frequently than its discrete-time Kalman filter equivalent.

The prediction and smoothing problems are also described and the section concludes with a brief discussion of the computational aspects of estimation, a topic of considerable practical importance.

8.4.1. Discrete-time Kalman filter

The discrete-time optimal filter was derived by Kalman [1] and is called the Kalman filter. If the output of a continuous-time plant is sampled and if the control-signal inputs are maintained constant during the sample interval then the system can be represented by a discrete-time model. Alternatively, the signal-generating process may be implemented by a discrete device. In either case, the following vector difference equation model might be employed to describe the state response:

$$x(t+1) = \mathbf{A}(t)x(t) + \mathbf{B}(t)u(t) + \mathbf{D}(t)\omega(t). \tag{8.150}$$

The observation model or output equation becomes:

$$z(t) = y(t) + v(t) = \mathbf{C}(t)x(t) + v(t). \tag{8.151}$$

The process $\{\omega(t)\}$ and measurement $\{v(t)\}$ noise sequences are assumed to be zero-mean and white. That is, the noise statistics are given as:

$$\operatorname{cov}[\omega(t), \omega(k)] = \mathbf{Q}(t)\ \delta_{tk} \tag{8.152}$$

$$\operatorname{cov}[v(t), v(k)] = \mathbf{R}(t)\ \delta_{tk} \tag{8.153}$$

$$\operatorname{cov}[\omega(t), v(t)] = \mathbf{O} \tag{8.154}$$

where δ_{tk} denotes the Kronecker delta function

$$(\delta_{tk} = 1 \text{ for } t = k \qquad \delta_{tk} = 0 \text{ for } t \neq k)$$

and

$$E\{\omega(t)\} = \mathbf{o}, \qquad E\{v(t)\} = \mathbf{o} \tag{8.155}$$

where $\mathbf{Q}(t) \geq \mathbf{O}$ and $\mathbf{R}(t) > \mathbf{O}$ for all $t \geq 0$ are symmetric covariance matrices. Note that the cross-covariance matrix need not be taken as zero but this assumption simplifies the results slightly. The initial condition vector is assumed to be uncorrelated with the noise processes and the expected value $\mathbf{m}_0$ and initial covariance $\mathbf{P}_0$ are assumed known where,

$$\mathbf{m}_0 = E\{x_0\} \quad \text{and} \quad \mathbf{P}_0 = E\{(x(0) - \mathbf{m}_0)(x(0) - \mathbf{m}_0)^T\}.$$

The optimal discrete linear filter is required to minimize the estimation error criterion:

$$\langle \tilde{x}(t \mid t-1), \tilde{x}(t \mid t-1) \rangle_{H_n} = \text{trace } E\{\tilde{x}(t \mid t-1)\tilde{x}^T(t \mid t-1)\} \tag{8.156}$$

where the estimation error:

$$\tilde{x}(t \mid t-1) = x(t) - \hat{x}(t \mid t-1). \tag{8.157}$$

The solution of this optimal estimation problem was presented in Kalman's classic paper [1]. The Kalman estimator, from this paper, has the form:

$$\hat{x}(t+1 \mid t) = \mathbf{A}(t)\hat{x}(t \mid t-1) + \mathbf{K}(t)(z(t) - \mathbf{C}(t)\hat{x}(t \mid t-1)) \tag{8.158}$$

where $\mathbf{K}(t)$ represents the Kalman gain matrix. There is a subtle difference between the estimation equation (8.158) and that given in various estimation books. The above expression defines what is often called a *single-stage predictor* since the current state estimate (at $t+1$) depends only upon past observations. The discrete-time Kalman filter algorithm may also be defined to process the current observation. In this case the estimation error criterion to be minimized becomes:

$$J = \langle \tilde{x}(t \mid t), \tilde{x}(t \mid t) \rangle_{H_n} = E\{\tilde{x}^T(t \mid t)\tilde{x}(t \mid t)\} \tag{8.159}$$

where

$$\tilde{x}(t \mid t) = x(t) - \hat{x}(t \mid t). \tag{8.160}$$

Sage and Melsa [17], for example, make such a distinction. The respective Kalman filter algorithms are summarized below.

Algorithm 8.1. *Discrete single-stage predictor*

Process model: $x(t+1) = \mathbf{A}(t)x(t) + \mathbf{B}(t)u(t) + \mathbf{D}(t)\omega(t)$.

Observation model: $z(t) = \mathbf{C}(t)x(t) + v(t)$.

Predictor: $\hat{x}(t+1 \mid t) = \mathbf{A}(t)\hat{x}(t \mid t-1) + \mathbf{B}(t)u(t) + \mathbf{K}(t)\nu(t)$.

(8.161)

Innovations: $\boldsymbol{\nu}(t) = z(t) - \hat{y}(t \mid t-1)$.

Output predictor: $\hat{y}(t \mid t-1) = \mathbf{C}(t)\hat{x}(t \mid t-1)$.

Gain:
$$\mathbf{K}(t) = \mathbf{A}(t)\mathbf{P}(t \mid t-1)\mathbf{C}^T(t) \times [\mathbf{C}(t)\mathbf{P}(t \mid t-1)\mathbf{C}^T(t) + \mathbf{R}(t)]^{-1}. \tag{8.162}$$

A priori covariance:
$$\mathbf{P}(t+1 \mid t) = \mathbf{A}(t)\mathbf{P}(t \mid t-1)\mathbf{A}^T(t) + \mathbf{D}(t)\mathbf{Q}(t)\mathbf{D}^T(t) - \mathbf{A}(t)\mathbf{P}(t \mid t-1)\mathbf{C}^T(t)[\mathbf{C}(t)\mathbf{P}(t \mid t-1)\mathbf{C}^T(t) + \mathbf{R}(t)]^{-1}\mathbf{C}(t)\mathbf{P}(t \mid t-1)\mathbf{A}^T(t) \tag{8.163}$$

Initial conditions: $\hat{x}(1 \mid 0) = \mathbf{A}(0)\mathbf{m}_0$
$\mathbf{P}(1 \mid 0) = \mathbf{A}(0)\mathbf{P}_0\mathbf{A}^T(0) + \mathbf{D}(0)\mathbf{Q}(0)\mathbf{D}^T(0)$. □

Note that the covariance update equation (8.163) may be given in several different forms for computational purposes. One alternative for the *a priori* covariance update is:

$$\mathbf{P}(t+1 \mid t) = (\mathbf{A}(t) - \mathbf{K}(t)\mathbf{C}(t))\mathbf{P}(t \mid t-1)(\mathbf{A}(t) - \mathbf{K}(t)\mathbf{C}(t))^T + \mathbf{D}(t)\mathbf{Q}(t)\mathbf{D}^T(t) + \mathbf{K}(t)\mathbf{R}(t)\mathbf{K}'(t).$$

Although this requires more computational operations than (8.163), it has the advantage over (8.163) that $\mathbf{P}(t \mid t-1) \geq \mathbf{O}$ will tend to give $\mathbf{P}(t+1 \mid t) \geq \mathbf{O}$ even in the presence of round-off errors. This is because the above equation comprises the sum of symmetric matrix terms rather than the difference of symmetric matrix terms as in (8.163).

The optimal estimator given by Kwakernaak and Sivan [4] is of the above single-stage predictor type.

Algorithm 8.2. *Discrete Kalman filter*

Process model: $x(t+1) = \mathbf{A}(t)x(t) + \mathbf{B}(t)u(t) + \mathbf{D}(t)\omega(t)$.

Observation model: $z(t) = \mathbf{C}(t)x(t) + v(t)$.

Kalman filter:
$$\hat{x}(t \mid t) = \mathbf{A}(t-1)\hat{x}(t-1 \mid t-1) + \mathbf{B}(t-1)u(t-1) + \mathbf{K}(t)\boldsymbol{\nu}(t) \tag{8.164}$$
$\boldsymbol{\nu}(t) = z(t) - \hat{y}(t \mid t-1)$
$\hat{y}(t \mid t-1) = \mathbf{C}(t)\mathbf{A}(t-1)\hat{x}(t-1 \mid t-1)$.

Gain:
$$\mathbf{K}(t) = \mathbf{P}(t \mid t-1)\mathbf{C}^T(t)[\mathbf{C}(t)\mathbf{P}(t \mid t-1)\mathbf{C}^T(t) + \mathbf{R}(t)]^{-1}. \tag{8.165}$$

A priori covariance:
$$\mathbf{P}(t+1 \mid t) = \mathbf{A}(t)\mathbf{P}(t \mid t-1)\mathbf{A}^T(t) + \mathbf{D}(t)\mathbf{Q}(t)\mathbf{D}^T(t). \tag{8.166}$$

A posteriori covariance: $\mathbf{P}(t \mid t) = \mathbf{P}(t \mid t-1) - \mathbf{K}(t)\mathbf{C}(t)\mathbf{P}(t \mid t-1)$.

Initial conditions: $\hat{x}(0 \mid 0) = \mathbf{m}_0$ and $\mathbf{P}(0 \mid 0) = \mathbf{P}_0$. □

The discrete optimal filter equations described by Meditch [19] have the above structure.

An examination of the Kalman filter equations reveals that they may be written in a predictor corrector form. The algorithm is given in this form below. Recall that if the system includes known inputs these should be input to the filter at the corresponding point they enter the system. Sage and Melsa [17] show that if the noise signals include bias terms then these must also be input to the filter following this principle. Let $\bar{v}(t) \triangleq E\{v(t)\}$ and $\bar{\omega}(t) = E\{\omega(t)\}$ then the algorithm becomes as follows:

Algorithm 8.3. *Discrete Kalman filter: Predictor-corrector form*

Process model: $x(t+1) = \mathbf{A}(t)x(t) + \mathbf{B}(t)u(t) + \mathbf{D}(t)\omega(t)$.

Observation model: $z(t) = \mathbf{C}(t)x(t) + v(t)$.

Predictor:
$$\hat{x}(t+1 \mid t) = \mathbf{A}(t)\hat{x}(t \mid t) + \mathbf{B}(t)u(t) + \mathbf{D}(t)\bar{\omega}(t). \quad (8.167)$$

Kalman filter:
$$\hat{x}(t+1 \mid t+1) = \hat{x}(t+1 \mid t) + \mathbf{K}(t+1)(z(t+1) - \mathbf{C}(t+1)\hat{x}(t+1 \mid t) - \bar{v}(t)). \quad (8.168)$$

Gain:
$$\mathbf{K}(t+1) = \mathbf{P}(t+1 \mid t)\mathbf{C}^T(t+1) \times [\mathbf{C}(t+1)\mathbf{P}(t+1 \mid t)\mathbf{C}^T(t+1) + \mathbf{R}(t+1)]^{-1}. \quad (8.169)$$

A priori covariance:
$$\mathbf{P}(t+1 \mid t) = \mathbf{A}(t)\mathbf{P}(t \mid t)\mathbf{A}^T(t) + \mathbf{D}(t)\mathbf{Q}(t)\mathbf{D}^T(t). \quad (8.170)$$

A posteriori covariance:
$$\mathbf{P}(t+1 \mid t+1) = \mathbf{P}(t+1 \mid t) - \mathbf{K}(t+1)\mathbf{C}(t+1)\mathbf{P}(t+1 \mid t). \quad (8.171)$$

Initial conditions: $\hat{x}(0 \mid 0) = \mathbf{m}_0$ and $\mathbf{P}(0 \mid 0) = \mathbf{P}_0$.

Bias terms: $\bar{\omega}(t) = E\{\omega(t)\}, \bar{v}(t) = E\{v(t)\}$. □

Stationary discrete-time Kalman filter

If the plant or message generating process is stable and the noise sources are stationary the Kalman filter, in the steady-state ($t_0 \to -\infty$), reduces to a time-invariant filter. This is equivalent to an optimal discrete filter which in transfer-function form is known as a Wiener filter (see Chapter 9). In this situation a return-difference matrix may be defined which has similarities to those discussed in Chapters 2 and 3 in the dual optimal control situation. The equivalent of the control return-difference, spectral optimal factorization result, of Corollary 2.3 may be obtained by the approach of Arcasoy [42]. However, the results are not identical to those of Arcasoy who considered the infinite–time case of the single-stage predictor algorithm (this involves a single-step delay between estimates and observations).

The return-difference results for Algorithm 8.2 have not previously been established and are therefore considered below.

Note. When evaluating Wiener or finite-time filters in the following chapter, the definition of the causal transform $\{\cdot\}_+$ includes the z^0 or constant terms. This implies that these estimators do not include a single-step delay. Thus, this filter is equivalent to a steady-state version of Algorithm 8.2. Chen [43] has shown that the estimation error is, as expected, lower for the case where the observations $\{z(t), z(t-1), \ldots\}$ are processed rather than for the first type of algorithm which rely upon $[z(t-1), z(t-2, \ldots$.

Steady-state Riccati equation

The solution of the Riccati differential equation for continuous-time systems converges to a constant matrix under the appropriate conditions (Chapters 1 and 2). The solution can be obtained from the steady-state or algebraic Riccati equation, when the infinite-time ($t_0 \to -\infty$) estimation problem is considered and the plant is time invariant, with stationary noise sources.

The algebraic Riccati equation for discrete systems may be derived formally by substituting $\mathbf{P}(t+1 \mid t) = \mathbf{P}_\infty$ in Algorithm 8.2, and by making a straightforward substitution for $\mathbf{P}(t \mid t)$. The positive-definite symmetric matrix $\mathbf{P}_\infty$ satisfies the steady-state matrix. Riccati equation:

$$\mathbf{P}_\infty = \mathbf{A}\mathbf{P}_\infty\mathbf{A}^T - \mathbf{A}\mathbf{P}_\infty\mathbf{C}^T(\mathbf{R} + \mathbf{C}\mathbf{P}_\infty\mathbf{C}^T)^{-1}\mathbf{C}\mathbf{P}_\infty\mathbf{A}^T + \mathbf{D}\mathbf{Q}\mathbf{D}^T. \tag{8.172}$$

The Kalman filter-gain matrix becomes:

$$\mathbf{K} = \mathbf{P}_\infty\mathbf{C}^T(\mathbf{R} + \mathbf{C}\mathbf{P}_\infty\mathbf{C}^T)^{-1}. \tag{8.173}$$

The return-difference matrix for the filter is defined as:

$$\mathbf{F}(z^{-1}) = \mathbf{I}_r + \mathbf{C}\mathbf{\Phi}(z^{-1})\mathbf{A}\mathbf{K} \tag{8.174}$$

where

$$\mathbf{\Phi}(z^{-1}) \triangleq (z\mathbf{I}_n - \mathbf{A})^{-1}. \tag{8.175}$$

This definition can be justified, and the matrix $\mathbf{F}(z^{-1})$ identified as a return-difference matrix, as follows. For simplicity, take $\boldsymbol{u}(t) = \mathbf{o}$ and then from Algorithm 8.2:

$$\hat{x}(t+1 \mid t+1) = \mathbf{A}\hat{x}(t \mid t) + \mathbf{K}(z(t+1) - \hat{y}(t+1 \mid t))$$

hence

$$\hat{x}(t \mid t) = \mathbf{\Phi}(z^{-1})\mathbf{K}(z(t+1) - \hat{y}(t+1 \mid t)) \tag{8.176}$$

but

$$\begin{aligned}\hat{y}(t+1 \mid t) &= \mathbf{C}\mathbf{A}\hat{x}(t \mid t)\\ &= \mathbf{C}\mathbf{A}\mathbf{\Phi}(z^{-1})\mathbf{K}(z(t+1) - \hat{y}(t+1 \mid t))\end{aligned}$$

giving

$$(\mathbf{I}_r + \mathbf{C A \Phi}(z^{-1})\mathbf{K})\hat{y}(t \mid t-1) = \mathbf{C A \Phi}(z^{-1})\mathbf{K} z(t). \tag{8.177}$$

This equation describes the closed-loop response of the estimated output $\hat{y}(t \mid t-1)$ given inputs $z(t-1)$, $z(t-2)$, ... The matrix $(\mathbf{I}_r + \mathbf{C A \Phi}(z^{-1})\mathbf{K})$ is the return-difference matrix since $\mathbf{A\Phi}(z^{-1}) = \mathbf{\Phi}(z^{-1})\mathbf{A}$.

The innovations signal for the discrete Kalman filter, follows from (8.176) as $z(t+1) - \hat{y}(t+1 \mid t)$,

or:

$$\boldsymbol{\nu}(t) = z(t) - \mathbf{C A}\hat{x}(t-1 \mid t-1) \tag{8.178}$$

but from (8.150) and (8.151),

$$\boldsymbol{\nu}(t) = v(t) + \mathbf{C}(\mathbf{A}\tilde{x}(t-1 \mid t-1) + \mathbf{D}\omega(t-1)) \tag{8.179}$$

where the estimation error:

$$\tilde{x}(t \mid t) \triangleq x(t) - \hat{x}(t \mid t).$$

Since the signals that form $z(t)$ above are orthogonal, the covariance matrix for the innovations signal follows using (8.166) as:

$$\mathbf{R}_\varepsilon = \mathbf{R} + \mathbf{C P}_\infty \mathbf{C}^T. \tag{8.180}$$

If $\mathbf{R} > \mathbf{O}$ then $\mathbf{R}_\varepsilon > \mathbf{O}$ and is non-singular.

Corollary 8.2. ***Return-difference, spectral factorization relationship.*** *Let* $\mathbf{W}(z^{-1}) \triangleq \mathbf{C\Phi}(z^{-1})\mathbf{D}$ *then the return difference* $\mathbf{F}(z^{-1})$ *matrix, for the optimal discrete-time filter, is related to the spectral factor* $\mathbf{Y}(z^{-1})$ *of the equivalent Wiener filter by the following relationship:*

$$\mathbf{F}(z^{-1})\mathbf{R}_\varepsilon \mathbf{F}^T(z) = \mathbf{W}(z^{-1})\mathbf{Q}\mathbf{W}^T(z) + \mathbf{R} = \mathbf{Y}(z^{-1})\mathbf{Y}^T(z). \tag{8.181}$$

Proof. From the Riccati equation (8.172) adding and subtracting terms:

$$\mathbf{D Q D}^T = (z\mathbf{I} - \mathbf{A})\mathbf{P}_\infty(z^{-1}\mathbf{I} - \mathbf{A}^T) + \mathbf{A P}_\infty(z^{-1}\mathbf{I} - \mathbf{A}^T)$$

$$+ (z\mathbf{I} - \mathbf{A})\mathbf{P}_\infty \mathbf{A}^T + \mathbf{A K C P}_\infty \mathbf{A}^T$$

hence

$$\mathbf{C}\boldsymbol{\Phi}(z^{-1})\mathbf{D}\mathbf{Q}\mathbf{D}^T\boldsymbol{\Phi}^T(z)\mathbf{C}^T = \mathbf{C}\mathbf{P}_\infty\mathbf{C}^T + \mathbf{C}\boldsymbol{\Phi}(z^{-1})\mathbf{A}\mathbf{P}_\infty\mathbf{C}^T + \mathbf{C}\mathbf{P}_\infty\mathbf{A}^T\boldsymbol{\Phi}^T(z)\mathbf{C}^T + \mathbf{C}\boldsymbol{\Phi}(z^{-1})\mathbf{A}\mathbf{K}\mathbf{C}\mathbf{P}_\infty\mathbf{A}^T\boldsymbol{\Phi}^T(z)\mathbf{C}^T \tag{8.182}$$

or

$$\begin{aligned}\mathbf{W}(z^{-1})\mathbf{Q}\mathbf{W}^T(z) + \mathbf{R} &= \mathbf{R} + \mathbf{C}\mathbf{P}_\infty\mathbf{C}^T + \mathbf{C}\boldsymbol{\Phi}(z^{-1})\mathbf{A}\mathbf{P}_\infty\mathbf{C}^T + \mathbf{C}\mathbf{P}_\infty\mathbf{A}^T\boldsymbol{\Phi}^T(z)\mathbf{C}^T \\ &\quad + \mathbf{C}\boldsymbol{\Phi}(z^{-1})\mathbf{A}\mathbf{K}(\mathbf{R} + \mathbf{C}\mathbf{P}_\infty\mathbf{C}^T)\mathbf{K}^T\mathbf{A}^T\boldsymbol{\Phi}^T(z)\mathbf{C}^T \\ &= (\mathbf{I} + \mathbf{C}\boldsymbol{\Phi}(z^{-1})\mathbf{A}\mathbf{K})(\mathbf{R} + \mathbf{C}\mathbf{P}_\infty\mathbf{C}^T)(\mathbf{I} + \mathbf{C}\boldsymbol{\Phi}(z)\mathbf{A}\mathbf{K})^T. \\ &= \mathbf{F}(z^{-1})\mathbf{R}_\varepsilon\mathbf{F}(z)^T. \end{aligned} \tag{8.183}$$

□

The return-difference matrix properties of the optimal linear estimator in the discrete-time singular case (where the measurement noise covariance matrix is singular) have been considered by O'Reilly [55].

The return-difference matrix provides a simple way of examining the behaviour of the stationary Kalman filter. For example, the paper by Arcasoy [42] establishes the necessary condition for optimality which can be derived in terms of the loci of the return-difference matrix.

8.4.2. Prediction and smoothing

The filtering problem is concerned with estimating the state (or output) of a system at time t given all measurements up to time t. This estimate is denoted by $\hat{x}(t \mid t)$. Prediction involves estimation of the state at some future time $t_1 > t$ and the corresponding estimate is denoted by $\hat{x}(t_1 \mid t)$. This is normally easily calculated from the Kalman filter estimate $\hat{x}(t \mid t)$. For the system described in Section 8.3.2 (with zero control input) the predicted state-vector becomes:

$$\hat{x}(t_1 \mid t) = \boldsymbol{\Phi}(t_1, t)\hat{x}(t \mid t). \tag{8.184}$$

This result may be proven by a simple extension of the previous theory and is intuitively justifiable, since given the state of the system $\boldsymbol{x}(t)$ at time t, the free response of the system gives: $\boldsymbol{x}(t_1) = \boldsymbol{\Phi}(t_1, t)\boldsymbol{x}(t)$. Thus, since $\hat{x}(t, t)$ is the best estimate of $\boldsymbol{x}(t)$ the result (8.184) follows.

The smoothing or interpolation problem involves the estimation of the system state (or output) at some time in the past. That is, a record of information up to time t is available and the estimate $\hat{x}(t_1 \mid t)$ is required at some time $t_1 < t$. For example, the initial state $\hat{x}(0 \mid t)$ of the system might be estimated. The predicted, filtered and smoothed linear estimates are derived using the projection of $x(\cdot)$ onto H_t^z and become $\mathscr{P}_t^z x(t+\tau)$, $\mathscr{P}_t^z x(t)$ and $\mathscr{P}_t^z x(t-\tau)$, respectively.

The early work on smoothing by Wiener [5], Bode [52] and Truxal [53], was based upon an infinite-time interval ($t_0 \to -\infty$) and solutions were obtained in the frequency domain. This type of smoother was not successful due to the infinite interval assumption and problems with realization and stability. A recursive solution to the smoothing problem was first obtained by Bryson and Frazier [47]. Such solutions were more difficult to obtain than those for the filtering problem and they came after Kalman and Bucy's work on filtering. Kailath and Frost [48] derived most of the smoothing formulae using an innovations approach. In these cases, the derivations and the forms of the solutions were simpler than those given previously. They also gave a useful adjoint-filter interpretation for the smoothing filter solution.

There are many applications for smoothing filters in communications, telemetry (Gelb [3]) and control systems (Lindquist [49]). Smoothing filters have also been used in ships in inertial navigation systems (Mehra and Bryson [50]) and in the control of aircraft using terrain-following radar (Anderson [51]). The smoother delay τ must not be too large, otherwise the performance of the control system will suffer. In this regard, it is helpful to note that if the delay is of the order of two or three times the dominant time-constant of the optimal filter, then effectively as much smoothing improvement, as it is possible to obtain, will be achieved.

A smoothing filter can be used to estimate the state of process $x(t_1)$, given the set of observations, $\{\dot{z}(\sigma) : 0 \le \sigma \le t\}$ where $0 \le t_1 < t$. The problem has been divided into three separate classes:

1. Fixed-interval smoothing: the observation interval is fixed (t is constant) and t_1 is varied $t_1 \in [0, t]$.
2. Fixed-point smoothing: the estimation time ($t_1 = T_1$) is fixed and the observation interval is varied for $t \ge t_1$.
3. Fixed-lag smoothing: the difference $\tau = t - t_1$ is fixed; t and t_1 are varied for $t \ge \tau$.

It is well known that smoothing generally improves the performance of estimates at a cost of increased computation and of course a delay of τ seconds. The number of applications of smoothing filters in control systems are limited but there are a large number of research papers on smoothing. This is probably because the solutions of the smoothing problem are all relatively complex compared with the simple Kalman filter equations. Recent work on finite-time smoothing problems has involved the use of time-invariant estimators to simplify implementation (Grimble [21]). This type of smoother has the added advantage that it also applies to some non-stationary systems (Kailath [22], Levy *et al.* [23]).

A more detailed introduction on smoothing problems is included in Maybeck [63]).

8.4.3. Computational problems in estimation

Errors in the computed state estimates can arise from several sources:

(a) Errors in the assumed system parameters and noise covariance matrices (Heffes [24]).
(b) Linearization errors.
(c) Neglected coloured-noise components in noise terms.
(d) Neglect of biases.
(e) Round-off errors.

One class of error phenomena is called divergence and this exists when the actual error covariance becomes large relative to the theoretically predicted covariance matrix $\mathbf{P}(t) = \mathbf{P}(t \mid t)$. Divergence can result from several causes including:

(a) Low-process noise.
(b) Signal models which are not asymptotically stable.
(c) Bias errors.

Divergence tends to arise from modelling rather than computational errors (Fitzgerald [25]). Such problems can be mitigated by using exponential data weighting or a fixed memory filter [26–29]. The former involves multiplying the noise covariance matrices by a term of the form α^{-2k} where $|\alpha| > 1$ (Anderson and Moore [30]). This scalar enters the covariance matrix update equations. Fixed memory filters have also been shown to reduce divergence problems (Jazwinski [39]). New forms of this type of filter have recently been proposed. Increasing the process-noise covariance often has the same effect as exponential data weighting, that is, a better degree of stability is obtained.

Square-root filtering

In the calculation of the error-covariance matrix using the usual Kalman filter equations the error-covariance matrix $\mathbf{P}(t)$ can fail to be non-negative definite and symmetric. This is more likely if some measurements are almost noise free. This difficulty can be reduced by propagating the error-covariance matrix in square-root form. Let $\mathbf{P}(t) = \mathbf{N}(t)\mathbf{N}(t)^T$ where $\mathbf{N}(t)$ is the square root of $\mathbf{P}(t)$. The Kalman filter equations may be written in terms of $\mathbf{N}(t)$ with two advantages:

(a) Since $\mathbf{N}(t)\mathbf{N}(t)^T$ is always non-negative definite the calculation of $\mathbf{P}(t)$

as $\mathbf{N}(t)\mathbf{N}(t)^T$ cannot lead to a matrix which fails to be non-negative definite.

(b) The numerical conditioning of $\mathbf{N}(t)$ is generally better than that for $\mathbf{P}(t)$.

A useful approximate rule cited by Bierman [41] is that square-root algorithms require only half the word length required by conventional non-square root methods. However, square-root filtering generally requires more computation but the additional accuracy may be essential in some cases. The book by Maybeck [14] includes a useful chapter on a square-root filtering.

A related technique for improving numerical reliability has been developed by Thornton and Bierman [31]. This involves the covariance factorization $\mathbf{P} = \mathbf{UDU}^T$ where $\mathbf{U}$ is upper triangular with unit diagonals and $\mathbf{D}$ is diagonal. A useful comparison of discrete linear filtering algorithms has been presented by Bierman [41].

Recent work by Kailath and co-workers at Stanford University has resulted in the so-called fast algorithms for calculating the gain matrix for time-invariant systems [32]–[35]. Ths usual Riccati equations are replaced by Chandrasekhar equations. For a limited class of system these algorithms can be more attractive computationally. Shaked and Priel [45]) have also obtained some simple expressions which enable the Kalman gain and the estimation error-covariance matrix to be calculated in terms of the zero structure of the measurement power spectrum.

8.5. CONCLUSIONS

The number of applications of Kalman filters in industry is increasing rapidly. For example, in the steel industry, Kalman filters have been considered for use in electric-arc furnace state estimation (Boland and Nicholson [36]), back-up-roll eccentricity filtering (Grimble [37]), shape-control systems (Grimble [38], Grimble and Patton [54]) and in centre temperature estimation in soaking pits. There are several reasons for their success:

(i) Easily implemented on digital computers, using a recursive algorithm.

(ii) Robust in the sense that they often work reasonably well because of their feedback structure, particularly when models and noise data are inaccurate.

(iii) Versatile and are applicable in non-stationary noise and time-varying signal model situations.

Thus, although the derivation of the Kalman filter involves lengthy mathematical analysis the result is a filter which has many practical advantages.

8.6. PROBLEMS

1. Let $\{x(t), t \geq 0\}$ denote a stationary orthogonal increments process with $x(0) = 0$ and $E\{x(t)\} = 0$. If the variance function $v(t) = E\{x(t)^2\}$ show that for stationary noise $v(t) = v(t-s) + v(s)$ or $v(t+s) = v(t) + v(s)$ and thence show that $v(t) = \sigma^2 t$ for some constant σ^2.

2. Consider the set S of step functions α: $[0, \infty) \to \mathbb{R}, \alpha \in S$ and assume there exist times $0 < t_1 < \cdots < t_n$ and constants $c_0, c_1, \ldots, c_{n-1}$ such that:

$$\alpha(t) = \begin{cases} c_i & t \in [t_i, t_{i+1}) \\ 0 & t \geq t_n \end{cases} \qquad i = 0, \ldots, n-1.$$

The Wiener integral of α with respect to $\{x(t)\}$ is denoted by: $I(\alpha) = \int_0^\infty \alpha(t)\, dx(t) = \sum_{i=0}^{n-1} c_i(x(t_i + 1) - x(t_i))$. Suppose that $\{x(t), t \geq 0\}$ represents a standard Wiener process. Show that $E\{I(\alpha)I(\beta)\} = \int_0^\infty \alpha(t)\beta(t)\, dt$.

3. Show that the matrix Riccati differential equation (8.99) has a unique solution $\mathbf{P}(\tau) \geq \mathbf{O}$ for all $\tau \in [0, t]$. Use the facts that $\mathbf{P}_0 \geq \mathbf{O}, \mathbf{P}(t) = \text{cov}[\tilde{\mathbf{x}}(t), \tilde{\mathbf{x}}(t)]$ and that $\{\mathbf{x}(t)\}$ is quadratic mean continuous on $[0, t]$.

4. Let $z(t)$, $t \geq 0$, denote a zero-mean independent increments process and define a new process $y(t)$ as $y(t) = z(t) - z(0)$, $t \geq 0$. Show that $y(t)$ is also an independent increments process, has the same increments as $x(t)$ does, and satisfies $P\{y(0) = 0\} = 1$.

5. Describe how stochastic processes are used to model disturbances in physical systems.

(a) Define a Gaussian stochastic process and describe a state equation model which includes measurement and noise and process noise inputs.

(b) Explain the difference between a stationary random process and a wide sense stationary process.

(c) Consider a sine wave $x(t) = \sin(\omega t + \phi)$ where ϕ is a random variable uniformly distributed over the interval $-\pi$ to π. Determine whether this process is wide sense stationary.

(d) If the phase ϕ is a random variable uniformly distributed over the interval 0 to π determine whether the process is wide-sense stationary and hence, whether the process is stationary.

8.7. REFERENCES

[1] Kalman, R. E. 1960. A new approach to linear filtering and prediction problems, *Journal of Basic Engineering*, **82**, 35–45, March.

[2] Kalman, R. E. 1961. New methods in Wiener filtering theory, *Proc. of the Symposium on Engineering Application of Random Function Theory and Probability*, pp. 270–388.

[3] Gelb, A. 1974. *Applied Optimal Estimation*, p. 142, M.I.T. Press.

[4] Kwakernaak, H. and Sivan, R. 1972. *Linear Optimal Control Systems*, Wiley-Interscience, New York.

[5] Weiner, N. 1949. *Extrapolation, Interpolation and Smoothing of Stationary Time Series, with Engineering Applications*, Technology Press and Wiley, New York. (Originally issued in Feb. 1942 as a classified National Defense Res. Council Rep.).

[6] Kalman, R. E. and Bucy, R. S. 1961. New results in linear filtering and prediction theory, *Journal of Basic Engineering*, **83**, 95–100, March.

[7] Jazwinski, A. H. 1970. *Stochastic Processes and Filtering Theory*, Academic Press, London.

[8] Curtain, R. F. and Pritchard, A. J. 1977. *Functional Analysis in Modern Applied Mathematics*, Academic Press, London.

[9] Davis, M. H. A. 1977. *Linear Estimation and Stochastic Control*, Chapman and Hall, London.

[10] Luenberger, D. G. 1969. *Optimization by Vector Space Methods*, Wiley, London.

[11] Melsa, J. L. and Sage, S. P. 1973. *An Introduction to Probability and Stochastic Processes*, pp. 206 and 343, Prentice-Hall, New Jersey.

[12] Doob, J. L. 1953. *Stochastic Processes*, Wiley, London.

[13] Astrom, K. J. 1970. *Introduction to Stochastic Control Theory*, Academic Press, London.

[14] Maybeck, P. S. 1979. *Stochastic Models, Estimation and Control*, Vol. 1, Academic Press, London.

[15] Rosenbrock, H. H. and Storey, C. 1970. *Mathematics of Dynamical Systems*, p. 195, Nelson, London.

[16] Moir, T. J. and Grimble, M. J. 1984. Optimal self-tuning filtering, prediction and smoothing for discrete multivariable processes, *IEEE Trans. Aut. Control*, **29**(2), 128–137.

[17] Sage, A. P. and Melsa, J. L. 1971. *Estimation Theory with Applications to Communications and Control*, p. 268. McGraw-Hill.

[18] Kailath, T. 1976. *Lectures on Linear Least-Squares Estimation*, Springer-Verlag, New York.

[19] Meditch, J. G. 1969. *Stochastic Optimal Linear Estimation and Control*, p. 296, McGraw-Hill.

[20] Grimble, M. J. 1978. Solution of the linear estimation problem in the *s*-domain, *Proc. IEE*, **125**(6), 541–549, June.

[21] Grimble, M. J. 1980. A new finite-time linear smoothing filter, *Int. J. Systems Sci.*, **11**(10), 1189–1212.

[22] Kailath, T. 1973. Some new algorithms for recursive estimation in constant linear systems, *IEEE Trans, on Inform. Theory*, **IT-19**(6), 750–760, November.

[23] Levy, B., Kailath, T., Ljung, L. and Morf, M. 1979. Fast time-invariant implementations for linear least-squares smoothing filters, *IEEE Trans. on Auto. Contr.* **AC-24**(5), 770–774, October.

[24] Heffes, H. 1966. The effect of erroneous models on Kalman filter response, *IEEE Trans. on Auto. Control.* **AC-11**, 541–543, July.

[25] Fitzgerald, R. J. 1971. Divergence of the Kalman filter, *IEEE Trans. on Auto. Control.*, **AC-16**(6), 736–747.

[26] Grimble, M. J. 1980. A finite-time linear filter for discrete-time systems, *International Journal of Control*, **31**(3), 413–432.

[27] Bierman, G. J. 1975. Fixed memory least-squares filtering, *IEEE Trans. Inf. Theory*, **IT-21**, 690–692.

[28] Buxbaum, P. J. 1974. Fixed-memory recursive filters, *IEEE Trans. Inf. Theory*, **IT-20**, 113–115, June.

[29] Lee, R. C. 1967. The 'moving window' approach to the problems of estimation and identification, Air Force Report No. SAMSO-TR-68-78, Los Angeles Air Force Station, Los Angeles, California.

[30] Anderson, B. D. O. and Moore, J. B. 1979. *Optimal Filtering*, Prentice-Hall.

[31] Thornton, C. L. and Bierman, G. J. 1978. Filtering and error analysis via the $\mathbf{UDU}^T$ covariance factorization, *IEEE Trans. on Auto. Cont.*, **AC-23**(5), 901–907, October.

[32] Kailath, T., Levy, B. C., Ljung, L. and Morf, M. 1978. Fast time-invariant implementation of Gaussian signal detectors, *IEEE Trans. on Inform. Theory*, **IT-24**(4), 469–477, July.

[33] Ljung, L. and Kailath, T. 1977. Efficient change of initial conditions, dual Chandraskhar equations and some applications, *IEEE Transactions on Automatic Control*, **AC-22**, 443–446, June.

[34] Kailath, T., Ljung, L. and Morf, M. 1978. *Generalized Krein–Levinson Equations for Efficient Calculation of Fredholm Resolvents of Non-Displacement Kernels*, pp. 169–184, Academic Press.

[35] Friedlander, B., Kailath, T., Morf, M., and Ljung, L. 1978. Extended Levinson and Chandrasekhar equations for general discrete-time linear estimation problems, *IEEE Trans. on Aut. Cont.*, **AC-23**(4), 653–659.

[36] Boland, F. M. and Nicholson, H. 1977. Estimation of the states during refining in electric-arc-furnace steelmaking, *Proc. IEE Cont. and Science*, **124**(2), 161–166, February.

[37] Grimble, M. J. 1982. Frequency domain properties of Kalman filters, Sheffield City Polytechnic, Research Report, EEE/42/1979; IEE Colloquium Savoy Place, February.

[38] Grimble, M. J. 1981. Development of models for shape control system design, published in Part II, *Modelling of Dynamical Systems*, edited by H. Nicholson, Peter Peregrinus Ltd.

[39] Jazwinski, A. H. 1968. Limited memory optimal filtering, *IEEE Trans. on Automatic Control*, **AC-13**, 558–563, October.

[40] McGarty, T. P. 1974. *Stochastic Systems and State Estimation*, Wiley, London.

[41] Bierman, G. J. 1973. A comparison of discrete linear filtering algorithms, *IEEE Trans. on Aerospace and Electronic Systems*, **AES**, **9**(1), 28–37.

[42] Arcasoy, C. C. 1971. Return difference-matrix properties for optimal stationary discrete Kalman filter, *Proc. IEE*, **118**(12), 1831–1834.

[43] Chen, Chi-Tsong. 1976. On digital Wiener filters, *Proc. IEEE*, **64**, 1736–1737, December.

[44] Wong, E. 1971. *Stochastic Processes in Information and Dynamical Systems*, McGraw-Hill Book Co., New York.

[45] Shaked, U. and Priel, B. 1982. 'Explicit solutions to the linear optimal estimation problem of discrete time invariant linear processes', *Int. J. Control*, **36**(5), 725–745.

[46] Abramowitz, N. and Segun, I. A. 1968. *Handbook of Mathematical Functions*, Dover Publications, Inc., New York.

[47] Bryson, A. E. and Frazier, M. 1963. Smoothing of linear and non-linear dynamic systems, Aeronautical System Division, Wright Patterson Air Force Base, Ohio, Report TDR 63-119.

[48] Kailath, T. and Frost, P. 1968. An innovations approach to least-squares estimation, Part II: Linear smoothing in additive white noise, *IEEE Trans. on Auto. Contr.* **AC-13**(6), 655–660, Dec.

[49] Lindquist, A. 1968. On optimal stochastic control with smoothed information, *Inform Sciences*, **1**, 43–54.

[50] Mehra, R. D. and Bryson, A. E. 1968. Linear smoothing using measurements containing correlated noise for an application to inertial navigation, *IEEE Trans. Aut. Cont.* **AC-13**(5), 496–503, October.

[51] Anderson, B. D. O. 1969. Properties of optimal linear smoothing, *IEEE Trans. on Auto. Cont.*, **AC-114**, 114–115, February.

[52] Bode, H. W. and Shannon, C. E. 1950. A simplified derivation of linear least-square smoothing and prediction theory, *Proc. of IRE*, **38**, 417–425, April.

[53] Truxal, J. G. 1955. *Automatic Feedback Control System Synthesis*, p. 465. McGraw-Hill.

[54] Grimble, M. J. and Patton, R. J. 1981. The design of shape control systems using stochastic optimal control theory, IEE International Conf. on Control and its Applications, University of Warwick, Mar.

[55] O'Reilly, J. 1982. A finite-time linear filter for discrete-time systems in singular case, *Int. J. Systems Sci.*, **13**(3), 257–263.

[56] Rhodes, J. B. 1971. A tutorial introduction to estimation and filtering, *IEEE Trans. on Automatic Control*, **AC-16**(6), 688–706.

[57] Grimble, M. J. 1979. Solution of the Kalman filtering problem, for stationary noise and finite data records, *Int. J. Systems Sci.*, **10**(2), 177–196.

[58] Soong, T. T. 1973 *Random Differential Equations in Science and Engineering*, Academic Press, New York and London.

[59] Taylor, S. J. 1973. *Introduction to Measure and Integration*, Cambridge University Press.

[60] Thomas, J. B. 1971. *An Introduction to Applied Probability and Random Processes*, John Wiley and Sons, Inc., New York.

[61] Assefi, T. 1980. *Stochastic Processes and Estimation Theory with Applications*, Wiley Interscience.

[62] Grimble, M. J. 1982. *Generalized Weiner and Kalman filters for uncertain systems*, IEEE Conf. On Dec. and Contr., Orlando, pp. 221–227, December.

[63] Maybeck, P. S. 1982. *Stochastic Models, Estimation and Control*, Vol. 2., Academic Press, New York.

CHAPTER 9

Frequency Domain Analysis of Filtering and Smoothing Problems

9.1. INTRODUCTION

The continuous-time Kalman filtering problem of the previous chapter is investigated further using a transform analysis. The infinite time or steady-state problem ($t_0 \to -\infty$) is reconsidered first from a frequency-response viewpoint. Engineers are usually very familiar with these concepts and thus an engineering appreciation of these characteristics of Kalman filters is obtained.

In most control applications, filters are now implemented in digital form and this requires an analysis in discrete equation form. Thus, the second part of the chapter concentrates on the z-domain analysis of filtering and smoothing problems. Here, the opportunity is taken to introduce a new type of finite-time estimator derived from the dual of the finite-time-contol problems considered in Chapter 2. For restricted set of problems, this estimator offers some advantages over the time-varying Kalman estimator usually applied.

Steady-state filtering problem

The Kalman filter (Kalman and Bucy [1], Kalman [2]) has assumed a role of ever increasing importance over the last twenty years and most aspects of its performance and application have been extensively reviewed. However, few authors have considered the frequency-domain performance of such filters. Recent work on the definition of the zeros of multivariable systems enables the performance of the filter to be re-examined. The need for such a study was emphasized when several industrial filtering problems were found to have the same basic structure and to employ the same classical solution; namely a notch filter. For such systems it is possible to define an industrial canonical Kalman filtering problem (Grimble and Åström [41]).

To introduce the basic frequency-domain properties of Kalman–Bucy filters, three very simple filtering problems are considered. In each problem the filter is found to have a low-frequency gain of unity. A more general multivariable situation is then considered and the conditions under which the gain becomes equal to the identity matrix at zero frequency are established (Section 9.2). The system structure is based upon that found in the dynamic ship-positioning control problem (Chapter 13). The presence of integrators in the system is shown to lead to a forward path gain of unity.

It is well known that the closed-loop poles of the Kalman–Bucy filter are determined by a return difference relationship (MacFarlane [3], Grimble [4]) involving the system noise covariance matrices and the system transfer-function matrix. The zeros of the optimal filter are not so well understood. However, from the simple examples referred to above it is shown that the zeros of the filter are related to the poles of the noise model. In Section 9.2.3 this relationship is investigated and generalized, using recent definitions of the zeros for multivariable systems. Also evident from the simple examples is a relationship between the transfer function for the filter and the signal-to-signal plus noise ratio. This is developed in more detail for the multivariable system description in Section 9.2.4.

The solution of the infinite-time optimal filtering problem was first obtained by Norbert Wiener [5] using an s-domain approach. The steady-state Kalman–Bucy filter is equivalent to a Wiener filter. However, the Kalman–Bucy filter depends upon a state-space system model whereas the Wiener filter is based upon a transfer-function system description. A Wiener analysis is used in Section 9.2.5 to confirm the transfer-function results of the previous sections.

In Section 9.2.6, a brief discussion of four industrial examples which exhibit the frequency-domain properties of the Kalman–Bucy filter is presented. The ship-positioning control problem is described first. This is followed by a problem which involves the estimation of the lateral and longitudinal motions of an aircraft. The state equations for the elastic modes filtering problem are very similar in structure to those of the ship-positioning problem. Two filtering problems arising from the control of steel-rolling mills are also discussed. These concern filters to remove back-up roll eccentricity signals and filters required in strip shape-control systems. Each problem involves a system with the same underlying structure and a predominantly sinusoidal disturbance.

The frequency-domain properties discussed here (first noted by Karl Åström of Lund Institute of Technology) were observed in some of the simple examples which open the chapter.

Finite-time filtering problem

The remaining sections are concerned with a new approach to finite-time estimation. The usual solution to the finite-time problem involves a Kalman–Bucy filter with a time-varying gain. Such a filter is not often used because of the complexity of implementation. The infinite-time or steady-state filter which involves a fixed gain matrix, is often employed instead, even though the time interval of interest is relatively short. The new time-invariant estimator described here, to some extent, achieves the best of both situations. it is time-invariant and can be implemented in transfer-function matrix form. It is also derived from a finite-time estimation problem.

9.2. FREQUENCY DOMAIN PROPERTIES OF KALMAN FILTERS

Useful insight into the operation of the steady-state Kalman–Bucy filter may be obtained by considering its frequency domain performance. The conditions under which the zero-frequency gain of the filter is unity are established and the pole-zero locations for the filter are determined for a class of industrial systems. The effect of the signal to signal-plus-noise ratio, on the gain of the filter, is discussed.

9.2.1. Kalman–Bucy filtering examples

In this section three simple examples of Kalman–Bucy filtering are considered. The relationship between the frequency domain properties of the Kalman–Bucy filter and of the signal and noise models are explored.

Example 9.1. *Single-integrator process.* Consider a signal process $\{x\}$ described by

$$dx = d\omega$$

where $\{\omega\}$ is a Wiener process with incremental covariance $q\ dt$ (Åström [6], McGarty [7]). Let the observation process $\{z\}$ be given by

$$dz = x\ dt + dv$$

where $\{v\}$ *is a Wiener process with incremental covariance r dt.* The Kalman–Bucy filter is given by:

$$d\hat{x} = k(dz - \hat{x}\ dt)$$

where the filter gain is given by $k = p/r$ and p is the solution to the Riccati differential equation:

$$\frac{dp}{dt} = \frac{-p^2}{r} + q.$$

The steady-state solution is $p_\infty = (qr)^{1/2}$ and $k = (q/r)^{1/2}$. The transfer function of the

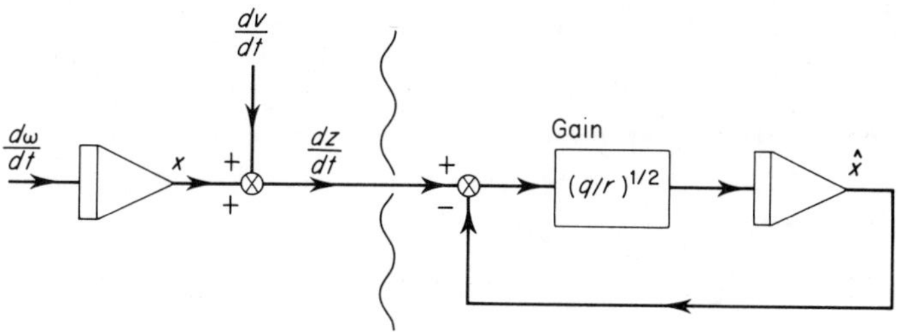

Fig. 9.1. System and filter for Example 8.1

filter (shown in Fig. 9.1) is thus:

$$H(s) = \frac{k}{(s+k)} = \frac{(q/r)^{1/2}}{(s+(q/r)^{1/2})}.$$

The steady-state or infinite-time filter has the same transfer function as the Wiener filter. Observe that:

(a) The zero-frequency gain $H(0)$ of the filter equals unity.
(b) The corner frequency equals $(q/r)^{1/2}$ so that the higher the measurement noise the lower the cut-off frequency.
(c) $|H(j\omega)|^2 = (q/\omega^2)/(r + q/\omega^2)$ where q/ω^2 and r are the signal and the noise powers at the frequency ω. The magnitude of the transfer function of the filter is thus proportional to the square root of the signal-to-signal plus noise ratio.

Example 9.2: *Coloured-measurement noise.* In this example, the same signal process used in Example 9.1 is considered. However, the measurement noise will be assumed to be generated by the following model:

$$dn = dv + dn_2$$

where v is a Wiener process and dn_2/dt is obtained by passing white noise, with covariance $q_2\,\delta(t)$, through a shaping filter (see Fig. 9.2), with the transfer function:

$$G_n(s) = \frac{s}{(s+a)} = 1 - \frac{a}{(s+a)}.$$

The total measurement noise spectral-density is therefore:

$$\Phi_{nn}(s) = r + \frac{(-s^2)}{(-s^2+a^2)}\,q_2$$

as shown in Fig. 9.3 for $s = j\omega$.

A state variable x_2 may be introduced to represent the dynamics of the shaping filter

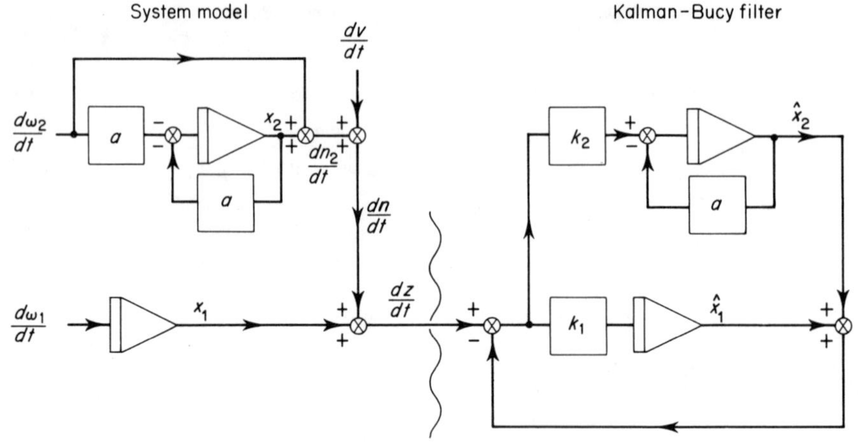

Fig. 9.2. System and Kalman–Bucy filter for Example 8.2

and thus the signal and noise processes can be represented by the following model:

$$dx_1 = d\omega_1$$

$$dx_2 = -ax_2\ dt - ad\omega_2$$

$$dz = x_1\ dt + x_2\ dt + dv + d\omega_2$$

where the Wiener processes $\{\omega_1\}$, $\{\omega_2\}$ and $\{v\}$ are independent with the incremental

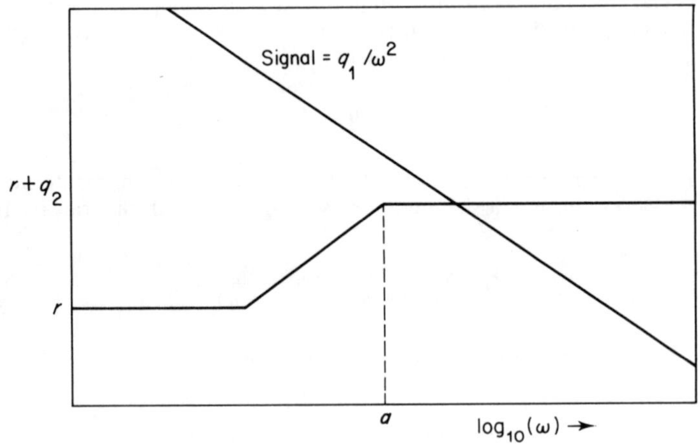

Fig. 9.3. Approximate spectral densities of the signal and noise used in Example 2.

$$\Phi_{nn}(j\omega) = \frac{\omega^2(r+q_2)+a^2r}{\omega^2+a^2}$$

covariances $q_1\,dt$, $q_2\,dt$ and $r\,dt$ respectively. The Kalman–Bucy filter then becomes:

$$d\hat{x} = \begin{bmatrix} 0 & 0 \\ 0 & -a \end{bmatrix} \hat{x}\,dt + \begin{bmatrix} k_1 \\ k_2 \end{bmatrix} [dz - \hat{x}_1\,dt - \hat{x}_2\,dt]$$

where the steady-state gains k_1 and k_2 are given by:

$$\mathbf{K} = (\mathbf{P}_\infty \mathbf{C}^T + \mathbf{DG})\mathbf{R}^{-1}$$

and

$$\mathbf{Q} = \text{diag}\{q_1, q_2\},\ \mathbf{R} = r + q_2 \text{ and } \mathbf{G}^T = [0 \quad q_2].$$

The matrix Riccati equation gives:

$$(\mathbf{A} - \mathbf{DGR}^{-1}\mathbf{C})\mathbf{P}_\infty + \mathbf{P}_\infty(\mathbf{A} - \mathbf{DGR}^{-1}\mathbf{C})^T - \mathbf{P}_\infty\mathbf{C}^T\mathbf{R}^{-1}\mathbf{CP}_\infty + \mathbf{D}(\mathbf{Q} - \mathbf{GR}^{-1}\mathbf{G})\mathbf{D}^T = \mathbf{0}$$

where

$$\mathbf{A} = \begin{bmatrix} 0 & 0 \\ 0 & -a \end{bmatrix}, \qquad \mathbf{D} = \begin{bmatrix} 1 & 0 \\ 0 & -a \end{bmatrix}, \qquad \mathbf{C} = [1 \quad 1].$$

Setting $\mathbf{P}_\infty = \{p_{ij}\}_{i=1, j=1}^{2\ \ 2}$ and noting $p_{ij} = p_{ji}$, the algebraic Riccati equation may be expanded to obtain:

$$q_1(r + q_2) - (p_{11} + p_{12})^2 = 0$$

$$a(q_2 p_{11} - r p_{12}) - (p_{11} + p_{12})(p_{12} + p_{22}) = 0$$

$$2a(q_2 p_{12} - r p_{22}) + a^2 q_2 r - (p_{12} + p_{22})^2 = 0$$

and the filter gain becomes:

$$k_1 = (p_{11} + p_{12})/(r + q_2) = (q_1/(r + q_2))^{1/2}$$

$$k_2 = (p_{12} + p_{22} - a q_2)/(r + q_2).$$

The transfer function of the filter (see Fig. 9.2) from measurement to the signal estimate $\hat{x}_1(t)$ is given by:

$$H_1(s) = [1 \quad 0] \begin{bmatrix} s + k_1 & k_1 \\ k_2 & s + a + k_2 \end{bmatrix}^{-1} \begin{bmatrix} k_1 \\ k_2 \end{bmatrix}$$

$$= \frac{k_1(s + a)}{(s^2 + s(k_1 + k_2 + a) + a k_1)}.$$

Recall that $\hat{x}_1(t)$ represents an estimate of the signal state and $\hat{x}_2(t)$ represents an estimate of the noise model state. Observe that:

(a) The low-frequency gain $H_1(0)$ equals unity.
(b) At high frequencies $H_1(s) \simeq k_1/s = ((q_1/s^2)/(r + q_2))^{1/2}$. As in Example 8.1, the

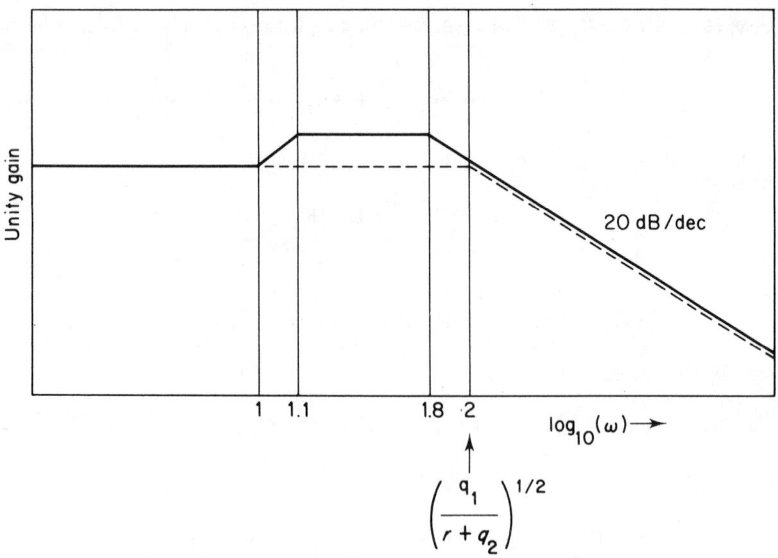

Fig. 9.4. Bode amplitude diagram for Example 8.2

high-frequency gain is proportional to the square root of the signal-to-signal plus noise ratio.

(c) The transfer function $H_1(s)$ has a zero at $s = -a$ which corresponds to a pole of the noise model.

The Bode diagram for the filter is shown in Fig. 9.4. For comparison, the Bode diagram for the filter of Example 9.1 is also shown in dotted lines and the noise levels are chosen so that the signal-to-noise ratios are the same at high frequencies. The diagram is drawn for the parameters $a = 1$, $q_1 = 8$, $q_2 = 1$ and $r = 1$. The corresponding Riccati equation solution becomes $p_{11} = 3.662$, $p_{22} = 0.49285$ and $p_{12} = 0.3381$ and the gains become $k_1 = 2$ and $k_2 = -0.0845$. The form of the Bode diagram in Fig. 9.4 may be explained by noting that at low frequencies the signal power is high and the noise power is low but increasing. The optimal gain $H_1(0)$ is therefore unity. At high frequencies the signal power is decreasing whilst the noise power comes constant. Thus, in this region the filter gain is decreasing.

Example 9.3. *Discrete-time system.* A discrete-time system will be considered in this example. Assume that the signal and noise processes can be described by:

$$x_1(t+1) = x_1(t) + \omega_1(t), \qquad t = \{t_0, t_0+1, t_0+2, \ldots\}$$

$$x_2(t+1) = -ax_2(t) + \omega_2(t), \qquad |a| \leq 1$$

$$z(t) = x_1(t) + x_2(t) + v(t)$$

where $\{\omega_1(t)\}$, $\{\omega_2(t)\}$ and $\{v(t)\}$ are mutually independent sequences of white noise with covariances q_1, q_2 and r respectively. The Kalman filter is given by:

$$\hat{\mathbf{x}}(t+1) = \begin{bmatrix} 1 & 0 \\ 0 & -a \end{bmatrix} \hat{\mathbf{x}}(t) + \begin{bmatrix} k_1 \\ k_2 \end{bmatrix} [z(t) - \hat{x}_1(t) - \hat{x}_2(t)].$$

The filter gains k_1 and k_2 are given by:

$$\mathbf{K} = \mathbf{A}\mathbf{P}_\infty\mathbf{C}^T(\mathbf{C}\mathbf{P}_\infty\mathbf{C}^T + \mathbf{R})^{-1}$$

where

$$\mathbf{P}_\infty = \mathbf{A}\mathbf{P}_\infty\mathbf{A}^T + \mathbf{Q} - \mathbf{A}\mathbf{P}_\infty\mathbf{C}^T(\mathbf{C}\mathbf{P}_\infty\mathbf{C}^T + \mathbf{R})^{-1}\mathbf{C}\mathbf{P}_\infty\mathbf{A}^T.$$

Setting $\mathbf{P}_\infty = \{p_{ij}\}_{i=1}^{2}{}_{j=1}^{2}$ and noting $p_{ij} = p_{ji}$, the matrix Riccati equation yields:

$$p_{11} = p_{11} + q_1 - (p_{11} + p_{12})^2/(r + p_{11} + 2p_{12} + p_{22})$$

$$p_{12} = -ap_{12} - (p_{11} + p_{12})(p_{12} + p_{22})/(r + p_{11} + 2p_{12} + p_{22})$$

$$p_{22} = a^2p_{22} + q_2 - a^2(p_{12} + p_{22})^2/(r + p_{11} + 2p_{12} + p_{22})$$

and the filter gains are given by:

$$k_1 = (p_{11} + p_{12})/(r + p_{11} + 2p_{12} + p_{22})$$

$$k_2 = -a(p_{12} + p_{22})/(r + p_{11} + 2p_{12} + p_{22}).$$

The transfer function from the observations input to the estimate $\hat{x}_1$ (see Fig. 9.5) is

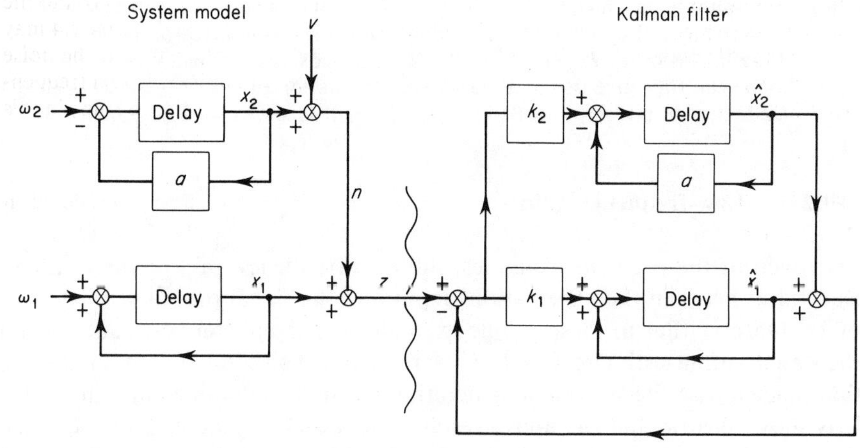

Fig. 9.5. System model and filter for Example 8.3

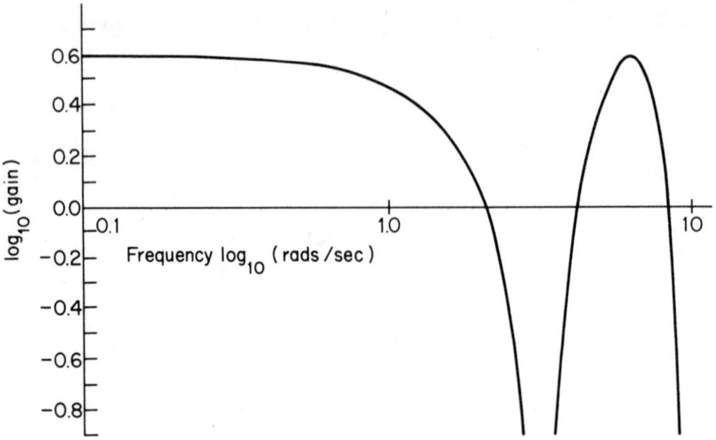

Fig. 9.6. Filter zero polynomial frequency response plot

given by:

$$H_1(z^{-1}) = [1 \quad 0]\begin{bmatrix} z-1+k_1 & k_1 \\ k_2 & z+a+k_2 \end{bmatrix}^{-1}\begin{bmatrix} k_1 \\ k_2 \end{bmatrix}$$

$$= \frac{k_1(1+az^{-1})z^{-1}}{(1+(a+k_1+k_2-1)z^{-1}+(ak_1-a-k_2)z^{-2}}.$$

Observe the following properties of the filter:

(a) The transfer function of the filter has a zero $z = -a$, corresponding to the pole $z = -a$ of the noise model.

(b) The zero-frequency response of the filter is unity that is $\lim_{|z| \to 1} H_1(z^{-1}) = 1$.

(c) For suitable parameter values, the zero can cause a significant damping of the disturbances. For example, the magnitude of the numerator polynomial in $H_1(z^{-1})$ becomes $H_{1n}(e^{-j\omega T}) = 1 + a^2 + 2a \cos(\omega T)$ when $k_1 = 1$. The zero alone can thus give rise to a maximum attenuation of $\alpha = (1-a)^2/(1+a)^2$. A frequency-response plot of $\log_{10}|H_{1n}(e^{-j\omega T})|$ is shown in Fig. 9.6. for $a = 1$.

9.2.2. Low-frequency gain

In each of the previous examples, the zero-frequency gain of the Kalman-Bucy filter between the observations input and the signal output estimate was unity. Observe that in these simple examples the signal state-estimate $\hat{x}_1$ and the signal output-estimate $\hat{y}_1 = C_1\hat{x}_1$ were identical since $C_1 = 1$. Also observe that in each case the subsystem generating the signal contained an integrator. It is shown below that the above result holds in more general cases. That is, the zero-frequency gain is equal to the identity matrix, providing the signal source contains integrator terms associated with certain of the state variables which form the outputs. This class of system will be illustrated by considering an industrial example.

Kalman–Bucy filtering techniques are now being used in dynamic ship-positioning control systems (Chapter 13). The vessel dynamics are represented by state equations of the following form:

$$\begin{bmatrix} dx_1 \\ dx_2 \\ dx_3 \end{bmatrix} \quad \begin{bmatrix} \mathbf{O} & \mathbf{A}_{12} & \mathbf{O} \\ \mathbf{O} & \mathbf{A}_{22} & \mathbf{O} \\ \mathbf{O} & \mathbf{O} & \mathbf{A}_{33} \end{bmatrix} \begin{bmatrix} x_1 \\ x_2 \\ x_3 \end{bmatrix} dt + \begin{bmatrix} D_1\ d\omega_1 \\ D_2\ d\omega_2 \\ D_3\ d\omega_3 \end{bmatrix} + \begin{bmatrix} \mathbf{o} \\ \mathbf{B}_2 \\ \mathbf{o} \end{bmatrix} u\ dt \tag{9.1}$$

$$dz = [\mathbf{C}_1 \quad \mathbf{O} \quad \mathbf{C}_3]x\ dt + dv, \qquad dz(t) \in \mathbb{R}^r \tag{9.2}$$

where the state vectors $\boldsymbol{x}_1$ and $\boldsymbol{x}_2$ represent the low-frequency ship motions and x_3 represents the high-frequency motions. The low-frequency position of the vessel must be controlled and this is represented by the state vector $\boldsymbol{x}_1$ (or $y_1 = \mathbf{C}_1\boldsymbol{x}_1$ in different coordinates). The low-frequency position states are obtained by integrating the velocity states contained in $\boldsymbol{x}_2$ and thus $\mathbf{A}_{11} = \mathbf{O}$. These position states do not influence the state variables $\boldsymbol{x}_2$ and thus $\mathbf{A}_{21} = \mathbf{O}$. The high-frequency subsystem $(\mathbf{C}_3, \mathbf{A}_{33}, \mathbf{D}_3)$ is by the nature of the model, decoupled from the low-frequency subsystem. The matrix $\mathbf{C}_1$ represents a coordinate transformation and is full-rank.

The above partitioning may be used to rewrite the ship equations in a canonical form which applies in several industrial situations:

$$\begin{bmatrix} dx_1 \\ dx_2 \end{bmatrix} = \begin{bmatrix} \mathbf{O} & \mathbf{A}_{12} \\ \mathbf{O} & \mathbf{A}_{22} \end{bmatrix} \begin{bmatrix} x_1 \\ x_2 \end{bmatrix} dt + \begin{bmatrix} \mathbf{D}_1\ d\omega_1 \\ \mathbf{D}_2\ d\omega_2 \end{bmatrix} \tag{9.3}$$

$$dz = [\mathbf{C}_1 \quad \mathbf{C}_2]x\ dt + dv \tag{9.4}$$

where the vectors and matrices have now been redefined and the thruster control signal u is neglected. The matrix $\mathbf{C}_1$ is square and is assumed full rank, as in the above system. It follows that the subsystem $(\mathbf{A}_{11}, \mathbf{C}_1)$ is observable. The Kalman–Bucy filter for this system is illustrated in Fig. 9.7.

The transfer function (between dz/dt and y) of the Kalman–Bucy filter may be calculated as follows:

$$\mathbf{H}(s) = \mathbf{C}(\mathbf{I} + \mathbf{\Phi}(s)\mathbf{KC})^{-1}\mathbf{\Phi}(s)\mathbf{K}$$

$$\mathbf{\Phi}(s) = \begin{bmatrix} s\mathbf{I} & -\mathbf{A}_{12} \\ \mathbf{O} & s\mathbf{I} - \mathbf{A}_{22} \end{bmatrix}^{-1} = \begin{bmatrix} \mathbf{I}/s & \mathbf{A}_{12}\phi_{22}/s \\ \mathbf{O} & \phi_{22} \end{bmatrix}$$

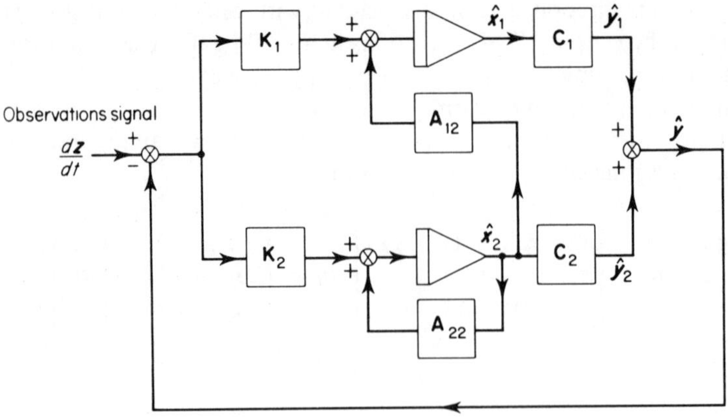

Fig. 9.7. Kalman–Bucy filter for the canonical system description

where $\boldsymbol{\phi}_{22}(s) \triangleq (s\mathbf{I} - \mathbf{A}_{22})^{-1}$. Also define $\mathbf{V}(s) = \mathbf{K}_1 + \mathbf{A}_{12}\boldsymbol{\phi}_{22}\mathbf{K}_2$, then

$$(\mathbf{I} + \boldsymbol{\Phi}(s)\mathbf{KC})^{-1} = \begin{bmatrix} \mathbf{I} + \mathbf{VC}_1/s & \mathbf{VC}_2/s \\ \boldsymbol{\phi}_{22}\mathbf{K}_2\mathbf{C}_1 & \mathbf{I} + \boldsymbol{\phi}_{22}\mathbf{K}_2\mathbf{C}_2 \end{bmatrix}^{-1}$$

$$= \begin{bmatrix} \mathbf{V}_1^{-1} + \mathbf{V}_1^{-1}\mathbf{VC}_2\mathbf{W}\boldsymbol{\phi}_{22}\mathbf{K}_2\mathbf{C}_1\mathbf{V}_1^{-1}/s & -\mathbf{V}_1^{-1}\mathbf{VC}_2\mathbf{W}/s \\ -\mathbf{W}\boldsymbol{\phi}_{22}\mathbf{K}_2\mathbf{C}_1\mathbf{V}_1^{-1} & \mathbf{W} \end{bmatrix}$$

where

$$\mathbf{W} \triangleq (\mathbf{I} + \boldsymbol{\phi}_{22}\mathbf{K}_2\mathbf{C}_2 - \boldsymbol{\phi}_{22}\mathbf{K}_2\mathbf{C}_1\mathbf{V}_1^{-1}\mathbf{VC}_2/s)^{-1} \quad \text{and} \quad \mathbf{V}_1 \triangleq (\mathbf{I} + \mathbf{VC}_1/s).$$

Define

$$\mathbf{M}(s) = (\mathbf{I}_n + \boldsymbol{\Phi}(s)\mathbf{KC})^{-1}\boldsymbol{\Phi}(s)\mathbf{K} = \begin{bmatrix} \mathbf{M}_1(s) \\ \mathbf{M}_2(s) \end{bmatrix}$$

where

$$\mathbf{M}_1(s) \triangleq \left(\mathbf{V}_1^{-1}\frac{\mathbf{V}}{s} + \mathbf{V}_1^{-1}\frac{\mathbf{V}}{s}\mathbf{C}_2\mathbf{W}\boldsymbol{\phi}_{22}\mathbf{K}_2\left(\mathbf{C}_1\mathbf{V}_1^{-1}\frac{\mathbf{V}}{s} - \mathbf{I}_r\right)\right) \tag{9.5}$$

and

$$\mathbf{M}_2(s) \triangleq \mathbf{W}\boldsymbol{\phi}_{22}\mathbf{K}_2\left(\mathbf{I}_r - \mathbf{C}_1\mathbf{V}_1^{-1}\frac{\mathbf{V}}{s}\right). \tag{9.6}$$

The transfer function between the observations signal dz/dt and the signal output estimate may now be defined as $\mathbf{H}_1(s) = \mathbf{C}_1\mathbf{M}_1(s)$. Similarly, $\mathbf{H}_2(s) \triangleq \mathbf{C}_2\mathbf{M}_2(s)$ represents the transfer-function between the observations signal and the disturbance output estimate.

Existence of an inverse matrix for $\mathbf{V}(0)$

It is shown below that the matrix $\mathbf{V}(s) = \mathbf{K}_1 + \mathbf{A}_{12}\boldsymbol{\phi}_{22}\mathbf{K}_2$ is non-singular at the origin of the s-plane. It may be easily shown from Fig. 9.7 that the filter may be drawn as two separate feedback systems with a forward path gain of $\mathbf{C}_1\boldsymbol{\phi}_{11}(\mathbf{K}_1 + \mathbf{A}_{12}\boldsymbol{\phi}_{22}\mathbf{K}_2)$ in one of the subsystems. The return difference matrix for this subsystem is obtained as

$$\mathbf{F}(s) = \mathbf{I}_r + \mathbf{C}_1 \frac{\mathbf{I}}{s} (\mathbf{K}_1 + \mathbf{A}_{12}\boldsymbol{\phi}_{22}\mathbf{K}_2) = (s\mathbf{I}_r + \mathbf{C}_1(\mathbf{K}_1 + \mathbf{A}_{12}\boldsymbol{\phi}_{22}\mathbf{K}_2))/s.$$

The filter is assumed to be asymptotically stable, so that

$$\lim_{|s| \to 0} \rho_0(s) \det \mathbf{F}(s) \neq 0$$

where $\rho_0(s)$ is the open-loop characteristic polynomial. This polynomial includes the term s^r so that $\rho_0(s)$ may be written in the form $s^r\rho_{22}(s)$. Thus, the following limit obtains:

$$\lim_{|s| \to 0} \rho_{22}(s) \det(s\mathbf{I}_r + \mathbf{C}_1(\mathbf{K}_1 + \mathbf{A}_{12}\boldsymbol{\phi}_{22}\mathbf{K}_2)) \neq 0$$

and hence

$$\lim_{|s| \to 0} \det \mathbf{C}_1(\mathbf{K}_1 + \mathbf{A}_{12}\boldsymbol{\phi}_{22}\mathbf{K}_2) \neq 0$$

which implies that the matrix $\mathbf{V}(0)$ is non-singular. It follows that:

$$\lim_{|s| \to 0} \mathbf{C}_1\mathbf{V}_1^{-1}\mathbf{V}/s = \lim_{|s| \to 0} \mathbf{C}_1(s\mathbf{I} + \mathbf{V}\mathbf{C}_1)^{-1}\mathbf{V} = \mathbf{I}_r$$

then

$$\mathbf{W}(0) = \lim_{|s| \to 0} \left(\mathbf{I} + \boldsymbol{\phi}_{22}\mathbf{K}_2\mathbf{C}_2 - \boldsymbol{\phi}_{22}\mathbf{K}_2\left(\mathbf{C}_1\mathbf{V}_1^{-1}\frac{\mathbf{V}}{s}\right)\mathbf{C}_2\right) = \mathbf{I}.$$

If the disturbance process does not contain low-frequency components, then the transfer function $\mathbf{C}_2\boldsymbol{\phi}_{22}\mathbf{K}_2$ remains finite as $|s| \to 0$. In this case the zero-frequency gains follow from (9.5) and (9.6) as:

$$\mathbf{H}_1(0) = \mathbf{I}_r, \qquad \mathbf{H}_2(0) = \mathbf{O} \tag{9.7}$$

and the overall transfer function becomes $\mathbf{H}(0) = \mathbf{I}$. These observations confirm the results obtained in the simple examples. That is, the signal estimate forward path gain is unity, whereas the disturbance estimate forward path gain is zero at zero frequency. These gains can be justified physically since the signal source contains a zero-frequency component whereas the disturbance was assumed to contain high-frequency components only. This applies in both the dynamic ship-positioning and the aircraft elastic modes filtering problems. For further justification, note that if the system contains integrators the input to the filter may have a constant non-zero mean value. Thus, if the filter is to be unbiased the zero frequency gain must be unity.

For reference later (Section 9.2.5), note that the inverse of the return difference matrix is zero at zero frequency. This may be shown as follows:

$$\mathbf{F}(s) = \mathbf{I}_r + \mathbf{C}\boldsymbol{\Phi}(s)\mathbf{K} = \mathbf{I}_r + \mathbf{C}_1\frac{\mathbf{V}}{s} + \mathbf{C}_2\boldsymbol{\phi}_{22}\mathbf{K}_2 \tag{9.8}$$

Let $\mathbf{F}_{22}(s) \triangleq \mathbf{I}_r + \mathbf{C}_2\boldsymbol{\phi}_{22}\mathbf{K}_2$, then since the filter is assumed asymptotically stable det $\mathbf{F}_{22}(0) \neq 0$ and $\mathbf{F}_{22}(0)$ is non-singular.

Thus,

$$\mathbf{F}(s)^{-1} = s(\mathbf{C}_1\mathbf{V})^{-1} - s(\mathbf{C}_1\mathbf{V})^{-1}(\mathbf{F}_{22}^{-1} + s(\mathbf{C}_1\mathbf{V})^{-1})^{-1}s(\mathbf{C}_1\mathbf{V})^{-1}$$

and

$$\lim_{|s| \to 0} \mathbf{F}(s)^{-1} = \mathbf{O}. \tag{9.9}$$

9.2.3. Zeros of the optimal filter

In all the previous examples, where the disturbance was modelled as coloured noise, the transfer function $\tilde{\mathbf{H}}_1(s)$ between the observations signal dz/dt and the signal estimate $\hat{x}_1$ contained zeros at the poles of the disturbance model. (Note that the tilde on $\tilde{\mathbf{H}}_1(s)$ signifies that the transfer function is to the state rather than output estimate.) The blocking action of zeros is well understood and the presence of zeros in the filter transfer function, corresponding to the poles of the disturbance model might have been expected. In this section these results are demonstrated for a more general system.

Consider the ship positioning problem and note that the **A** matrix in equation (9.1) may be partitioned differently to obtain a block diagonal matrix. The ship dynamics may then be represented by the following partitioned state equations:

$$\begin{bmatrix} dx_1 \\ dx_2 \end{bmatrix} = \begin{bmatrix} \mathbf{A}_{11} & \mathbf{O} \\ \mathbf{A}_{21} & \mathbf{A}_{22} \end{bmatrix} \begin{bmatrix} x_1 \\ x_2 \end{bmatrix} dt + \begin{bmatrix} \mathbf{D}_1 \; d\omega_1 \\ \mathbf{D}_2 \; d\omega_2 \end{bmatrix} + \begin{bmatrix} \mathbf{o} \\ \mathbf{B}_2 \end{bmatrix} u \; dt \tag{9.10}$$

$$dz = [\mathbf{C}_1 \quad \mathbf{C}_2] x \; dt + dv \tag{9.11}$$

where the matrix $\mathbf{A}_{21}$ is non-zero for more generality. Notice that the matrix $\mathbf{A}_{12}$ is assumed null since the disturbance must not influence the signal states if the above results are to hold. That is, the filter transfer function $\tilde{\mathbf{H}}_1(s)$ would not contain blocking zeros, equal to the disturbance subsystem poles, if the signal itself contained modes at these frequencies.

The Kalman filter for the above system is shown in Fig. 9.8 with the signal feedback loop open. It is well known that the set of transmission zeros of a system are invariant under a range of state and output feedback transformations (MacFarlane *et al.* [8], [9], [39]). It is therefore convenient to consider the open-loop system shown in Fig. 9.8. to demonstrate that some of the transmission zeros of $\tilde{\mathbf{H}}_1(s)$ are contained in the set of zeros of the disturbance model pole polynomial $\det(s\mathbf{I} - \mathbf{A}_{22})$. The Kalman–Bucy filter with $\mathbf{C}_1 = \mathbf{O}$ becomes:

$$d\hat{x} = \begin{bmatrix} \mathbf{A}_{11} & \mathbf{O} \\ \mathbf{A}_{21} & \mathbf{A}_{22} \end{bmatrix} \hat{x} \; dt + \begin{bmatrix} \mathbf{K}_1 \\ \mathbf{K}_2 \end{bmatrix} [dz - \mathbf{C}_2 \hat{x}_2 \; dt] \tag{9.12}$$

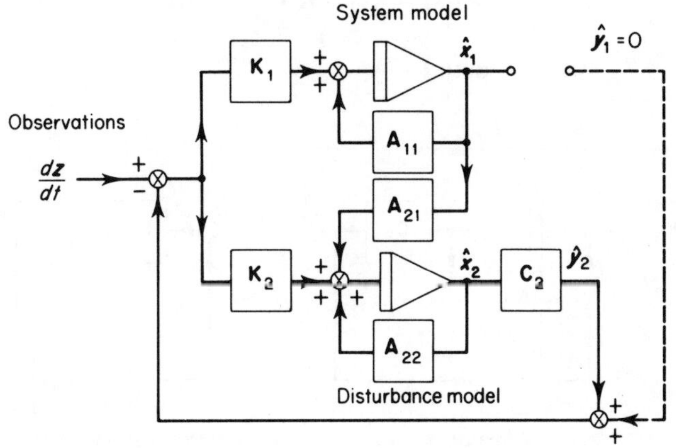

Fig. 9.8. Kalman–Bucy filter with the signal feedback loop open

Laplace transforming the filter equation (defining $\tilde{\mathbf{z}} = \mathscr{L}_1(dz/dt)$) gives:

$$\hat{\mathbf{x}}_1 = \boldsymbol{\phi}_{11}\mathbf{K}_1(\tilde{\mathbf{z}} - \mathbf{C}_2\hat{\mathbf{x}}_2)$$

$$\hat{\mathbf{x}}_2 = \boldsymbol{\phi}_{2c}\mathbf{A}_{21}\hat{\mathbf{x}}_1 + \boldsymbol{\phi}_{2c}\mathbf{K}_2\tilde{\mathbf{z}}$$

where

$$\boldsymbol{\phi}_{2c} \triangleq (s\mathbf{I} - \mathbf{A}_{22} + \mathbf{K}_2\mathbf{C}_2)^{-1}.$$

From these equations,

$$\hat{\mathbf{x}}_1 = (\mathbf{I} + \boldsymbol{\phi}_{11}\mathbf{K}_1\mathbf{C}_2\boldsymbol{\phi}_{2c}\mathbf{A}_{21})^{-1}\boldsymbol{\phi}_{11}\mathbf{K}_1(\mathbf{I} - \mathbf{C}_2\boldsymbol{\phi}_{2c}\mathbf{K}_2)\tilde{\mathbf{z}} \tag{9.13}$$

$$\hat{\mathbf{x}}_2 = (\mathbf{I} + \boldsymbol{\phi}_{2c}\mathbf{A}_{21}\boldsymbol{\phi}_{11}\mathbf{K}_1\mathbf{C}_2)^{-1}\boldsymbol{\phi}_{2c}(\mathbf{A}_{21}\boldsymbol{\phi}_{11}\mathbf{K}_1 + \mathbf{K}_2)\tilde{\mathbf{z}}. \tag{9.14}$$

The transfer-function matrix defined by (9.13) may be considered as representing two cascaded systems as shown in Fig. 9.9. Davison and Wang [10] have shown that the set of transmission zeros of a cascaded system S are contained in the set of zeros of the first subsystem S_1 together with the set of zeros of S_2. Note that the definition of transmission zeros used here is that due to Davison and Wang in their 1976 paper (Davison and Wang [11]). MacFarlane and Karcanias [8], have noted that there is a difference between their definition of transmission zeros and that due to these authors (Davison and Wang [11], Wang and Davison [12]).

The first subsystem S_1 shown in Fig. 9.9 is square and in this case the transmission zeros are the zeros of the polynomial:

$$\det\begin{bmatrix} s\mathbf{I} - \mathbf{A} & \mathbf{B} \\ -\mathbf{C} & \mathbf{D} \end{bmatrix} = \det\begin{bmatrix} s\mathbf{I} - \mathbf{A}_{22} + \mathbf{K}_2\mathbf{C}_2 & \mathbf{K}_2 \\ \mathbf{C}_2 & \mathbf{I} \end{bmatrix}$$

$$= \det(s\mathbf{I} - \mathbf{A}_{22} + \mathbf{K}_2\mathbf{C}_2)\cdot\det(\mathbf{I} - \mathbf{C}_2\boldsymbol{\phi}_{2c}\mathbf{K}_2)$$

$$= \det(s\mathbf{I} - \mathbf{A}_{22}). \tag{9.15}$$

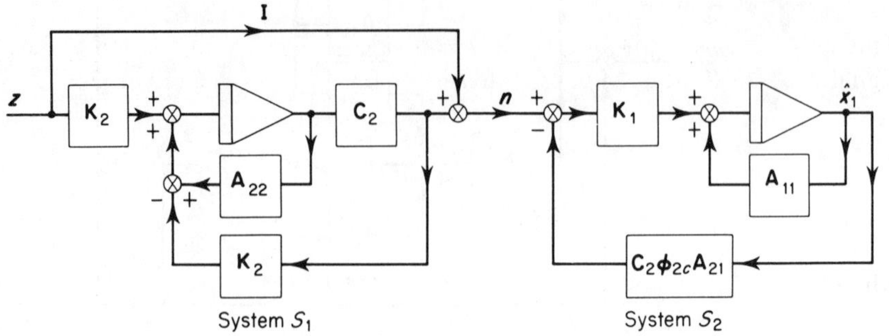

Fig. 9.9. Kalman–Bucy filter shown as two cascaded systems

The zeros so defined are referred to as invariant zeros by MacFarlane and Karcanias [8]. From the above results, it follows that the transmission zeros of $\tilde{\mathbf{H}}_1(s)$ are contained in the set of zeros of $\det(s\mathbf{I} - \mathbf{A}_{22})$ together with the set of transmission zeros of the subsystem S_2. The desired result has therefore been demonstrated since the polynomial (9.15) represents the pole polynomial of the disturbance model.

9.2.4. Signal to signal-plus-noise ratio

In the previous examples, a relationship was noted between the filter transfer function $\mathbf{H}_1(s)$ and the signal to signal-plus-noise ratio. This relationship is derived below for the system considered in Section 9.2.3 (where $\mathbf{A}_{21} = \mathbf{O}$). Let the transfer-function matrix $\mathbf{H}_0(s)$, for the Kalman filter, be defined as:

$$\mathbf{H}_0(s) = \mathbf{F}(s)^{-1}\mathbf{C}\mathbf{\Phi}(s) \tag{9.16}$$

where the filter input point comes after the gain matrix $\mathbf{K} \triangleq [\mathbf{K}_1^T \quad \mathbf{K}_2^T]^T$. Note that $\mathbf{H}_0(s)$ may be written in the form:

$$\mathbf{H}_0(s) = [\mathbf{H}_{01}(s) \ \mathbf{H}_{02}(s)] \tag{9.17}$$

where

$$\mathbf{H}_{01}(s) \triangleq \mathbf{F}(s)^{-1}\mathbf{C}_1\mathbf{\Phi}_{11}(s) \quad \text{and} \quad \mathbf{H}_{02}(s) \triangleq \mathbf{F}(s)^{-1}\mathbf{C}_2\mathbf{\Phi}_{22}(s)$$

and that $\mathbf{H}_1(s) = \mathbf{H}_{01}(s)\mathbf{K}_1$. The following return-difference relationship (MacFarlane [3], Arcasoy [13]) may be derived using the steady-state Riccati equation for the optimal filter:

$$\mathbf{F}(s)\mathbf{R}\mathbf{F}^T(-s) = \mathbf{G}(s)\mathbf{Q}\mathbf{G}^T(-s) + \mathbf{R}. \tag{9.18}$$

This represents the continuous-time equivalent of (8.181) of Chapter 8. The signal plus noise power spectral-density matrix $\mathbf{\Phi}_{nn}(s)$ is defined as:

$$\begin{aligned}\mathbf{\Phi}_{nn}(s) &= \mathbf{G}(s)\mathbf{Q}\mathbf{G}^T(-s) + \mathbf{R} \\ &= \mathbf{G}_{11}(s)\mathbf{Q}_1\mathbf{G}_{11}^T(-s) + \mathbf{G}_{22}(s)\mathbf{Q}_2\mathbf{G}_{22}^T(-s) + \mathbf{R}\end{aligned} \tag{9.19}$$

where

$$\mathbf{G}_{11}(s) \triangleq \mathbf{C}_1\boldsymbol{\phi}_{11}(s)\mathbf{D}_1 \quad \text{and} \quad \mathbf{G}_{22}(s) \triangleq \mathbf{C}_2\boldsymbol{\phi}_{22}(s)\mathbf{D}_2.$$

Using spectral-factorization (Grimble [14]):

$$\mathbf{\Phi}_{nn}(s) = \mathbf{N}(s)\mathbf{N}^T(-s) \tag{9.20}$$

where from (9.18) and (9.20) $\mathbf{N}(s) = \mathbf{F}(s)\mathbf{R}^{1/2}$ and is of normal full rank. The signal power-spectral density matrix $\mathbf{\Phi}_{ss}(s)$ is defined as:

$$\mathbf{\Phi}_{ss}(s) = \mathbf{G}_{11}(s)\mathbf{Q}_1\mathbf{G}_{11}^T(-s) \tag{9.21}$$

and this matrix may be spectrally factored into the form:

$$\mathbf{\Phi}_{ss}(s) = \mathbf{M}(s)\mathbf{M}^T(-s). \tag{9.22}$$

The desired relationship between the transfer function ($\mathbf{H}_{01}$) signal ($\mathbf{M}$) and signal plus noise ($\mathbf{N}$) matrices follows from equations (9.21) and (9.22) (note that $\mathbf{M}(s)^* \triangleq \mathbf{M}^T(-s)$ and for a constant matrix, then $\mathbf{D}_1^* = \mathbf{D}_1^T$):

$$\mathbf{N}^{-1}\mathbf{M}\mathbf{M}^*(\mathbf{N}^*)^{-1} = \mathbf{N}^{-1}\mathbf{G}_{11}\mathbf{Q}_1\mathbf{G}_{11}^*(\mathbf{N}^*)^{-1} = \mathbf{N}^{-1}\mathbf{C}_1\mathbf{\phi}_{11}\mathbf{D}_1\mathbf{Q}_1\mathbf{D}_1^*\mathbf{\phi}_{11}^*\mathbf{C}_1^*(\mathbf{N}^*)^{-1}$$

and from (9.17):

$$\mathbf{N}^{-1}\mathbf{M}\mathbf{M}^*(\mathbf{N}^*)^{-1} = \mathbf{R}^{-1/2}\mathbf{H}_{01}\mathbf{D}_1\mathbf{Q}_1\mathbf{D}_1^*\mathbf{H}_{01}^*\mathbf{R}^{-1/2}. \tag{9.23}$$

In multivariable problems, a scalar measure of the signal to signal-plus-noise ratio may be obtained using the trace operation. From (9.23), using the identity $\text{trace}\{\mathbf{AB}\} = \text{trace}\{\mathbf{BA}\}$:

$$\text{trace}\{\mathbf{\Phi}_{ss}(s)\mathbf{\Phi}_{nn}(s)^{-1}\} = \text{trace}\{\mathbf{H}_{01}(s)\mathbf{D}_1\mathbf{Q}_1\mathbf{D}_1^T\mathbf{H}_{01}^T(-s)\mathbf{R}^{-1}\} \tag{9.24}$$

thus, $\text{trace}\{\mathbf{\Phi}_{ss}(j\omega)\mathbf{\Phi}_{nn}(j\omega)^{-1}\}$ can be considered as a measure for the ratio of signal to signal-plus-noise power density at frequency ω.

If it is assumed that $\mathbf{Q}_1 > \mathbf{O}$ and $\mathbf{D}_1$ is full rank, then from (9.23):

$$\mathbf{N}(s)^{-1}\mathbf{M}(s) = \mathbf{R}^{-1/2}\mathbf{H}_{01}(s)\mathbf{D}_1\mathbf{Q}_1^{1/2} \tag{9.25}$$

or

$$\mathbf{H}_1(s) = \mathbf{R}^{1/2}\mathbf{N}(s)^{-1}\mathbf{M}(s)\mathbf{Q}_1^{-1/2}\mathbf{D}_1^{-1}\mathbf{K}_1 \tag{9.26}$$

where $\mathbf{M}(s) = \mathbf{G}_{11}(s)\mathbf{Q}_1^{1/2}$. Clearly, as in a scalar system, the gain of the optimal filter is directly proportional to the square root of the signal to signal-plus-noise ratio.

9.2.5. Analysis of the filtering problem in the s-domain

It is well known that the infinite-time or steady-state Kalman filter is identical to the Wiener filter for a given system and noise description (Hutchinson [15], Singer and Frost [16], Kučera [17]). In this section the Wiener filter will be analysed and it will be shown that many of the previous results may be demonstrated easily using this approach. The Wiener filter is given by the following expression (Grimble [18]):

$$\mathbf{H}(s) = \{\mathbf{G}(s)\mathbf{Q}\mathbf{G}^T(-s)\mathbf{Y}^T(-s)^{-1}\}_+\mathbf{Y}(s)^{-1} \tag{9.27}$$

where the generalized spectral factor $\mathbf{Y}(s)$ (see Chapter 2) is calculated using:

$$\mathbf{Y}(s)\mathbf{Y}^T(-s) \triangleq \mathbf{G}(s)\mathbf{Q}\mathbf{G}^T(-s) + \mathbf{R} = \mathbf{F}(s)\mathbf{R}\mathbf{F}^T(-s). \tag{9.28}$$

An alternative expression for $\mathbf{H}(s)$ was obtained by Barrett [19] and by Shaked [20]:

$$\mathbf{H}(s) = \mathbf{I} - \mathbf{F}(s)^{-1} \tag{9.29}$$

where $\mathbf{F}(s)$ is the return-difference matrix.

In Section 9.2.2 it was shown that if the system contains integrators associated with the output then the zero-frequency gain of the filter is unity. From equation (9.29) it follows that a necessary and sufficient condition for the filter gain to be unity (or identity matrix in the multivariable case) is that,

$$\lim_{|s|\to 0} \mathbf{F}(s)^{-1} = \mathbf{O}.$$

This condition is satisfied (equation (9.9)) for the system considered in Section 9.2.2, confirming the results obtained.

In Section 9.2.3 it was shown that the transmission zeros of the filter transfer-function matrix $\tilde{\mathbf{H}}_1(s)$ are contained in a set which includes the zeros of the disturbance polynomial. This result may be confirmed for the special case of the ship system (where the partitioned $\mathbf{A}$ matrix is block diagonal). Let $\mathbf{\Phi}_h(s)$ represent the sea-wave power spectral density. This disturbance signal is added to the system output in the same way as the measurement system noise and $\mathbf{\Phi}_h(s) = \mathbf{G}_{22}(s)\mathbf{Q}_2\mathbf{G}_{22}^T(-s)$. The system transfer-function matrix can be expressed in the form:

$$\mathbf{G}(s) = \mathbf{C}\tilde{\mathbf{G}}(s)$$

where

$$\mathbf{C} \triangleq [\mathbf{C}_1 \quad \mathbf{C}_2] \quad \text{and} \quad \tilde{\mathbf{G}}(s) \triangleq \begin{bmatrix} \tilde{\mathbf{G}}_{11} & \mathbf{O} \\ \mathbf{O} & \tilde{\mathbf{G}}_{22}(s) \end{bmatrix}.$$

From (9.27) the Wiener filter is given by:

$$\mathbf{H}(s) = [\mathbf{C}_1 \quad \mathbf{C}_2]\left\{\begin{bmatrix} \tilde{\mathbf{G}}_{11}(s)\mathbf{Q}_1 & \mathbf{O} \\ \mathbf{O} & \tilde{\mathbf{G}}_{22}(s)\mathbf{Q}_2 \end{bmatrix}\begin{bmatrix} \mathbf{G}_{11}^T(-s) \\ \mathbf{G}_{22}^T(-s) \end{bmatrix}\mathbf{Y}^T(-s)^{-1}\right\}_+ \mathbf{Y}(s)^{-1}$$

The wave filter transfer function between the observation input and estimate $\hat{x}_1$ becomes:

$$\tilde{\mathbf{H}}_1(s) = \{\tilde{\mathbf{G}}_{11}(s)\mathbf{Q}_1\mathbf{G}_{11}^T(-s)\mathbf{Y}^T(-s)^{-1}\}_+\mathbf{Y}(s)^{-1} \tag{9.30}$$

where

$$\mathbf{Y}(s)\mathbf{Y}^T(-s) = \mathbf{G}_{11}(s)\mathbf{Q}_1\mathbf{G}_{11}^T(-s) + \mathbf{G}_{22}(s)\mathbf{Q}_2\mathbf{G}_{22}^T(-s) + \mathbf{R}$$

and $\mathbf{G}_{11}(s)$ represents the low-frequency ship dynamics.

Let $\rho_1(s)$ represent the pole polynomial for the matrix $\mathbf{G}_{11}(s)$ and let $\rho_2(s)$ represent the stable pole polynomial of the power spectral density matrix $\mathbf{\Phi}_h(s)$. Also use spectral factorization (Youla [21], Davis [22]) to define the square polynomial matrix $\mathbf{Y}_0(s)$ where

$$\mathbf{Y}_0(s)\mathbf{Y}_0^T(-s) = \rho_1(s)\rho_1(-s)\rho_2(s)\rho_2(-s)\mathbf{Y}(s)\mathbf{Y}^T(-s)$$

and $\mathbf{Y}_0(s)$ has a stable inverse (Shaked [20]). From equation (9.30)

$$\tilde{\mathbf{H}}_1(s) = (\{\tilde{\mathbf{G}}_{11}(s)\mathbf{Q}\mathbf{G}_{11}^T(-s)\mathbf{Y}^T(-s)^{-1}\}_+\rho_1(s))\mathbf{Y}_0(s)^{-1}\rho_2(s). \tag{9.31}$$

Cancellation of terms is possible within this expression, however, the transmission zeros of $\tilde{\mathbf{H}}_1(s)$ are clearly determined by a set of zeros which include the zeros of the disturbance pole polynomial $\rho_2(s)$. Note that the poles of the term $\mathbf{C}_1\{\cdot\}_+$ are determined by the zeros of the polynomial $\rho_1(s)$. Also notice that the zeros of the polynomial $\det(\mathbf{Y}_0(s))$ determine the closed-loop characteristic frequencies of the Kalman–Bucy filter (MacFarlane [3]). The above results apply to any system where the low- and high-frequency dynamics may be separated as in a ship-positioning problem (Grimble [23]) or the aircraft elastic modes filtering problem to be discussed (Moore *et al.* [38]).

9.2.6. Industrial examples

The basic form of the ship-positioning control and filtering problem is very similar to that in several other industrial applications. The class of problem is that for which a notch filter might be appropriate. The ship-state equations may be partitioned into a low-frequency subsystem which is to be controlled and a high-frequency subsystem which represents the wave-motion disturbances. The wave filter is required to remove these high-frequency disturbances so that control of the low-frequency motions can be achieved. The problems in designing the wave filters are compounded by the overlap of the high-and low-frequency bands. The structure and basic form of the ship-positioning problem is very similar to an aircraft elastic modes filtering problem and a rolling mill, back-up roll eccentricity filtering problem. The ship-positioning and back-up roll eccentricity filtering problems are given further detailed exposition in Chapter 13.

Example 9.4. *Dynamic ship positioning.* The linearized equations of motion for a dynamically positioned vessel separate into low- and high-frequency subsystems (see Section 13.2.6). The position of the vessel ($\dot{z}$) may normally be measured with reasonable accuracy. However, only the low-frequency motions (y_l) are to be controlled. The filtering problem is therefore to obtain estimates of the low-frequency position y_l given the observations signal $\dot{z}$. For state estimate feedback schemes the low-frequency states x_l must also be estimated.

The linearized equations for the dynamically positioned vessel *Wimpey Sealab* have the form:

$$\begin{bmatrix} dx_l \\ dx_h \end{bmatrix} = \begin{bmatrix} \mathbf{A}_l & \mathbf{O} \\ \mathbf{O} & \mathbf{A}_h \end{bmatrix} \begin{bmatrix} x_l \\ x_h \end{bmatrix} dt + \begin{bmatrix} \mathbf{B}_l \\ \mathbf{O} \end{bmatrix} u\, dt + \begin{bmatrix} \mathbf{D}_l\, d\omega_l \\ \mathbf{D}_h\, d\omega_h \end{bmatrix} \tag{9.32}$$

$$dz = [\mathbf{C}_l \quad \mathbf{C}_h] \begin{bmatrix} x_l \\ x_h \end{bmatrix} dt + dv \tag{9.33}$$

where for sway and yaw motions:

$$\mathbf{A}_l = \begin{bmatrix} -0.0546 & 0 & 0.0016 & 0 & 0.5435 & 0 \\ 1 & 0 & 0 & 0 & 0 & 0 \\ 0.0573 & 0 & -0.0695 & 0 & 0 & 9.785 \\ 0 & 0 & 1 & 0 & 0 & 0 \\ 0 & 0 & 0 & 0 & -1.55 & 0 \\ 0 & 0 & 0 & 0 & 0 & -1.55 \end{bmatrix}$$

$$\mathbf{B}_l = \begin{bmatrix} 0 & 0 \\ 0 & 0 \\ 0 & 0 \\ 0 & 0 \\ 1.55 & 0 \\ 0 & 1.55 \end{bmatrix} \quad \mathbf{D}_l = \begin{bmatrix} 0.5435 & 0 \\ 0 & 0 \\ 0 & 9.785 \\ 0 & 0 \\ 0 & 0 \\ 0 & 0 \end{bmatrix} \quad \mathbf{C}_l^T = \begin{bmatrix} 0 & 0 \\ 1 & 0 \\ 0 & 0 \\ 0 & 1 \\ 0 & 0 \\ 0 & 0 \end{bmatrix}.$$

The matrix $\mathbf{A}_h$ is block diagonal with the following sway and yaw submatrices:

$$\mathbf{A}_h^s = \mathbf{A}_h^y = \begin{bmatrix} 0 & 3.104 & 0 & 0 \\ 0 & 0 & 3.104 & 0 \\ 0 & 0 & 0 & 3.104 \\ -0.15 & -1.884 & -2.44 & -8.555 \end{bmatrix}.$$

Similarly,

$$\mathbf{D}_h^s = \begin{bmatrix} 0 \\ 0 \\ 0 \\ 0.088 \end{bmatrix}, \qquad \mathbf{D}_h^y = \begin{bmatrix} 0 \\ 0 \\ 0 \\ 0.04 \end{bmatrix}$$

and

$$\mathbf{C}_h = \begin{bmatrix} 0 & 0 & 1 & 0 & 0 & 0 & 0 & 0 \\ 0 & 0 & 0 & 0 & 0 & 0 & 1 & 0 \end{bmatrix}.$$

The step response of the Kalman-Bucy filter for the ship-positioning system is illustrated in Figs. 9.10 to 9.13. The sway and yaw responses of the filter for a unit step

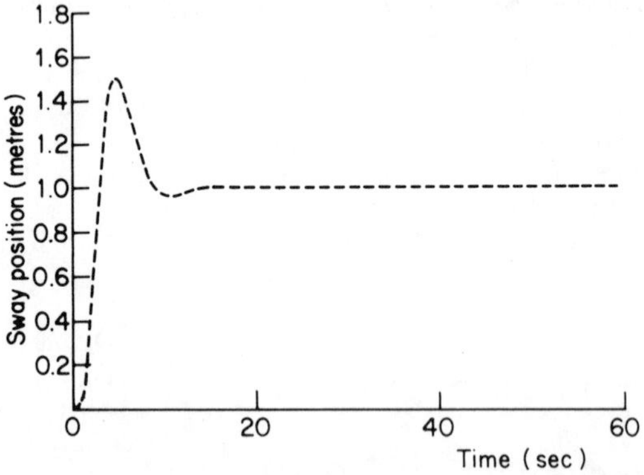

Fig. 9.10. Step response of the Kalman filter for a unit step into sway

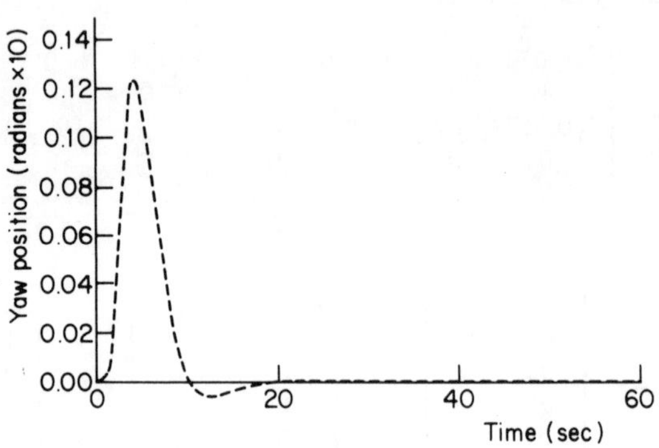

Fig. 9.11. Yaw response of the Kalman filter for a step in sway

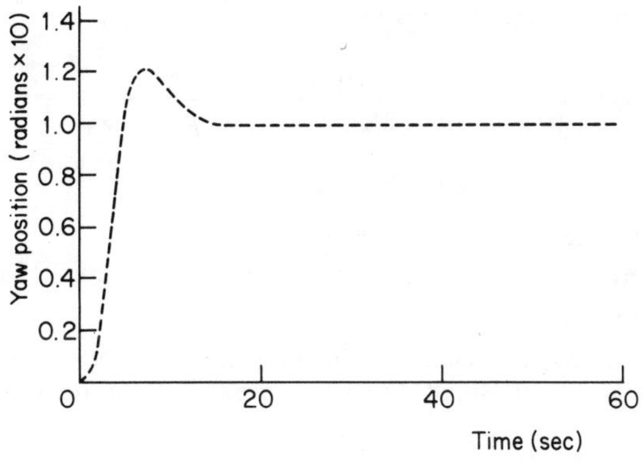

Fig. 9.12. Step response of the Kalman filter for a unit step into yaw

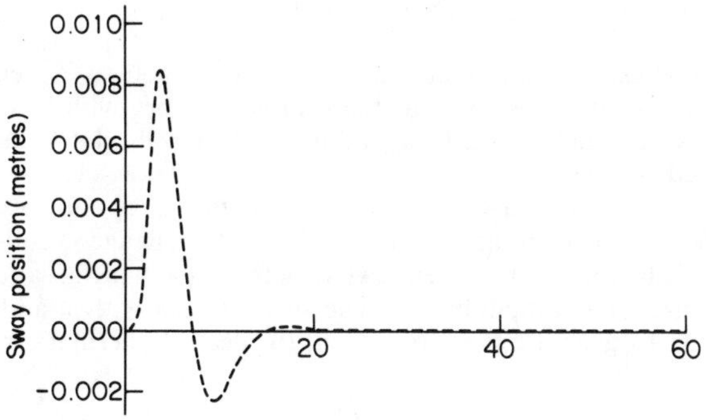

Fig. 9.13. Sway response for a Kalman filter for step into yaw

into the sway input are shown in Figs. 9.10 and 9.11 respectively. Recall that such a system was shown to have zero frequency gain equal to the identity matrix. The steady-state values of the outputs, shown in these figures, clearly accord with this result.

Example 9.5. *Aircraft elastic modes filter.* In this example, the signal is again assumed to be a Wiener process but the noise process is assumed to be generated by passing white noise through an oscillatory shaping filter. This example resembles the situation obtained in aircraft where the elastic mode disturbances are to be removed by filtering (Bryson [24], Healey [25]). The characteristic equation of the oscillatory mode subsystem is of the form $(s^2 + 2\xi\omega s + \omega^2)$. The Kalman–Bucy filter for this problem is defined as:

$$d\hat{x} = \begin{bmatrix} 0 & 0 & 0 \\ 0 & -\xi\omega & \omega_1 \\ 0 & -\omega_1 & -\xi\omega \end{bmatrix} \hat{x}\, dt + \begin{bmatrix} k_1 \\ k_2 \\ k_3 \end{bmatrix} [dz - \hat{x}_1\, dt - \hat{x}_2 dt]$$

where $\omega_1 \triangleq \omega_2(1 - \xi^2)$. The filter has the following transfer-function between the signal estimate $\hat{x}_1$ and the observations signal $\dot{z}$:

$$H_1(s) = \frac{k_1(s^2 + 2\xi\omega s + \omega^2)}{((s + k_1)(s^2 + 2\xi\omega s + \omega^2) + s(sk_2 + \omega_1 k_3 + \xi\omega k_2))}.$$

Notice the following properties of the filter:

(a) The low-frequency gain $H_1(0)$ equals unity.

(b) The numerator polynomial in $H_1(s)$ equals the denominator polynomial of the noise model.

Also notice that the complex poles of the filter are close to the complex zeros if the magnitudes of k_2 and k_3 are small. In such a case the system resembles a regular body bending modes filter.

Example 9.6. *Steel mill back-up roll eccentricity filtering problems.* A common problem in rolling mills where the gauge-control system is based upon roll force signals, is the adverse influence of roll eccentricity on the control system components and performance. The primary cause of eccentricity components in the roll-force signal in a four-high mill is the eccentricity of the back-up rolls. This problem is discussed in more detail in Chapter 13.

The BISRA Gaugemeter automatic gauge-control system is dependent upon the elastic deformation characteristics of a stand (in accordance with Hooke's Law). The spring constant for a stand is called the mill modulus and it relates the roll separating force to the stand stretch.

Consider the operation of a Gaugemeter automatic gauge-control system. A roll-force signal is fed into a set point circuit to calculate the change in roll-force ΔF. This signal is multiplied by the roll-force sensitivity coefficient k_f which is as close as possible to the effective mill stretch modulus k_m. The signal $k_f \Delta F$ is then added to the screw position signal Δs to generate the error signal (neglecting thermal errors):

$$e = k_f \Delta F + \Delta s.$$

This error signal is amplified and fed into the mill screwdown motor in a direction to reduce the error signal. An increase in strip hardness then causes an increase in ΔF and the screws are driven downwards to compensate for the mill stretch. The performance of this system is dependent upon the accuracy of the roll-force sensitivity factor and upon the magnitude of the gain factor as well as the response time of the primary screw actuator.

Unfortunately, the roll-force signal typically includes a fundamental periodic component caused by eccentricity in the mill rolls. In fact, the signal may contain significant distortion and higher-order harmonics.

The filtering problem is to separate out the low-frequency component of force which is to be controlled from the total force signal. The basic problem is clearly very similar to that in the ship-positioning problem and thus a Kalman–Bucy filter may be employed with characteristics similar to those discussed previously.

Example 9.7. *Steel-mill shape control filtering problem.* A Kalman filtering scheme has recently been proposed (Grimble [26], Grimble and Patton [27]) for use in the control of strip shape in cold-rolling mills. The shape of the material does not refer to its gauge profile but to its internal stress distribution (see Chapter 13). The object is to roll strip with good shape. This is strip which when cut into sections lies flat on a level

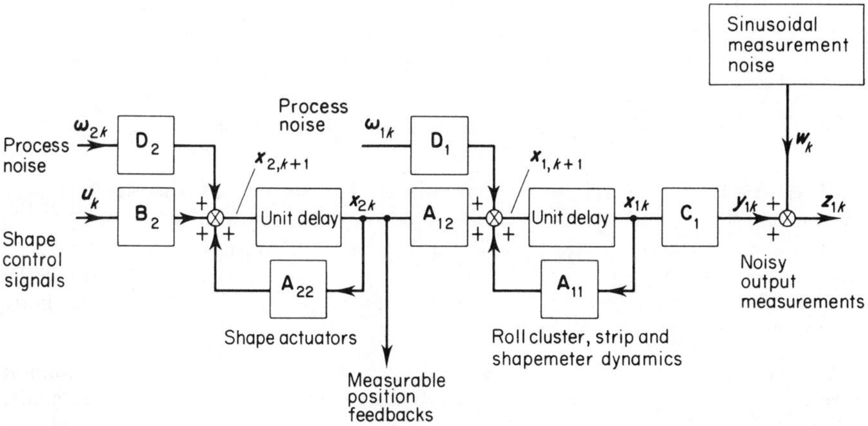

Fig. 9.14. Block diagram for a Sendzimir mill (shape actuators and measurements)

surface. A block diagram of the open-loop plant is shown in Fig. 9.14. The sinusoidal measurement noise is proportional to mill speed and is added to the strip shape measurements. The system and measurement noise clearly have the same form as in the previous examples.

Thus, the four industrial filtering problems have the same feature of a low-frequency controllable system and an output corrupted by noise with a predominantly sinusoidal component.

9.3. FINITE-TIME ESTIMATION

The problem of filtering over a finite-time interval was first solved by Kalman and Bucy [1]. This time-domain solution involved the use of a time-varying Kalman filter gain. More recently, *s*-domain and *z*-domain solutions to this problem have been presented by Grimble [18], [28], [29], [30], [31]. These new transfer-function time-invariant estimators are easier to implement. Furthermore, estimates at particular times (for example, at the end of the estimation interval) are identical to those obtained from the time-varying Kalman filter. However, the new time-invariant estimates suffer from some computational complexity but this has been alleviated by the development of computer algorithms which depend upon well-established numerical techniques as described in the sequel (Grimble [32]).

The solution to the finite-time optimal control problem was presented in Chapter 2. The method used was to embed the finite-time problem in an equivalent infinite-time problem which was easier to solve. The same general technique is used here for the finite-time filtering problem. The infinite-time and finite-time filtering problems are shown to be equivalent by demonstrating that the cost function being minimized is the same in each case. The solution for the filtering, smoothing and prediction problems is obtained by specializing the results.

9.3.1. Discrete-time system and estimator description

It is assumed that the measurements of the output of the discrete-time system are corrupted by white noise. The system is time invariant and linear and can be represented by the discrete time state equations:

$$\boldsymbol{x}(t+1) = \mathbf{A}\boldsymbol{x}(t) + \mathbf{D}\boldsymbol{\omega}(t) + \boldsymbol{u}(t) \tag{9.34}$$

$$z(t) = \mathbf{C}\boldsymbol{x}(t) + \boldsymbol{v}(t) \tag{9.35}$$

where

$$\boldsymbol{x}(t) \in \mathbb{R}^n, \qquad \boldsymbol{\omega}(t) \in \mathbb{R}^q, \qquad z(t) \in \mathbb{R}^r, \qquad \boldsymbol{u}(t) \in \mathbb{R}^n \quad \text{and}$$

$$t \in \{\ldots, -2, -1, 0, 1, 2, \ldots\}.$$

The system is assumed to be completely reconstructible (Kwakernaak and Sivan [40]) and to be in operation from time $t_0 = -\infty$. The process noise $\{\boldsymbol{\omega}(t)\}$ and measurement noise $\{\boldsymbol{v}(t)\}$ are stationary, zero-mean, white-noise sequences with the following covariances:

$$\begin{aligned} \operatorname{cov}[\boldsymbol{\omega}(t), \boldsymbol{\omega}(k)] &= \mathbf{Q}\,\delta(t-k) \\ \operatorname{cov}[\boldsymbol{v}(t), \boldsymbol{v}(k)] &= \mathbf{R}\,\delta(t-k) \\ \operatorname{cov}[\boldsymbol{\omega}(t), \boldsymbol{v}(k)] &= \mathbf{O} \end{aligned} \tag{9.36}$$

where $\delta(t-k)$ is the Kronecker delta function (defined as $\delta(t-k) = 1$ if $t = k$ and $\delta(t-k) = 0$ if $t \neq k$) and $\mathbf{Q} \geq \mathbf{O}, \mathbf{R} > \mathbf{O}$.

The above infinite-time situation ($t_0 = -\infty$) is related to the usual finite-time problem where $\mathbf{m}_0 = E\{\boldsymbol{x}(0)\}$ and $\boldsymbol{\Sigma}_0 = \operatorname{cov}[\boldsymbol{x}(0), \boldsymbol{x}(0)]$ by assuming that the system has a Kronecker delta-function input at time $t = 0$ of magnitude $\boldsymbol{x}_1' = \mathbf{A}\boldsymbol{x}_0'$. That is,

$$\boldsymbol{u}(t) = \delta(t)\boldsymbol{x}_1'$$

where the random variable $\boldsymbol{x}_0'$ has mean $\mathbf{m}_0$ and covariance $\boldsymbol{\Sigma}_0$ and $\boldsymbol{x}_0'$ is assumed to be uncorrelated with the noise signals. The following system z-transfer-function matrices are required:

$$\Phi(z^{-1}) = (z\mathbf{I}_n - \mathbf{A})^{-1}$$

$$\mathbf{W}(z^{-1}) = \mathbf{C}\Phi(z^{-1})\mathbf{D}$$

$$\mathbf{W}_0(z^{-1}) = \mathbf{\Phi}(z^{-1})\mathbf{D}. \tag{9.37}$$

System response

In calculating the system-state trajectory, particular attention is given to the time-interval containing zero because of the input to the plant at time zero. The state trajectory for the system (9.34) is:

$$\boldsymbol{x}(t) = \sum_{j=-\infty}^{t-1} \boldsymbol{\phi}(t-1-j)(\mathbf{D}\boldsymbol{\omega}(j) + \delta(j)\boldsymbol{x}_1'), \qquad \text{for all } t \in (-\infty, \infty)$$

$$= \mathbf{\Phi}(t)\boldsymbol{x}_1' + \sum_{j=-\infty}^{t-1} \mathbf{\Phi}(t-j)\mathbf{D}\boldsymbol{\omega}(j), \qquad \text{for } t > 0$$

where

$$\mathbf{\Phi}(t) \triangleq \mathbf{A}^{t-1}U(t-1),\ \boldsymbol{\phi}(t) \triangleq \mathbf{A}^{t}U(t)$$

and $U(t)$ is the unit step-function ($U(t) \triangleq 0$ for $t < 0$ and $U(t) \triangleq 1$ for $t \geq 0$).

The mean value of the system-state trajectory is given as:

$$E\{\boldsymbol{x}(t)\} = \mathbf{\Phi}(t)\mathbf{A}\mathbf{m}_0, \quad \text{for } t > 0$$

$$= \mathbf{O} \qquad\qquad \text{for } t \leq 0.$$

To simplify the notation $\boldsymbol{\phi}(t) \triangleq \mathbf{O}$ for $t < 0$ and $\mathbf{\Phi}(t) \triangleq \mathbf{O}$ for $t \leq 0$. The delta function input signal establishes the desired initial mean value and thus:

$$\boldsymbol{x}(t) - E\{\boldsymbol{x}(t)\} = \sum_{j=-\infty}^{t-1} \mathbf{\Phi}(t-j)(\mathbf{D}\boldsymbol{\omega}(j) + \delta(j)\mathbf{A}(\boldsymbol{x}_0' - \mathbf{m}_0)).$$

Define the vector $\boldsymbol{x}_0$ as:

$$\boldsymbol{x}_0 = \boldsymbol{x}_0' + \boldsymbol{x}(0)$$

then for $t > 0$ obtain:

$$x(t) = \phi(t)x_0 + (W_0\omega)(t)$$

where

$$(W_0\omega)(t) \triangleq \sum_{j=0}^{t-1} \Phi(t-j)\mathbf{D}\omega(j)$$

and $\mathscr{Z}_1(\phi(t)) \triangleq \phi(z^{-1}) = z\Phi(z^{-1})$, $\mathscr{Z}_1((W_0\omega)(t)) \triangleq \mathbf{W}_0(z^{-1})\omega(z^{-1})$.
The impulse-response matrix $\mathbf{W}(t) \triangleq \mathbf{C}\phi(t-1)\mathbf{D} = \mathbf{C}\Phi(t)\mathbf{D}$.

Time-invariant estimator

The time-invariant estimator is assumed to be linear and is represented in the convolution summation form:

$$\hat{x}(t_1 \mid t) = \sum_{j=-\infty}^{t} (\mathbf{H}(t-j; T, l)z(j) + \Psi(t-j+1; l)\delta(j-1)x_0')$$

for all $t \in (-\infty, \infty)$

$$= \Psi(t; l)x_0' + \sum_{j=-\infty}^{t} \mathbf{H}(t-j; T, l)z(j) \quad \text{for } t > 0 \tag{9.38}$$

where $\mathbf{H}(t; T, l) \triangleq \mathbf{O}$ for all $t < 0$. The estimator has an input x_0' at time zero which is required to optimize the zero-input response of the estimator. Recall that the system also has an input at time zero. The smoothing delay $l \triangleq t - t_1 > 0$ and for prediction $l < 0$, where $\min(t, t_1) \geq 0$. The dependence of the estimator on T (a fixed time interval) and l will be evident from the solution to be obtained. The state estimates are given as:

$$\hat{x}(t_1 \mid t) = d_0(t; T, l) + (Hz)(t), \quad \text{for all } t \in (-\infty, \infty) \tag{9.39}$$

where

$$d_0(t; T, l) = \Psi(t; l)x_0'$$

and

$$(Hz)(t) \triangleq \sum_{j=-\infty}^{t} \mathbf{H}(t-j; T, l)z(j)$$

the input–output response, for $t > 0$, follows from the operator:

$$(H_0 z)(t) \triangleq \sum_{j=1}^{t} \mathbf{H}(t-j; T, l)\mathbf{z}(j) = \sum_{\tau=0}^{t-1} \mathbf{H}(\tau, T, l)\mathbf{z}(t-\tau)$$

hence for $\min(t, t_1) > 0$ the mean estimator response:

$$E\{\hat{x}(t \mid t)\} = \boldsymbol{\Psi}(t; l)\mathbf{x}_0' + (\mathbf{H}_0\mathbf{C}\boldsymbol{\phi})(t)\mathbf{m}_0, \quad \text{for } t > 0$$

$$= \mathbf{o} \quad \text{for } t \leq 0. \tag{9.40}$$

9.3.2. Time-invariant estimation problem

The finite-interval linear estimation problem to be considered is not the same as the Kalman estimation problem since the estimator is required to be time invariant. Also, the estimation error is only required to be a minimum at the end point of a time interval of length T.

Problem 9.1. *Time-invariant estimator.* Find the linear time-invariant estimator which minimizes the estimation error at the end of a fixed time interval of length T. The system descriptions are those of Section 9.3.1. The error criterion to be minimized is therefore:

$$J(\tilde{x}(T_1 \mid T)) = \langle \tilde{x}(T_1 \mid T), \tilde{x}(T_1 \mid T)\rangle_H \triangleq E\{\langle \tilde{x}(T_1 \mid T), \tilde{x}(T_1 \mid T)\rangle_{E_n}\} \tag{9.41}$$

where the estimation error is given as:

$$\tilde{x}(T_1 \mid T) = x_i(T_1) - \hat{x}(T_1 \mid T) \tag{9.42}$$

and $x_i(T_1)$ denotes the ideal output at time T_1. □

The problem is illustrated in Fig. 9.15 and the solution is that of the finite-time problem of interest. It is derived by the use of the Wiener–Hopf equation.

Theorem 9.1 ***Wiener–Hopf equation.*** *The necessary and sufficient conditions for $\hat{x}(T_1 \mid T)$ to be the minimum variance estimator for $\mathbf{x}_i(T_1)$ are:*

(*i*) *Matrix function $\mathbf{H}(T-\tau; T, l)$ satisfies the Wiener–Hopf equation:*

$$\operatorname{cov}[x_i(T_1), z(m)] = \sum_{j=-\infty}^{T} \mathbf{H}(T-j; T, l)\operatorname{cov}[z(j), z(m)]$$

$$\textit{for all } m \in (-\infty, T] \tag{9.43}$$

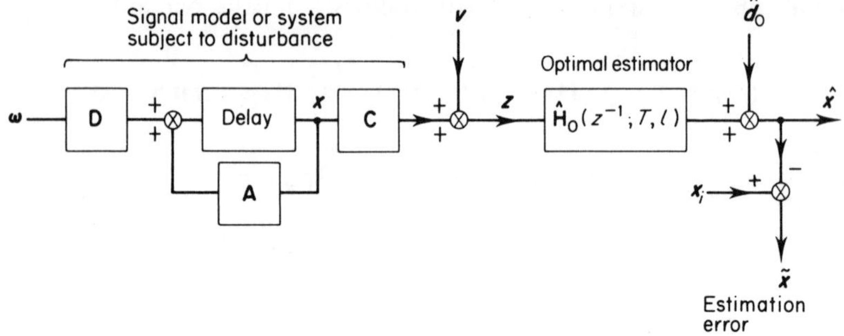

Fig. 9.15. The optimal estimation problem

and

(ii) *Vector* $\mathbf{d}_0(T; T, l)$ *satisfies:*

$$\mathbf{d}_0(T; T, l) = E\{x_i(T_1)\} - \sum_{j=-\infty}^{T} \mathbf{H}(T-j; T, l)E\{z(j)\} \tag{9.44}$$

giving $E\{\hat{x}(T_1 \mid T)\} = E\{x_i(T_1)\}$.

Proof. The necessary and sufficient condition is obtained from the orthogonal projection theorem (see Chapter 8, Theorem 8.2). Let $\mathscr{X}$ denote the space generated by the random vectors $\{x_i(T_1)\}$ and let $\mathscr{U} \subset \mathscr{X}$ represent the subspace generated by the set of observations $\{z(t)\}$ where elements $u(T_1) \in \mathscr{U}$ are given by,

$$u(T_1) = \mathbf{d}_0(T; T, l) + \sum_{m=-\infty}^{T} \mathbf{H}(T-m; T, l)z(m)$$

and $\mathbf{d}_0$ is an unknown deterministic vector. The projection theorem states $\| x_i - u \|_H$ is minimized by $\hat{x} \in \mathscr{U}$ if and only if $(x_i - \hat{x}) \perp \mathscr{U}$. That is:

$$\langle x_i(T) - \hat{x}(T_1 \mid T), u(t_1)\rangle_H = 0 \quad \text{for all } u \in \mathscr{U}.$$

Expanding this equation gives:

$$E\{x_i(T) - \hat{x}(T_1 \mid T), u(T_1)\rangle\} = \text{trace}\{E\{(x_i(T) - \hat{x}(T_1 \mid T))u^T(T_1)\}\} = 0.$$

Sufficiency

Substituting for $u(T_1)$ in the optimality condition yields

$$\text{trace}\, E\{(\boldsymbol{x}_i(T) - \hat{\boldsymbol{x}}(T_1 \mid T))\boldsymbol{u}^T(T_1)\}$$

$$= \text{trace}\, E\left\{(\boldsymbol{x}_i(T) - \hat{\boldsymbol{x}}(T_1 \mid T))(\mathbf{d}_0(T; T, l) + \sum_{m=-\infty}^{T} \mathbf{H}(T-m; T, l)\boldsymbol{z}(m))^T\right\}$$

$$= \text{trace}\, E\{(\boldsymbol{x}_i(T) - \hat{\boldsymbol{x}}(T_1 \mid T))\, \mathbf{d}_0^T(T; T, l)\}$$

$$+ \text{trace}\, E\left\{(\boldsymbol{x}_i(T) - \hat{\boldsymbol{x}}(T_1 \mid T)\left(\sum_{m=-\infty}^{T} \mathbf{H}(T-m; T, l)\boldsymbol{z}(m)\right)^T\right\}$$

$$= \text{trace}\, E\{(\boldsymbol{x}_i(T) - \hat{\boldsymbol{x}}(T_1 \mid T)\}\mathbf{d}_0^T(T; T, l)$$

$$+ \text{trace} \sum_{m=-\infty}^{T} E\{(\boldsymbol{x}_i(T) - \hat{\boldsymbol{x}}(T_1 \mid T))\boldsymbol{z}^T(m)\}\mathbf{H}^T(T-m; T, l). \tag{9.45}$$

Optimality is therefore attained if,

$$\text{(i)} \quad E\{(\boldsymbol{x}_i(T) - \hat{\boldsymbol{x}}(T_1 \mid T))\} = \mathbf{o} \tag{9.46}$$

and

$$\text{(ii)} \quad E\{(\boldsymbol{x}_i(T) - \hat{\boldsymbol{x}}(T_1 \mid T))\boldsymbol{z}^T(m)\} = \mathbf{O} \text{ for all } m \in (-\infty, T]. \tag{9.47}$$

Necessity

Select vector $\mathbf{d}_0$ and matrix function $\mathbf{H}$ so that:

$$\mathbf{d}_0(T; T, l) = E\{\boldsymbol{x}_i(T)\} - E\{\hat{\boldsymbol{x}}(T_1 \mid T)\} \tag{9.48}$$

and

$$\mathbf{H}(T-m; T, l) = E\{(\boldsymbol{x}_i(T) - \hat{\boldsymbol{x}}(T_1 \mid T))\boldsymbol{z}^T(m)\}. \tag{9.49}$$

Substituting into the optimality condition (9.45) gives:

$$\text{trace}\, E\{(\boldsymbol{x}_i(T) - \hat{\boldsymbol{x}}(T_1 \mid T))(\boldsymbol{x}_i(T) - \hat{\boldsymbol{x}}(T_1 \mid T))^T\}$$

$$+ \text{trace} \sum_{m=-\infty}^{T} E\{(\boldsymbol{x}_i(T) - \hat{\boldsymbol{x}}(T_1 \mid T))\boldsymbol{z}^T(m)\boldsymbol{z}(m)(\boldsymbol{x}_i(T) - \hat{\boldsymbol{x}}(T_1 \mid T))^T\} = 0. \tag{9.50}$$

Since the matrix terms in (9.50) are all positive definite, the trace relationship and Sylvester's Theorem implies that all these terms are zero, thus proving necessity.

Equation (9.46) demonstrates that the estimator is unbiased, and from equation (9.47):

$$E\{x_i(T_1)z^T(m)\} = \mathbf{d}_0(T; T, l)E\{z^T(m)\} + \sum_{j=-\infty}^{T} \mathbf{H}(T-j; T, l)E\{z(j)z^T(m)\}$$

thus

$$\operatorname{cov}[x_i(T_1), z(m)] = \sum_{j=-\infty}^{T} \mathbf{H}(T-j; T, l)\operatorname{cov}[z(j), z(m)]$$

$$\text{for all } m \in (-\infty, T]. \quad (9.51)$$

□

Equivalence between the infinite and finite-time problems

The estimator is assumed to be in operation from time $t_0 = -\infty$, however, by defining the ideal output appropriately the optimal estimate at time T, will involve only the system signals for times greater than zero. The resulting estimator will then provide the required solution to the finite-time problem. Let the ideal output be defined as:

$$x_i(T_1) = x(T_1) - \mathbf{\Psi}(T; l)x(0) \quad (9.52)$$

where if the estimator is to be unbiased from (9.42)

$$\mathbf{\Psi}(T; l) \triangleq \boldsymbol{\phi}(T_1) - (\mathbf{H}_0\mathbf{C}\boldsymbol{\phi})(T) \quad \text{and} \quad x_0' = \mathbf{m}_0. \quad (9.53)$$

Consider the expression for the estimation error:

$$\hat{x}(T_1 \mid T) = \boldsymbol{\phi}(T_1)x_0' + (W_0\omega)(T_1) + (\mathbf{H}_0\mathbf{C}\boldsymbol{\phi})(T)x(0) - \hat{x}(T_1 \mid T)$$

$$= x'(T_1) - (\mathbf{\Psi}(T; l)x_0 + (\mathbf{H}_0(v + \mathbf{C}x'))(T))$$

where

$$x'(T_1) \triangleq \boldsymbol{\phi}(T_1)x_0' + (W_0\omega)(T_1).$$

This expression for the estimation error $\tilde{x}(T_1 \mid T)$ depends only upon the system inputs for $t > 0$ and the vector x_0'. The estimation error is therefore the same as that for a finite-time estimation problem in which the initial state of the system is x_0'. Recall that x_0' was assigned the statistics $(\mathbf{m}_0, \mathbf{\Sigma}_0)$ of the initial state vector in the finite-time problem of interest. It follows that the optimal estimator for this problem depends only upon the observations in the interval $[l, T]$ and the vector x_0'. The estimator obtained by solving the above infinite-time problem therefore yields the desired solution to the finite-time problem. The impulse response for the optimal estimator is found to satisfy the property that $\mathbf{H}(t; T, l) = \mathbf{O}$ for $t \geq T$. That is, this estimator has a finite memory of T sample points.

9.3.3. Necessary and sufficient condition for optimality

The Wiener–Hopf equation must be expanded and simplifed before the z-domain solution may be obtained. A number of preliminary results are required.

Lemma 9.1. *Expression for* $\operatorname{cov}[\boldsymbol{x}_i(T_1), z(m)]$

$$\operatorname{cov}[\boldsymbol{x}_i(T_1), z(m)] = \operatorname{cov}[\boldsymbol{x}(T_1), \boldsymbol{x}(m)]\mathbf{C}^T - \boldsymbol{\Psi}(T; l)\boldsymbol{\phi}_{x_0 x}(-m)\mathbf{C}^T$$

where $\boldsymbol{\phi}_{x_0 x}(-m) = E\{\boldsymbol{x}(0)\boldsymbol{x}^T(m)\}$.

Proof. From the definition of the covariance function:

$$\begin{aligned}\operatorname{cov}[\boldsymbol{x}_i(T_1), z(m)] &= E\{\boldsymbol{x}_i(T_1)z^T(m)\} - E\{\boldsymbol{x}_i(T_1)\}E\{z^T(m)\} \\ &= \operatorname{cov}[\boldsymbol{x}_i(T_1), \boldsymbol{x}(m)]\mathbf{C}^T\end{aligned}$$

since $E\{\boldsymbol{x}_i(T_1)\boldsymbol{v}^T(m)\} = \mathbf{O}$. Recall that,

$$\boldsymbol{x}(T_1) = \sum_{j=-\infty}^{T_1-1} \boldsymbol{\phi}(T_1 - j - 1)(\mathbf{D}\boldsymbol{\omega}(j) + \delta(j)\mathbf{A}\boldsymbol{x}_0')$$

and that $\boldsymbol{x}_0'$ is assumed to be uncorrelated with $\boldsymbol{v}$. The term $\operatorname{cov}[\boldsymbol{x}_i(T_1), \boldsymbol{x}(m)]$ may now be simplifed using $E\{\boldsymbol{x}(0)\} = \mathbf{o}$ as:

$$\operatorname{cov}[\boldsymbol{x}_i(T_1), \boldsymbol{x}(m)] = \operatorname{cov}[\boldsymbol{x}(T_1), \boldsymbol{x}(m)] - \boldsymbol{\Psi}(T; l)E\{\boldsymbol{x}(0)\boldsymbol{x}^T(m)\}$$

Let $\boldsymbol{\phi}_{x_0 x}(-m) \triangleq E\{\boldsymbol{x}(0)\boldsymbol{x}^T(m)\}$

then,

$$\operatorname{cov}[\boldsymbol{x}_i(T_1), z(m)] = \operatorname{cov}[\boldsymbol{x}(T_1), \boldsymbol{x}(m)]\mathbf{C}^T - \boldsymbol{\Psi}(T; l)\boldsymbol{\phi}_{x_0 x}(-m)\mathbf{C}^T \qquad \square$$

Lemma 9.2. *Expression for* $\operatorname{cov}[\boldsymbol{x}(t), \boldsymbol{x}(m)]$

$$\operatorname{cov}[\boldsymbol{x}(t), \boldsymbol{x}(m)] = \sum_{j=-\infty}^{\min(t,m)-1} \boldsymbol{\Phi}(t-j)\mathbf{DQD}^T\boldsymbol{\Phi}^T(m-j) + \boldsymbol{\Phi}(t)\mathbf{A}\boldsymbol{\Sigma}_0\mathbf{A}^T\boldsymbol{\Phi}^T(m). \tag{9.54}$$

Proof

$$\begin{aligned}\operatorname{cov}[\boldsymbol{x}(t), \boldsymbol{x}(m)] &= E\{(\boldsymbol{x}(t) - E\{\boldsymbol{x}(t)\})(\boldsymbol{x}(m) - E\{\boldsymbol{x}(m)\})^T\} \\ &= \sum_{i=-\infty}^{t-1} \sum_{j=-\infty}^{m-1} (\boldsymbol{\Phi}(t-i)\mathbf{DQD}^T\boldsymbol{\Phi}^T(m-j)\delta(j-i) \\ &\quad + \delta(i)\boldsymbol{\Phi}(n-1)\mathbf{A}\boldsymbol{\Sigma}_0\mathbf{A}^T\boldsymbol{\Phi}^T(m-j)\delta(j)) \\ &= \sum_{j=-\infty}^{\min(t,m)-1} (\boldsymbol{\Phi}(t-j)\mathbf{DQD}^T\boldsymbol{\Phi}^T(m-j)) \\ &\quad + \boldsymbol{\Phi}(t)\mathbf{A}\boldsymbol{\Sigma}_0\mathbf{A}^T\boldsymbol{\Phi}^T(m).\end{aligned}$$

□

Lemma 9.3. *Expression for* $\mathbf{H}(T-j; T, l)\operatorname{cov}[z(j), z(m)]$

$$\begin{aligned}&\sum_{j=-\infty}^{T} \mathbf{H}(T-j; T, l)\operatorname{cov}[z(j), z(m)] \\ &\quad = \sum_{j=-\infty}^{T} \mathbf{H}(T-j; T, l) \sum_{i=-\infty}^{j-1} \sum_{k=-\infty}^{\min(m,j)-1} \mathbf{W}(j-i)\mathbf{Q}\delta(i-k)\mathbf{W}^T(m-k) \\ &\quad + \sum_{j=-\infty}^{T} \mathbf{H}(T-j; T, l)\mathbf{R}\,\delta(j-m) \\ &\quad + \sum_{j=-\infty}^{T} \mathbf{H}(T-j; T, l)\mathbf{C}\boldsymbol{\Phi}(j)\mathbf{A}\boldsymbol{\Sigma}_0\mathbf{A}^T\boldsymbol{\Phi}^T(m)\mathbf{C}^T.\end{aligned} \tag{9.55}$$

Proof.

$$\begin{aligned}\operatorname{cov}[z(j), z(m)] &= \mathbf{C}\operatorname{cov}[\boldsymbol{x}(j), \boldsymbol{x}(m)]\mathbf{C}^T + \boldsymbol{\phi}_{vv}(j-m) \\ &\quad + \mathbf{C}E\{\boldsymbol{x}(j)\boldsymbol{v}^T(m)\} + E\{\boldsymbol{v}(j)\boldsymbol{x}^T(m)\}\mathbf{C}^T\end{aligned}$$

where $\boldsymbol{\phi}_{vv}(j-m) \triangleq E\{\boldsymbol{v}(j)\boldsymbol{v}^T(m)\}$. The expression for the covariance matrix may now be simplified since the process and measurement noise terms are uncorrelated, thus:

$$\operatorname{cov}[z(j), z(m)] = \mathbf{C} \operatorname{cov}[\boldsymbol{x}(j), \boldsymbol{x}(m)]\, \mathbf{C}^T + \boldsymbol{\phi}_{vv}(j-m). \tag{9.56}$$

Substituting from (9.56) and using the system impulse response matrix $\mathbf{W}(i)$ obtain:

$$\sum_{j=-\infty}^{T} \mathbf{H}(T-j; T, l)\operatorname{cov}[z(j), z(m)]$$

$$= \sum_{j=-\infty}^{T} \mathbf{H}(T-j; T, l) \sum_{i=-\infty}^{j-1} \sum_{k=-\infty}^{\min(m,j)-1} \mathbf{W}(j-1)\mathbf{Q}\,\delta(i-k)\mathbf{W}^T(m-k)$$

$$+ \sum_{j=-\infty}^{T} \mathbf{H}(T-j; T, l)(\mathbf{R}\,\delta(j-m))$$

$$+ \sum_{j=-\infty}^{T} \mathbf{H}(T-j; T, l)\mathbf{C}\boldsymbol{\Phi}(j)\mathbf{A}\boldsymbol{\Sigma}_0\mathbf{A}^T\boldsymbol{\Phi}^T(m)\mathbf{C}^T. \qquad \square$$

Lemma 9.4. ***Expansion of the Wiener–Hopf equation.*** *On expansion and simplification, the Wiener–Hopf equation becomes:*

$$\mathbf{G}_0(t; T, l) = \mathbf{O} \quad \text{for all integers } t \geq 0$$

where

$$\mathbf{G}_0(t; T, l) = \boldsymbol{\Psi}(T; l)[\boldsymbol{\Sigma}_0\mathbf{A}^T\boldsymbol{\Phi}^T(T-t)\mathbf{C}^T - \boldsymbol{\phi}_{x_0x}(t-T)\mathbf{C}^T]$$

$$- \sum_{j=-\infty}^{T} \mathbf{H}(T-j; T, l)\left[\sum_{i=-\infty}^{j-1} \sum_{k=-\infty}^{\min(j, T-t)-1} \mathbf{W}(j-i)\mathbf{Q}\delta(i-k)\mathbf{W}^T(T-t-k)\right.$$

$$\left. + \mathbf{R}\delta(j+t-T)\right]$$

$$+ \sum_{j=-\infty}^{\min(T_1, T-t)-1} \boldsymbol{\Phi}(T_1-j)\mathbf{D}\mathbf{Q}\mathbf{W}^T(T-t-j).$$

Proof. From the above results the left-hand side of (9.51) becomes:

$$\operatorname{cov}[\boldsymbol{x}_i(T_1), z(m)] = \sum_{j=-\infty}^{\min(T_1, m)-1} (\boldsymbol{\Phi}(T_1-j)\mathbf{D}\mathbf{Q}\mathbf{D}^T\boldsymbol{\Phi}^T(m-j)\mathbf{C}^T)$$

$$+ \boldsymbol{\Phi}(T_1)\mathbf{A}\boldsymbol{\Sigma}_0\mathbf{A}^T\boldsymbol{\Phi}^T(m)\mathbf{C}^T - \boldsymbol{\Psi}(T; l)\boldsymbol{\phi}_{x_0x}(-m)\mathbf{C}^T \tag{9.57}$$

for $m \in (-\infty, T]$. Similarly the right-hand side becomes:

$$\sum_{j=-\infty}^{T} \mathbf{H}(T-j; T, l)\operatorname{cov}[z(j), z(m)]$$

$$= \sum_{j=-\infty}^{T} \mathbf{H}(T-j; T, l)\Bigg[\sum_{i=-\infty}^{j-1} \sum_{k=-\infty}^{\min(m,j)-1} \mathbf{W}(j-1)\mathbf{Q}\,\delta(i-k)\mathbf{W}^T(m-k)$$

$$+\mathbf{R}\,\delta(j-m) + \mathbf{C}\mathbf{\Phi}(j)\mathbf{A}\mathbf{\Sigma}_0\mathbf{A}^T\mathbf{\Phi}^T(m)\mathbf{C}^T\Bigg]. \tag{9.58}$$

From (9.57) and (9.58) the terms including $\mathbf{\Sigma}_0$ in the Wiener–Hopf equation (for $m > 0$) become:

$$\left[\boldsymbol{\phi}(T_1) - \sum_{j=1}^{T} \mathbf{H}(T-j; T, l)\mathbf{C}\boldsymbol{\phi}(j)\right]\mathbf{\Sigma}_0\mathbf{A}^T\mathbf{\Phi}^T(m)\mathbf{C}^T = \mathbf{\Psi}(T; l)\mathbf{\Sigma}_0\mathbf{A}^T\mathbf{\Phi}^T(m)\mathbf{C}^T \tag{9.59}$$

From the above results (9.51) becomes:

$$\mathbf{\Psi}(T; l)[\mathbf{\Sigma}_0\mathbf{A}^T\mathbf{\Phi}^T(m)\mathbf{C}^T - \boldsymbol{\phi}_{x_0x}(-m)\mathbf{C}^T]$$

$$+\sum_{j=-\infty}^{T} \mathbf{H}(T-j; T, l)\Bigg[\sum_{i=-\infty}^{j-1} \sum_{k=-\infty}^{\min(m,j)-1} \mathbf{W}(j-1)\mathbf{Q}\,\delta(i-k)\mathbf{W}^T(m-k)$$

$$+\mathbf{R}\,\delta(j-m)\Bigg]$$

$$+\sum_{j=-\infty}^{\min(T_1, m)-1} \mathbf{\Phi}(t_1-j)\mathbf{D}\mathbf{Q}\mathbf{W}^T(m-j) = \mathbf{O} \quad \text{for all } m \in (-\infty, T].$$

Now substitute $m = T - t$, $l = T - T_1$ then the Wiener–Hopf equation becomes:

$$\mathbf{\Psi}(T; l)[\mathbf{\Sigma}_0\mathbf{A}^T\mathbf{\Phi}^T(T-t)\mathbf{C}^T - \boldsymbol{\phi}_{x_0x}(t-T)\mathbf{C}^T]$$

$$-\sum_{j=-\infty}^{T} \mathbf{H}(T-j; T, l)\Bigg[\sum_{i=-\infty}^{j-1} \sum_{k=-\infty}^{\min(j,T-t)-1} \mathbf{W}(j-i)\mathbf{Q}\,\delta(i-k)\mathbf{W}^T(T-t-k)$$

$$+\mathbf{R}\,\delta(j+t-T)\Bigg] + \sum_{j=-\infty}^{\min(T_1, T-t)-1} \mathbf{\Phi}(T_1-j)\mathbf{D}\mathbf{Q}\mathbf{W}^T(T-t-j) = \mathbf{O} \tag{9.60}$$

for all $t \in [0, \infty)$. Denote the equation (9.60) now as:

$$\mathbf{G}_0(t; T, l) = \mathbf{O} \text{ for all integers } t \geq 0. \tag{9.61}$$

□

These lemmas now permit the Wiener–Hopf equation to be transformed into the z-domain:

9.3.4. Solution for the discrete-time estimator in the z-domain

The solution for the time-invariant, discrete-time estimator may now be derived. However, the z-transfer functions of terms in the Wiener–Hopf equation (9.60) are required.

Discrete-time sequences: z-domain description

From the definition of $\boldsymbol{\phi}(t)$ the two-sided z-transform gives:

$$\mathscr{Z}_2(\boldsymbol{\phi}(t-1-l)) = z^{-1-l}\boldsymbol{\phi}(z^{-1}) = z^{-l}\boldsymbol{\Phi}(z^{-1})$$

$$\mathscr{Z}_2(\boldsymbol{\phi}(t-T-1)) = z^{-1-T}\boldsymbol{\phi}(z^{-1}) = z^{-T}\boldsymbol{\Phi}(z^{-1})$$

$$\mathscr{Z}_2(\boldsymbol{\phi}^T(T-t-1)) = \sum_{i=-\infty}^{T-1} \boldsymbol{\phi}^T(T-i-1)z^{-i} = \sum_{m=1}^{\infty} \boldsymbol{\Phi}^T(m)z^{-(T-m)}$$

$$= \boldsymbol{\Phi}^T(z)z^{-T} = \boldsymbol{\phi}^T(z)z^{-(T-1)}. \tag{9.62}$$

From (9.54) and the definition of $\boldsymbol{\phi}_{x_0x} = E\{\mathbf{x}(0)\mathbf{x}^T(m)\}$ obtain:

$$\boldsymbol{\phi}_{x_0x}(-m) = \sum_{j=-\infty}^{-1} \sum_{i=-\infty}^{m-1} \boldsymbol{\Phi}(-j)\mathbf{DQD}^T\,\delta(j-i)\boldsymbol{\Phi}^T(m-i) \tag{9.63}$$

$$\boldsymbol{\phi}_{x_0x}(t-T) = \sum_{j=-\infty}^{-1} \sum_{i=-\infty}^{T-t-1} \boldsymbol{\Phi}(-j)\mathbf{DQD}^T\,\delta(j-i)\boldsymbol{\Phi}^T(T-t-i) \tag{9.64}$$

$$\mathscr{Z}_2(\boldsymbol{\phi}_{x_0x}(t-T)) = \sum_{t=-\infty}^{\infty} \sum_{j=-\infty}^{-1} \sum_{i=-\infty}^{T-t-1} \boldsymbol{\Phi}(-j)\mathbf{DQD}^T\,\delta(j-1)\boldsymbol{\Phi}^T(T-t-i)z^{-t}$$

$$= \boldsymbol{\Phi}(z^{-1})\mathbf{DQD}^T\boldsymbol{\Phi}(z)z^{-T}. \tag{9.65}$$

The transformations of the summation terms may also be obtained as follows:

$$\sum_{j=-\infty}^{T} \mathbf{H}(T-j;T,l) \sum_{i=-\infty}^{j-1} \sum_{k=-\infty}^{\min(j,\,T-t)-1} \mathbf{W}(j-i)\mathbf{Q}\,\delta(i-k)\mathbf{W}^T(T-t-k)$$

$$= \sum_{j=-\infty}^{T} \mathbf{H}(T-j;T,l) \sum_{k=-\infty}^{\min(j,\,T-t)-1} \mathbf{W}(j-k)\mathbf{Q}\mathbf{W}^T(T-t-k)$$

$$\sum_{t=-\infty}^{\infty} \sum_{j=-\infty}^{T} \mathbf{H}(T-j;T,l) \sum_{k=-\infty}^{\min(j,\,T-t)-1} \mathbf{W}(j-k)\mathbf{Q}\mathbf{W}^T(T-t-k)z^{-t}$$

$$= \mathbf{H}(z^{-1};T,l)\mathbf{W}(z^{-1})\mathbf{Q}\mathbf{W}^T(z) \tag{9.66}$$

where

$$\mathbf{H}(z^{-1};T,l) \triangleq \mathscr{Z}_1(\mathbf{H}(t;T,l)) \text{ and } \mathbf{W}(z^{-1}) \triangleq \mathscr{Z}_1(\mathbf{W}(t))$$

$$\sum_{t=-\infty}^{\infty} \sum_{j=-\infty}^{T} \mathbf{H}(T-j;T,l)[\mathbf{R}\,\delta(j+t-T)z^{-t}] = \mathbf{H}(z^{-1};T,l)\mathbf{R} \tag{9.67}$$

$$\sum_{t=-\infty}^{\infty} \sum_{j=-\infty}^{\min(T_1,\,T-t)-1} \mathbf{\Phi}(T_1-j)\mathbf{D}\mathbf{Q}\mathbf{W}^T(T-t-j)z^{-t} = \mathbf{\Phi}(z^{-1})\mathbf{D}\mathbf{Q}\mathbf{W}^T(z)z^{-l}. \tag{9.68}$$

Lemma 9.5. *The time-invariant estimator—a z-domain solution. The time-invariant estimator is obtained as:*

$$\mathbf{H}(z^{-1};T,l) = (\mathbf{N}_1(z^{-1},l) - \mathbf{\Psi}(T;l)(\mathbf{N}_1(z^{-1},T) - \mathbf{N}_2(z^{-1},T)))\mathbf{Y}(z^{-1})^{-1} \tag{9.69}$$

where

$$\mathbf{N}_1(z^{-1},l) \triangleq \{\mathbf{\Phi}(z^{-1})\mathbf{D}\mathbf{Q}\mathbf{W}^T(z)\mathbf{Y}^T(z)^{-1}z^{-l}\}_+ \tag{9.70}$$

$$\mathbf{N}_2(z^{-1},T) \triangleq \mathbf{\Sigma}_0\{\mathbf{M}^{\mathrm{T}}(z)z^{-T}\}_+ \tag{9.71}$$

and

$$\mathbf{M}(z^{-1}) \triangleq \mathbf{Y}(z^{-1})^{-1}\mathbf{C}\mathbf{\Phi}(z^{-1})\mathbf{A}. \tag{9.72}$$

Proof. Transforming the Wiener–Hopf equation (9.61) into the

z-domain, using the above results, gives:

$$\mathbf{G}_0(z^{-1}; T, l) = \mathbf{\Psi}(T; l)[\mathbf{\Sigma}_0\mathbf{A}^T\mathbf{\Phi}^T(z)\mathbf{C}^T - \mathbf{\Phi}(z^{-1})\mathbf{DQW}^T(z)]z^{-T}$$

$$+ \mathbf{\Phi}(z^{-1})\mathbf{DQW}^T(z)z^{-l} - \mathbf{H}(z^{-1}: \mathbf{T}, l)(\mathbf{W}(z^{-1})\mathbf{QW}^T(z) + \mathbf{R}).$$

Now introduce the generalized spectral factor $\mathbf{Y}(z^{-1})$ (having a pole polynomial equal to the system pole polynomial) from the discrete spectral density matrix of the observations:

$$\mathbf{Y}(z^{-1})\mathbf{Y}^T(z) = \mathbf{W}(z^{-1})\mathbf{QW}^T(z) + \mathbf{R} \tag{9.73}$$

hence

$$\mathbf{G}_0(z^{-1}, T)\mathbf{Y}^T(z)^{-1} = \mathbf{\Psi}(T; l)[\mathbf{\Sigma}_0\mathbf{A}\mathbf{\Phi}^T(z)\mathbf{C}^T - \mathbf{\Phi}(z^{-1})\mathbf{DQW}^T(z)]z^{-T}$$

$$+ \mathbf{\Phi}(z^{-1})\mathbf{DQW}^T(z)z^{-l}\mathbf{Y}^T(z)^{-1} - \mathbf{H}(z^{-1}; T, l)\mathbf{Y}(z^{-1}).$$

If the transforms of the positive time signals are equated, the desired time-invariant estimator: (equations 9.69 to 9.72) is obtained to prove the lemma. □

To evaluate the above expression for the time-invariant estimator the unknown matrix $\mathbf{\Psi}(T; l)$ must be calculated and this requires an expression for the zero-input response, namely:

$$\mathbf{d}_0(T; T, l) = \mathbf{\Psi}(T; l)x_0'.$$

thus from (9.44)

$$\mathbf{d}_0(T; T, l) = (\mathbf{\phi}(T_1) - \sum_{j=1}^{T} \mathbf{H}(T-j; T, l)\mathbf{C}\mathbf{\phi}(j))\mathbf{m}_0 \qquad \text{for } T > 0.$$

If the state estimates are to unbiased, for all t, let:

$$\mathbf{d}_0(t; T; l) = (\mathbf{\phi}(t_1) - (\mathbf{H}_0\mathbf{C}\mathbf{\phi})(t))m_0$$

where $t_1 \triangleq t - l$.

Lemma 9.6. *Expression for the matrix* $\mathbf{\Psi}(T; l)$. *The unknown matrix* $\mathbf{\Psi}(T; l)$ *is given by:*

$$\mathbf{\Psi}(T; l) = \mathbf{\phi}(T_1) - \sum_{j=0}^{T-1} \mathbf{N}_1(j; l)\mathbf{M}(T-j) + \mathbf{\Psi}(T; l)\left(\sum_{j=0}^{T-1} \mathbf{N}_1(j, T)\mathbf{M}(T-j)\right.$$

$$\left. - \mathbf{\Sigma}_0\left(\sum_{j=0}^{T} \mathbf{M}^T(T-j)\mathbf{M}(T-j)\right)\right).$$

Proof. $\boldsymbol{\Psi}(T; l)$ can be calculated as:

$$\boldsymbol{\Psi}(T; l) = \boldsymbol{\phi}(T_1) - \sum_{j=1}^{T} \mathbf{H}(T-j; T, l)\mathbf{C}\boldsymbol{\phi}(j) \tag{9.74}$$

$$= \boldsymbol{\phi}(T_1) - \sum_{j=0}^{T-1} \mathbf{H}(j; T, l)\mathbf{C}\boldsymbol{\phi}(T-j-1)\mathbf{A}.$$

Recall the following equations:

$$\mathbf{H}(z^{-1}; T, l)\mathbf{C}\boldsymbol{\Phi}(z^{-1})\mathbf{A} = (N_1(z^{-1}, l) + \boldsymbol{\Psi}(T; l)(-\mathbf{N}_1(z^{-1}, T) + \mathbf{N}_2(z^{-1}, T)))\mathbf{M}(z^{-1})$$

$$\{\mathbf{M}^T(z)z^{-T}\}_+ = \mathcal{Z}_1(\mathbf{M}^T(T-t)U(T-t)) = \sum_{n=0}^{T-1} \mathbf{M}^T(T-t)z^{-t}.$$

Giving

$$\boldsymbol{\Psi}(T; l) = \boldsymbol{\phi}(T_1) - \sum_{j=0}^{T-1} \mathbf{N}_1(j, l)\mathbf{M}(T-j) + \boldsymbol{\Psi}(T; l)\left(\sum_{j=0}^{T-1} \mathbf{N}_1(j, T)\mathbf{M}(T-j) - \Sigma_0\left(\sum_{j=0}^{T} \mathbf{M}^T(T-j)\mathbf{M}(T-j)\right)\right). \quad \square$$

9.3.5. Time-invariant estimator: Theorem and examples

The main theorem defining the time-invariant predictor, filter or smoother, now follows.

Theorem 9.2. ***Time-invariant estimator.*** *Consider the discrete-time finite-interval estimation problem described in Section* 9.3.2. *The optimal time-invariant estimator is given as:*

$$\mathbf{H}(z^{-1}; T, l) = (\mathbf{N}_1(z^{-1}, l) - \boldsymbol{\Psi}(T; l)(\mathbf{N}_1(z^{-1}, T) - \mathbf{N}_2(z^{-1}, T)))\mathbf{Y}(z^{-1})^{-1} \tag{9.75}$$

where

$$\mathbf{N}_1(z^{-1}, l) = \{\boldsymbol{\Phi}(z^{-1}\mathbf{DQW}^T(z)\mathbf{Y}^T(z)^{-1}z^{-l}\}_+ \tag{9.76}$$

$$\mathbf{N}_2(z^{-1}, T) = \boldsymbol{\Sigma}_0\{\mathbf{M}^T(z)z^{-T}\}_+ \tag{9.77}$$

and

$$\mathbf{M}(z^{-1}) = \mathbf{Y}(z^{-1})^{-1}\mathbf{C}\mathbf{\Phi}(z^{-1})\mathbf{A}. \tag{9.78}$$

The matrix $\mathbf{\Psi}(T; l)$ *may be calculated using:*

$$\mathbf{\Psi}(T; l) = (\boldsymbol{\phi}(T_1) - \mathbf{I}_1(T, l))(\mathbf{I}_n - \mathbf{I}_1(T, T) + \mathbf{\Sigma}_0\mathbf{S}(T))^{-1} \tag{9.79}$$

where

$$\mathbf{I}_1(T, l) = \sum_{j=0}^{T-1} \mathbf{N}_1(j, l)\mathbf{M}(T-j) \tag{9.80}$$

$$\mathbf{S}(T) = \sum_{j=0}^{T-1} \mathbf{M}^T(T-j)\mathbf{M}(T-j). \tag{9.81}$$

Proof. Follows by collecting the results of the previous Lemmas 9.1 to 9.6. □

Special cases

Wiener estimators. The first term in the expression for the estimator transfer-function matrix represents the solution to the usual infinite time ($T_0 = -\infty$) estimation problem. In the continuous-time case this corresponds to the Wiener estimator. The first term in (9.75) may be written as:

$$\mathbf{H}(z^{-1}; \infty, l) = \mathbf{N}_1(z^{-1}, l)\mathbf{Y}(z^{-1})^{-1} \tag{9.82}$$

(since $\mathbf{\Psi}(T; l) \to \mathbf{O}$ as $T \to \infty$ for an asymptotically stable filter). If the initial state of the system is known exactly $\mathbf{\Sigma}_0 = \mathbf{O}$, $\mathbf{N}_2(z^{-1}, T) = \mathbf{O}$ and the estimator is given by the first two terms in equation (9.75):

$$\mathbf{H}'(z^{-1}; T, l) = (\mathbf{N}_1(z^{-1}, l) - \mathbf{\Psi}(T; l)\mathbf{N}_1(z^{-1}, T))(\mathbf{Y}(z^{-1}))^{-1}.$$

If alternatively $\mathbf{\Sigma}_0 \to \infty$, then from (9.75) and (9.77) the estimator depends upon the first and third terms in (9.75):

$$\mathbf{H}''(z^{-1}; T, l) = (\mathbf{N}_1(z^{-1}, l) + \mathbf{\Psi}(T; l)\mathbf{N}_2(z^{-1}, T))(\mathbf{Y}(z^{-1}))^{-1}.$$

Singular measurement noise. O'Reilly [33] has considered the case where the measurement noise covariance matrix is singular and has derived the finite-time filter for this case. The conclusion was that the transfer-function matrix for the filter is almost as easy to implement as the infinite-time Brammer filter.

Wiener computation procedure. The calculation of the time-invariant estimators is considerably simplified in later sections. Two lemmas are derived by which standard algorithms may be used to implement the smoothers or predictors. The simplifications to the estimator equations are achieved using relationships derived from the steady state Riccati equation (Shaked [20]). These relationships enable the causal transforms to be calculated by a standard approach. However, the results of Theorem 9.2 are used to evaluate the filter in the following example.

Example 9.8. *Discrete-time estimation problem.* Consider a stable signal generating process defined in state-space form:

$$x(t+1) = ax(t) + \omega(t)$$

$$z(t) = x(t) + v(t)$$

where $|a| \leq 1$. Assume that the noise processes and $x(0)$ are independent. The random signals $\omega(t)$ and $v(t)$ are assumed to be zero mean and to have the covariances q and r respectively. The initial state vector is assumed to be zero mean and to have a covariance $\Sigma_0 = \sigma_0^2$.

Solution. The finite-time estimator is derived below in its general form and the results are then specialized to obtain the optimal predictor $l < 0$, smoother $l > 0$ or filter $l = 0$.

$$\Phi(z^{-1}) = \frac{z^{-1}}{1 - az^{-1}}, \qquad W(z^{-1}) = \frac{cz^{-1}}{1 - az^{-1}}$$

$$\Phi(t) = a^{t-1}U(t-1) = \phi(t-1)U(t-1).$$

The spectral factor $Y(z^{-1})$ can be identified as follows:

$$W(z^{-1})QW^T(z) + R = \frac{c^2q + r + ra^2 - ar(z^{-1} + z)}{(z-a)(z^{-1} - a)}.$$

Let $\xi \triangleq (c^2q + r + ra^2)/(2ar)$ then,

$$W(z^{-1})QW^T(z) + R = \frac{ar(2\xi - z^{-1} - z)}{(z-a)(z^{-1} - a)}$$

$$= \frac{ar(\alpha - z)(\alpha - z^{-1})}{\alpha(z-a)(z^{-1} - a)}$$

hence

$$Y(z^{-1}) = \left(\frac{ar}{\alpha}\right)^{1/2} \frac{(1 - \alpha z^{-1})}{(1 - az^{-1})}$$

where $\alpha = \xi \pm (\xi^2 - 1)^{1/2}$ and $|\alpha| < 1$.

The transfer functions directly involved in the estimator may now be calculated:

$$M(z^{-1}) = Y(z^{-1})^{-1}C\Phi(z^{-1})A = \left(\frac{\alpha}{ar}\right)^{1/2} \frac{ca}{z-\alpha}$$

$$N_1(z^{-1}, l) = \{\Phi(z^{-1})DQW^T(z)Y^{-T}(z)z^{-l}\}_+.$$

There are two solutions for N_1 depending on whether $l > 0$ or $l < 0$. The solutions coincide when $l = 0$.

$$N_1(z^{-1}l) = \beta\left\{z^{-l+1}\left(\frac{1}{z-a} - \frac{\alpha}{\alpha z - 1}\right)\right\}_+$$

where

$$\beta \triangleq qc\left(\frac{\alpha}{ar}\right)^{1/2} \frac{1}{1-a\alpha}.$$

Case (i) $l \geq 0$

$$N_1(z^{-1}, l) = \beta\,\frac{z^{-l+1}}{z-a} - \beta\left\{\frac{\alpha z^{-l+1}}{\alpha z - 1}\right\}_+$$

$$= \beta\,\frac{z^{-l+1}}{z-a} + \beta\alpha\left\{\frac{z^{-l+1}}{1-\alpha z}\right\}_+$$

By expanding the term $z^{-l+1}/(1-\alpha z)$ in terms of powers of αz and summing the constant and terms with negative powers obtain:

$$N_1(z^{-1}, l) = \beta\left(\frac{z^{-l+1}}{z-a} + \frac{\alpha z^{-l+1} - z\alpha^{l+1}}{1-\alpha z}\right). \tag{9.83}$$

Case (ii) $l \leq 0$

$$N_1(z^{-1}, l) = \beta\left\{\frac{z^{-l+1}}{z-a}\right\}_+ = \beta a^{-l}\,\frac{z}{z-a}. \tag{9.84}$$

The term $N_2(z^{-1}, T)$ does not depend upon l and is obtained as:

$$N_2(z^{-1}, T) = \sigma_0^2\{M^T(z)z^{-T}\}_+$$

$$= \sigma_0^2\left(\frac{\alpha}{ar}\right)^{1/2} \text{ca}\,\frac{(z^{-T+1} - z\alpha^T)}{1-az}. \tag{9.85}$$

The matrix $\Psi(T;l)$ can be calculated using (9.79) where $\phi(T_1)=a^{T-l}$ and for $l\geq 0$:

$$I_1(T,l)=\beta\left(\frac{\alpha}{ar}\right)^{1/2}\left(\text{ca}\,\frac{(a^{T-l}-\alpha^{T-l})}{a-\alpha}+\frac{(1-\alpha^{2l})}{a-\alpha^2}\,\alpha^{1-l+T}\right) \tag{9.86}$$

or for $l\leq 0$:

$$I_1(T,l)=\beta\left(\frac{\alpha}{ar}\right)^{1/2}\text{ca}\,\frac{(a^{T}-\alpha^{T})}{a-\alpha}\,\alpha^{-l}. \tag{9.87}$$

Also

$$I_1(T,T)=\beta\left(\frac{\alpha}{ar}\right)^{1/2}\text{ca}\;\alpha\,\frac{(1-\alpha^{2T})}{(1-\alpha^2)}$$

$$S(T)=\frac{c^2a\alpha}{r}\,\frac{(1-\alpha^{2T})}{(1-\alpha^2)}. \tag{9.88}$$

It is clear that $\Psi(T;l)$ can easily be calculated from the above results for particular system parameters.

The expressions for the optimal estimators now follow.

Optimal smoother or filter $l\geq 0$

$$\begin{aligned}H(z^{-1};T,l)=\beta\left(\frac{\alpha}{ar}\right)^{1/2}\Bigg\{&\frac{z^{-l+1}}{z-\alpha}+\frac{(z-a)(\alpha z^{-l+1}-z\alpha^{l+1})}{(1-\alpha z)(z-\alpha)}\\&+\Psi(T;l)\left[\frac{-z^{-T+1}}{z-\alpha}-\frac{(z-a)(\alpha z^{-T+1}-z\alpha^{T+1})}{(1-\alpha z)(z-\alpha)}\right.\\&\left.+\frac{\sigma_0^2}{\beta}\left(\frac{\alpha}{ar}\right)^{1/2}\text{ca}\,\frac{(z^{-T+1}-z\alpha^{T})(z-a)}{(1-\alpha z)(z-\alpha)}\right]\Bigg\}.\end{aligned} \tag{9.89}$$

Optimal predictor or filter $l\leq 0$

By inverse tranforming the above equations, for $l\geq 0$,

$$N_1(t,l)=\beta(a^{t-l}-\alpha^{-(t-l)})U(t-l)+\beta\alpha^{l-t}U(t)$$

and for $l\leq 0$,

$$N_1(t,l)=\beta a^{t-l}U(t)$$

and for all l,

$$N_1(t, T) = \beta(a^{t-T} - \alpha^{-(t-T)})U(t-T) + \beta\alpha^{T-t}U(t)$$

$$N_2(t; T) = \sigma_0^2\left(\frac{\alpha}{ar}\right)^{1/2} ca\ \alpha^{T-t-1}(U(t) - U(t-T))$$

$$M(t) = \left(\frac{\alpha}{ar}\right)^{1/2} ca\ \alpha^{t-1}U(t-1).$$

The optimal finite-time predictor or filter becomes:

$$H(z^{-1}; T, l) = \beta\left(\frac{\alpha}{ar}\right)^{1/2}\left\{\frac{a^{-l}z}{z-\alpha} + \Psi(T; l)\left[\frac{-z^{-T+1}}{z-\alpha} - \frac{(z-a)(\alpha z^{-T+1} - z\alpha^{T+1})}{(1-\alpha z)(z-\alpha)} + \frac{\sigma_0^2}{\beta}\left(\frac{\alpha}{ar}\right)^{1/2} ca\ \frac{(z^{-T+1} - z\alpha^{T})(z-a)}{(1-\alpha z)(z-\alpha)}\right]\right\}. \tag{9.90}$$

Comments

(a) If the estimators are to be used only within the finite interval $[0, T]$ then all terms involving z^{-T} can be neglected since they are inactive within the interval. The above solution and estimators are considerably simplified in this case.

(b) The impulse response may be shown to be zero outside the interval $[0, T]$.

(c) For a stable estimator $\Psi(T; l) \to 0$ as $T \to \infty$ and hence the infinite-time smoothers and predictors are given by the first terms in (9.89), (9.90).

Simulation results

Simulation results for the finite-interval ($T = 100$) time-invariant estimator are shown in Figs. 9.16 to 9.19. The system parameters were defined as: $T = 100$, $a = 0.99$, $\sigma_0^2 = 1$, $q = 0.01$ and $r = 2$. The observations signal is very noisy as may be seen in Fig. 9.16. The smoothing filter with impulse response shown in Fig. 9.17 (zero for $t \geq T$) gives the estimates shown in Fig. 9.18. Note that there is a delay of $l = 50$ between the signal and its estimate. The major trends are clearly recovered by the smoothed estimate. Finally, the filtered estimate is shown in Fig. 9.19. Note, that in both of these cases the estimator employs a fixed memory length outside of the interval of length $T = 100$. The smoothed estimates are normally superior to the filtered estimates since information on both sides of the time point of interest is available in the smoothing problem.

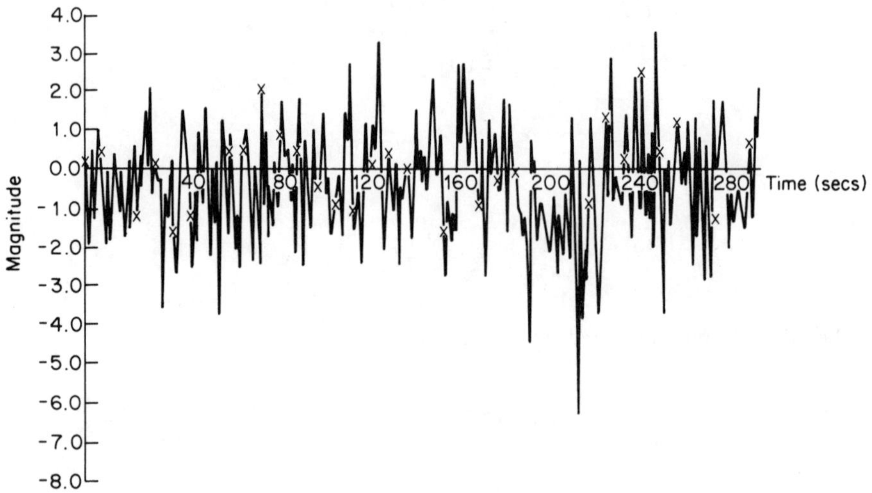

Fig. 9.16. Signal and measurement noise signal

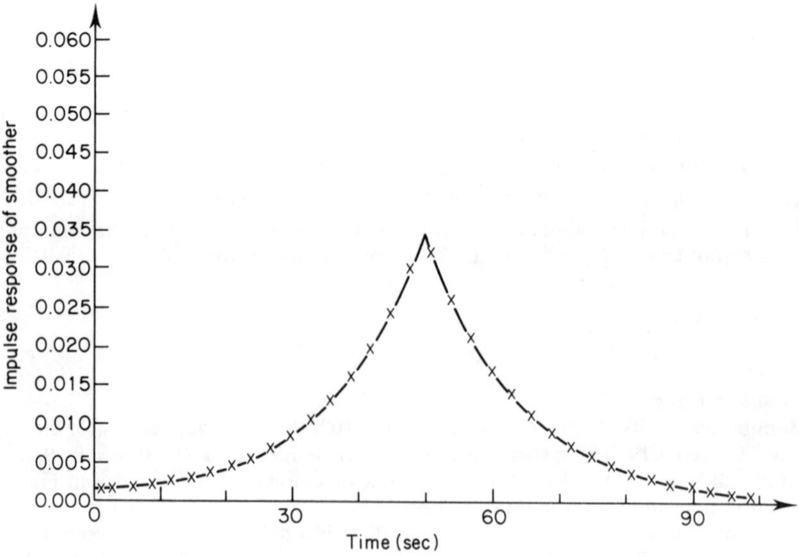

Fig. 9.17. Smoothing filter impulse response ($T + 100$, Delay = 50 seconds)

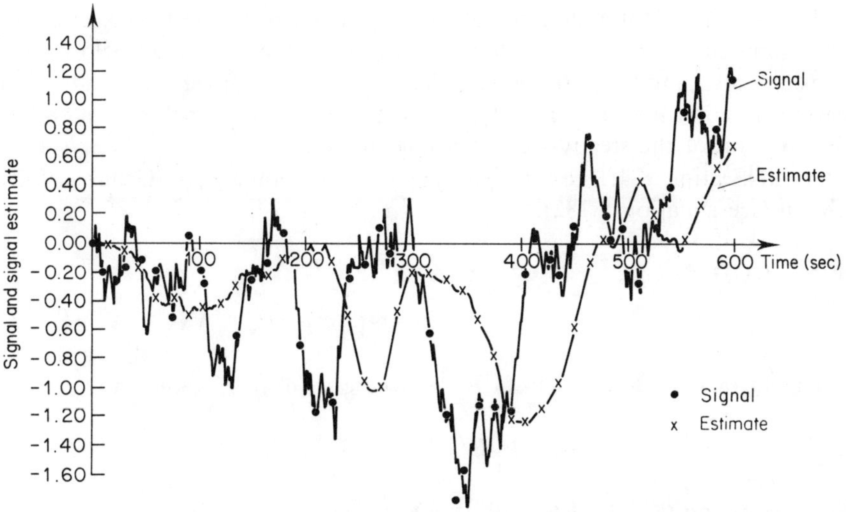

Fig. 9.18. Smoothing estimate $T = 100$, Delay = 50 seconds

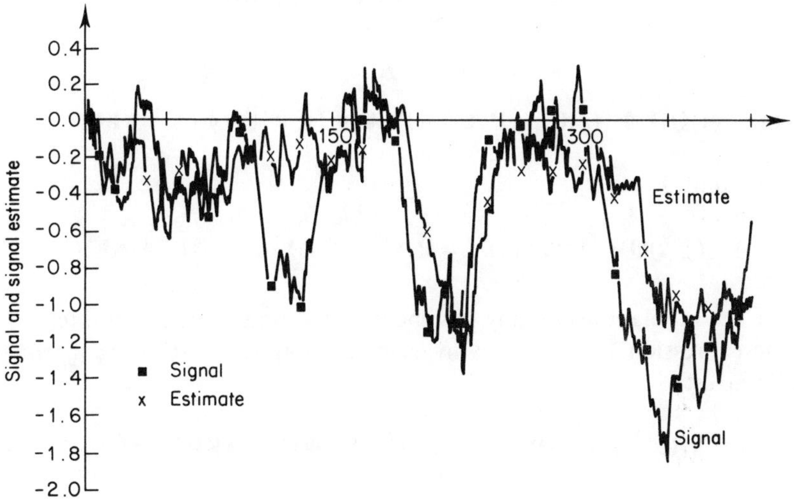

Fig. 9.19. Filtered estimate ($T = 100$ seconds)

9.3.6. Calculation of the time-invariant estimator

The time-invariant finite-interval estimator described in Section 9.3.5 cannot be computed by routine calculations. An algorithm is derived below which considerably simplifies these calculations. The algorithm depends upon various relationships established between the expression for the estimator and the solution of the steady-state Riccati equation.

As in Section 8.4.1, by simple algebraic manipulations (Grimble [42]) obtain (see equation 8.182):

$$\mathbf{\Phi}(z^{-1})\mathbf{DQD}^T\mathbf{\Phi}^T(z)\mathbf{C}^T = \mathbf{P}_\infty\mathbf{C}^T + \mathbf{P}_\infty\mathbf{A}^T\mathbf{\Phi}^T(z)\mathbf{C}^T$$
$$+\,\mathbf{\Phi}(z^{-1})(\mathbf{AP}_\infty + \mathbf{AKCP}_\infty\mathbf{A}^T\mathbf{\Phi}^T(z))\mathbf{C}^T$$

but from (8.173): $\mathbf{KR}_\varepsilon^{1/2} = \mathbf{P}_\infty\mathbf{C}^T\mathbf{R}_\varepsilon^{-1/2}$ where from (8.173) the gain:

$$\mathbf{K} \triangleq \mathbf{P}_\infty\mathbf{C}^T(\mathbf{R} + \mathbf{CP}_\infty\mathbf{C}^T)^{-1}.$$

From equation (8.181): $\mathbf{F}(z^{-1})\mathbf{R}_\varepsilon^{1/2} = \mathbf{Y}(z^{-1})$ or

$$\mathbf{KR}_\varepsilon^{1/2}\mathbf{Y}^T(z) = \mathbf{KR}_\varepsilon\mathbf{F}^T(z) = \mathbf{KR}_\varepsilon(\mathbf{I}_r + \mathbf{K}^T\mathbf{A}^T\mathbf{\Phi}^T(z)\mathbf{C}^T)$$
$$= \mathbf{P}_\infty\mathbf{C}^T + \mathbf{KCP}_\infty\mathbf{A}^T\mathbf{\Phi}^T(z)\mathbf{C}^T$$

thus

$$\mathbf{\Phi}(z^{-1})\mathbf{DQD}^T\mathbf{\Phi}^T(z)\mathbf{C}^T = \mathbf{P}_\infty\mathbf{C}^T + \mathbf{P}_\infty\mathbf{A}^T\mathbf{\Phi}^T(z)\mathbf{C}^T + \mathbf{\Phi}(z^{-1})\mathbf{AKR}_\varepsilon^{1/2}\mathbf{Y}^T(z)$$

hence from (9.73)

$$\mathbf{W}_0(z^{-1})\mathbf{QW}^T(z)\mathbf{Y}^T(z)^{-1} = \mathbf{P}_\infty\mathbf{C}^T + \mathbf{P}_\infty\mathbf{M}^T(z) + \mathbf{\Phi}(z^{-1})\mathbf{AKR}_\varepsilon^{1/2}.$$

The above relationship may now be used to compute the causal term $\{\cdot\}_+$ in the expression for the time-invariant estimator. Substituting in equation (9.71) gives:

$$\mathbf{N}_1(z^{-1}, l) = \{(\mathbf{P}_\infty\mathbf{C}^T + \mathbf{P}_\infty\mathbf{M}^T(z) + \mathbf{\Phi}(z^{-1})\mathbf{AKR}_\varepsilon^{1/2})z^{-l}\}_+. \tag{9.91}$$

The $\{\cdot\}_+$ terms involves the negative powers of the delay operator: $z^0, z^{-1}, z^{-2}, \ldots$. If $l = T$ the first and last terms do not contribute to the filter impulse response within the interval $[1, T]$ and thus:

$$\mathbf{N}_1(z^{-1}, T) = \mathbf{P}_\infty\{\mathbf{M}^T(z)z^{-T}\}_+ \tag{9.92}$$

If $0 \leq l < T$, then since the first and last terms in (9.91) are causal:

$$\mathbf{N}_1(z^{-1}, l) = \mathbf{P}_\infty \mathbf{C}^T z^{-l} + \mathbf{P}_\infty \{\mathbf{M}^T(z) z^{-1}\}_+ + \mathbf{\Phi}(z^{-1}) \mathbf{AKR}_\varepsilon^{1/2} z^{-l}.$$

Similarly, consider the case where $l < 0$, then from (8.91):

$$\begin{aligned}\mathbf{N}_1(z^{-1}, l) &= \{\mathbf{\Phi}(z^{-1}) \mathbf{AKR}_\varepsilon^{1/2} z^{-l}\}_+ \\ &= \mathscr{Z}_1\{\mathbf{A}^{t-1-l} \mathbf{U}(t-1-l)\} \mathbf{AKR}_\varepsilon^{1/2} \\ &= \mathbf{A}^{-1-l} \boldsymbol{\phi}(z^{-1}) \mathbf{AKR}_\varepsilon^{1/2}\end{aligned}$$

To simplify the above expressions, let the following positive-time transform be defined, for $\sigma > 0$, as:

$$\begin{aligned}\mathbf{L}(z^{-1}, \sigma) &\triangleq \{\mathbf{M}^T(z) z^{-\sigma}\}_+ = \mathscr{Z}_1\{\mathbf{M}^T(\sigma - t)\} \\ &= \sum_{t=0}^{\sigma-1} \mathbf{M}^T(\sigma - t) z^{-t}\end{aligned} \tag{9.93}$$

where $\mathbf{M}(t) \triangleq \mathscr{Z}_1^{-1}(\mathbf{M}(z^{-1}))$ and $\mathbf{M}(t) \triangleq \mathbf{O}$ for $t \leq 0$.

Note that $\mathbf{L}(z^{-1}, \sigma) = \mathbf{O}$ for $\sigma \leq 0$. Also define,

$$\begin{aligned}\mathbf{V}(z^{-1}, \sigma) &= (\mathbf{P}_\infty \mathbf{C}^T + \mathbf{\Phi}(z^{-1}) \mathbf{AKR}_\varepsilon^{1/2}) z^{-l}, \quad &\text{for } 0 \leq l \leq T \\ &= \mathbf{A}^{-1-l} \boldsymbol{\phi}(z^{-1}) \mathbf{AKR}_\varepsilon^{1/2}, &\text{for } l < 0.\end{aligned}$$

From (9.75) and noting (9.92), the expression for the time-invariant estimator becomes:

$$\begin{aligned}\mathbf{H}(z^{-1}; T, l) = (&\mathbf{P}_\infty \mathbf{L}(z^{-1}, l) + \mathbf{V}(z^{-1}, l) \\ &+ \mathbf{\Psi}(T; l)(\mathbf{\Sigma}_0 - \mathbf{P}_\infty) \mathbf{L}(z^{-1}, T)) \mathbf{Y}(z^{-1})^{-1}.\end{aligned} \tag{9.94}$$

The calculation of the matrix $\mathbf{\Psi}$ is also simplified. From (9.79):

$$\mathbf{\Psi}(T; l) = (\boldsymbol{\phi}(T_1) - \mathbf{\Psi}_1(T, l))(\mathbf{I}_n + (\mathbf{\Sigma}_0 - \mathbf{P}_\infty) \mathbf{S}(T))^{-1}. \tag{9.95}$$

Corollary 9.1. *Finite-interval prediction* (l_0 = *prediction interval*). *The optimal time-invariant predictor* ($l < 0, l_0 = -l$) *for the system* (9.34) *and criterion* (9.41), *becomes:*

$$\begin{aligned}\mathbf{H}(z^{-1}; T, l_0) = \mathbf{A}^{l_0 - 1}(&\boldsymbol{\phi}(z^{-1}) \mathbf{AKR}_\varepsilon^{1/2} \\ &+ \mathbf{\Psi}(T; -1)(\mathbf{\Sigma}_0 - \mathbf{P}_\infty) \mathbf{L}(z^{-1}, T)) \mathbf{Y}(z^{-1})^{-1}\end{aligned} \tag{9.96}$$

where

$$\boldsymbol{\Psi}(T; -1) = (\mathbf{A}^{T+1} - \mathbf{I}_1(T, -1))(\mathbf{I}_n + (\boldsymbol{\Sigma}_0 - \mathbf{P}_\infty)\mathbf{S}(T))^{-1} \tag{9.97}$$

and

$$\mathbf{I}_1(T, -1) = \sum_{j=0}^{T-1} \boldsymbol{\phi}(j)\mathbf{AKR}_\varepsilon^{1/2}\mathbf{M}(T-j). \tag{9.98}$$

The summations $\mathbf{L}(z^{-1}, T)$ *and* $\mathbf{S}(T)$ *may be evaluated using* (9.93) *and* (9.81), *respectively.*

Proof. From the definitions of $\mathbf{N}_1$ and $\mathbf{I}_1$:

$$\mathbf{N}_1(z^{-1}, -1) = \boldsymbol{\phi}(z^{-1})\mathbf{AKR}_\varepsilon^{1/2}$$

$$\mathbf{N}_1(z^{-1}, l) = \mathbf{A}^{-1-l}\mathbf{N}_1(z^{-1}, -1)$$

$$\mathbf{I}_1(T, l) = \sum_{j=0}^{T-1} \mathbf{A}^{-1-l}\mathbf{N}_1(j, -1)\mathbf{M}(T-j).$$

$$= \mathbf{A}^{-1-l}\mathbf{I}_1(T, -1)$$

and substituting in the expression for $\boldsymbol{\Psi}(T; l)$ completes the proof. □

Corollary 9.2. ***Finite-interval filtering and smoothing*** ***(l = smoothing interval).*** *The optimal time-invariant filter* ($l = 0$) *and smoothing filter* ($l > 0$) *for the system* (9.34) *and criterion* (9.41) *become:*

$$\mathbf{H}(z^{-1}; T, l) = (\mathbf{P}_\infty\mathbf{L}(z^{-1}, l) + \mathbf{V}(z^{-1}, l) + \boldsymbol{\Psi}(T; l)(\boldsymbol{\Sigma}_0 - \mathbf{P}_\infty)\mathbf{L}(z^{-1}, T))\mathbf{Y}(z^{-1})^{-1} \tag{9.99}$$

where

$$\boldsymbol{\Psi}(T; l) = (\boldsymbol{\phi}(T_1) - \mathbf{I}_1(T, l))(\mathbf{I}_n + (\boldsymbol{\Sigma}_0 - \mathbf{P}_\infty)\mathbf{S}(T))^{-1} \tag{9.100}$$

and

$$\mathbf{I}_1(T, l) = \mathbf{P}_\infty\mathbf{C}^T\mathbf{M}(T-l) + \mathbf{P}_\infty \sum_{j=0}^{l-1} \mathbf{M}^T(l-j)\mathbf{M}(T-j)$$

$$+ \sum_{j=l+1}^{T-1} \boldsymbol{\Phi}(j-l)\mathbf{AKR}_\varepsilon^{1/2}\mathbf{M}(T-j) \tag{9.101}$$

$$l \triangleq T - T_1.$$

Proof. The expression for the time-invariant estimator was obtained as (9.94), and the matrix $\boldsymbol{\Psi}(T; l)$ was defined by (9.95). Recall the definition of $\mathbf{I}_1(T, l)$ from (9.80):

$$\mathbf{I}_1(T, l) = \sum_{j=0}^{T-1} \mathbf{N}_1(j, l)\mathbf{M}(T-j)$$

and note, using (9.93):

$$\mathbf{N}_1(z^{-1}, l) = \mathbf{P}_\infty \mathbf{C}^T z^{-l} + \mathbf{P}_\infty \sum_{t=0}^{l-1} \mathbf{M}^T(l-t)z^{-t} + \boldsymbol{\Phi}(z^{-1})\mathbf{AKR}_\varepsilon^{1/2} z^{-l}$$

thus,

$$\mathbf{I}_1(T, l) = \sum_{j=0}^{T-1} \mathbf{P}_\infty \mathbf{C}^T\, \delta(j-l)\mathbf{M}(T-j) + \sum_{j=0}^{l-1} \mathbf{P}_\infty \mathbf{M}^T(l-j)\mathbf{M}(T-j)$$

$$+ \sum_{j=l-1}^{T-1} \boldsymbol{\Phi}(j-l)\mathbf{AKR}_\varepsilon^{1/2}\mathbf{M}(T-j). \qquad \square$$

Special cases

The finite-time discrete filter is a special case of these results obtained when $l = 0$:

$$\mathbf{H}(z^{-1}; T, 0) = (\mathbf{P}_\infty \mathbf{C}^T + \boldsymbol{\Phi}(z^{-1})\mathbf{AKR}_\varepsilon^{1/2} + \boldsymbol{\Psi}(T; 0)(\boldsymbol{\Sigma}_0 - P_\infty)\mathbf{L}(z^{-1}, T))\mathbf{Y}(z^{-1})^{-1}$$

where

$$\boldsymbol{\Psi}(T; 0) = (\boldsymbol{\phi}(T) - \mathbf{I}_1(T, 0))(\mathbf{I}_n + (\boldsymbol{\Sigma}_0 - \mathbf{P}_\infty)\mathbf{S}(T))^{-1}$$

and

$$\mathbf{I}_1(T, 0) = \mathbf{P}_\infty \mathbf{C}^T + \sum_{j=1}^{T-1} \boldsymbol{\Phi}(j)\mathbf{AKR}_\varepsilon^{1/2}\mathbf{M}(T-j).$$

The smoothing filter to estimate the initial condition is obtained by setting $l = T$. In this case $\mathbf{I}_1(T, T) = \mathbf{P}_\infty \mathbf{S}(T)$. Note that if $\boldsymbol{\Sigma}_0 = \mathbf{O}$, $\boldsymbol{\Psi}(T; T) = \mathbf{I}_n$ and from (9.99) $\mathbf{H}(z^{-1}; T, T) = \mathbf{O}$ within the interval. This result might be expected since when the initial state is known exactly the best estimate is given by $\hat{\mathbf{x}}(0 \mid T) = \mathbf{m}_0$.

Algorithms and structure

Algorithms to implement the above estimators may be constructed directly from the above corollaries. These expressions reveal the structure of the z-domain estimators with more clarity than was previously possible. For example, let $\mathbf{H}_w$ denote the optimal Wiener filter and let $\mathbf{G}$ represent a constant matrix which depends upon $\boldsymbol{\Psi}$ and the initial covariance matrix $\boldsymbol{\Sigma}_0$, then from (9.99):

$$\mathbf{H}(z^{-1}; T, l) = \mathbf{H}_w(z^{-1}; \infty, l) + \mathbf{G}\mathbf{L}(z^{-1}, T)\mathbf{Y}(z^{-1})^{-1}.$$

The final term in this expression represents the correction term which occurs when finite-time rather than infinite-time problems are considered. The advantage of the Kalman filter is that it is in state-space form and the structure is fixed. The structure of the z-domain estimators is not predetermined to the same extent but there is more structural information than is available for estimators which are defined only via their impulse response matrices.

Shaked and Priel [34] have developed a solution to the optimal estimation problem for linear discrete time invariant processes. This also involves a mixed z- and time-domain approach but is more related to the Kalman filtering problem than to the above finite-time problem.

9.3.7. Relationship to the Kalman filter

A simple relationship exists between the time-invariant filter and the Kalman filter and this is discussed below. The impulse response of the optimal time invariant filter is zero for $t \geq T$. This property is a consequence of the way the ideal output of the filter was defined in Section 9.2.3. The weighting-sequence matrix for the filter problem ($T_1 = T$) satisfies the following Wiener–Hopf equation:

$$\operatorname{cov}[\mathbf{x}_i(T), \mathbf{z}(m)] = \sum_{j=1}^{T} \mathbf{H}(T-j; T, 0)\operatorname{cov}[\mathbf{z}(j), \mathbf{z}(m)] \quad \text{for all } m \in (-\infty, T].$$

Substituting for $\mathbf{x}_i(T)$ from (9.52) and for $\boldsymbol{\Psi}(T; 0)$ from (9.53) the equation (for all $m \in [1, T]$) becomes:

$$\operatorname{cov}[\mathbf{x}'(T), \mathbf{z}'(m)] = \sum_{j=1}^{T} \mathbf{H}(T-j; T, 0)\operatorname{cov}[\mathbf{z}'(j), \mathbf{z}'(m)] \tag{9.102}$$

where $\mathbf{x}'(T)$ and $\mathbf{z}'(m)$ are the state and observations for a system with initial states vector $\mathbf{x}'_0$ at time zero. The weighting sequence for the discrete-time Kalman filter $\mathbf{k}(T, j)$ must satisfy a similar Wiener–Hopf equation, at time

$t = T$:

$$\operatorname{cov}[\mathbf{x}'(T), z'(m)] = \sum_{j=1}^{T} \mathbf{k}(T, j)\operatorname{cov}[z'(j), z'(m)] \tag{9.103}$$

for all $m \in [1, T]$. The Wiener–Hopf equation represents a necessary and sufficient condition for optimality in these problems and since the covariance matrices in (9.102) and (9.103) are identical, it follows that:

$$\mathbf{H}(T - j; T, 0) = \mathbf{k}(T, j) \text{ for all integers } j \in [1, T] \tag{9.104}$$

but

$$\mathbf{k}(T, j) = \mathbf{\Psi}_k(T, j)\mathbf{K}(j) \tag{9.105}$$

where $\mathbf{\Psi}_k(T, j)$ is the transition matrix for the Kalman filter and $\mathbf{K}(j)$ is the Kalman gain matrix. From (9.66) and (9.67):

$$\mathbf{K}(T) = \mathbf{H}(0; T, 0) \tag{9.106}$$

thus

$$\mathbf{K}(t) = \mathbf{H}(0; t, 0)$$

and from the z-transform initial value theorem:

$$\mathbf{K}(t) = \lim_{|z| \to \infty} \mathbf{H}(z; t, 0) \quad \text{for all } t > 0. \tag{9.107}$$

From the second condition for optimality both filters must be unbiased and the transition matrices satisfy:

$$\mathbf{\Psi}(t; 0) = \mathbf{\Psi}_k(t, 0) \quad \text{for all } t > 0. \tag{9.108}$$

From equation (9.104) the weighting sequence for the time-invariant filter is the same as that for the Kalman filter at time T. Similarly, the matrix $\mathbf{\Psi}(T; 0)$ is the same as the Kalman filter transition matrix as time T. It follows that given inputs from the same observation signal and $\mathbf{m}_0$ the two filters will give identical state estimates at time T. The Kalman filter gain matrix can be calculated directly from the transfer-function matrix from the time-invariant filter using equation (9.107). This matrix and the system equations completely define the Kalman solution to the filtering problem.

Advantages of digital filtering techniques

Digital filtering techniques are now employed in many important areas such as speech signal processing, digital telephony and communications, image processing, radar and sonar systems, biomedicine, space research, geoscience signal processing and several others (Cappellini *et al.* [37]). The digital representation of signals has several advantages:

(i) Transmission and manipulation of digital signals can be achieved with a greater accuracy and better noise immunity than analogue signals.
(ii) Ease of implementing digital filters using modern computing devices.
(iii) Decreasing cost of hardware due to the production of very reliable large-scale integrated (LSI) circuits.

9.3.8. Continuous-time s-domain results

The time-invariant estimators for continuous-time systems can also be written in an algorithmic form. Using an equivalent notation to the preceding z-domain analysis the s-domain algorithms are as follows.

Algorithm 9.1. *Time-invariant predictor* (τ_0 = prediction interval)

(i) Calculate $\mathbf{Y}(s)$ using a spectral factorization algorithm which gives $\mathbf{P}_\infty$ at an intermediate stage.
(ii) Calculate $\mathbf{M}(s)=\mathbf{Y}(s)^{-1}\mathbf{C}\mathbf{\Phi}(s)$ and $\mathbf{M}(t)=\mathscr{L}_{\mathrm{I}}^{-1}(\mathbf{M}(s))$.
(iii) Calculate $\mathbf{S}(T)=\int_0^T\mathbf{M}^T(T-\tau)\mathbf{M}(T-\tau)\,d\tau$ and $\mathbf{I}_1(T,0)=\int_0^T\mathbf{\Phi}(\tau)\mathbf{K}\mathbf{R}^{1/2}\mathbf{M}(T-\tau)\,d\tau$.
(iv) Evaluate $\mathbf{L}(s,T)=\int_0^\infty\mathbf{M}^T(T-t)e^{-st}\,dt$.
(v) Calculate $\mathbf{\Psi}(T;0)=(\mathbf{\Phi}(T)-\mathbf{\Psi}_1(t,0))(\mathbf{I}_n+(\mathbf{\Sigma}_0-\mathbf{P}_\infty)\mathbf{S}(T))^{-1}$.
(vi) Obtain $\mathbf{H}(s;T,\tau_0)=\mathbf{\Phi}(\tau_0)(\mathbf{\Phi}(s)\mathbf{K}\mathbf{R}^{1/2}$
$+\mathbf{\Psi}(T;0)(\mathbf{\Sigma}_0-\mathbf{P}_\infty)\mathbf{L}(s,T))\mathbf{Y}(s)^{-1}$. □

Algorithm 9.2. *Time-invariant filter and smoothing filter* (τ_1 = smoothing delay)

(i) Calculate $\mathbf{Y}(s)$ and $\mathbf{P}_\infty$.
(ii) Calculate $\mathbf{M}(s)=\mathbf{Y}(s)^{-1}\mathbf{C}\mathbf{\Phi}(s)$ and $\mathbf{m}(t)$.
(iii) Calculate $\mathbf{S}(T)=\int_0^T\mathbf{M}^T(T-\tau)\mathbf{M}(T-\tau)\,d\tau$ and
$\mathbf{I}_1(T,\tau_1)=\int_0^{\tau_1}\mathbf{P}_\infty\mathbf{M}^T(\tau_1-\tau)\mathbf{M}(T-\tau)\,d\tau$
$+\int_{\tau_1}^T\mathbf{\Phi}(\tau-\tau_1)\mathbf{K}\mathbf{R}^{1/2}\mathbf{M}(T-\tau)\,d\tau$.
(iv) Evaluate $\mathbf{L}(s,\tau_1)=\int_0^{\tau_1}\mathbf{M}^T(\tau_1-T)e^{-st}\,dt$ and $\mathbf{L}(s,T)$ (neglecting terms in e^{-sT} which are effective outside the interval).

(v) Evaluate $\mathbf{\Psi}(T;\tau_1) = (\mathbf{\Phi}(T_1) - \mathbf{I}_1(T;\tau_1))(\mathbf{I} + (\mathbf{\Sigma}_0 - \mathbf{P}_\infty)\mathbf{S}(T))^{-1}$.

(vi) Obtain $\mathbf{H}(s;T;\tau_1) = (\mathbf{P}_\infty \mathbf{L}(s,\tau_1) + \mathbf{\Phi}(s)\mathbf{K}\mathbf{R}^{1/2}\, e^{-s\tau_1}$
$+ \mathbf{\Psi}(T;\tau_1)(\mathbf{\Sigma}_0 - \mathbf{P}_\infty)\mathbf{L}(s,T))\mathbf{Y}(s)^{-1}$. □

Moving window estimation property

The impulse response of the time-invariant estimator is zero for $\tau \geq T$. In the continuous time-case this is a consequence of the e^{-sT} terms in the expression for the estimator transfer-function matrix. These operator terms represent time delays which cannot easily be implemented in continuous-time systems. To overcome this problem the continuous-time estimator is only used to give state estimates within the initerval $[0, T]$. Any e^{-sT} terms in the transfer-function solution for the estimator may therefore be neglected since they do not affect the state estimates within this interval. This implementation problem does not arise in discrete-time systems and the finite memory property can be exploited. The state estimate at any time $t \geq T$ will then depend upon the observations in a moving window $[t-T, t]$.

Simulation results

The finite-interval filter or smoother can be used outside the interval $[0, T]$ as shown in Fig. 9.20. In this case, the optimal estimate is obtained at the end of each T seconds interval. The signal generating process for this example was a single integrator and the noise covariances were set at unity. A typical impulse-response for the continuous-time smoothing filter is shown in Fig. 9.21. Further results, for this example, are given in Grimble [31].

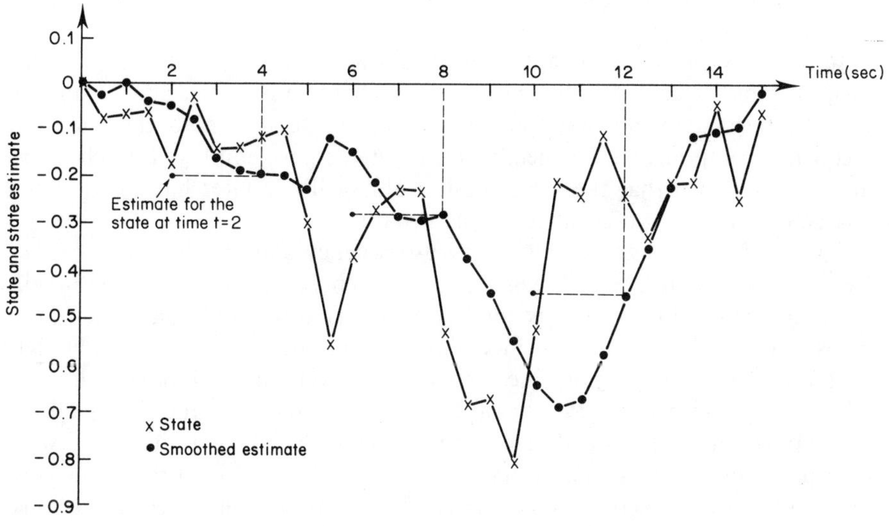

Fig. 9.20. State and smoothed estimate ($q = r = 1$, $T = 4$, $T_1 = 2$, $\Sigma = 1$)

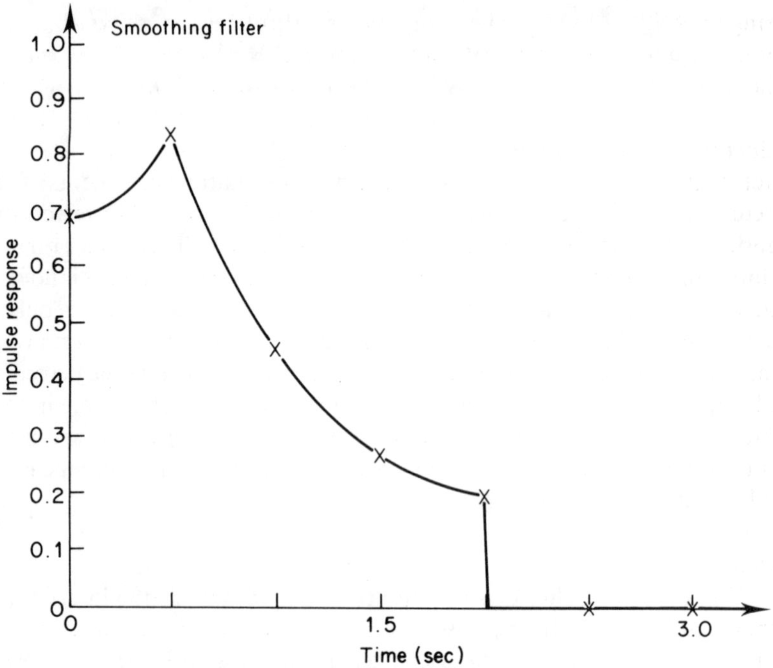

Fig. 9.21. Smoother impulse response

9.4. CONCLUSIONS

In the early part of the chapter, the discussions centred upon a range of industrial filtering problems having the same characteristic of a low-frequency controllable system and an output corrupted by noise of dominantly sinusoidal frequency components. Classically, such systems have employed notch filters and it was shown that the steady-state Kalman–Bucy filter has some of the characteristics of a notch filter in this situation.

The conditions under which the zero-frequency gain of the filter is unity were discussed and it was shown that the presence of an output integrator in the process model leads to this condition. It was also noted that some of the zeros of the optimal filter coincided with the poles of the output noise model. This result is not surprising since zeros have a well-known blocking action.

A relationship between the forward path gain of the filter and the signal-to-signal plus noise ratio was derived. It was found that the gain increased with the square root of the signal to signal-plus-noise ratio. Many of the frequency-domain properties of the Kalman–Bucy filter were demonstrated most easily using the equivalent Wiener filter. There are probably many other industrial filtering problems which have a common structure and for which general conclusions, regarding the performance of optimal filters, may be drawn.

The later sections of the chapter described a time-invariant filter which will give the same state estimate, at some fixed time T, as that which would be obtained from a time-varying Kalman–Bucy filter. The state estimates within the interval $[0, T]$ can have a lower error variance than is obtained from a discrete Wiener filter. The filter may be obtained in s- or z-transfer function, weighting sequence or difference equation forms. If the filter is to be used within the finite interval $[0, T]$ only, the filter is almost as easy to implement as the infinite-time (constant gain) Kalman filter. The number of calculations at each sample point can therefore be reduced compared with the time-varying Kalman filter. Related time-invariant smoothing filters and predictors have also been introduced to produce a new class of linear estimators.

The time-invariant estimator is particularly useful when the minimum variance estimate is required at one point in time. Such an application arises in fixed interval prediction where the state estimate is required at the end of an interval of specified length T. In this case, the Kalman–Bucy and time-invariant predictors give identical estimates of the future plant states at time $t = T$. The implementation of the continuous-time Kalman–Bucy estimator is complicated by the analogue system which generates the time-varying gain matrix and the multipliers which are necessary. However, the time invariant estimator is in transfer-function matrix form and may be implemented using standard techniques.

The main disadvantage of the s-and z-domain estimators has been the difficulty of calculating the transfer-function matrices. The results presented in Sections 9.2.6 and 9.2.7 enable this calculation procedure to be simplified and formalized. This procedure involves the solution of the algebraic Riccati equation for which there are well established numerical techniques. Note that this is not an on-line calculation and there is no requirement that the algorithm should be numerically efficient. If the time-invariant filter is used for $t > T$ the filter behaves as a finite memory filter and only uses the previous T seconds worth of observations. This property follows from the fact that the filter weighting sequence is zero for $t \geq T$ which in turn is a result of the z^{-T} terms in the filter transfer-function matrix. These terms are not involved when the filter is used within the interval $[0, T]$. The difference equation form of the filter includes an additional term in the delayed observations $z(t - T)$ for operation outside the fixed interval. Limited memory filters have been used to reduce problems with divergence and the time-invariant filter provides one method of generating such a filter with well-defined properties (Jazwinski [35]).

The Kalman type estimators have proven their worth in many industrial and other applications and the need for new estimators may be questioned. However, it is only in the control literature that the Kalman–Bucy filter has achieved such a dominant position. The Levinson filter was developed in 1947 and is now used in fast algorithms for solving Toeplitz equations in many applications areas including geophysics and speech processing (Friedlander *et al.* [36]). Thus, there could be restricted role for the new *s*- and *z*-domain estimators.

9.5. PROBLEMS

1. *Discrete-time filtering problem.* Consider the following discrete-time scalar system:

$$x(n+1) = x(n) + \omega(n)$$

$$z(n) = x(n) + v(n).$$

Assume that the noise processes and $x(0)$ are independent. The random signals $\omega(n)$ and $v(n)$ are zero mean and have covariances $q = 25$ and $r = 15$, respectively. The initial state vector is assumed to have a zero-mean and a covariance $\sigma_0^2 = 100$. Calculate the optimal time-invariant filter which will give a minimum mean-square estimation error at time at $N = 50$. Show that,

(a) $$\Psi(N;0) = \alpha^N/1 - \frac{\alpha}{1+\alpha}(1-\alpha^{2N}) + \sigma_0^2 \frac{\alpha}{r}\frac{(1-\alpha^{2N})}{(1-\alpha^2)},$$

(b) $$H(z^{-1}; N, 0) = \frac{\alpha^{1/2}}{r}\frac{\beta z}{z-\alpha}(1 - \Psi(N;0)z^{-N})$$

$$+ \Psi(N;0)\left(\left(\frac{\alpha}{r}\right)^{1/2}\beta - \frac{\sigma_0^2}{r}\frac{\alpha z(z-1)}{(1-\alpha z)(z-\alpha)}\frac{(\alpha^N - z^{-N})}{r}\right),$$

(c) $H(n; N, 0) =$

$$\left(\left(\frac{\alpha}{r}\right)^{1/2}\beta\alpha^n + \Psi(N;0)\left(\left(\frac{\alpha}{r}\right)^{1/2}\beta - \frac{\sigma_0^2}{r}\right)\frac{(-\alpha^{N-n} + \alpha^{n+N+1})}{1+\alpha}\right)U(n)$$

$$+ \phi(N;0)\left(-\left(\frac{\alpha}{r}\right)^{1/2}\beta\alpha^{n-N} + \left(\left(\frac{\alpha}{r}\right)^{1/2}\beta - \frac{\sigma_0^2}{r}\right)\right.$$

$$\left.\left(\frac{(\alpha^{N-n} + \alpha^{n+N+1})}{1+\alpha}\right)\right)U(n-N),$$

and verify that $H(n; N, 0) = 0$ for all $n \geq N$. Obtain an expression for the

Kalman filter gain matrix and show that $K(1) = 0.8929$, $K(2) = 0.7191$, $k(3) = 0.7046$, $K(4) = 0.7034$.

2. *Finite-interval filter.* Consider the following single integrator plant:

$$\dot{x} = \omega$$

$$z = x + v, \qquad E\{x(0)\} = 0$$

where the zero-mean white-noise signals ω and v are uncorrelated and $Q = q$, $R = r$, and $\Sigma_0 = \sigma_0^2$. Show the time-invariant finite-interval filter is given as:

$$H(s; T, 0) = \frac{\beta}{s+\beta} + \Psi(T; 0)\,\frac{(\sigma_0^2 - \beta)}{r}\,\frac{s\,e^{-\beta T}}{(s^2 - \beta^2)}$$

where

$$\Psi(T; 0) = \frac{\beta}{(\beta \cosh \beta T + (\sigma_0^2/r) \sinh \beta T)}.$$

3. *Finite-interval smoother.* Consider the system described in the previous example and assume that the time-invariant smoothing filter is required. Show the smoothing filter may be expressed in the form:

$$H(s\colon T, \tau_1) = \left(e^{-\beta\tau_1} - \frac{\beta}{s}\,e^{-s\tau_1} + \Psi(T; \tau_1)((\sigma_0^2/r\beta) - 1)\,e^{-\beta T}\right)\frac{s\beta}{(s^2 - \beta^2)}$$

where

$$\Psi(T; \tau_1) = \frac{\cosh \beta\tau_1}{(\cosh \beta T + (\sigma_0^2/\sqrt{(qr)}) \sinh \beta T)}.$$

4. The transfer-function matrix for the Kalman filter can be obtained from (8.164) as:

$$\mathbf{H}(z^{-1}\colon \infty, 0) = (\mathbf{I} - \mathbf{A}z^{-1} + \mathbf{KCA}z^{-1})^{-1}\mathbf{K}.$$

It follows that the Kalman filter gain can be found from the transfer function using:

$$\mathbf{K} = \lim_{|z| \to \infty} \mathbf{H}(z^{-1}; \infty, 0).$$

Using the results of Corollary 9.2 show that the filter gain is given by (8.173).

9.6. REFERENCES

[1] Kalman, R. E. and Bucy. R. S. 1961. New Results in linear filtering and prediction theory, *Trans ASME, Journal of Basic Engineering*, **83**, 95–108, March.

[2] Kalman, R. E. 1961. New methods in Wiener filtering theory, *Proc. of the Symposium on Engineering Applications of Random Function Theory and Probability,* 270–388.

[3] MacFarlane, A. G. J. 1971. Return-difference matrix properties for optimal stationary Kalman–Bucy filter, *Proc. IEE,* **118**(2), 373–376.

[4] Grimble, M. J. 1977. The optimal return-difference matrix, *Proc. IEE,* **124** (2), 167–168.

[5] Wiener, N. 1949. *Extrapolation, Interpolation and Smoothing of Stationary Time Series, with Engineering Applications,* Technology Press, Cambridge, Mass.

[6] Astrom, K. J. 1970. *Introduction to Stochastic Control Theory,* p. 20. Academic Press.

[7] McGarty, T. P. 1974. *Stochastic Systems and State Estimation,* p. 99, Wiley.

[8] MacFarlane, A. G. J. and Karcanias, N. 1976. Poles and zeros of linear multivariable systems: a survey of the algebraic, geometric and complex-variable theory, Research Report, CUED/F-Control/TR 105, Cambridge University, Dept of Engineering.

[9] Kouvaritakis, B. and MacFarlane, A. G. J. 1976. Geometric approach to analysis and synthesis of system zeros, Part 1: Square systems, *Int. J. Control,* **23**(2), 149–166.

[10] Davison, E. J. and Wang, S. H. 1974. Properties and calculation of transmission zeros of linear multivariable systems, *Automatica,* **10**, 643–658.

[11] Davison, E. J. and Wang, S. H. 1976. Remark on multiple transmission zeros of a system, *Automatica,* **12**, 195.

[12] Wang, S. H. and Davison, E. J. 1978. An algorithm for the calculation of transmission zeros of the system (C, A, B, D) using high-gain output feedback, *IEEE Trans. on Auto. Cont.,* **AC-23**(4), 738–741, August.

[13] Arcasoy, C. C. 1971. Return-difference matrix properties for optimal stationary discrete Kalman filter, *Proc. IEEE,* **118**(12), 1831–1834. December.

[14] Grimble, M. J. 1978. Factorization procedure for a class of rational matrices, *Int. J. Control,* **28**(1), 105–111.

[15] Hutchinson, C. E. 1961. An example of the equivalence of the Kalman and Wiener filters, *IEEE Trans. on Auto. Control.,* **AC-6**, 324, April.

[16] Singer, R. A. and Frost, P. A. 1969. On the relative performance of the Kalman and Wiener filters, *IEEE Trans. of Auto. Control,* **AC-14**, 390–394, August.

[17] Kucera, V. 1978. Transfer-function solution of the Kalman–Bucy Filtering Problem, *Kybernetika,* **14**(2), 110–121.

[18] Grimble, M. J. 1980. Finite-time linear filtering, prediction and control, Inst. of Mathematics and its Applications Conf. on the Analysis and Optimization of Stochastic Systems, University of Oxford, September, 1978, edited by O. L. R. Jacobs, M. H. A. Davis, M. A. H. Dempster, C. J. Harris, and P. C. Parks, Academic Press.

[19] Barrett, J. F. 1977. Construction of Wiener filters using the return-difference, *Int. J. Control,* **26**(5), 797–803.

[20] Shaked, U. 1976. A general transfer function approach to linear stationary

filtering and steady-state optimal control problems, *Int. J. Control,* **24**(6), 741–770.
[21] Youla, D. C. 1961. On the factorization of rational matrices, *IEEE Trans. Information Theory,* **IT-7**, 172–189, July.
[22] Davis, M. G. 1963. Factoring the spectral matrix, *IEEE Trans Automatic Control,* 296–305.
[23] Grimble, M. J. 1978. Relationship between Kalman and notch filters used in dynamic ship positioning systems, *Electronics Letters,* **14**(13), 399–400, June.
[24] Bryson, A. E. 1978. Kalman filter divergence and aircraft motion estimators, *J. Guidance and Control,* **1**(1), 71–79, Jan–Feb.
[25] Healey, M. 1974 Optimal control theories in the design of aircraft stability augmentation and control systems, *Measurement and Control,* **7**, 387–395, Oct.
[26] Grimble, M. J. 1980, *Development of Models for Shape Control System Design in Modelling of Dynamical Systems,* Edited by H. Nicholson, published by Peter Peregrinus Limited.
[27] Grimble, M. J. and Patton, R. J. 1981. The design of shape-control systems using stochastic optimal control theory, IEE Conference on Control and Its Applications, University of Warwick, March.
[28] Grimble, M. J. 1978. Solution of the linear estimation problem in the *s*-domain, *Proc. IEE.* **125**(6), 541–549, June.
[29] Grimble, M. J. 1979. Solution of the Kalman filtering problem for stationary noise and finite data records, *Int. J. Systems Science,* **10**(2), 177–196.
[30] Grimble, M. J. 1980. A finite-time linear filter for discrete-time systems, *Int. J. Control,* **31**(3), 413–432.
[31] Grimble, M. J. 1980. A new finite-time linear smoothing filter, *Int. J. Systems Sci.,* **11**(10), 1189–1212.
[32] Grimble M. J. 1980. 'New finite-time linear filters and smoothers, *Proc. of Third IMA Conf. on Contr. Theory,* pp. 547–565, University of Sheffield, Academic Press.
[33] O'Reilly, J. 1982. A finite-time linear filter for discrete-time systems in singular case, *Int. J. Systems Sci.,* **13**(3), 257–263.
[34] Shaked, U. and Priel, B. 1982. Explicit solutions to the linear optimal estimation problem of discrete time invariant linear processes, *Int. J. Control,* **36**(5), 725–745.
[35] Jazwinski, A. H. 1970. *Stochastic Processes and Filtering Theory,* Academic Press.
[36] Eriedlander, B., Kailath, T., Morf, M. and Ljunf, L. 1978. Extended Levinson and Chandrasekhar equations for general discrete time linear estimation problems, *IEEE Trans. on Auto. Cont.,* **AC-23**(4), 653–659.
[37] Cappellini, V., Constantinides, A. G. and Emiliani, P. 1978. *Digital Filters and their Applications,* Academic Press, London.
[38] Moore, J. B., Hotz, A. F., and Gangsaas, D. 1982. Adaptived flutter suppression as a complement to LQG based aircraft control, 6th IFAC Symposium, Arlington, Virginia USA on Identification and System Parameter Estimation, Vol. 2, pp. 1604–1609.
[39] Kouvaritakis, B. and MacFarlane, A. G. J. 1976. Geometric approach to analysis and synthesis of system zeros, Part 2: Non-square systems, *Int. J. Control,* **23**(2), 67–181.
[40] Kwakernaak, H. and Sivan, R. 1972. *Linear Optimal Control,* John Wiley and Sons.

[41] Grimble, M. J. and Åström, K. J. 1987. Frequency-domain properties of Kalman filters, *Int. J. Control*, **45**(3), 907–925.

[42] Grimble, M. J. 1980, A reduced order optimal controller for discrete-time stochastic systems, *Proc. IEE,* Pt. D, **2**, 55–63.

CHAPTER 10

Time-domain Analysis of the Optimal Stochastic Linear Control Problem

10.1. INTRODUCTION

No treatise on optimal control and stochastic estimation is complete without a treatment of the separation principle of stochastic optimal control theory. In this chapter, two types of analyses are presented, one for the continuous time-domain formulation (Section 10.2) and one for discrete-time systems (Section 10.3). The level of the two presentations is slightly different so that one gives full rigour to the derivation (the continuous-time case) and the other (the discrete-time case) subsumes a number of preliminary assumptions to simplify the analysis.

The final section of the chapter (Section 10.4) considers a number of ways in which the linear stochastic regulator problem has been extended and discusses the usefulness of these results.

10.2. THE CONTINUOUS-TIME LINEAR STOCHASTIC REGULATOR PROBLEM

The analysis of the optimal linear stochastic regulator problem for systems described in the continuous time domain has followed two routes in the literature. The seminal treatment (Wonham [1]) of the separation principle was based upon dynamic programming. Davis [2] also adopted this scheme in a presentation which was designed to be as close as possible to the analysis used in non-linear stochastic problems.

Alternative derivations (albeit for more general inifinite dimensional systems) are the functional analytic treatments reported by Brooks [3], Balakrishnan [4], Lindquist [5], Bensoussan and Viot [6] and Curtain and Ichikawa [7]. The approach of this section follows the functional analytic route.

10.2.1. Abstract probability theory

The triple $(\Omega, \mathscr{F}, P)$ denotes a fixed probability space where Ω is the space of sample points, $\omega \in \Omega$, $\mathscr{F}$ is a non-empty class of sets in Ω, forming a σ-field of subsets of Ω and P is a probability measure, mapping $E \in \mathscr{F}$ to $\mathbb{R}$.

DEFINITION 10.1. *σ-field and sub σ-field.* (i) Let $\mathscr{F}$ be a non-empty class of sets in Ω, then $\mathscr{F}$ forms a σ-field if:

(a) $\Omega \in \mathscr{F}$.

(b) $\mathscr{F}$ forms a ring.

(c) If $\{E_1, E_2, \ldots\} \in \mathscr{F}$ then $E_i^c = \Omega - E_i \in \mathscr{F}$ for all i and $\{\cup_{i-1}^{\infty} E_i\} \in \mathscr{F}$.

(ii) Let γ be a class of sets in $\mathscr{F}$, then $\mathscr{F}_1 = \sigma(\gamma)$ is called a sub σ-field of $\mathscr{F}$, if it is the smallest σ-field of sets which contains γ. Since $\mathscr{F}$ is a σ-field then $\mathscr{F}_1 \subset \mathscr{F}$. □

DEFINITION 10.2. *Probability measure.* The probability measure $P\colon \mathscr{F} \to \mathbb{R}$ satisfies:

(i) $P(E) \geq 0$ $(E \in \mathscr{F})$.

(ii) $P(\Omega) = 1$ $(\Omega \in \mathscr{F})$.

(iii) $P(\cup_i E_i) = \Sigma_i P(E_i)$ provided $E_i \in \mathscr{F}$ and $E_i \cap E_j = \phi$ for all i, j. □

The triple $(\Omega, \mathscr{F}, P)$ is regarded as a complete probability space in the sense that the probability measure is complete with respect to sets and subsets of measure zero (see Doob [8]). The concept of a Borel field is described in considerable detail by Loeve [9] and the definition given here is due to Naylor and Sell [10].

DEFINITION 10.3. *Borel fields.* (i) The Borel field in $\mathbb{R}$ is defined to be the smallest σ-field of sets from $\mathbb{R}$ which contains all the open intervals (a, b).

(ii) The Borel field into $\mathbb{R}^n$ is the smallest σ-field of sets from $\mathbb{R}^n$ which contains all the open rectangles:

$$\{(x_1, \ldots, x_n);\ a_i < x_i < b_i,\ i = 1, 2, \ldots, n\}.$$ □

The Borel field is denoted by β (in this text) and the sets in a Borel field are often termed Borel sets (Loeve [9]). The Borel fields of $\mathbb{R}$, and $\mathbb{R}^n$ are of importance in the construction of the Lebesgue measure and the Lebesgue integral; see the discussion in Naylor and Sell [9]. In this context where $\Omega \in \mathscr{F}$, a measurable set is simply a set in the σ-field $\mathscr{F}$, and a real valued function $f\colon (\Omega, \mathscr{F}) \to (\mathbb{R}, \beta)$ is measurable if and only if $f^{-1}(B) \in \mathscr{F}$ for every Borel set $B \in \beta$ (see Taylor [11]).

DEFINITION 10.4. *Random variable.* A random variable x is a real valued measurable function $x: (\Omega, \mathscr{F}) \rightarrow (\mathbb{R}, \beta)$ where $\mathscr{F}$ is a σ-field of the non-empty class of events in Ω and β is the Borel σ-field of $\mathbb{R}$. □

Thus for a random variable x, every Borel set $B \in \beta$ corresponds to an event $x^{-1}(B)$ such that $x^{-1}(B) \in \mathscr{F}$. If $\mathscr{F}_x$ is the minimal σ-field in $\mathscr{F}$ which contains all the events of the form $x^{-1}(B)$ $(B \in \beta)$ then $\mathscr{F}_x$ is called the sub σ-field generated by the random variable x. (The inverse map $x^{-1}(\cdot)$ is taken to mean $x^{-1}(B) \triangleq \{\omega \in \Omega$ such that $x(\omega) \in B\}$.)

DEFINITION 10.5. *Closure of* $\mathscr{F}_1$. Let $\mathscr{F}_1$ denote a collection of sets in $\mathscr{F}$, then $\mathscr{F}_1$ is closed under finite intersections if $F_a \cap F_b \in \mathscr{F}_1$ for all $F_a, F_b \in \mathscr{F}_1$. □

DEFINITION 10.6. *Independent events.* Let $(\Omega, \mathscr{F}, P)$ be a complete probability space and F_1, F_2 two collection of events such that $F_1, F_2 \in \mathscr{F}$ then the collection of events are said to be independent if:

$$P(F_{1i} \cap F_{2j}) = P(F_{1i})P(F_{2j}) \quad \text{for all } F_{1i} \in F_1 \quad \text{and} \quad F_{2j} \in F_2. \qquad \square$$

Theorem 10.1. ***Independent random variables.*** *(i) If collections $F_1, F_2 \in \mathscr{F}$ are each closed under finite intersections and are independent then $\mathscr{F}_1 \triangleq \sigma(F_1)$ and $\mathscr{F}_2 \triangleq \sigma(F_2)$ are independent sub σ-fields of $\mathscr{F}$.*

(ii) Random variables x, and y are independent if they generate independent sub σ-fields $\mathscr{F}_x$ and $\mathscr{F}_y$.

Proof (Loeve [9]). □

A result of importance in the sequel for the classes of admissible feedback is as follows.

Theorem 10.2. ***Induced measurable function.*** *Let $(\Omega, \mathscr{F}, P)$ be a complete probability space, and let $y: (\Omega, \mathscr{F}_y) \rightarrow (\Omega_1, \mathscr{G})$ be a random variable where $\mathscr{F}_y \triangleq \sigma(y^{-1}(\mathscr{G}))$. Let x be a random variable, then x is $\mathscr{F}_y$-measurable if and only if there is a measurable function $g: \Omega_1 \rightarrow \mathbb{R}$ such that $x(\omega) = g(y(\omega))$.*

Proof (Davis [2]). □

A very useful diagram is given by Davis [2] which appears here as Fig. 10.1. In the applications to follow, random variable y represents an observation which, via the sub σ-field $\mathscr{F}_y$, generates a measurable subspace $(\Omega, \mathscr{F}_y) \subset (\Omega, \mathscr{F})$. The methods require a feedback to be constructed to be $\mathscr{F}_y$-measurable. Theorem 10.2 demonstrates that for such a map there exists a (non-linear) dependence of the form $x(\omega) = g(y(\omega))$.

To complete the discussion of the abstract probability theory required in the sequel, the concept of conditional expectation is considered more fully.

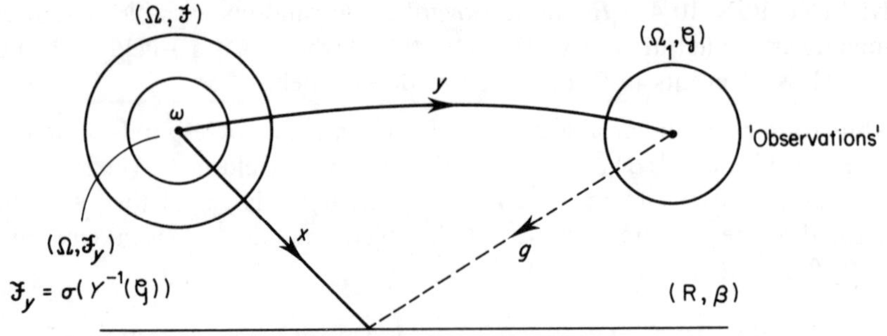

Fig. 10.1. Induced measurable functions and random variables (after Davis [2])

Introduce the specific probability spaces:

$$\text{(i)} \quad L_1(\Omega, \mathscr{F}, P) = \{x\colon E\{\,|x|\,\} \triangleq \int_\Omega |x(\omega)|\, dP(\omega) < \infty\}$$

and

$$\text{(ii)} \quad L_2(\Omega, \mathscr{F}, P) = \{x\colon E\{x^2(\omega)\} \triangleq \int_\Omega x^2\, dP(\omega) < \infty\}.$$

The latter is the familiar Hilbert space of random variables (Chapter 8) where $H = L_2(\Omega, \mathscr{F}, P)$ and its subspace $H^0 \triangleq \{x \in H$ such that $E\{x\} = 0\}$. Other useful subspace relations include:

(i) $H = L_2(\Omega, \mathscr{F}, P) \subset L_1(\Omega, \mathscr{F}, P)$

and

(ii) if $\mathscr{F}_1$ is a sub σ-field of $\mathscr{F}$ define:

$$H(\mathscr{F}_1) \triangleq \{y \in H \text{ such that } y \text{ is } \mathscr{F}_1\text{-measurable}\}$$

then

$$H(\mathscr{F}_1) = L_2(\Omega, \mathscr{F}_1, P) \subset H(\mathscr{F}) = L_2(\Omega, \mathscr{F}, P).$$

Hence let $\mathscr{F}_1$ be a sub σ-field of $\mathscr{F}$, and let $x \in L_1(\Omega, \mathscr{F}, P)$ and define $\mu(F_1)$, where $F_1 \in \mathscr{F}_1$, by:

$$\mu(F_1) = \int_{F_1} x(\omega)\; dP(\omega) \qquad F_1 \in \mathscr{F}_1.$$

Thus, $\mu(\cdot)$ defines a measure on $\mathscr{F}_1$, and by the Radon–Nikodym Theorem

(Naylor and Sell [10]) there is a unique random variable $E\{x \mid \mathscr{F}_1\}$ which is $\mathscr{F}_1$-measurable, such that:

$$\mu(\mathscr{F}_1) = \int_{F_1} E\{x \mid \mathscr{F}_1\}\ dP_{\mathscr{F}_1}(\omega) = \int_{F_1} x(\omega)\ dP(\omega) \quad \text{for all } F_1 \in \mathscr{F}_1.$$

The random variable $E\{x \mid \mathscr{F}_1\}$ is called the conditional expectation of x with respect to the sub σ-field $\mathscr{F}_1$. Note that whilst the random variable x is $\mathscr{F}$-measurable (namely $x^{-1}(B) \in \mathscr{F}$ for all $B \in \beta$), the random variable $E\{x \mid \mathscr{F}_1\}$ is measurable with respect to $\mathscr{F}_1$ (namely $(E\{x \mid \mathscr{F}_1\})^{-1}(B) \in \mathscr{F}_1$ for all $B \in \beta$).

DEFINITION 10.7. *Conditional expectation* $E\{x \mid y\}$. Let $x, y \in L_1(\Omega, \mathscr{F}, P)$ and let $\mathscr{F}_y$ be the sub σ-field generated by the random variable y(viz. $\mathscr{F}_y = c(y^{-1}(B))$ for all $B \in \beta$) then:

$$E\{x \mid y\} \triangleq E\{x \mid \mathscr{F}_y\}. \tag{10.1}$$

□

Lemma 10.1. ***Properties of conditional expectation.*** *Let* $x \in L_1(\Omega, \mathscr{F}, P)$ *and let* $\mathscr{F}_1$ *be a sub* σ*-field of* $\mathscr{F}$ *then:*

(i) $E\{x \mid \mathscr{F}_1\}$: $L_1(\Omega, \mathscr{F}, P) \to L_1(\Omega, \mathscr{F}_1, P)$.
(ii) $E\{E\{x \mid \mathscr{F}_1\}\} = E\{x\}$,
(iii) $E\{x \mid \mathscr{F}_1\} = x$ *if* x *is* $\mathscr{F}_1$-measurable,
(iv) $E\{ax + by \mid \mathscr{F}_1\} = aE\{x \mid \mathscr{F}_1\} + bE\{y \mid \mathscr{F}_1\}$,
(v) $E\{x \mid \mathscr{F}_1\} = E\{x\}$ *if* x *is independent of* $\mathscr{F}_1$,
(vi) $E\{yx \mid \mathscr{F}_1\} = yE\{x \mid \mathscr{F}_1\}$ if y is $\mathscr{F}_1$-measurable.
(vii) *If* $\mathscr{F}_2$ *is a sub* σ-field of $\mathscr{F}_1$ *which is a sub* σ*-field of* $\mathscr{F}$ *(viz.* $\mathscr{F}_2 \subset \mathscr{F}_1 \subset \mathscr{F}$) *then:* $E\{(x \mid \mathscr{F}_1) \mid \mathscr{F}_2\} = E\{x \mid \mathscr{F}_2\}$.

Theorem 10.3. ***Orthogonal projection.*** *If* $x \in H = L_2(\Omega, \mathscr{F}, P)$ *and* $\mathscr{F}_1$ *is a sub* σ*-field of* $\mathscr{F}$ *then random variable* $E\{x \mid \mathscr{F}_1\}$ *is the orthogonal projection of* x *onto* $H(\mathscr{F}_1)$ *and so is the least squares best* $\mathscr{F}_1$*-measurable estimator for* x.

Proof (Davis [2]). □

10.2.2. Stochastic processes revisited

The underlying probability space is given by the triple $(\Omega, \mathscr{F}, P)$ and Definition 10.4 gave a random variable to be a real valued measurable function x: $(\Omega, \mathscr{F}) \to (\mathbb{R}, \beta)$. The vector version of this function is $\boldsymbol{x}$: $(\Omega, \mathscr{F}^n) \to (\mathbb{R}^n, \beta^n)$ where the notation $\mathscr{F}^n$, and β^n reflect the multivariable nature of the random variable $\boldsymbol{x}$. A real stochastic process subsumes a time dimension into this definition. Introduce a finite time interval $\mathcal{T}_T = [0, T]$ and let $\beta(T)$ be the σ-field of Borel sets of $\mathcal{T}_T$, then a real stochastic process is the real valued measurable function $\boldsymbol{x}: (\Omega_T, \mathcal{F}_T) \to (\mathbb{R}^n, \beta^n)$ where $\Omega_T \triangleq \mathcal{T}_T \times \Omega$ and $\mathcal{F}_T \triangleq \beta(T) \times \mathcal{F}^n$.

DEFINITION 10.8 *Hilbert spaces and stochastic processes.* Let the underlying probability space be $H = L_2(\Omega, \mathcal{F}, P)$ then

(i) $H^x \triangleq \{\boldsymbol{x}: (\Omega_T, \mathcal{F}_T) \to (\mathbb{R}^n, \beta^n)$ and

$$\int_0^T E\{\langle \boldsymbol{x}(t;\omega), \boldsymbol{x}(t;\omega)\rangle_{E_n}\}\, dt < \infty\} \tag{10.2}$$

(ii) $H^0 \triangleq \{\boldsymbol{x} \in H^x: E\{\boldsymbol{x}\} = \mathbf{o} \in \mathbb{R}^n \text{ for all } t \in [0, T]\}.$ (10.3)

□

Hence for fixed $t \in [0, T]$, $x(\cdot, \omega)$ is a random variable, measurable with respect to $\mathcal{F}_T$ and if ω is fixed, then $\boldsymbol{x}(t: \omega)$ is Lebesgue measurable with respect to $t \in \mathcal{T}_T$.

DEFINITION 10.9. *$\mathcal{F}_t$-adapted process.* Let $\{\mathcal{F}_t\}$ be a family of nondecreasing σ-fields (namely $\mathcal{F}_s \subseteq \mathcal{F}_t \subseteq \mathcal{F}_T$, $s \le t \le T$) then the stochastic process $\boldsymbol{x}(t;\omega)$ is adapted to $\{\mathcal{F}_t\}$, if it is $\mathcal{F}_t$ measurable for every $t \in [0, T]$.

□

It is also useful to recall the definitions of some particular stochastic processes.

DEFINITION 10.10. *Centred (or zero-mean) stationary orthogonal increments process.* A stationary orthogonal increments process $\{\boldsymbol{w}(t), t \in [0, T]\}$, taking values in H^0 is given by:

$$\boldsymbol{w}(t) = \sum_{i=1}^{n} w_i(t)\mathbf{e}_i$$

where $\mathbf{e}_i \in \mathbb{R}^n$ is the usual ith unit basis vector for $\mathbb{R}^n$ and $\{w_i(t), t \in [0, T]\}_{i=1}^n$ are real orthogonal increment processes satisfying:

(i) Zero mean:

$$E\{w_i(t)\} = 0 \quad \text{for all } t \in [0, T\}, \; i = 1, \dots, n.$$

(ii) Orthogonal increments:

$$E\{(w_i(t_4) - w_i(t_3))(w_j(t_2) - w_j(t_1))\} = 0$$

with $0 \le t_1 < t_2 < t_3 < t_4 \le T$
and $(i, j) \in (1, \dots, n) \times (1, \dots, n)$.

(iii) Incremental covariance function:

$$E\{(w(t)-w(s))(w(t)-w(s))^T\}=(t-s)\mathbf{R}_0,\ \mathbf{R}_0=\mathbf{R}_0^T>\mathbf{O}\in\mathbb{R}^{n\times n}.\quad\square$$

DEFINITION 10.11. *Wiener process.* A centred stationary orthogonal increments process is the Wiener process $\{w(t), t\in[0,T]\}$:

$$w(t)=\sum_{i=1}^{n} w_i(t)\mathbf{e}_i$$

where for $i=1,\dots,n$: $\mathbf{e}_i\in\mathbb{R}^n$ is the usual ith unit basis vector and $\{w_i(t)\}$ are mutually independent Wiener processes with incremental covariances $q_i\in\mathbb{R}$.

$\square$

The multivariable Wiener process has independent increments and further details can be found in Chapter 8, see Section 8.2.4. To be able to solve the stochastic differential equation which represents the system description it is necessary to recall the stochastic integral theorem from Chapter 8.

Theorem 10.4. ***Stochastic integrals for orthogonal increments (zero-mean) processes.*** *Let $\{w(t); t\in[0,T]\}$ be a stationary zero-mean orthogonal increments process with covariance matrix:*

$$\text{cov}[w(t),w(\tau)]=\mathbf{R}(t,\tau)=(t\wedge\tau)\mathbf{Q}$$

where $\mathbf{Q}\in\mathbb{R}^{n\times n}$ is a constant diagonal matrix. Let two transformations of $\{w(t); t\in[0,T]\}$ be defined:

$$\left\{x(t), t\in[0,T]: x(t)=\int_0^t \mathbf{M}(\tau)\,dw(\tau),\ \int_0^T m_{ij}^2(\tau)\,d\tau<\infty \right.$$

$$\left. \text{and } i=1,\dots,m;\ j=1,\dots,n\right\}$$

$$\left\{y(t), t\in[0,T]: y(t)=\int_0^t \mathbf{N}(\tau)\,dw(\tau),\ \int_0^T n_{ij}^2(\tau)\,d\tau<\infty \right.$$

$$\left. \text{and } i=1,\dots,r;\ j=1,\dots,n\right\}$$

then:

(*i*) $E\{x(t)\}=\mathbf{o}\in\mathbb{R}^m$ *for all* $t\in[0,T]$

$E\{y(t)\}=\mathbf{o}\in\mathbb{R}^r$ *for all* $t\in[0,T]$, *(ii)* $\operatorname{cov}[\boldsymbol{x}(t), \boldsymbol{y}(\tau)] = \int_0^{t \wedge \tau} \mathbf{M}(u)\mathbf{Q}\mathbf{N}^T(u)\, du \in \mathbb{R}^{m \times r}$,

(iii) *If* $\mathbf{G}(t)$: $[0, T] \to \mathbb{R}^{n \times n}$ *and* $\sum_{i=1}^{n} \sum_{j=1}^{n} \int_0^t g_{ij}^2(\tau)\, d\tau < \infty$

for any $t \in [0, T]$ *then* $\boldsymbol{p}(t) \triangleq \int_0^t \mathbf{G}(\tau)\, d\boldsymbol{w}(\tau)$

satisfies $\{\boldsymbol{p}(s); s \in [0, t]\} \in H_t^w$.

Proof (see Proposition 8.3, Chapter 8). □

Utilizing the stochastic integrals of Theorem 10.4, the solution of the following stochastic differential equation is considered:

$$d\boldsymbol{x}(t, \omega) = \mathbf{A}(t)\boldsymbol{x}(t, \omega)\, dt + \mathbf{B}(t)\boldsymbol{u}(t, \omega)\, dt + \mathbf{D}(t)\, d\boldsymbol{w}(t, \omega)$$

$$\boldsymbol{x}(0) = \boldsymbol{x}_0 \tag{10.4}$$

where

$\mathbf{A}(\cdot) \in \mathbb{R}^{n \times n}$ is continuous on $[0, T]$,
$\mathbf{B}(\cdot) \in \mathbb{R}^{n \times m}$ and $\mathbf{D}(\cdot) \in \mathbb{R}^{n \times l}$ are piecewise continuous,
$u \in H^u$, $\omega \in H^0$ is a centred orthogonal increments process,

and

$x_0 \in L_2^n(\Omega, \mathscr{F}, P)$.

The solution is written in convolution form as:

$$\boldsymbol{x}(t, \omega) = \mathbf{\Phi}(t, 0)\boldsymbol{x}_0 + \int_0^t \mathbf{\Phi}(t, \tau)\mathbf{B}(\tau)\boldsymbol{u}(\tau, \omega)\, dt + \int_0^t \mathbf{\Phi}(t, \tau)\mathbf{D}(\tau)\, d\boldsymbol{w}(\tau, \omega) \tag{10.5}$$

where $\mathbf{\Phi}(\cdot, \cdot) \in \mathbb{R}^{n \times n}$ is the transition matrix associated with the homogeneous differential equation: $\dot{\mathbf{x}}(t) = \mathbf{A}(t)\mathbf{x}(t)$, $t \in [0, T]$. (The properties of the transition matrix can be found in Theorems 1.2, and 1.3 of Chapter 1). Extensions of these ideas to evolution operators and the related stochastic evolution equations can be found in Curtain and Pritchard [12] and Curtain and Ichikawa [7].

10.2.3. The linear stochastic regulator problem

In this section the details of the linear stochastic regulators problem are presented. The two sections which then follow detail the solution procedures for the problems with complete and incomplete observations, culminating in the famed separation principle. Incremental forms for the system, output and observations models are presented first.

System equation

$$d\boldsymbol{x}(t,\omega) = \mathbf{A}(t)\boldsymbol{x}(t,\omega)\ dt + \mathbf{B}(t)\boldsymbol{u}(t,\omega)\ dt + \mathbf{D}(t)\ d\boldsymbol{w}(t,\omega). \tag{10.6}$$

Output equations

$$\boldsymbol{y}(t,\omega) = \mathbf{C}(t)\boldsymbol{x}(t,\omega). \tag{10.7}$$

Observation equation

$$d\boldsymbol{z}(t,\omega) = d\boldsymbol{y}(t,\omega) + d\boldsymbol{v}(t,\omega). \tag{10.8}$$

Measurement noise process

$$d\boldsymbol{v}(t,\omega) = \mathbf{F}(t)\ d\boldsymbol{v}_1(t,\omega). \tag{10.9}$$

System parameters

$\mathbf{A}(\cdot) \in \mathbb{R}^{n\times n}$, assumed continuous on $[0, T]$,

$\mathbf{B}(\cdot) \in \mathbb{R}^{n\times m}$, $\mathbf{C}(\cdot) \in \mathbb{R}^{r\times n}$, $\mathbf{D}(\cdot) \in \mathbb{R}^{n\times l}$ and $\mathbf{F}(\cdot) \in \mathbb{R}^{r\times r}$

are assumed piecewise continuous on interval $[0, T]$.

Process noise characteristics

The process noise $\boldsymbol{w} \triangleq \{\boldsymbol{w}(t,\omega), t \in [0, T]\}$ is a l-vector, centered, stationary orthogonal increments process, hence $\boldsymbol{w} \in H^0$ with

$$\operatorname{cov}[\boldsymbol{w}(t,\omega), \boldsymbol{w}(s,\omega)] = (t \wedge s)\mathbf{Q} \qquad (\mathbf{Q}^T = \mathbf{Q} \geq \mathbf{O} \in \mathbb{R}^{l\times l}).$$

The initial condition: $\boldsymbol{x}_0 \in L_2^n(\Omega, \mathscr{F}, P)$ with $E\{\boldsymbol{x}_0\} = \mathbf{m}_0$,

$$\operatorname{cov}[\boldsymbol{x}_0, \boldsymbol{x}_0] = \mathbf{X}_0 \in \mathbb{R}^{n\times n} \quad \text{and} \quad \boldsymbol{x}_0 \perp H^0.$$

Measurement noise characteristics

The measurement process driving noise

$$\boldsymbol{v}_1 \triangleq \{\boldsymbol{v}_1(t,\omega), t \in [0, T]\} \in H^{v_1} \subseteq H^0\}$$

is assumed to be a r-vector centered stationary orthogonal increments process such that:

$$\operatorname{cov}[\mathbf{v}_1(t,\omega), \mathbf{v}_1(\tau,\omega)] = (t \wedge \tau)\mathbf{I}_r, \qquad t, \tau \in [0, T].$$

Hence the relation (10.9) gives:

$$\mathbf{v}(t,\omega) = \int_0^t \mathbf{F}(\tau)\, d\mathbf{v}_1(\tau,\omega)$$

and (Theorem 10.4) the measurement noise

$$\mathbf{v} \triangleq \{\mathbf{v}(t,\omega),\ t \in [0,T]\} \in H^v \subseteq H^0$$

is a r-vector centred orthogonal increments process such that:

$$\operatorname{cov}[\mathbf{v}(t,\omega), \mathbf{v}(\tau,\omega)] = \int_0^{(t\wedge\tau)} \mathbf{F}(u)\mathbf{F}^T(u)\, du = \mathbf{R}(t\wedge\tau), \qquad t, \tau \in [0, T]$$

where

$$\mathbf{R}(t)\colon [0,T] \to \mathbb{R}^{l\times l},\ \mathbf{R}^T(t) = \mathbf{R}(t) > \mathbf{O} \quad \text{for all } t \in [0,T]$$

and it assumed that

$$\int_0^t \|\mathbf{R}(\tau)\|_E^2\, d\tau < \infty \quad \text{for all } t \in [0,T].$$

Hilbert space of controls

The control process $\mathbf{u}$: $\{\mathbf{u}(t,\omega),\ t \in [0,T]\}$ is assumed to derive from a Hilbert space of control

$$H^u \triangleq \left\{\mathbf{u}\colon (\Omega_T, \mathscr{F}_T) \to (\mathbb{R}^m, \beta^m) \text{ and } \int_0^T E\{\langle \mathbf{u}(t,\omega), \mathbf{u}(t,\omega)\rangle_{E_m}\}\, dt < \infty\right\} \tag{10.10}$$

Hence for fixed $t \in [0,T]$, $\mathbf{u}(t,\omega)$ is a random variable measurable with respect to the σ-field $\mathscr{F}_T$, and if ω is fixed then $\mathbf{u}(t,\omega)$ is Lebesgue measurable on $[0,T]$.

If $\mathbf{u} \in H^u$ then equations (10.6) to (10.9) defines a quadratic mean continuous process, such that $\mathbf{x} \in H^x$ which may be given a convolution form:

$$\mathbf{x}(t,\omega) = \mathbf{\Phi}(t,0)\mathbf{x}_0 + \int_0^t \mathbf{\Phi}(t,\tau)\mathbf{B}(\tau)\mathbf{u}(\tau,\omega)\, d\tau + \int_0^t \mathbf{\Phi}(t,\tau)\mathbf{D}(\tau)\, d\mathbf{w}(\tau,\omega) \tag{10.11}$$

with observation process:

$$z(t,\omega)=\int_0^t \mathbf{C}(\tau)x(\tau,\omega)\;d\tau+\int_0^t dv(\tau,\omega). \tag{10.12}$$

Stochastic regulator cost functional

The cost functional to be optimized is the stochastic version of the finite-time state regulator problem (Chapter 1), namely:

$$J(u)=E\left\{\int_0^T (\langle x(t,\omega),\mathbf{Q}_1(t)x(t,\omega)\rangle_{E_n}+\langle u(t,\omega),\mathbf{R}_1(t)u(t,\omega)\rangle_{E_m})\;dt\right\}$$
$$+E\{\langle x(T,\omega),\mathbf{Q}_1(T)x(T,\omega)\rangle_{E_n}\} \tag{10.13}$$

where the cost-functional weightings satisfy:

$$\mathbf{Q}_1(t)=\mathbf{Q}_1^T(t)\in\mathbb{R}^{n\times n},\qquad \mathbf{Q}_1(t)\geq\mathbf{O}\quad\text{for all } t\in[0,T]$$

$$\mathbf{R}_1(t)=\mathbf{R}_1^T(t)\in\mathbb{R}^{m\times m},\qquad \mathbf{R}_1(t)>\mathbf{O}\quad\text{for all } t\in[0,T].$$

The linear stochastic regulator problem is therefore:

$$\min_{\text{w.r.t. } u\in\mathcal{U}_{ad}\subset H^u}\{J(u)\}$$

and the main problems arise in the specification of $\mathcal{U}_{ad}$, the set of admissible controls. The first analysis to be presented is that for the case of complete state observations, that is, where complete state feedback free of measurement noise may be used.

10.2.4. The stochastic linear regulator with complete state observations

The case of complete state observations is considered first, thus the only noise entering the problem is the process noise. The problem statement of the previous section becomes:

$$\min_{\text{w.r.t. } u\in\mathcal{U}_{ad}\subset H^u}\{J(u)\}$$

where

$$J(u)=E\left\{\int_0^T (\langle x(t,\omega),\mathbf{Q}_1(t)x(t,\omega)\rangle_{E_n}+\langle u(t,\omega),\mathbf{R}_1(t)u(t,\omega)\rangle_{E_m})\;dt\right\}$$
$$+E\{\langle x(T,\omega),\mathbf{Q}_1(T)x(T,\omega)\rangle_{E_n}\}. \tag{10.14}$$

Subject to the process equation:

$$x(t,\omega) = \Phi(t,0)x_0 + \int_0^t \Phi(t,\tau)\mathbf{B}(\tau)u(\tau,\omega)\,d\tau + \int_0^t \Phi(t,\tau)\mathbf{D}(\tau)\,dw(\tau,\omega) \tag{10.15}$$

where w is a l-vector centred stationary orthogonal increments process with

$$w \in H^0 \text{ and } \operatorname{cov}[w(t,\omega), w(s,\omega)] = (t \wedge s)\mathbf{Q}$$

and initial condition $x_0 \in L_2^n(\Omega, \mathscr{F}, P)$ with $E\{x_0\} = \mathbf{m}_0$,

$$\operatorname{cov}[x_0, x_0] = \mathbf{X}_0 \in \mathbb{R}^{n\times n} \quad \text{and} \quad x_0 \perp H^0.$$

Recall that on the time interval $[0,t]$, the restriction of the stochastic process w, satisfies w_t: $(\Omega_T, \mathscr{F}_t^w) \to (\mathbb{R}^l, \beta^l)$ where $\mathscr{F}_t^w$ is the sub σ-field generated by the stochastic process $\{w(s,\omega)\colon 0 \le s \le t \le T\}$. The set of admissible controls $\mathscr{U}_{ad}$ is then selected so that the restriction of the stochastic process is measurable with respect to the sub σ-field $\mathscr{F}_t^w$, viz.: u_t: $(\Omega_T, \mathscr{F}_t^w) \to (\mathbb{R}^m, \beta^m)$. Theorem 10.2 ensures that if u is $\mathscr{F}_t^w$ measurable then there exists a (non-linear) measurable function $\psi: \mathbb{R}^l \to \mathbb{R}^m$ where $u(t,\omega) = \psi(t, w(t,\omega))$. The situation in Fig. 10.2 pertains. Thus, the set of admissible controls which are $\mathscr{F}_t^w$ measurable lie in a subspace $H^{u,w} \subset H^u$, where

$$H^{u,w} \triangleq \left\{u_t\colon (\Omega_T, \mathscr{F}_t^w) \to (\mathbb{R}^m, \beta^m) \text{ and } \int_0^t E\{\langle u(t,\omega), u(t,\omega)\rangle_{E_m}\right\} dt < \infty$$

$$\text{for all } t \in [0,T]\} \subset H^u. \tag{10.16}$$

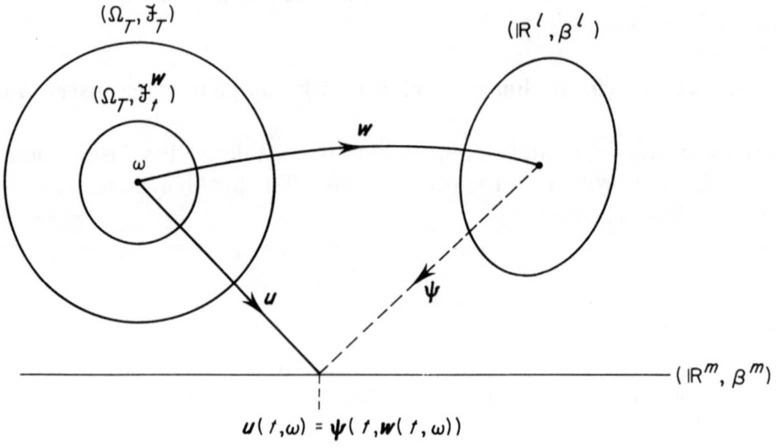

Fig. 10.2. A $\mathscr{F}_t^w$ – measurable control map u: $(\Omega_T, \mathscr{F}_t^w) \to (\mathbb{R}^m, \beta^m)$

To facilitate the proof of the main theorem introduce some important definitions and some linear operators:

(i) *Inner product on H^x:*

$$\langle x, y\rangle_{H^x} \triangleq E\left\{\int_0^T \langle x(t,\omega), y(t,\omega)\rangle_{E_n}\, dt\right\} \tag{10.17}$$

where

$$x, y \in H^x.$$

(ii) *Inner product on $H^{u,w}$:*

$$\langle u, v\rangle_{H^{u,w}} \triangleq E\left\{\int_0^T \langle u(t,\omega), v(t,\omega\rangle_{E_m}\, dt\right\}. \tag{10.18}$$

where

$$u, v \in H^{u,w}.$$

(iii) *Inner product on $H_n \triangleq L_2^n(\Omega, \mathscr{F}, P)$*

$$\langle x, y\rangle_{H_n} \triangleq E\{\langle x(\omega), y(\omega)\rangle_{E_n}\}. \tag{10.19}$$

(iv) *Define W: $H^{u,w} \to H^x$*

where

$$(Wu)(t,\omega) \triangleq \int_0^t \Phi(t,\tau)\mathbf{B}(\tau)u(t,\omega)\, d\tau. \tag{10.20}$$

(v) *Define W_T: $H^{u,w} \to H_n$*

$$(W_T u)(\omega) \triangleq \int_0^T \Phi(T,s)\mathbf{B}(\tau)u(\tau,\omega)\, d\tau. \tag{10.21}$$

Lemma 10.2. *Adjoint of W: $H^{u,w} \to H^x$. Let*

$$W: H^{u,w} \to H^x \quad \textit{where } (Wu)(t,\omega) \triangleq \int_0^t \Phi(t,\tau)\mathbf{B}(\tau)u(\tau,\omega)\, d\tau$$

then

$$W^* x: H^x \to H^{u,w}$$

is given by:

$$(W^*\boldsymbol{x})(t,\omega) = \int_t^T \mathbf{B}^T(t)\boldsymbol{\Phi}^T(\tau,t)E\{\boldsymbol{x}(\tau,\omega)\mid \mathscr{F}_t^w\}\ d\tau. \tag{10.22}$$

Proof.

$$\langle \boldsymbol{x}, W\boldsymbol{u}\rangle_{H^x} \triangleq E\left\{\int_0^T \langle \boldsymbol{x}(t,\omega), \int_0^t \boldsymbol{\Phi}(t,\tau)\mathbf{B}(\tau)\boldsymbol{u}(\tau,\omega)\ d\tau\rangle_{E_n}\ dt\right\}.$$

See Section 1.2.2, Chapter 1, to obtain:

$$= E\left\{\int_0^T \langle \int_t^T \mathbf{B}^T(t)\boldsymbol{\Phi}^T(\tau,t)\boldsymbol{x}(\tau,\omega)\ d\tau, \boldsymbol{u}(t,\omega)\rangle_{E_n}\ dt\right\} \tag{10.23}$$

Define

$$\tilde{\boldsymbol{u}}(t,\omega) \triangleq \int_t^T \mathbf{B}^T(t)\boldsymbol{\Phi}^T(\tau,t)\boldsymbol{x}(\tau,\omega)\ d\tau$$

and recall that $\boldsymbol{u}$ is $\mathscr{F}_t^w$-measurable thus:

$$\begin{aligned} E\{\langle \tilde{\boldsymbol{u}}(t,\omega), \boldsymbol{u}(t,\omega)\rangle_{E_m}\} &= E\{E\{\langle \tilde{\boldsymbol{u}}(t,\omega), \boldsymbol{u}(t,\omega)\rangle_{E_m} \mid \mathscr{F}_t^w\}\} \\ &= E\{\langle E\{\tilde{\boldsymbol{u}}(t,\omega)\mid \mathscr{F}_t^w\}, \boldsymbol{u}(t,\omega)\rangle_{E_m}\}. \end{aligned}$$

Hence from (10.23):

$$\begin{aligned} \langle \boldsymbol{x}, W\boldsymbol{u}\rangle_{H^x} &= E\left\{\int_0^T \langle \int_t^T \mathbf{B}^T(t)\boldsymbol{\Phi}^T(\tau,t)E\{\boldsymbol{x}(\tau,\omega)\mid \mathscr{F}_t^w\}\ d\tau, \boldsymbol{u}(t,\omega)\right\}\ dt\rangle \\ &= \langle W^*\boldsymbol{x}, \boldsymbol{u}\rangle_{H^{u,w}} \end{aligned}$$

and W^*: $H^x \to H^{u,w}$ where

$$(W^*\boldsymbol{x})(t,\omega) \triangleq \int_t^T \mathbf{B}^T(t)\boldsymbol{\Phi}^T(\tau,t)E\{\boldsymbol{x}(\tau,\omega)\mid \mathscr{F}_t^w\}\ d\tau. \qquad \square$$

Lemma 10.3. *Adjoint of* W_T: $H^{u,w} \to H_n$. *Let*

W_T^*: $H^{u,w} \to H_n$ *where* $(W_T\boldsymbol{u})(\omega) = \int_0^T \boldsymbol{\Phi}(T,t)\mathbf{B}(t)\boldsymbol{u}(t,\omega)\ dt$

then

$W_T^*: H_n \to H^{u,w}$ *is given by:*

$$(W_T^* \boldsymbol{x})(t, \omega) = \mathbf{B}^T(t)\mathbf{\Phi}^T(T, t)E\{\boldsymbol{x}(T, \omega) \mid \mathscr{F}_t^w\}.$$

Proof

$$\langle \boldsymbol{x}, W_T \boldsymbol{u} \rangle_{H_n} \triangleq E\left\{\langle \boldsymbol{x}(T, \omega), \int_0^T \mathbf{\Phi}(T, t)\mathbf{B}(t)\boldsymbol{u}(t, \omega)\ dt \rangle_{E_n}\right\}$$

$$= E\left\{\int_0^T \langle \mathbf{B}^T(t)\mathbf{\Phi}^T(T, t)\boldsymbol{x}(T, \omega), \boldsymbol{u}(t, \omega)\rangle_{E_m}\ dt\right\}. \quad (10.24)$$

Define

$$\tilde{\boldsymbol{u}}(t, \omega) = \mathbf{B}^T(t)\mathbf{\Phi}^T(T, t)\boldsymbol{x}(T, \omega)$$

and recall that $\boldsymbol{u}$ is $\mathscr{F}_t^w$ measurable thus:

$$E\{\langle \tilde{\boldsymbol{u}}(t, \omega), \boldsymbol{u}(t, \omega)\rangle_{E_m}\} = E\{E\{\langle \tilde{\boldsymbol{u}}(t, \omega), \boldsymbol{u}(t, \omega)\rangle_{E_m} \mid \mathscr{F}_t^w\}\}$$

$$= E\{\langle E\{\tilde{\boldsymbol{u}}(t, \omega) \mid \mathscr{F}_t^w\}, \boldsymbol{u}(t, \omega)\rangle_{E_m}\}.$$

Hence in (10.24):

$$\langle \boldsymbol{x}, W_T \boldsymbol{u}\rangle_{H_n} = E\left\{\int_0^T \langle \mathbf{B}^T(t)\mathbf{\Phi}^T(T, t)E\{\boldsymbol{x}(T, \omega) \mid \mathscr{F}_t^w\}, \boldsymbol{u}(t, \omega)\rangle_{E_m}\ dt\right\}$$

$$= \langle W_T^* \boldsymbol{x}, \boldsymbol{u}\rangle_{H^{u,w}} \quad \text{and} \quad W_T^*: H_n \to H^{u,w}$$

where

$$(W_T^* \boldsymbol{x})(t, \omega) \triangleq \mathbf{B}^T(t)\mathbf{\Phi}^T(T, t)E\{\boldsymbol{x}(T, \omega) \mid \mathscr{F}_t^w\}.$$

□

The above discussion can be used to transform the linear stochastic regulator problem to a form closely related to the Hilbert space optimization problems of Chapter 1. The problem specified in equations (10.14) and (10.15) become:

$$\min_{\text{w.r.t. } \boldsymbol{u} \in H^{u,w}} \{J(\boldsymbol{u})\} \quad (10.25)$$

where

$$J(\boldsymbol{u}) = \langle \boldsymbol{x}, Q_1 \boldsymbol{x}\rangle_{H^x} + \langle \boldsymbol{u}, R_1 \boldsymbol{u}\rangle_{H^u} + \langle \boldsymbol{x}_T, Q_T \boldsymbol{x}_T\rangle_{H_n} \quad (10.26)$$

subject to the linear system equation (10.15) which may be written in operator form as:

$$\boldsymbol{x} = \boldsymbol{d} + W\boldsymbol{u} \tag{10.27}$$

where

$$\boldsymbol{d} \in H^x,$$

$$(\boldsymbol{d})(t, \omega) = \boldsymbol{\Phi}(t, 0)\boldsymbol{x}_0 + \int_0^t \boldsymbol{\Phi}(t, \tau)\mathbf{D}(\tau)\ d\boldsymbol{w}(\tau, \omega) \tag{10.28}$$

$$W\colon H^{u,w} \to H^x$$

such that

$$(W\boldsymbol{u})(t, \omega) = \int_0^T \boldsymbol{\Phi}(t, \tau)\mathbf{B}(\tau)\boldsymbol{u}(\tau, \omega)\ d\tau \quad \text{and} \quad (t, \omega) \in \mathscr{T}_T \times \Omega. \tag{10.29}$$

Also

$$\boldsymbol{x}_T = \boldsymbol{d}_T + W_T\boldsymbol{u} \tag{10.30}$$

where

$$\boldsymbol{d}_T \in H_n \quad \boldsymbol{d}_T(\omega) = \boldsymbol{d}(T, \omega) \tag{10.31}$$

$$W_T\colon H^{u,w} \to H_n$$

with

$$(W_T\boldsymbol{u})(\omega) = (W\boldsymbol{u})(T, \omega). \tag{10.32}$$

Theorem 10.5. ***Optimal control for complete observations linear stochastic regulator.*** *The optimal control to minimize the quadratic cost functional* (10.26) *over the admissible controls $H^{u,w}$, subject to the linear system constraints* (10.27) *and* (10.30) *is given by:*

$$\boldsymbol{u}^0(t, \omega) = -\mathbf{R}_1^{-1}(t)\mathbf{B}^T(t)[\boldsymbol{\Phi}^T(T, t)\mathbf{Q}_1(T)E\{\boldsymbol{x}^0(T, \omega) \mid \mathscr{F}_t^w\}$$

$$+ \int_t^T \boldsymbol{\Phi}^T(\tau, t)\mathbf{Q}_1(t)E\{\boldsymbol{x}^0(\tau, \omega) \mid \mathscr{F}_t^w\}\ d\tau] \tag{10.33}$$

here $\boldsymbol{x}^0(t, \omega)$ derives from (10.15) *using $\boldsymbol{u}^0(t, \omega)$ prescribed by* (10.33).

Proof

$$\min_{\text{w.r.t. } u \in H^{u,w}} J(u) = \langle x, Q_1 x\rangle_{H^x} + \langle u, R_1 u\rangle_{H^u} + \langle x_T, Q_T x_T\rangle_{H_n}$$

subject to

$$x = d + Wu \text{ and } x_T = d_T + W_T u.$$

Substituting for the linear constraint equations, obtain:

$$J(u) = \langle d + Wu, Q_1(d + Wu)\rangle + \langle u, R_1 u\rangle + \langle d_T + W_T u, Q_T(d_T + W_T u)\rangle.$$

The usual variational necessary and sufficient condition is:

$$W^* Q_1(d + Wu^0) + R_1 u^0 + W_T^* Q_T(d_T + W_T u^0) = o.$$

Define

$$x^0 = d + Wu^0 \quad \text{and} \quad x_T^0 = d_T + W_T u^0,$$

then

$$R_1 u^0 + W_T^* Q_T x_T^0 + W^* Q_1 x^0 = o.$$

Use the adjoint definitions of Lemmas 10.2 and 10.3 to construct a $\mathcal{F}_t^W$-measurement control map $u \in H^{u,w} \subset H^u$ and equation (10.33) follows. □

Theorem 10.6. ***Feedback form of complete observations linear stochastic regulator.*** *The optimal control* (10.33) *may be expressed in the feedback form:*

$$u^0(t, \omega) = -\mathbf{R}_1^{-1}(t)\mathbf{B}^T(t)\mathbf{P}_1(t)x^0(t, \omega) \tag{10.34}$$

here for each $t \in [0, T]$, $\mathbf{P}_1(t) \in \mathbb{R}^{n\times n}$ *satisfies:*

$$\frac{d\mathbf{P}_1}{dt} + \mathbf{A}^T(t)\mathbf{P}_1(t) + \mathbf{P}_1(t)\mathbf{A}(t) - \mathbf{P}_1(t)\mathbf{B}(t)\mathbf{R}_1^{-1}(t)\mathbf{B}^T(t)\mathbf{P}_1(t) + \mathbf{Q}_1(t) = \mathbf{O}$$

and $\mathbf{P}_1(T) = \mathbf{Q}_1(T)$. (10.35)

Proof. Define an adjoint state:

$$\boldsymbol{\lambda}^0(t,\omega) = \int_t^T \boldsymbol{\Phi}^T(\tau,t)\mathbf{Q}_1(\tau)\boldsymbol{x}^0(\tau,\omega)\ d\tau + \boldsymbol{\Phi}^T(T,t)\mathbf{Q}_1(T)\boldsymbol{x}^0(T,\omega) \quad (10.36)$$

then (10.33) becomes:

$$u^0(t,\omega) = -\mathbf{R}_1^{-1}(t)\mathbf{B}^T(t)E\{\boldsymbol{\lambda}^0(t,\omega) \mid \mathscr{F}_t^w\}. \quad (10.37)$$

From equation (10.15) obtain:

$$\boldsymbol{x}^0(\tau,\omega) = \boldsymbol{\Phi}(\tau,t)\boldsymbol{x}^0(t,\omega) - \int_t^T \boldsymbol{\Phi}(\tau,s)\mathbf{B}(s)\mathbf{R}_1^{-1}(s)\mathbf{B}^T(s)E\{\boldsymbol{\lambda}^0(s,\omega) \mid \mathscr{F}_s^w\}\ ds$$

$$+ \int_t^T \boldsymbol{\Phi}(\tau,s)\mathbf{D}(s)\ dw(s,\omega), \qquad (\tau \geq t). \quad (10.38)$$

Introduce the notation:

$$\hat{\boldsymbol{\lambda}}^0(s) \triangleq E\{\boldsymbol{\lambda}^0(s,\omega) \mid \mathscr{F}_s^w\}, \qquad s \in [0, T] \quad (10.39)$$

and

$$\tilde{\boldsymbol{\lambda}}^0(\tau) \triangleq E\{\boldsymbol{\lambda}^0(\tau,\omega) \mid \mathscr{F}_t^w\}, \qquad \tau \geq t \quad (10.40)$$

thus

(i) $\hat{\boldsymbol{\lambda}}^0(t) = \tilde{\boldsymbol{\lambda}}^0(t)$ (10.41)

(ii) $E\{\hat{\boldsymbol{\lambda}}^0(\sigma) \mid \mathscr{F}_t^w\} = \tilde{\boldsymbol{\lambda}}^0(\sigma), \qquad \sigma \geq t.$ (10.42)

From equation (10.38) obtain:

$$\tilde{\boldsymbol{x}}^0(\tau) = \boldsymbol{\Phi}(\tau,t)\boldsymbol{x}^0(t,\omega) - \int_t^T \boldsymbol{\Phi}(\tau,s)\mathbf{B}(s)\mathbf{R}_1^{-1}(s)\mathbf{B}^T(s)\tilde{\boldsymbol{\lambda}}^0(s)\ ds$$

where $t \leq \tau \leq T$. (10.43)

From equation (10.36):

$$\hat{\boldsymbol{\lambda}}^0(t) = \tilde{\boldsymbol{\lambda}}^0(t) = \int_t^T \boldsymbol{\Phi}^T(\tau,t)\mathbf{Q}_1(\tau)\tilde{\boldsymbol{x}}^0(\tau)\ d\tau + \boldsymbol{\Phi}^T(T,t)\mathbf{Q}_1(T)\tilde{\boldsymbol{x}}^0(T). \quad (10.44)$$

Using (10.43) in (10.44) yields:

$$\tilde{\boldsymbol{\lambda}}^0(t) = \left[\int_t^T \boldsymbol{\Phi}^T(\tau, t)\mathbf{Q}_1(\tau)\boldsymbol{\Phi}(\tau, t)\, d\tau + \boldsymbol{\Phi}^T(T, t)\mathbf{Q}_1(T)\boldsymbol{\Phi}(T, t)\right] \boldsymbol{x}^0(t, \omega)$$

$$- \left[\int_t^T \boldsymbol{\Phi}^T(\tau, t)\mathbf{Q}_1(\tau) \int_t^\tau \boldsymbol{\Phi}(\tau, s)\mathbf{B}(s)\mathbf{R}_1^{-1}(s)\mathbf{B}^T(s)\tilde{\boldsymbol{\lambda}}^0(s)\, ds\, d\tau\right.$$

$$\left. + \boldsymbol{\Phi}^T(T, t)\mathbf{Q}_1(T) \int_t^T \boldsymbol{\Phi}(T, \tau)\mathbf{B}(\tau)\mathbf{R}_1^{-1}(\tau)\mathbf{B}^T(\tau)\tilde{\boldsymbol{\lambda}}^0(\tau)\, d\tau\right]. \tag{10.45}$$

This is a linear integral equation similar in form to a Volterra equation of the second kind. The solution of (10.45) may be written (Taylor [11]):

$$\tilde{\boldsymbol{\lambda}}^0(t) = \mathbf{P}_1(t)\boldsymbol{x}^0(t, \omega) \tag{10.46}$$

where $\mathbf{P}_1(t)$ satisfies the usual matrix Riccati equation, so that:

$$\mathbf{P}_1(T) = \mathbf{Q}(T), \qquad \mathbf{P}_1^T(t) = \mathbf{P}_1(t) \geq 0, \qquad \mathbf{P}_1(t) \in \mathbb{R}^{n \times n}$$

and

$$\frac{d\mathbf{P}_1}{dt} + \mathbf{A}^T(t)\mathbf{P}_1(t) + \mathbf{P}_1(t)\mathbf{A}^T(t) - \mathbf{P}_1(t)\mathbf{B}(t)\mathbf{R}_1^{-1}(t)\mathbf{B}^T(t)\mathbf{P}_1(t) + \mathbf{Q}_1(t) = \mathbf{O}.$$

Using (10.41) and (10.46):

$$\tilde{\boldsymbol{\lambda}}^0(t) = \hat{\boldsymbol{\lambda}}^0(t) = E\{\boldsymbol{\lambda}^0(t, \omega) \mid \mathcal{F}_t^w\} = \mathbf{P}_1(t)\boldsymbol{x}^0(t, \omega) \tag{10.47}$$

and hence substituting (10.47) in (10.37) proves (10.34) as desired. □

This proof which is based on that in Curtain and Ichikawa [7] also has strong similarities to the analysis in Section 1.4.2, Chapter 1.

Theorem 10.7. ***Optimal cost for the complete state observations linear stochastic regulator.*** *The value of the optimal cost functional, on the interval* $[t, T]$, *for the complete state observations linear stochastic regulator is given by:*

$$J_t^0 = E\{\langle \boldsymbol{x}^0(t, \omega), \mathbf{P}_1(t)\boldsymbol{x}^0(t, \omega)\rangle_{E_n}\} + \operatorname{trace}\left\{\mathbf{Q}\int_t^T \mathbf{D}^T(s)\mathbf{P}_1(s)\mathbf{D}(s)\, ds\right\} \tag{10.48}$$

where $t \in [0, T]$, $\mathbf{P}_1(t)$ *is the matrix Riccati solution* (*equation* (10.35)) *and*

Q *is the incremental covariance matrix associated with the l-vector centred stationary orthogonal increments process,* $\boldsymbol{w}$.

Proof. From (10.4) the incremental form for the optimal system is:

$$d\boldsymbol{x}^0(t,\omega) = \mathbf{A}(t)\boldsymbol{x}^0(t,\omega)\ dt + \mathbf{B}(t)\boldsymbol{u}^0(t,\omega)\ dt + \mathbf{D}(t)\ d\boldsymbol{w}(t,\omega).$$

Using (10.34) in the form

$$\boldsymbol{u}^0(t,\omega) = -\mathbf{K}_c(t)\boldsymbol{x}^0(t,\omega)$$

where

$$\mathbf{K}_c(t) \triangleq \mathbf{R}_1^{-1}(t)\mathbf{B}^T(t)\mathbf{P}_1(t)$$

yields:

$$d\boldsymbol{x}^0(t,\omega) = (\mathbf{A}(t) - \mathbf{B}(t)\mathbf{K}_c(t))\boldsymbol{x}^0(t,\omega)\ dt + \mathbf{D}(t)\ d\boldsymbol{w}(t,\omega) \qquad (10.49)$$

In convolution form, (10.49) gives:

$$\boldsymbol{x}^0(\tau,\omega) = \boldsymbol{\Psi}(\tau,t)\boldsymbol{x}^0(t,\omega) + \int_t^T \boldsymbol{\Psi}(\tau,s)\mathbf{D}(s)\ d\boldsymbol{w}(s,\omega) \qquad (10.50)$$

where $\boldsymbol{\Psi}$: $[0,T]\times[0,T]\to\mathbb{R}^{n\times n}$ is the transition matrix associated with the optimal closed-loop matrix $\mathbf{A}(t)-\mathbf{B}(t)\mathbf{K}_c(t)$ and $0\le t\le\tau\le T$. The optimal cost-function over the interval $[t, T\}$ is considered:

$$J_t{}^0 = E\left\{\int_t^T (\langle \boldsymbol{x}^0(\tau,\omega), \mathbf{Q}_1(\tau)\boldsymbol{x}^0(\tau,\omega)\rangle_{E_n} + \langle \boldsymbol{u}^0(\tau,\omega), \mathbf{R}_1(\tau)\boldsymbol{u}^0(\tau,\omega)\rangle_{E_m})\ dt\right\}$$
$$+ E\{\langle \boldsymbol{x}_T^0(\omega), \mathbf{Q}_1(T)\boldsymbol{x}_T^0(\omega)\rangle_{E_n}\}.$$

Using equation (10.34), this becomes:

$$J_t^0 = E\left\{\int_t^T \langle \boldsymbol{x}^0(\tau,\omega), \boldsymbol{\varkappa}(\tau)\boldsymbol{x}^0(\tau,\omega)\rangle_{E_n}\ d\tau\right\} + E\{\langle \boldsymbol{x}_T^0(\omega), \mathbf{Q}_1(T)\boldsymbol{x}_T^0(\omega)\rangle_{E_n}\}$$

where

$$\boldsymbol{\varkappa}(\tau) = \mathbf{Q}_1(\tau) + \mathbf{K}_c^T(\tau)\mathbf{R}_1(\tau)\mathbf{K}_c(\tau) \in \mathbb{R}^{n\times n}, \qquad \tau\in[0,T].$$

From (10.50) write:

$$\boldsymbol{x}^0(\tau,\omega) = \boldsymbol{a} + \boldsymbol{b} \quad \text{and} \quad \boldsymbol{x}_T^0(\omega) = \boldsymbol{a}_T + \boldsymbol{b}_T$$

where

$$\left.\begin{aligned} a &\triangleq \boldsymbol{\Psi}(\tau, t)\boldsymbol{x}^0(t, \omega), \qquad & a_T &= \boldsymbol{\Psi}(T, t)\boldsymbol{x}^0(t, \omega) \\ b &\triangleq \int_t^T \boldsymbol{\Psi}(\tau, s)\mathbf{D}(s)\; dw(s, \omega), \qquad & b_T &= \int_t^T \boldsymbol{\Psi}(T, s)\mathbf{D}(s)\; dw(s, \omega). \end{aligned}\right\} \tag{10.51}$$

Thus,

$$J_t^0 = E\left\{\int_t^T (\langle (a+b), \boldsymbol{\varkappa}(\tau)(a+b)\rangle_{E_n}\; d\tau\right\}$$

$$+ E\{\langle (a_T + b_T), \mathbf{Q}_1(T)(a_T + b_T)\rangle_{E_n} = T_1 + T_2 + T_3 \tag{10.52}$$

where

$$T_1 = E\left\{\int_t^T \langle a, \boldsymbol{\varkappa}(\tau) a\rangle_{E_n}\; d\tau + \langle a_T, \mathbf{Q}_1(T) a_T\rangle_{E_n}\right\} \tag{10.53}$$

$$T_2 = 2E\left\{\int_t^T \langle a, \boldsymbol{\varkappa}(\tau) b\rangle_{E_n}\; d\tau + 2\langle a_T, \mathbf{Q}_1(T) b_T\rangle_{E_n}\right\} \tag{10.54}$$

and

$$T_3 = E\left\{\int_t^T \langle b, \boldsymbol{\varkappa}(\tau) b\rangle_{E_n}\; d\tau + \langle b_T, \mathbf{Q}_1(T) b_T\rangle_{E_n}\right\}. \tag{10.55}$$

Each term T_1, T_2, T_3 is examined in turn. Equations (10.53) and (10.51) yield:

$$\begin{aligned} T_1 &= E\left\{\int_t^T \langle \boldsymbol{\Psi}(\tau, t)\boldsymbol{x}^0(t, \omega), \boldsymbol{\varkappa}(\tau)\boldsymbol{\Psi}(\tau, t)\boldsymbol{x}^0(t, \omega)\rangle_{E_n}\; d\tau\right\} \\ &\quad + E\{\langle \boldsymbol{\Psi}(T, t)\boldsymbol{x}^0(t, \omega), \mathbf{Q}_1(T)\boldsymbol{\Psi}(T, t)\boldsymbol{x}^0(t, \omega)\rangle_{E_n}\} \\ &= E\{\langle \boldsymbol{x}^0(t, \omega), \mathbf{P}_1(t)\boldsymbol{x}^0(t, \omega)\rangle_{E_n}\} \end{aligned} \tag{10.56}$$

where

$$\mathbf{P}_1(t) = \int_t^T \boldsymbol{\Psi}^T(\tau, t)\boldsymbol{\varkappa}(\tau)\boldsymbol{\Psi}(\mathrm{T}, t)\; d\tau + \boldsymbol{\Psi}^T(t, t)\mathbf{Q}_1(T)\boldsymbol{\Psi}(T, t).$$

(See Section 1.4.2, Chapter 1 for verification that $\mathbf{P}_1(t)$ solves the matrix Riccati equation (10.35).)

Equations (10.54) and (10.51) yield:

$$T_2 = 2E\left\{\int_t^T \langle \boldsymbol{\Psi}(\tau, t)\boldsymbol{x}^0(t, \omega), \boldsymbol{\varkappa}(\tau) \int_t^\tau \boldsymbol{\Psi}(\tau, s)\mathbf{D}(s)\ dw(s, \omega)\rangle_{E_n}\ d\tau\right\}$$

$$+ 2E\left\{\langle \boldsymbol{\Psi}(T, t)\boldsymbol{x}^0(t, \omega), \mathbf{Q}_1(T) \int_t^T \boldsymbol{\Psi}(T, s)\mathbf{D}(s)\ dw(s, \omega)\rangle_{E_n}\right\}.$$

Since w is a centred process and $\boldsymbol{x}^0(t, \omega)$ is $\mathscr{F}_t^w$-measurable,

$$T_2 = 0 \text{ (use properties (ii) and (vi) Lemma 10.1).} \tag{10.57}$$

Equations (10.55) and (10.51) yield:

$$T_3 = E\left\{\int_t^T \left\langle \int_t^\tau \boldsymbol{\Psi}(\tau, s)\mathbf{D}(s)\ dw(s, \omega), \boldsymbol{\varkappa}(\tau) \int_t^\tau \boldsymbol{\Psi}(\tau, s)\mathbf{D}(s)\ dw(s, \omega)\right\rangle_{E_n} d\tau\right\}$$

$$+ E\left\{\left\langle \int_t^T \boldsymbol{\Psi}(T, s)\mathbf{D}(s)\ dw(s, \omega), \mathbf{Q}_1(T) \int_t^T \boldsymbol{\Psi}(T, s)\mathbf{D}(s)\ dw(s, \omega)\right\rangle_{E_n}\right\}.$$

Use the trace identity to obtain:

$$T_3 = E\left\{\int_t^T \operatorname{trace}\left(\boldsymbol{\varkappa}(\tau)\left(\int_t^\tau \boldsymbol{\Psi}(\tau, s)\mathbf{D}(s)\ d\omega(s, \omega)\right)\left(\int_t^\tau \boldsymbol{\Psi}(\tau, s)\mathbf{D}(s)\ d\omega(s, \omega)\right)^T\right) d\tau\right\}$$

$$+ E\left\{\operatorname{trace}\left(\mathbf{Q}_1(T)\left(\int_t^T \boldsymbol{\Psi}(T, s)\mathbf{D}(s)\ d\omega(s, \omega)\right)\left(\int_t^T \boldsymbol{\Psi}(T, s)\mathbf{D}(s)\ d\omega(s, \omega)\right)^T\right)\right\}.$$

Note that w is a centred stationary orthogonal increments process with incremental covariance matrix $\mathbf{Q}$. Apply Theorem 10.4 to obtain:

$$T_3 = \int_t^T \operatorname{trace}\left(\boldsymbol{\varkappa}(\tau) \int_t^\tau \boldsymbol{\Psi}(\tau, s)\mathbf{D}(s)\mathbf{Q}\mathbf{D}^T(s)\boldsymbol{\Psi}^T(\tau, s)\ ds\right) d\tau$$

$$+ \operatorname{trace}\left(\mathbf{Q}_1(T) \int_t^T \boldsymbol{\Psi}(T, s)\mathbf{D}(s)\mathbf{Q}\mathbf{D}^T(s)\boldsymbol{\Psi}^T(T, s)\ ds\right)$$

$$= \operatorname{trace}\left(\mathbf{Q}\int_t^T \mathbf{D}^T(s)\left(\int_s^T \boldsymbol{\Psi}^T(\tau, s)\boldsymbol{\varkappa}(\tau)\boldsymbol{\Psi}(\tau, s)\ d\tau + \boldsymbol{\Psi}^T(T, s)\mathbf{Q}_1(T)\boldsymbol{\Psi}(T, s)\right)\mathbf{D}(s)\ ds\right)$$

$$T_3 = \operatorname{trace}\left(\mathbf{Q}\int_t^T \mathbf{D}^T(s)\mathbf{P}_1(s)\mathbf{D}(s)\ ds\right). \tag{10.58}$$

Finally collate (10.52), with (10.56), (10.57) and (10.58) to obtain (10.48) as required □

Corollary 10.1. ***Optimal cost value on* $[0, T]$.** *If the initial statistics of the state vector satisfy:* $E\{x_0\} = \mathbf{m}_0$, $E\{x_0 x_0^T\} = \mathbf{X}_0 + \mathbf{m}_0\mathbf{m}_0^T$, *then:*

$$J_0^0 = \langle \mathbf{m}_0, \mathbf{P}_1(0)\mathbf{m}_0 \rangle_{E_n} + \mathrm{trace}(\mathbf{P}_1(0)\mathbf{X}_0) + \mathrm{trace}\left\{\mathbf{Q} \int_0^T \mathbf{D}^T(s)\mathbf{P}_1(s)\mathbf{D}(s)\; ds\right\}. \tag{10.59}$$

The above cost expression (10.59) demonstrates how additional terms are added to the deterministic optimal cost value to account for the stochastic setting. One term, trace $(\mathbf{P}_1(0)\mathbf{X}_0)$, arises from the uncertainty in the knowledge of the initial condition and the final trace term reflects the introduction of the process noise. The interesting feature of Theorem 10.6 is the retention of the deterministic optimal control feedback strategy despite the stochastic disturbances which are present.

The theorems of this section are required for the derivation of the linear stochastic regulator with incomplete observations. Indeed, the route followed in the next section is to use the analysis for the Kalman filter of Chapter 8 and reduce the incomplete observations problem to one which can be solved by the application of the above results. The concept of an innovations process plays an important role in this derivation.

10.2.5. The stochastic linear regulator with incomplete observations

The statement of the stochastic linear regulator with incomplete observations may now be given as:

$$\min_{\text{w.r.t. } u \in \mathscr{U}_{\text{ad}} \subset H^u} \{J(u)\} \tag{10.60}$$

where $J(u)$ is given by (10.13):

$$J(u) = E\left\{\int_0^T \langle x(t,\omega), \mathbf{Q}_1(t)x(t,\omega)\rangle_{E_n} + \langle u(t,\omega), \mathbf{R}_1(t)u(t,\omega)\rangle_{E_m}\; dt\right\}$$
$$+ E\{\langle x(T,\omega), \mathbf{Q}_1(T)x(T,\omega)\rangle_{E_n}\} \tag{10.61}$$

and the linear system constraints in convolution form are as follows.

State equation

$$x(t,\omega) = \mathbf{\Phi}(t,0)x_0 + \int_0^t \mathbf{\Phi}(t,\tau)\mathbf{B}(\tau)u(\tau,\omega)\; d\tau + \int_0^t \mathbf{\Phi}(t,\tau)\mathbf{D}(\tau)\; dw(\tau,\omega). \tag{10.62}$$

Observation equation

$$z(t, \omega) = \int_0^t \mathbf{C}(\tau)\boldsymbol{x}(\tau, \omega)\, d\tau + \boldsymbol{v}(t, \omega). \tag{10.63}$$

Measurement process

$$\boldsymbol{v}(t, \omega) = \int_0^t \mathbf{F}(\tau)\, d\boldsymbol{v}_1(\tau, \omega). \tag{10.64}$$

where

(i) $\{\boldsymbol{w}: (\Omega_T, \mathscr{F}^w) \to (\mathbb{R}^l, \beta^l)\}$ is a Wiener process, namely $\boldsymbol{w}$ is a stationary centred orthogonal increments process whose increments are independent Gaussian random variables, and

$$\operatorname{cov}[\boldsymbol{w}(t, \omega), \boldsymbol{w}(\tau, \omega)] = (t \wedge \tau)\mathbf{Q} \in \mathbb{R}^{l \times l} \tag{10.65}$$

(ii) $\boldsymbol{x}_0$ is Gaussian, $\boldsymbol{x}_0 \in L_2^n(\Omega, \mathscr{F}, P)$

(iii) $\{\boldsymbol{v}_1: (\Omega_T, \mathscr{F}^{v_1}) \to (\mathbb{R}^r, \beta^r)\}$ is a Wiener process

where

$$\operatorname{cov}[\boldsymbol{v}_1(t, \omega), \boldsymbol{v}_1(\tau, \omega)] = (t \wedge \tau)\mathbf{I}_r. \tag{10.66}$$

(iv) $\boldsymbol{v}: (\Omega_T, \mathscr{F}^v) \to (\mathbb{R}^r, \beta^r)\}$ is a stationary centred orthogonal increments process whose incremental covariance is:

$$\operatorname{cov}[\boldsymbol{v}(t, \omega), \boldsymbol{v}(\tau, \omega)] = \int_0^{(t \wedge \tau)} \mathbf{F}(u)\mathbf{F}^T(u)\, du = \mathbf{R}(t \wedge \tau) \in \mathbb{R}^{r \times r} \tag{10.67}$$

where

$$\mathbf{R}\colon [0, T] \to \mathbb{R}^r, \mathbf{R}^T(t) = \mathbf{R}(t) > \mathbf{O}$$

and

$$\int_0^t \| \mathbf{R}(\tau) \|_E^2\, d\tau < \infty \quad \text{for all } t \in [0, T].$$

(v)
$$\boldsymbol{x}_0 \perp H^w \perp H^{v_1}. \tag{10.68}$$

Let $\mathscr{F}_t^z$ be the σ-field generated by the restriction of the stochastic process

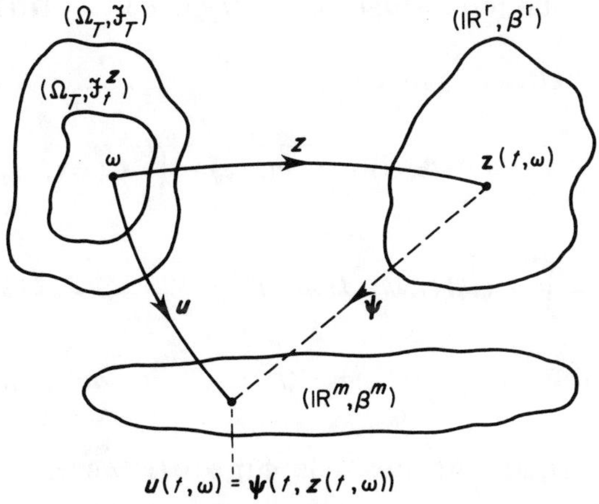

Fig. 10.3. Relationship for u $\mathcal{F}_t^z$-measurable

v to the time interval $[0, t]$. In selecting the control process $u \in H^u$ interest centres on processes which are $\mathcal{F}_t^z$ measurable, hence $u \in H^{u,z}$. Thus using Theorem 10.2 there exist a measurable map

$$\psi \colon (\mathbb{R}^r, \beta^r) \to (\mathbb{R}^m \beta^m)$$

such that $u(t, \omega) = \psi(t, z(t, \omega))$ as a control law (see Fig. 10.3). For any $u \in H^{u,z}$ define:

$$x_u(t, \omega) = \int_0^t \mathbf{\Phi}(t, \tau)\mathbf{B}(\tau)u(\tau, \omega)\, d\tau \tag{10.69}$$

and

$$y_u(t, \omega) = \int_0^t \mathbf{C}(\tau)x_u(\tau, \omega)\, d\tau. \tag{10.70}$$

Introduce two new stochastic processes:

$$\xi(t, \omega) \triangleq x(t, \omega) - x_u(t, \omega) \tag{10.71}$$

$$= \mathbf{\Phi}(t, 0)x_0 + \int_0^t \mathbf{\Phi}(t, \tau)\mathbf{D}(\tau)\, dw(\tau, \omega).$$

(The incremental form is $d\boldsymbol{\xi}(t,\omega)=\mathbf{A}(t)\boldsymbol{\xi}(t,\omega)\;dt+\mathbf{D}(t)\;dw(t,\omega)$.)

$$\boldsymbol{\eta}(t,\omega)\triangleq z(t,\omega)-y_u(t,\omega) \tag{10.72}$$

$$=\int_0^t \mathbf{C}(\tau)\boldsymbol{x}(\tau,\omega)\;d\tau+\boldsymbol{v}(t,\omega)-\int_0^t \mathbf{C}(\tau)\boldsymbol{x}_u(\tau,\omega)\;d\tau$$

$$=\int_0^t \mathbf{C}(\tau)\boldsymbol{\xi}(\tau,\omega)\;d\tau+\boldsymbol{v}(t,\omega).$$

(The incremental form is $d\boldsymbol{\eta}(t,\omega)=\mathbf{C}(t)\boldsymbol{\xi}(t,\omega)\;dt+d\boldsymbol{v}(t,\omega)$.) Collating:

$$\boldsymbol{\xi}(t,\omega)=\boldsymbol{\Phi}(t,0)\boldsymbol{x}_0+\int_0^t \boldsymbol{\Phi}(t,\tau)\mathbf{D}(\tau)\;dw(\tau,\omega) \tag{10.73}$$

$$\boldsymbol{\eta}(t,\omega)=\int_0^t \mathbf{C}(\tau)\boldsymbol{\xi}(\tau,\omega)\;d\tau+\boldsymbol{v}(t,\omega). \tag{10.74}$$

Consider the problem of estimating $\boldsymbol{\xi}(t,\omega)$ given the observation process $\boldsymbol{\eta}(t,\omega)$. Denote the σ-field generated by the process $\boldsymbol{\eta}(\tau,\omega)$ by $\mathscr{F}_t^\eta$. Define this estimate as:

$$\hat{\boldsymbol{\xi}}(t,\omega)=E\{\boldsymbol{\xi}(t,\omega)\mid\mathscr{F}_t^\eta\}. \tag{10.75}$$

Suppose now that $\boldsymbol{u}_t$: $(\Omega_T,\mathscr{F}_t^z)\to(\mathbb{R}^m,\beta^m)$ is $\mathscr{F}_t^\eta$-measurable then

$$\hat{\boldsymbol{\xi}}(t,\omega)=E\{(\boldsymbol{x}(t,\omega)-\boldsymbol{x}_u(t,\omega))\mid\mathscr{F}_t^\eta\}$$

$$\hat{\boldsymbol{\xi}}(t,\omega)=\hat{\boldsymbol{x}}(t,\omega)-\boldsymbol{x}_u(t,\omega) \tag{10.76}$$

where

$$\hat{\boldsymbol{x}}(t,\omega)\triangleq E\{\boldsymbol{x}(t,\omega)\mid\mathscr{F}_t^\eta\}.$$

Differencing equations (10.71) and (10.76) yields:

$$\boldsymbol{e}(t,\omega)\triangleq\boldsymbol{\xi}(t,\omega)-\hat{\boldsymbol{\xi}}(t,\omega)=\boldsymbol{x}(t,\omega)-\hat{\boldsymbol{x}}(t,\omega) \tag{10.77}$$

where

$$E\{\boldsymbol{e}(t,\omega)\mid\mathscr{F}_t^\eta\}=\mathbf{o}. \tag{10.78}$$

The problem of minimizing $E\{\langle\boldsymbol{e}(t,\omega),\boldsymbol{e}(t,\omega)\rangle_{E_n}\}$, subject to the linear system and observation equations (10.73) and (10.74), is solved by the Kalman Filter Theorems of Chapter 8. In particular, the optimal filter is just (10.75), since by Definition 10.7:

$$\hat{\boldsymbol{\xi}}(t,\omega) \triangleq E\{\boldsymbol{\xi}(t,\omega) \mid \mathscr{F}_t^{\eta}\} = E\{\boldsymbol{\xi}(t,\omega) \mid \{\boldsymbol{\eta}(s,\omega), 0 \le s \le t\}\}$$

where

$$\hat{\boldsymbol{\xi}}(t,\omega) = \boldsymbol{\Phi}(t,0)\boldsymbol{x}_0 + \int_0^t \boldsymbol{\Phi}(t,\tau)\mathbf{K}_f(\tau)\, d\boldsymbol{\nu}(\tau,\omega) \tag{10.79}$$

$$\boldsymbol{\nu}(t,\omega) = \boldsymbol{\eta}(t,\omega) - \int_0^t \mathbf{C}(\tau)\hat{\boldsymbol{\xi}}(\tau,\omega)\, d\tau \tag{10.80}$$

with

$$\mathbf{K}_f(t) = \mathbf{P}(t)\mathbf{C}(t)\mathbf{R}^{-1}(t) \in \mathbb{R}^{n\times n}, \qquad t \in [0, T]$$

$$-\frac{d\mathbf{P}(t)}{dt} + \mathbf{A}(t)\mathbf{P}(t) + \mathbf{P}(t)\mathbf{A}^T(t) - \mathbf{P}(t)\mathbf{C}^T(t)\mathbf{R}^{-1}(t)\mathbf{C}(t)\mathbf{P}(t)$$

$$+ \mathbf{D}(t)\mathbf{Q}\mathbf{D}^T(t) = \mathbf{O}$$

$$\mathbf{P}^T(t) = \mathbf{P}(t) \ge \mathbf{O} \in \mathbb{R}^{n\times n} \quad \text{and} \quad \mathbf{P}(0) = \operatorname{cov}[\boldsymbol{x}_0, \boldsymbol{x}_0].$$

(The noise process statistics of $\boldsymbol{w}$ and $\boldsymbol{v}$ enter via the matrices $\mathbf{Q}$ and $\mathbf{R}$; see (10.65) and (10.67).) Note that the Gaussian assumption on $\boldsymbol{w}$ and $\boldsymbol{x}_0$ imply that

(i) $\hat{\boldsymbol{\xi}}(t,\omega)$ is the best global estimate of $\boldsymbol{\xi}(t,\omega)$, based on the observation process (10.74).

(ii) $\boldsymbol{e}(t,\omega)$ is independent of $\mathscr{F}_t^{\eta}$ and $E\{\boldsymbol{e}(t,\omega) \mid \mathscr{F}_t^{\eta}\} = \mathbf{o}$.

(iii) $\boldsymbol{e}(t,\omega)$ is also Gaussian where $\operatorname{cov}[\boldsymbol{e}(t,\omega), \boldsymbol{e}(t,\omega)] = \mathbf{P}(t)$.

Combining equations (10.76) and (10.79) yields:

$$\hat{\boldsymbol{x}}(t,\omega) = \boldsymbol{\Phi}(t,0)\boldsymbol{x}_0 + \int_0^t \boldsymbol{\Phi}(t,\tau)\mathbf{B}(\tau)\boldsymbol{u}(\tau,\omega)\, dt + \int_0^t \boldsymbol{\Phi}(t,\tau)\mathbf{K}_f(\tau)\, d\boldsymbol{\nu}(\tau,\omega) \tag{10.81}$$

where the innovation process satisfies (10.80). Note also:

$$\boldsymbol{\nu}(t,\omega)=\boldsymbol{\eta}(t,\omega)-\int_0^t \mathbf{C}(\tau)\hat{\boldsymbol{\xi}}(\tau,\omega)\ d\tau$$

$$=\boldsymbol{z}(t,\omega)-\boldsymbol{y}_u(t)-\int_0^t \mathbf{C}(\tau)[\hat{\boldsymbol{x}}(\tau,\omega)-\boldsymbol{x}_u(\tau,\omega)]\ d\tau$$

$$\boldsymbol{\nu}(t,\omega)=\boldsymbol{z}(t,\omega)-\int_0^t \mathbf{C}(\tau)\hat{\boldsymbol{x}}(\tau,\omega)\ d\tau. \tag{10.82}$$

From the Kalman filter results is the important property:

$$H_t^{\nu}=H_t^{\eta}$$

$$\mathscr{F}_t^{\nu}=\mathscr{F}_t^{\eta}$$

and (10.81) represents a process which is driven by the innovation process and $\hat{\boldsymbol{x}}(t,\omega)$ is a $\mathscr{F}_t^{\nu}$-measurable process. The stochastic properties of processes $\boldsymbol{\nu}$ and $\boldsymbol{\eta}$, are identical to those of $\boldsymbol{v}$ (see equation (10.67) and the description relating to $\boldsymbol{v}$ therein).

The last question to be resolved is the relationship between σ-fields $\mathscr{F}_t^z$ and $\mathscr{F}_t^{\eta}(=\mathscr{F}_t^{\nu})$. The following theorem due to Balakrishnan settles the issue.

Theorem 10.8. *The σ-fields $\mathscr{F}_t^z$ and $\mathscr{F}_t^{\nu}$.* (*i*) $\mathscr{F}_t^z=\mathscr{F}_t^{\nu}(=\mathscr{F}_t^{\eta})$. (10.83)
(*ii*) *Any control function $\boldsymbol{u}(t,\omega)\in H^u$ which is $\mathscr{F}_t^{\nu}$-measurable is also $\mathscr{F}_t^z$-measurable and $u\in H^{u,\eta}\subset H^{u,z}$.*

Proof (Balakrishnan [4]). □

These ideas can now be utilized to solve the problem of the linear stochastic regulator with incomplete observations. Consider the kernel of the cost function (10.61), property (ii) of Lemma 10.1 and the following facts:

(i) $\boldsymbol{u}(t,\omega)$ is $\mathscr{F}_t^z(=\mathscr{F}_t^{\nu})$ measurable
(ii) $\boldsymbol{x}(t,\omega)=\hat{\boldsymbol{x}}(t,\omega)+\boldsymbol{e}(t,\omega)$ (see (10.77))
(iii) $\hat{\boldsymbol{x}}(t,\omega)$ is $\mathscr{F}_t^z(=\mathscr{F}_t^{\nu})$ measurable (see (10.83)) and by definition)
(iv) $E\{\boldsymbol{e}(t,\omega)\,|\,\mathscr{F}_t^z\}=\mathbf{o}$ for all $t\in[0,T]$ (see (10.78) and (10.83))
(v) $E\{\boldsymbol{e}(t,\omega)\ \boldsymbol{e}^T(t,\omega)\}=\mathbf{P}(t)$

to obtain:

$$E\{\langle x(t,\omega)\mathbf{Q}_1(t)x(t,\omega)\rangle_{E_n} + \langle u(t,\omega), \mathbf{R}_1(t)u(t,\omega)\rangle_{E_m}\}$$

$$= E\{E\{(\langle x(t,\omega), \mathbf{Q}_1(t)x(t,\omega)\rangle_{E_n} + \langle u(t,\omega), \mathbf{R}_1(t)u(t,\omega)\rangle_{E_m}) \mid \mathscr{F}_t^z\}\}$$

$$= E\{\langle \hat{x}(t,\omega), \mathbf{Q}_1(t)\hat{x}(t,\omega)\rangle_{E_n} + \langle u(t,\omega), \mathbf{R}_1(t)u(t,\omega)\rangle_{E_m}\}$$

$$+ \operatorname{trace}(\mathbf{Q}_1(t)\mathbf{P}(t)) \tag{10.84}$$

and

$$E\{\langle x(T,\omega), \mathbf{Q}_1(T)x(T,\omega)\rangle_{E_n}\} = E\{E\{\langle x(T,\omega), \mathbf{Q}_1(T)x(T,\omega)\rangle_{E_n} \mid \mathscr{F}_t^z\}\}$$

$$= E\{\langle \hat{x}(T,\omega), \mathbf{Q}_1(T)\hat{x}(T,\omega)\rangle_{E_n}\} + \operatorname{trace}(\mathbf{Q}_1(T)\mathbf{P}(T)) \tag{10.85}$$

where the evolution of $\hat{x}(t,\omega)$ is given by (10.81).

Lemma 10.4. *Cost-functional equivalence. If* u: $(\Omega_T, \mathscr{F}_t^z) \to (\mathbb{R}^m, \beta^m)$ *then the restricted cost functionals on* $[t, T]$ *satisfy:*

$$J_t \triangleq E\left\{\int_t^T (\langle x(\tau,\omega), \mathbf{Q}_1(\tau)x(\tau,\omega)\rangle_{E_n} + \langle u(\tau,\omega), \mathbf{R}_1(\tau)u(\tau,\omega)\rangle_{E_m})\, d\tau\right\}$$

$$+ E\{\langle x(T,\omega), \mathbf{Q}_1(T)x(T,\omega)\rangle_{E_n}\}$$

$$= J_t^1 + \operatorname{trace}\left(\int_t^T \mathbf{Q}_1(\tau)\mathbf{P}(\tau)\, d\tau\right) + \operatorname{trace}(\mathbf{Q}_1(T)\mathbf{P}(T)) \tag{10.86}$$

where

$$J_t^1 = E\left\{\int_t^T \langle \hat{x}(\tau,\omega), \mathbf{Q}_1(\tau)\hat{x}(\tau,\omega)\rangle_{E_n} + \langle u(\tau,\omega), \mathbf{R}_1(\tau)u(\tau,\omega)\rangle_{E_m})\, d\tau\right.$$

$$\left. + E\{\langle \hat{x}(T,\omega), \mathbf{Q}_1(T)\hat{x}(T,\omega)\}_{E_n}\right\}. \tag{10.87}$$

Proof. Substitute equations (10.84) and (10.85) into (10.61). □

Lemma 10.5. *Intermediate cost-functional optimization. The intermediate cost-functional optimization problem:*

$$\min_{w.r.t.\ u \in H^{u,z}} \{J_0^1(u)\}$$

where $J_0^1(u)$ *is given by* (10.87),

subject to the linear system constraint (10.81), *is solved by*

$$u^0(t, \omega) = -\mathbf{K}_c \hat{x}^0(t, \omega) \tag{10.88}$$

where the gain of the control feedback law is given by:

$$\mathbf{K}_c(t) = \mathbf{R}_1^{-1}(t)\mathbf{B}^T(t)\mathbf{P}_1(t) \tag{10.89}$$

and $\mathbf{P}_1^T(t) = \mathbf{P}_1(t) \geq \mathbf{O} \in R^{n\times n}$ *solves the control matrix Riccati equation given in Theorem* 10.5 *by* (10.35) *with the associated endpoint* $\mathbf{P}_1(T) = \mathbf{Q}_1(T)$. *The optimal cost value over the time interval* $[t, T]$ *is given by:*

$$J_t^1(u^0) = E\{\langle \hat{x}^0(t, \omega), \mathbf{P}_1(t)\hat{x}^0(t, \omega)\rangle_{E_m}\} + \operatorname{trace}\left(\int_t^T \mathbf{F}^T(\tau)\mathbf{K}_f^T(\tau)\mathbf{P}_1(\tau)\mathbf{K}_f(\tau)\mathbf{F}(\tau)\, d\tau\right).$$

Proof. Apply Theorem 10.6 and 10.7 to the optimization of $J_t^1(u)$; a problem which has the same structure of the complete observations problem. □

The analysis of the section may be collated to give a theorem which is essentially of summary nature.

Theorem 10.9. ***Linear stochastic regulator with incomplete observations.*** *The minimization of the cost function given in* (10.61), *subject to the state equation* (10.62), *the observation equation* (10.63) *and the measurement process* (10.64), *is achieved by:*

Feedback control law:

$$u^0(t, \omega) = -\mathbf{K}_c(t)\hat{x}^0(t, \omega)$$

where the control gain

$$\mathbf{K}_c(t) = \mathbf{R}_1^{-1}(t)\mathbf{B}^T(t)\mathbf{P}_1(t) \in \mathbb{R}^{m\times n} \quad \textit{and} \quad \mathbf{P}_1(t) \in \mathbb{R}^{n\times n}$$

satisfies the control matrix Riccati equation:

$$\frac{d}{dt}\mathbf{P}_1(t) + \mathbf{A}^T(t)\mathbf{P}_1(t) + \mathbf{P}_1(t)\mathbf{A}(t) - \mathbf{P}_1(t)\mathbf{B}(t)\mathbf{R}_1^{-1}(t)\mathbf{B}^T(t)\mathbf{P}_1(t) + \mathbf{Q}_1(t) = \mathbf{O}$$

$$\mathbf{P}_1(T) = \mathbf{Q}_1(T) \qquad t \in [0, T]$$

and the stochastic process $\hat{x}^0(t, \omega)$ *is given by:*

$$\hat{x}^0(t, \omega) = \mathbf{\Phi}(t, 0)x_0 + \int_0^t \mathbf{\Phi}(t, \tau)\mathbf{B}(\tau)u^0(\tau, \omega)\, d\tau + \int_0^t \mathbf{\Phi}(t, \tau)\mathbf{K}_f(\tau)\, d\nu^0(\tau, \omega)$$

with innovations process:

$$\nu^0(t, \omega) = z(\tau, \omega) - \int_0^t \mathbf{C}(\tau)\hat{x}^0(\tau, \omega)\, d\tau$$

where

$$\mathbf{K}_f(t) = \mathbf{P}(t)\mathbf{C}^T(t)\mathbf{R}^{-1}(t) \in \mathbb{R}^{n\times r}$$

and $\mathbf{P}(t) \in \mathbb{R}^{n\times n}$ *satisfies the filter matrix Riccati equation:*

$$-\frac{d}{dt}\mathbf{P}(t) + \mathbf{A}(t)\mathbf{P}(t) + \mathbf{P}(t)\mathbf{A}^T(t) - \mathbf{P}(t)\mathbf{C}^T(t)\mathbf{R}^{-1}(t)\mathbf{C}(t)\mathbf{P}(t) + \mathbf{D}(t)\mathbf{Q}\mathbf{D}^T(t) = \mathbf{O}$$

$$\mathbf{P}(0) = \operatorname{cov}[x_0, x_0].$$

The optimal cost value over time interval $[t, T]$ *is given by:*

$$J_t(u^0) = E\{\langle \hat{x}^0(t), \mathbf{P}_1(t)\hat{x}^0(t)\rangle_{E_n}\} + \operatorname{trace}\left(\int_t^T \mathbf{F}^T(\tau)\mathbf{K}_f^T(\tau)\mathbf{P}_1(\tau)\mathbf{K}_f(\tau)\mathbf{F}(\tau)\, d\tau\right)$$

$$+ \operatorname{trace}\left(\int_t^T \mathbf{Q}_1(\tau)\mathbf{P}(\tau)\, d\tau\right) + \operatorname{trace}(\mathbf{Q}_1(T)\mathbf{P}(T)). \quad (10.90)$$

Proof. The requirement for u to be $\mathscr{F}_t^z$-measurable motivates the investigation of the filtering problem. The fundamental result Theorem 8.7 where $\mathscr{F}_t^z = \mathscr{F}_t^\eta = \mathscr{F}_t^\nu$ enables the separation (10.77):

$$x(t, \omega) = \hat{x}(t, \omega) + e(t, \omega)$$

for u which is $\mathscr{F}_t^z$ measurable. The important features are $\hat{x}$ which is $\mathscr{F}_t^z$-measurable, e which is independent of $\mathscr{F}_t^z$ and the innovations representation for the process $\hat{x}$. Lemma 10.4 states a cost-functional equivalence so that the minimization of $J(u)$ becomes one of minimizing J^1 with respect to the $\hat{x}$ process. Lemma 10.5 solves this subproblem by reference to the complete observations theorem to give the main results of this theorem. The optimal cost value is obtained from the relation:

$$J_t^0 = J_t^1(\boldsymbol{u}^0) + \text{trace}\left(\int_t^T \mathbf{Q}_1(\tau)\mathbf{P}(\tau)\ d\tau\right) + \text{trace}(\mathbf{Q}_1(T)\mathbf{P}(T))$$

where $J_t^1(\boldsymbol{u}^0)$ obtains from Lemma 10.5. □

It is useful to give the results of Theorem 10.9 a more succinct presentation in incremental form:

System: $d\boldsymbol{x}^0(t) = \mathbf{A}(t)\boldsymbol{x}^0(t,\omega)\ dt + \mathbf{B}(t)\boldsymbol{u}^0(t,\omega)\ dt + \mathbf{D}(t)\ dw(t,\omega)$.

Feedback: $\boldsymbol{u}^0(t,\omega) = -\mathbf{K}_c(t)\hat{\boldsymbol{x}}^0(t,\omega)$.

Filter: $d\hat{\boldsymbol{x}}^0(t,\omega) = \mathbf{A}(t)\hat{\boldsymbol{x}}^0(t,\omega) + \mathbf{B}(t)\boldsymbol{u}^0(t,\omega) + \mathbf{K}_f(t)\ d\boldsymbol{\nu}^0(t,\omega)$.

Innovations: $d\boldsymbol{\nu}^0(t,\omega) = dz(t,\omega) - \mathbf{C}(t)\hat{\boldsymbol{x}}^0(t,\omega)$.

Gains: $\mathbf{K}_c = \mathbf{R}_1^{-1}(t)\mathbf{B}^T(t)\mathbf{P}_1(t) \in \mathbb{R}^{m\times n}$ (control)
$\mathbf{K}_f = \mathbf{P}(t)\mathbf{C}^T(t)\mathbf{R}^{-1}(t) \in \mathbb{R}^{n\times r}$ (filter).

These results form a celebrated theorem for the linear stochastic regulator designed to optimise a stochastic quadratic cost functional. There is a separating out of the problem of estimation without altering the framework of the underlying deterministic control problem; this occurs at equation (10.77) and is utilized in (10.84) and subsequently in Lemma 10.4.

Certainty equivalence and separation principles

The terms *certainty equivalence principle* and *separation principle* are often used interchangeably when discussing stochastic control problems. However, in general stochastic control problems there is a subtle difference. The *certainty equivalence principle* holds if the closed-loop control law is the same as the equivalent deterministic optimal control law when all the random variables have been replaced by their expected values. Thus, if the deterministic optimal control law is of the form:

$$\mathbf{u}_d^0(t) = \boldsymbol{\phi}(t, \mathbf{x}(t))$$

and the optimal control law for the stochastic problem is obtained with the same functional dependence, viz.:

$$\boldsymbol{u}_s^0(t) = \boldsymbol{\phi}(t, \hat{\boldsymbol{x}}(t))$$

then the certainty equivalence principle is deemed to hold. Clearly, the certainty equivalence principle holds for the LQG problem just solved. However, this is not the case for many problems, but it is often assumed since this considerably simplifies the computation of a control law. For example, in adaptive control designs, the optimal control law is calculated assuming that the estimates of parameter estimates represents true values. This is discussed further in Chapter 8.

The *separation principle* is weaker than certainty equivalence. It occurs if the complete observations problem gives a closed-loop control law:

$$\mathbf{u}_d^0(t) = \boldsymbol{\phi}(t, \mathbf{x}(t))$$

and the incomplete observations stochastic problem yields a (possibly) different functional dependence:

$$u_s^0(t) = \boldsymbol{\psi}(t, \hat{x}(t)).$$

Hence, separation of the estimation and control problems occurs but equivalence between the control laws $\boldsymbol{\phi}(\cdot\,,\,\cdot)$ and $\boldsymbol{\psi}(\cdot\,,\,\cdot)$ may not. In the LQG case both *separation* of the estimation and control problems and *equivalence* of the control laws occurs. This is a feature of the particular structure of the problem: quadratic cost functional and linear system dynamics. To conclude:

$$\text{certainty equivalence} \rightarrow \text{separation}$$

but not vice versa.

10.2.6. Time-invariant systems and steady-state cost-functionals

The results for infinite time-interval problems are cited without proof for a time-invariant system and stationary disturbance and measurement processes. Wonham [13] has given a definitive account of these results and Davis [2] recapitulates the main argument of the proofs. Consequently, the results are given here for completeness and in much abbreviated form. The time-invariant system and measurement process descriptions are given first:

System equation

$$dx(t, \omega) = \mathbf{A}x(t, \omega)\ dt + \mathbf{B}u(t, \omega)\ dt + \mathbf{D}\ dw(t, \omega) \qquad (10.91)$$

where $\mathbf{A} \in \mathbb{R}^{n\times n}$, $\mathbf{B} \in \mathbb{R}^{n\times m}$, $\mathbf{D} \in \mathbb{R}^{n\times l}$.

Output equation

$$y(t, \omega) = \mathbf{C}x(t, \omega) \qquad (10.92)$$

where $\mathbf{C} \in \mathbb{R}^{r\times n}$.

Observation equation

$$dz(t,\omega) = dy(t,\omega) + dv(t,\omega). \tag{10.93}$$

Process noise characteristics

The process noise w is a l-vector centred stationary orthogonal increments process with:

$$\operatorname{cov}[w(t,\omega), w(\tau,\omega)] = (t \wedge \tau)\mathbf{Q} \tag{10.94}$$

where $\mathbf{Q}^T = \mathbf{Q} \geq \mathbf{O}$, $\mathbf{Q} \in \mathbb{R}^{l \times l}$.

Measurement noise characteristics

The measurement noise v is a r-vector-centred stationary orthogonal increments process with:

$$\operatorname{cov}[v(t,\omega), v(\tau,\omega)] = (t \wedge \tau)\mathbf{R} \tag{10.95}$$

where $\mathbf{R}^T = \mathbf{R} > \mathbf{O}, \mathbf{R} \in \mathbb{R}^{r \times r}$.

The steady-state cost-function is obtained by a limit process, viz.:

$$J = \lim_{T \to \infty} \frac{1}{T} E\left\{ \int_0^T \langle x(t,\omega), \mathbf{Q}_1 x(t,\omega)\rangle_{E_n} + \langle u(t,\omega), \mathbf{R}_1 u(t,\omega)\rangle_{E_m}\, dt \right\} \tag{10.96}$$

where

$$\mathbf{Q}_1^T = \mathbf{Q}_1 \geq \mathbf{O}, \qquad \mathbf{Q}_1 \in \mathbb{R}^{n \times n}$$

and

$$\mathbf{R}_1^T = \mathbf{R}_1 > \mathbf{O}, \qquad \mathbf{R}_1 \in \mathbb{R}^{m \times m}.$$

Lemma 10.6. ***Control matrix Riccati equation-Steady state.*** *Given the finite time-control matrix Riccati equation:*

$$\dot{\mathbf{P}}_1(t) + \mathbf{A}^T\mathbf{P}_1(t) + \mathbf{P}_1(t)\mathbf{A} - \mathbf{P}_1(t)\mathbf{B}\mathbf{R}_1^{-1}\mathbf{B}^T\mathbf{P}_1(t) + \mathbf{Q}_1 = \mathbf{O}$$

$$\mathbf{P}_1(T) = \mathbf{Q}_1(T).$$

If the pair $(\mathbf{A}, \mathbf{B})$ *is stabilizable and the pair* $(\tilde{\mathbf{Q}}_1, \mathbf{A})$ *is detectable* $(\tilde{\mathbf{Q}}_1^T\tilde{\mathbf{Q}}_1 = \mathbf{Q}_1)$ *then there exists a unique positive semidefinite matrix* $\mathbf{P}_1 \in \mathbb{R}^{n \times n}$, *such that:*

(*i*) $\mathbf{P}_1(0) \to \mathbf{P}_1$ *as* $T \to \infty$, (*ii*) $\mathbf{P}_1$ *satisfies* $\mathbf{A}^T\mathbf{P}_1 + \mathbf{P}_1\mathbf{A} - \mathbf{P}_1\mathbf{B}\mathbf{R}_1^{-1}\mathbf{B}^T\mathbf{P}_1 + \mathbf{Q}_1 = \mathbf{O}$,

(*iii*) $\mathbf{A} - \mathbf{B}\mathbf{K}_c$ *is asymptotically stable where* $\mathbf{K}_c = \mathbf{R}_1^{-1}\mathbf{B}^T\mathbf{P}_1 \in \mathbb{R}^{m\times n}$.

Proof (Kwakernaak and Sivan [14]). □

Lemma 10.7. ***Filter matrix Riccati equation-Steady state.*** *Given the finite time-filter matrix Riccati equation:*

$$-\dot{\mathbf{P}}(t) + \mathbf{A}\mathbf{P}(t) + \mathbf{P}(t)\mathbf{A}^T - \mathbf{P}(t)\mathbf{C}^T\mathbf{R}^{-1}\mathbf{C}\mathbf{P}(t) + \mathbf{D}\mathbf{Q}\mathbf{D}^T = \mathbf{O}$$

$$\mathbf{P}(0) = \mathbf{X}_0 > \mathbf{O}.$$

If the pair $(\mathbf{A}^T, \mathbf{C}^T)$ *is stabilizable* ($(\mathbf{C}, \mathbf{A})$ *detectable*) *and the pair* $(\tilde{\mathbf{Q}}^T\mathbf{D}^T, \mathbf{A}^T)$ *is detectable* ($(\mathbf{A}, \mathbf{D}\tilde{\mathbf{Q}})$ *stabilizable*) *where* $\tilde{\mathbf{Q}}\tilde{\mathbf{Q}}^T = \mathbf{Q}$, *then there exists a unique positive semidefinite matrix* $\mathbf{P} \in \mathbb{R}^{n\times n}$ *such that:*

(*i*) $\mathbf{P}(T) \to \mathbf{P}$ *as* $T \to \infty$,

(*ii*) $\mathbf{P}$ *satisfies* $\mathbf{A}\mathbf{P} + \mathbf{P}\mathbf{A}^T - \mathbf{P}\mathbf{C}^T\mathbf{R}^{-1}\mathbf{C}\mathbf{P} + \mathbf{D}\mathbf{Q}\mathbf{D}^T = \mathbf{O}$,

(*iii*) $\mathbf{A}^T - \mathbf{C}^T\mathbf{K}_f^T$ *is asymptotically stable where* $\mathbf{K}_f = \mathbf{P}\mathbf{C}\mathbf{R}^{-1} \in \mathbb{R}^{n\times r}$.

Proof. Dual of Lemma 10.6. □

The above Lemmas give conditions under which the control and filter matrix Riccati equations have well-behaved solutions. They also facilitate the notation adopted in the subsequent theorems.

Theorem 10.10. ***Infinite-time/time-invariant stochastic linear regulator problem with complete state observation.*** *The cost-functional* (10.96), *subject to a system equation of the form* (10.91), *where the system conditions of Lemma* 10.6 *hold, is minimized by the control law:*

$$u^0(t, \omega) = -\mathbf{K}_c x^0(t, \omega)$$

where

$$\mathbf{K}_c = \mathbf{R}_1^{-1}\mathbf{B}^T\mathbf{P}_1 \in \mathbb{R}^{m\times n}.$$

(*This being optimal in a class of linear control laws.*) *The minimal cost is given by:*

$$J = \lim_{T\to\infty} \frac{1}{T} E\left\{\int_0^t \langle x^0(t, \omega), \mathbf{R}_1 x^0(t, \omega)\rangle_{E_n} + \langle u^0(t, \omega), \mathbf{Q}_1 u^0(t, \omega)\rangle_{E_m}\, dt\right\}$$

$$= \text{trace}(\mathbf{Q}\mathbf{D}^T\mathbf{P}_1\mathbf{D}) \quad (10.97)$$

Proof (Davis [2]). □

Theorem 10.11. *Infinite-time/time-invariant stochastic linear regulator problem with incomplete observations. The cost-functional* (10.96), *subject to the system equation* (10.91) *and observation equation* (10.93), *where the system conditions of Lemmas* 10.5 *and* 10.6 *apply, is minimized by the control law:*

$$u^0(t,\omega) = -\mathbf{K}_c\tilde{x}^0(t,\omega)$$

where

$$dx^0(t,\omega) = \mathbf{A}x^0(t,\omega)\ dt + \mathbf{B}u^0(t,\omega)\ dt + \mathbf{D}\ d\omega(t,\omega)$$

and

$$\mathbf{d}\tilde{x}^0(t,\omega) = \mathbf{A}\tilde{x}^0(t,\omega) + \mathbf{B}u^0(t,\omega)\ dt + \mathbf{K}_f(dz^0(t,\omega) - \mathbf{C}\tilde{x}^0(t,\omega))$$

where

$$\mathbf{K}_c = \mathbf{R}_1^{-1}\mathbf{B}^T\mathbf{P}_1 \quad \text{and} \quad \mathbf{K}_f = \mathbf{PCR}^{-1}.$$

The optimal value of the cost-functional is:

$$J^0 = \lim_{T\to\infty}\frac{1}{T}E\left\{\int_0^T \langle x^0(t,\omega), \mathbf{Q}_1x^0(t,\omega)\rangle_{E_n} + \langle u^0(t,\omega), \mathbf{R}_1u^0(t,\omega)\rangle_{E_m}\ dt\right\}$$

$$= \lim_{t\to\infty} E\{\langle x^0(t,\omega), \mathbf{Q}_1x^0(t,\omega)\rangle_{E_n} + \langle u^0(t,\omega), \mathbf{R}_1u^0(t,\omega)\rangle_{E_m}$$

$$= \text{trace}(\mathbf{RK}_f^T\mathbf{P}_1\mathbf{K}_f + \mathbf{Q}_1\mathbf{P}) = \text{trace}(\mathbf{P}_1\mathbf{Q} + \mathbf{PK}_c{}^T\mathbf{R}_1\mathbf{K}_c).$$

Proof (Davis [2]). □

10.3. THE DISCRETE-TIME LINEAR STOCHASTIC REGULATOR PROBLEM

The solution of the linear stochastic regulator problem for discrete-time systems follows the same pattern as that presented in Section 10.2 for continuous time-systems. The first problem to be solved is that for the complete state observation formulation. As in the equivalent continuous time problem of Section 10.2.4, the optimal law gain is exactly that of the deterministic discrete-time optimal linear regulator. The optimal control law gain $\mathbf{K}_c$

depends in this case upon the discrete-time control matrix Riccati difference equation.

Subsequently, the situation where the state is observed via a set of noisy outputs is pursued. The Kalman filter is introduced to provide a state estimate and the innovations form for the filter used to obtain the structure of the complete state observations problem. The demonstration of the results is essentially pragmatic to provide a little light relief from the measure-theoretic concepts and proofs utilized in the previous section.

10.3.1. Quadratic covariance summation and difference equations

It will be necessary in the sequel to evaluate quadratic summations of the form:

$$J_T = E\left\{\sum_{t=0}^{T-1} (\langle x(t), \mathbf{M}(t)x(t)\rangle_{E_n} + \langle x(T), \mathbf{P}_T x(T)\rangle_{E_n})\right\} \tag{10.98}$$

where

$$\mathbf{M}(t) = \mathbf{M}^T(t) > \mathbf{O} \quad \text{for all } t \in [0, T] \text{ and } \mathbf{P}_T \geq \mathbf{O},$$

and the linear system difference equation is:

$$x(t+1) = \mathbf{A}_c(t)x(t) + \mathbf{D}(t)\omega(t), \qquad x(t) \in \mathbb{R}^n \tag{10.99}$$

with white-noise sequence $\omega(t)$ such that $\text{cov}[\omega(t), \omega(k)] = \mathbf{Q}\ \delta_{tk}$ and initial condition $x_0 \in \mathbb{R}^n, E\{x_0\} = \mathbf{o}$, $\text{cov}[x_0, x_0] = \mathbf{P}_0$.

The convolution summation solution for linear difference equation (10.99) may be written as:

$$x(t) = \mathbf{\Phi}(t, 0)x_0 + \sum_{j=0}^{t-1} \mathbf{\Phi}(t, j+1)\mathbf{D}(j)\omega(j) \tag{10.100}$$

where $\mathbf{\Phi}(t, \tau)$ denotes the state-transition matrix for the above discrete system. The text by deRusso, Roy and Close [15] provides an excellent introduction to the representation of discrete systems using state equations and describes methods for calculating $\mathbf{\Phi}(t, 0)$.

DEFINITION 10.12. *Discrete-system state-transition matrix.* The discrete system transition matrix $\mathbf{\Phi}(\cdot, \cdot) \in \mathbb{R}^{n \times n}$ is defined by:

$$\begin{aligned}\mathbf{\Phi}(t, t_0) &= \mathbf{A}(t-1)\mathbf{A}(t-2)\ldots\mathbf{A}(t_0+1)\mathbf{A}(t_0), && t \geq t_0 + 1\\ &= \mathbf{I}_n. && t = t_0.\end{aligned}$$

Lemma 10.8. ***Discrete-system transition matrix-semigroup rules and other properties.*** *The discrete-system-transition matrix* $\mathbf{\Phi}(\cdot, \cdot) \in \mathbb{R}^{n \times n}$ *has the following properties:*

(*i*) $\mathbf{\Phi}(t_0, t_0) = \mathbf{I}_n$

(*ii*) $\mathbf{\Phi}(t, t_0) = \mathbf{\Phi}(t, s)\mathbf{\Phi}(s, t_0) \qquad t_0 \le s < t.$

Other properties:

(*i*) $\mathbf{\Phi}(t + 1, t_0) = \mathbf{A}(t)\mathbf{\Phi}(t, t_0) \qquad t \ge t_0.$

If $\mathbf{A}(\cdot) \in \mathbb{R}^{n \times n}$ *is independent of* $t = t_0, t_1, \ldots,$ *then:*

(*ii*) $\mathbf{\Phi}(t, t_0) = \mathbf{A}^{t - t_0} \triangleq \mathbf{\Phi}(t - t_0)$

(*iii*) $\mathbf{\Phi}(t + t_0) = \mathbf{\Phi}(t)\mathbf{\Phi}(t_0).$

If time-invariant $\mathbf{A}$ *is also invertible then:*

(*iv*) $\mathbf{\Phi}^{-1}(t) = \mathbf{\Phi}(-t).$

It is useful to note that if the discrete-system description (10.99) arises from a sampled (using a zero-order hold) continuous-system description then the matrix $\mathbf{A}(\cdot)$ has a prescribed form. If the underlying state-space system is time varying then $\mathbf{A}(t) = \mathbf{\Phi}(t + 1, t)$ $(t > t_0)$ where $\mathbf{\Phi}(\cdot, \cdot)$ is the continuous state-space system transition matrix (see Theorems 1.2 to 1.4, Chapter 1). This implies that additional properties may be ascribed to the discrete-system transition (see property (1.20) for example). If the underlying continuous state space system is time-invariant, then $\mathbf{A}_{\text{discrete}} = e^{\mathbf{A}\,\Delta T}$ where ΔT is the sampling interval. Hence, whatever the underlying open loop eigenvalue distribution of the state-space system $\mathbf{A}_{\text{discrete}} = e^{\mathbf{A}\,\Delta T}$ is non-singular $(\Delta T > 0)$, and property (iv) of Lemma 10.8 will be satisfied by the discrete system transition matrix.

The discrete convolution (10.100) may be given the more concise representation:

$$x(t) = \mathbf{\Phi}(t, 0)x_0 + \sum_{j=0}^{t-1} \mathbf{W}(t, j)\omega(j) \tag{10.101}$$

where the impulse matrix is defined by:

$$\begin{aligned} \mathbf{W}(t, j) &= \mathbf{\Phi}(t, j + 1)\mathbf{D}(j), \qquad t \ge j + 1; \qquad j = 0, 1, \ldots, \\ &= \mathbf{O}, \qquad\qquad\qquad\qquad t \le j. \end{aligned} \tag{10.102}$$

Using equation (10.101), the following theorem prescribes the value of the quadratic cost function given in (10.98).

Theorem 10.12. ***Quadratic covariance summation and associated difference equation.*** *For the linear difference equation* (10.99), *or equivalent* *from* (10.101), *the cost function* (10.98) *has the value:*

$$J_T = \text{trace}\left[\mathbf{P}_c(0)\mathbf{P}_0 + \sum_{j=0}^{T-1} \mathbf{D}(j)\mathbf{Q}(j)\mathbf{D}^T(j)\mathbf{P}_c(j+1)\right] \tag{10.103}$$

where

$$\mathbf{P}_c(j-1) = \mathbf{A}_c^T(j-1)\mathbf{P}_c(j)\mathbf{A}_c(j-1) + \mathbf{M}(j-1)$$

$$\mathbf{P}_c(T) = \mathbf{P}_T. \tag{10.104}$$

Proof. (i) Expansion of the cost functional. Using (10.101), substitute into (10.98):

$$J_T = E\left\{\sum_{t=0}^{T-1}\left(\left\langle \sum_{j=0}^{t-1} \mathbf{W}(t,j)\omega(j), \mathbf{M}(t)\sum_{j=0}^{t-1}\mathbf{W}(t,j)\omega(j)\right\rangle_{E_n}\right.\right.$$

$$\left. + \langle \mathbf{\Phi}(t,0)x_0(t), \mathbf{M}(t)\mathbf{\Phi}(t,0)x_0\rangle_{E_n}\right)$$

$$+ \left\langle \sum_{j=0}^{T-1} \mathbf{W}(T,j)\omega(j), \mathbf{P}_T\sum_{j=0}^{T-1}\mathbf{W}(T,j)\omega(j)\right\rangle_{E_n}$$

$$\left. + \langle \mathbf{\Phi}(T,0)x_0, \mathbf{P}_T\mathbf{\Phi}(T,0)x_0\rangle_{E_n}\right\}$$

$$= \text{trace}\left\{\left(\sum_{t=0}^{T-1} \mathbf{\Phi}^T(t,0)\mathbf{M}(t)\mathbf{\Phi}(t,0) + \mathbf{\Phi}^T(T,0)\mathbf{P}_T\mathbf{\Phi}(T,0)\right)\mathbf{P}_0\right.$$

$$\left. + \left[\sum_{t=0}^{T-1}\left(\sum_{j=0}^{t-1} \mathbf{Q}(j)\mathbf{W}^T(t,j)\mathbf{M}(t)\mathbf{W}(t,j)\right) + \sum_{j=0}^{T-1}\mathbf{Q}(j)\mathbf{W}^T(T,j)\mathbf{P}_T\mathbf{W}(T,j)\right]\right\}.$$

(10.105)

Using the result (for compatible matrices) that trace$\{\mathbf{XY}\}$ = trace$\{\mathbf{YX}\}$, interchanging the order of the summation and using (10.102) for $t \leqslant j$, obtain for the last two terms $[\cdot]$ of (10.105):

$$[\cdot] = \text{trace}\left\{\sum_{j=0}^{T-1} \mathbf{D}(j)\mathbf{Q}(j)\mathbf{D}^T(j)\left[\sum_{t=j+1}^{T-1} \mathbf{\Phi}^T(t,j)\mathbf{M}(t)\mathbf{\Phi}(t,j)\right.\right.$$

$$\left.\left. + \mathbf{\Phi}^T(T,j+1)\mathbf{P}_T\mathbf{\Phi}(T,j+1)\right]\right\}$$

The expression for J_T now becomes:

$$J_T = \text{trace}\left\{\mathbf{P}_c(0)\mathbf{P}_0 + \sum_{j=0}^{T-1} \mathbf{D}(j)\mathbf{Q}(j)\mathbf{D}^T(j)\mathbf{P}_c(j+1)\right\}$$

where the symmetric matrix $\mathbf{P}_c(j)$ satisfies:

$$\mathbf{P}_c(j+1) = \sum_{t=j+1}^{T-1} \mathbf{\Phi}^T(t, j+1)\mathbf{M}(t)\mathbf{\Phi}(t, j+1) + \mathbf{\Phi}^T(T, j+1)\mathbf{P}_T\mathbf{\Phi}(T, j+1)$$

or

$$\mathbf{P}_c(j) = \sum_{t=j}^{T-1} \mathbf{\Phi}^T(t, j)\mathbf{M}(t)\mathbf{\Phi}(t, j) + \mathbf{\Phi}^T(T, j)\mathbf{P}_T\mathbf{\Phi}(T, j)$$

where

$$\mathbf{P}_c(0) = \sum_{t=0}^{T-1} \mathbf{\Phi}^T(t, 0)\mathbf{M}(t)\mathbf{\Phi}(t, 0) + \mathbf{\Phi}^T(T, 0)\mathbf{P}_T\mathbf{\Phi}(T, 0).$$

This proves equation (10.103) of the theorem.

(ii) Difference equation. The associated difference equation follows as:

$$\mathbf{P}_c(j-1) = \mathbf{\Phi}^T(j, j-1) \sum_{t=j-1}^{T-1} \mathbf{\Phi}^T(t, j)\mathbf{M}(t)\mathbf{\Phi}(t, j)\mathbf{\Phi}(j, j-1)$$

$$+ \mathbf{\Phi}^T(j, j-1)\mathbf{\Phi}^T(T, j)\mathbf{P}_T\mathbf{\Phi}(T, j)\mathbf{\Phi}(j, j-1)$$

$$= \mathbf{\Phi}^T(j, j-1)\mathbf{P}_c(j)\mathbf{\Phi}(j, j-1) + \mathbf{M}(j-1).$$

From Lemma 10.8, the transition matrix satisfies the unforced system equation:

$$\mathbf{\Phi}(t+1, k) = \mathbf{A}_c(t)\mathbf{\Phi}(t, k) \qquad (t \geq k)$$

Hence

$$\mathbf{\Phi}(j, j-1) = \mathbf{A}(j-1)\mathbf{\Phi}(j-1, j-1) = \mathbf{A}_c(j-1)$$

giving

$$\mathbf{P}_c(j-1) = \mathbf{A}_c^T(j-1)\mathbf{P}_c(j)\mathbf{A}_c(j-1) + \mathbf{M}(j-1)$$

where

$$\mathbf{P}_c(T) = \mathbf{P}_T. \qquad \square$$

10.3.2. Discrete-time stochastic linear regulator problem with complete state observations

In preparation for the stochastic linear regulator problem, where the state information is obtained from the measurement of noise corrupted outputs, the regulator problem (with complete state observations) is solved.

An optimal state-feedback law is required which minimises the stochastic cost criterion:

$$J_T = E\left\{ \sum_{t=0}^{T-1} (\langle x(t), \mathbf{Q}_1(t)x(t)\rangle_{E_n} + \langle u(t), \mathbf{R}_1(t)u(t)\rangle_{E_m}) + \langle x(T), \mathbf{Q}_1(T)x(T)\rangle_{E_n} \right\} \tag{10.106}$$

where the linear stochastic system description is:

$$x(t+1) = \mathbf{A}(t)x(t) + \mathbf{B}(t)u(t) + \mathbf{D}(t)\omega(t) \tag{10.107}$$

and

(i) Weighting matrices $\mathbf{Q}_1(\cdot)$ and $\mathbf{R}_1(\cdot)$ satisfy:

$$\mathbf{Q}_1(\cdot) = \mathbf{Q}_1^T(\cdot) \geq \mathbf{O}, \qquad \mathbf{Q}_1(\cdot) \in \mathbb{R}^{n \times n}, \qquad 0 \leq t \leq T$$

$$\mathbf{R}_1(\cdot) = \mathbf{R}_1^T(\cdot) > \mathbf{O}, \qquad \mathbf{R}_1(\cdot) \in \mathbb{R}^{m \times m}, \qquad 0 \leq t < T-1.$$

(ii) Noise process $\{\omega(t)\}$ is zero mean and white with

$$\operatorname{cov}[\omega(t), \omega(k)] = \mathbf{Q}(t)\,\delta_{tk}$$

and the initial condition x_0 satisfies:

$$E\{x_0\} = \mathbf{o}, \quad \operatorname{cov}[x_0, x_0] = \mathbf{P}_0.$$

To distinguish between the gain matrices and the Riccati equations for the control and filter equations, the subscripts c and f are used respectively.

Theorem 10.13: ***Discrete-time optimal stochastic linear regulator with complete state observations.*** *The linear stochastic feedback law which optimizes the cost functional* (10.106) *subject to linear system description* (10.107) *is given by:*

$$u^0(t) = -\mathbf{K}_c(t)x(t) \tag{10.108}$$

where

$$\left.\begin{aligned} \mathbf{K}_c(t) &= \tilde{\mathbf{R}}_1^{-1}(t)\mathbf{B}^T(t)\mathbf{P}_c(t+1)\mathbf{A}(t) \\ &\text{and} \\ \tilde{\mathbf{R}}_1(t) &= \mathbf{R}_1(t) + \mathbf{B}^T(t)\mathbf{P}_c(t+1)\mathbf{B}(t). \end{aligned}\right\} \tag{10.109}$$

The associated matrix Riccati difference equation is:

$$\begin{aligned} \mathbf{P}_c(t) &= \mathbf{A}^T(t)\mathbf{P}_c(t+1)\mathbf{A}(t) \\ &\quad - \mathbf{A}(t)\mathbf{P}_c(t+1)\mathbf{B}(t)\tilde{\mathbf{R}}_1^{-1}(t)\mathbf{B}^T(t)\mathbf{P}_c(t+1)\mathbf{A}(t) + \mathbf{Q}_1(t) \\ &\qquad\qquad\qquad\qquad 0 \le t \le T-1 \\ \mathbf{P}_c(T) &= \mathbf{Q}_1(T). \end{aligned} \tag{10.110}$$

The optimal cost value is given by:

$$J_T^0 = \operatorname{trace}\left\{\mathbf{P}_c(0)\mathbf{P}_0 + \sum_{t=0}^{T-1} \mathbf{P}_c(t+1)\mathbf{D}(t)\mathbf{Q}(t)\mathbf{D}^T(t)\right\} \tag{10.111}$$

where $E\{x_0\} = \mathbf{o}$ *and* $\operatorname{cov}[x_0, x_0] = \mathbf{P}_0$.

Proof. Expansion of the cost functional. Let the linear stochastic feedback law be given as:

$$u(t) = -\mathbf{K}_c(t)x(t)$$

where $\mathbf{K}_c(t)$ denotes a non-dynamic, possibly time-varying feedback matrix. The closed-loop system equation becomes:

$$\begin{aligned} x(t+1) &= (\mathbf{A}(t) - \mathbf{B}(t)\mathbf{K}_c(t))x(t) + \mathbf{D}(t)\omega(t) \\ &= \mathbf{A}_c(t)x(t) + \mathbf{D}(t)\omega(t) \end{aligned} \tag{10.112}$$

where

$$\mathbf{A}_c(t) \triangleq \mathbf{A}(t) - \mathbf{B}(t)\mathbf{K}_c(t)$$

and the performance criterion (10.106) becomes:

$$J_T = E\left\{\sum_{t=0}^{T-1} (\langle x(t), \mathbf{Q}_1(t)x(t)\rangle_{E_n} + \langle u(t), \mathbf{R}_1(t)u(t)\rangle_{E_m}) + \langle x(T), \mathbf{Q}_1(T)x(T)\rangle_{E_n}\right\}$$

$$= E\left\{\sum_{t=0}^{T-1} (\langle x(t), (\mathbf{Q}_1(t) + \mathbf{K}_c^T(t)\mathbf{R}_1(t)\mathbf{K}_c(t))x(t)\rangle_{E_n} + \langle x(T), \mathbf{Q}_1(T)x(T)\rangle_{E_n}\right\}. \tag{10.113}$$

Define

$$\mathbf{M}(t) \triangleq \mathbf{Q}_1(t) + \mathbf{K}_c^T(t)\mathbf{R}_1(t)\mathbf{K}_c(t)$$

then (10.113) becomes:

$$J_T = E\left\{\sum_{t=0}^{T-1} (\langle x(t), \mathbf{M}(t)x(t)\rangle_{E_n} + \langle x(T), \mathbf{Q}_1(T)x(T)\rangle_{E_n}\right\}. \tag{10.114}$$

Clearly, the evaluation of (10.114) subject to (10.112) can be achieved by the application of the results of Theorem 10.12, viz:

$$J_T = \text{trace}\left\{\mathbf{P}_c(0)\mathbf{P}_0 + \sum_{t=0}^{T-1} \mathbf{P}_c(t+1)\mathbf{D}(t)\mathbf{Q}(t)\mathbf{D}^T(t)\right\} \tag{10.115}$$

where $\mathbf{P}_c(t)$ solves:

$$\mathbf{P}_c(t-1) = \mathbf{A}_c^T(t-1)\mathbf{P}_c(t)\mathbf{A}_c(t-1) + \mathbf{Q}_1(t-1) + \mathbf{K}_c^T(t-1)\mathbf{R}_1(t-1)\mathbf{K}_c(t-1)$$

and $\mathbf{P}_c(T) = \mathbf{Q}_1(T)$.

Hence, expanding $\mathbf{A}_c(\cdot)$ from (10.112) yields:

$$\begin{aligned}\mathbf{P}_c(t-1) &= (\mathbf{A}(t-1) - \mathbf{B}(t-1)\mathbf{K}_c(t-1))^T\mathbf{P}_c(t)(\mathbf{A}(t-1) - \mathbf{B}(t-1)\mathbf{K}_c(t-1))\\ &\quad + \mathbf{Q}_1(t-1) + \mathbf{K}_c^T(t-1)\mathbf{R}_1(t-1)\mathbf{K}_c(t-1)\\ &= \mathbf{A}^T(t-1)\mathbf{P}_c(t)\mathbf{A}(t-1) + \mathbf{Q}_1(t-1)\\ &\quad - \mathbf{K}_c^T(t-1)\mathbf{B}^T(t-1)\mathbf{P}_c(t)\mathbf{A}(t-1)\\ &\quad - \mathbf{A}^T(t-1)\mathbf{P}_c(t)\mathbf{B}(t-1)\mathbf{K}_c(t-1)\\ &\quad + \mathbf{K}_c^T(t-1)(\mathbf{R}_1(t-1) + \mathbf{B}^T(t-1)\mathbf{P}_c(t)\mathbf{B}(t-1))\mathbf{K}_c(t-1).\end{aligned}$$

Completing squares obtain:

$$\begin{aligned}\mathbf{P}_c(t-1) = {} & [\mathbf{A}^T(t-1)\mathbf{P}_c(t)\mathbf{A}(t-1) + \mathbf{Q}_1(t-1) \\ & - \mathbf{A}^T(t-1)\mathbf{P}_c(t)\mathbf{B}(t-1)\tilde{\mathbf{R}}_1^{-1}(t-1)\mathbf{B}^T(t-1)\mathbf{P}_c(t)\mathbf{A}(t-1)] \\ & + [\mathbf{A}^T(t-1)\mathbf{P}_c(t)\mathbf{B}(t-1)\tilde{\mathbf{R}}_1^{-1}(t-1) \\ & - \mathbf{K}_c^T(t-1)]\tilde{\mathbf{R}}_1(t-1)[\tilde{\mathbf{R}}_1^{-1}(t-1)\mathbf{B}^T(t-1)\mathbf{P}_c(t)\mathbf{A}(t-1) \\ & - \mathbf{K}_c(t-1)] \end{aligned} \qquad (10.116)$$

where

$$\tilde{\mathbf{R}}_1(t) \triangleq \mathbf{R}(t) + \mathbf{B}^T(t)\mathbf{P}_c(t+1)\mathbf{B}(t).$$

With an obvious identification of terms:

$$\mathbf{P}_c(t-1) = [\mathbf{V}(t-1)] + [\mathbf{U}(t-1)]^T\tilde{\mathbf{R}}_1(t-1)[\mathbf{U}(t-1)]. \qquad (10.117)$$

Note that $\mathbf{V}(t-1)$ is independent of the gain matrix $\mathbf{K}_c(\cdot)$ at $t-1$. The cost functional (10.115) may be written:

$$\begin{aligned} J_T = \operatorname{trace}\Bigg\{ & \left(\mathbf{V}(0)\mathbf{P}_0 + \sum_{t=0}^{T-2} \mathbf{V}(t+1)\mathbf{D}(t)\mathbf{Q}(t)\mathbf{D}^T(t)\right) \\ & + \mathbf{P}_c(T)\mathbf{D}(T-1)\mathbf{Q}(T-1)\mathbf{D}^T(T-1) \\ & + \left[\mathbf{U}^T(0)\tilde{\mathbf{R}}_1(0)\mathbf{U}(0) + \sum_{t=0}^{T-2} \mathbf{U}^T(t+1)\tilde{\mathbf{R}}_1(t+1)\mathbf{U}(t+1)\mathbf{D}(t)\mathbf{Q}(t)\mathbf{D}^T(t)\right]\Bigg\} \end{aligned} \qquad (10.118)$$

where the summation terms are both zero for the special case of a single-step horizon ($T = 1$).

Optimality analysis

Only the final term $[\cdot]$ in the cost functional (10.118) depends on the feedback gain matrix via its influence on $\mathbf{U}(t)$. Let $\mathbf{U}(t) = \mathbf{U}_0(t) + \varepsilon\tilde{\mathbf{U}}(t)$ where ε is a real scalar, $\mathbf{U}_0(t)$ denotes the optimal value of $\mathbf{U}(t)$ and $\tilde{\mathbf{U}}(t)$ denotes an arbitrary variation. Using the familiar gradient argument the necessary condition for optimality becomes:

$$\left.\frac{\partial J_T}{\partial \varepsilon}\right|_{\varepsilon=0} = 0$$

namely

$$\left.\frac{\partial J_T}{\partial \varepsilon}\right|_{\varepsilon=0} = \operatorname{trace}\left\{\mathbf{U}_0^T(0)\tilde{\mathbf{R}}_1(0)\tilde{\mathbf{U}}(0) + \tilde{\mathbf{U}}^T(0)\tilde{\mathbf{R}}_1(0)\mathbf{U}_0(0)\right.$$

$$+ \sum_{t=0}^{T-2} [\mathbf{U}_0^T(t+1)\tilde{\mathbf{R}}_1(t+1)\tilde{\mathbf{U}}(t+1)$$

$$\left. + \tilde{\mathbf{U}}^T(t+1)\tilde{\mathbf{R}}_1(t+1)\mathbf{U}_0(t+1)]\,\mathbf{D}(t)\mathbf{Q}(t)\mathbf{D}^T(t)\right\} = 0$$

which must be satisfied for any $\tilde{\mathbf{U}}(t)$ and arbitrary choice of matrix $\mathbf{D}(t)\mathbf{Q}(t)\mathbf{D}^T(t)$, hence the necessary condition for optimality becomes:

$$\mathbf{U}_0(t) = \mathbf{O} \quad \text{for all } t \in [0, T-1] \tag{10.119}$$

giving equations (10.109) as:

$$\mathbf{K}_c(t) = \tilde{\mathbf{R}}_1^{-1}(t)\mathbf{B}^T(t)\mathbf{P}_c(t+1)\mathbf{A}(t)$$

where

$$\tilde{\mathbf{R}}_1(t) = \mathbf{R}_1(t) + \mathbf{B}^T(t)\mathbf{P}_c(t+1)\mathbf{B}(t).$$

The second derivative obtains as:

$$\left.\frac{1}{2}\frac{\partial^2 J_T}{\partial \varepsilon^2}\right|_{\varepsilon=0} = \operatorname{trace}\left\{\tilde{\mathbf{U}}^T(0)\tilde{\mathbf{R}}_1(0)\tilde{\mathbf{U}}(0) + \right.$$

$$\left.\sum_{t=0}^{T-2} \tilde{\mathbf{U}}^T(t+1)\tilde{\mathbf{R}}_1(t+1)\tilde{\mathbf{U}}(t+1)\mathbf{D}(t)\mathbf{Q}(t)\mathbf{D}^T(t)\right\}.$$

Clearly, since $\tilde{\mathbf{R}}_1(0) = \mathbf{R}(0) + \mathbf{B}^T(0)\mathbf{P}_c(1)\mathbf{B}(0)$ and $\mathbf{R}(0) > \mathbf{O}$ then the above condition is also sufficient.

Optimal cost value

The condition of optimality (10.119) yields in equation (10.117):

$$\mathbf{P}_c(t) = \mathbf{V}(t) \qquad t \in [0, T-1]. \tag{10.120}$$

Thus the identification between equations (10.116) and (10.117) gives the desired recurrent relationship of (10.110). The use of (10.119) and (10.120) in

(10.118) gives the optimal cost:

$$J_T^0 = \text{trace}\left\{\mathbf{V}(0)\mathbf{P}_0 + \sum_{t=0}^{T-2} \mathbf{V}(t+1)\mathbf{D}(t)\mathbf{Q}(t)\mathbf{D}^T(t) + \mathbf{P}_c(T)\mathbf{D}(T-1)\mathbf{Q}(T-1)\mathbf{D}^T(T-1)\right\}$$

$$= \text{trace}\left\{\mathbf{P}_c(0)\mathbf{P}_0 + \sum_{t=0}^{T-1} \mathbf{P}_c(t+1)\mathbf{D}(t)\mathbf{Q}(t)\mathbf{D}^T(t)\right\}. \qquad \square$$

Corollary 10.2. *Alternative control matrix Riccati difference equation. An equivalent control matrix Riccati difference equation is:*

$$\mathbf{P}_c(t-1) = \mathbf{A}_c^T(t-1)\mathbf{P}_c(t)\mathbf{A}_c(t-1) + \mathbf{Q}_1(t-1) + \mathbf{K}_c^T(t-1)\mathbf{R}_1(t-1)\mathbf{K}_c(t-1)$$

$$\mathbf{P}_c(T) = \mathbf{Q}_1(T) \qquad 1 \le t \le T$$

where

$$\mathbf{K}_c(t) = \tilde{\mathbf{R}}_1^{-1}(t)\mathbf{B}^T(t)\mathbf{P}_c(t+1)$$

$$\mathbf{A}_c(t) = \mathbf{A}(t) - \mathbf{B}(t)\mathbf{K}_c(t)$$

and

$$\tilde{\mathbf{R}}_1(t) = \mathbf{R}_1(t) + \mathbf{B}^T(t)\mathbf{P}_c(t+1)\mathbf{B}(t).$$

Proof. In the proof of Theorem 10.13, note that the optimal cost follows directly from equation (10.115) which has its source in Theorem 10.12. The recurrent formula then derives directly from equation (10.103) and (10.104) with appropriate interpretation for $\mathbf{A}_c(\cdot)$ and $\mathbf{M}(\cdot)$. $\square$

10.3.3. Discrete-time stochastic linear regulator problem with incomplete state observations

To the problem of the stochastic linear regulator of the previous section is added the realistic constraint that the information about the state can only be obtained through a set of noisy output measurements. The problem becomes one of minimizing the stochastic cost-functional:

$$J_T = E\left\{\sum_{t=0}^{T-1} (\langle x(t), \mathbf{Q}_1(t)x(t)\rangle_{E_n} + \langle u(t), \mathbf{R}_1(t)u(t)\rangle_{E_m}) + \langle x(T), \mathbf{Q}_1(T)x(T)\rangle_{E_n}\right\} \quad (10.121)$$

subject to the process equation:

$$x(t+1) = \mathbf{A}(t)x(t) + \mathbf{B}(t)u(t) + \mathbf{D}(t)\omega(t) \tag{10.122}$$

and utilizing the measurement process:

$$z(t) = \mathbf{C}(t)x(t) + v(t) \tag{10.123}$$

where the white, zero-mean, process and measurement noise signals satisfy:

$$\mathrm{cov}\,[\omega(t), \omega(k)] = \mathbf{Q}(t)\ \delta_{tk}, \qquad \mathbf{Q}^T(t) = \mathbf{Q}(t) \geq \mathbf{O},$$

$$\mathrm{cov}\,[v(t), v(k)] = \mathbf{R}(t)\ \delta_{tk}, \qquad \mathbf{R}^T(t) = \mathbf{R}(t) > \mathbf{O},$$

and

$$\mathrm{cov}\,[\omega(t), v(t)] = \mathbf{O}.$$

These noise processes are assumed uncorrelated with the zero mean initial condition, x_0 where $\mathrm{cov}\,[x_0, x_0] = \mathbf{P}_0$.

Theorem 10.14. ***Discrete-time stochastic regulator with incomplete state observations.*** *Consider the problem of optimizing the cost-functional* (10.121) *subject to the linear discrete system* (10.122) *and the measurement process* (10.123). *The optimal state estimate feedback is given by:*

$$u^0(t) = -\mathbf{K}_c(t)\hat{x}(t \mid t-1) \tag{10.124}$$

where

$$\mathbf{K}_c(t) = \tilde{\mathbf{R}}_1^{-1}(t)\mathbf{B}^T(t)\mathbf{P}_c(t+1)\mathbf{A}(t) \tag{10.125}$$

$$\tilde{\mathbf{R}}_1(t) = \mathbf{R}_1(t) + \mathbf{B}^T(t)\mathbf{P}_c(t+1)\mathbf{B}(t) \tag{10.126}$$

and $\mathbf{P}_c(t)$ *satisfies the control matrix Riccati difference equation:*

$$\begin{aligned}\mathbf{P}_c(t) = {} & \mathbf{A}^T(t)\mathbf{P}_c(t+1)\mathbf{A}(t) \\ & - \mathbf{A}(t)\mathbf{P}_c(t+1)\mathbf{B}(t)\tilde{\mathbf{R}}_1^{-1}(t)\mathbf{B}^T(t)\mathbf{P}_c(t+1)\mathbf{A}(t) + \mathbf{Q}_1(t) \\ & 0 \leq t \leq T-1 \end{aligned} \tag{10.127}$$

$$\mathbf{P}_c(T) = \mathbf{Q}_1(T).$$

The optimal estimator $\hat{\mathbf{x}}(t \mid t-1)$ *satisfies:*

$$\hat{\mathbf{x}}(t+1 \mid t) = \mathbf{A}(t)\hat{\mathbf{x}}(t \mid t-1) + \mathbf{B}(t)\mathbf{u}^0(t) + \mathbf{K}_f(t)(\mathbf{z}(t) - \mathbf{C}(t)\hat{\mathbf{x}}(t \mid t-1)) \tag{10.128}$$

where $\hat{\mathbf{x}}(0 \mid -1) = \mathbf{o}$ *and*

$$\mathbf{K}_f(t) = \mathbf{A}(t)\mathbf{P}_f(t \mid t-1)\mathbf{C}^T(t)\tilde{\mathbf{R}}^{-1}(t) \tag{10.129}$$

$$\tilde{\mathbf{R}}(t) = \mathbf{R}(t) + \mathbf{C}(t)\mathbf{P}_f(t \mid t-1)\mathbf{C}^T(t) \tag{10.130}$$

and $\mathbf{P}_f(t \mid t-1)$ *satisfies the filter matrix Riccati difference equation:*

$$\begin{aligned}\mathbf{P}_f(t+1 \mid t) = {}& \mathbf{A}(t)\mathbf{P}_f(t \mid t-1)\mathbf{A}^T(t) \\ & - \mathbf{A}(t)\mathbf{P}_f(t \mid t-1)\mathbf{C}^T(t)\tilde{\mathbf{R}}^{-1}(t)\mathbf{C}(t)\mathbf{P}_f(t \mid t-1)\mathbf{A}(t) \\ & + \mathbf{D}(t)\mathbf{Q}(t)\mathbf{D}^T(t) \qquad 0 \le t \le T\end{aligned} \tag{10.131}$$

$$\mathbf{P}_f(0 \mid -1) = \mathbf{P}_0$$

The optimal value of the cost functional is given by:

$$J_T = \text{trace}\left\{\sum_{t=0}^{T} \mathbf{Q}_1(t)\mathbf{P}_f(t \mid t-1) + \mathbf{P}_c(0)\mathbf{P}_0 + \sum_{t=0}^{T-1} \mathbf{P}_c(t+1)\mathbf{K}_f(t)\mathbf{R}(t)\mathbf{K}_f^T(t)\right\}. \tag{10.132}$$

Proof. Simplification of the cost-functional. Recall from Algorithm 8.1 (Chapter 8), the discrete single-state predictor system:

$$\hat{\mathbf{x}}(t+1 \mid t) = \mathbf{A}(t)\hat{\mathbf{x}}(t \mid t-1) + \mathbf{B}(t)\mathbf{u}(t) + \mathbf{K}_f(t)\boldsymbol{\nu}(t)$$

where innovations process $\{\boldsymbol{\nu}(t)\}$ is given by:

$$\boldsymbol{\nu}(t) = \mathbf{z}(t) - \mathbf{C}(t)\hat{\mathbf{x}}(t \mid t-1)$$

$$\mathbf{K}_f(t) = \mathbf{A}(t)\mathbf{P}_f(t \mid t-1)\mathbf{C}^T(t)\tilde{\mathbf{R}}^{-1}(t)$$

$$\tilde{\mathbf{R}}(t) = \mathbf{R}(t) + \mathbf{C}(t)\mathbf{P}_f(t \mid t-1)\mathbf{C}^T(t)$$

$$\begin{aligned}\mathbf{P}_f(t+1 \mid t) = {}& \mathbf{A}(t)\mathbf{P}(t \mid t-1)\mathbf{A}^T(t) + \mathbf{D}(t)\mathbf{Q}(t)\mathbf{D}^T(t) \\ & - \mathbf{A}(t)\mathbf{P}(t \mid t-1)\mathbf{C}^T(t)\tilde{\mathbf{R}}^{-1}(t)\mathbf{C}(t)\mathbf{P}(t \mid t-1)\mathbf{A}^T(t)\end{aligned}$$

and

$$\mathbf{P}_f(0 \mid -1) = \mathbf{P}_0.$$

Define the estimation error (which is independent of $\{\boldsymbol{u}(t)\}$):

$$\tilde{\boldsymbol{x}}(t \mid t-1) \triangleq \boldsymbol{x}(t) - \hat{\boldsymbol{x}}(t \mid t-1) \tag{10.133}$$

where

$$\mathbf{P}_f(t \mid t-1) = \operatorname{cov}[\tilde{\boldsymbol{x}}(t \mid t-1), \tilde{\boldsymbol{x}}(t \mid t-1)].$$

Equation (10.133) rearranges as:

$$\boldsymbol{x}(t) = \tilde{\boldsymbol{x}}(t \mid t-1) + \hat{\boldsymbol{x}}(t \mid t-1)$$

hence from cost functional (10.121) the state terms become:

$$E\left\{\sum_{t=0}^{T} \langle \boldsymbol{x}(t), \mathbf{Q}_1(t)\boldsymbol{x}(t)\rangle_{E_n}\right\}$$

$$= E\left\{\sum_{t=0}^{T} \langle \tilde{\boldsymbol{x}}(t \mid t-1) + \hat{\boldsymbol{x}}(t \mid t-1), \mathbf{Q}_1(t)(\tilde{\boldsymbol{x}}(t \mid t-1) + \hat{\boldsymbol{x}}(t \mid t-1))\rangle_{E_n}\right\}$$

$$= \sum_{t=0}^{T} \operatorname{trace}\{\mathbf{Q}_1(t)\mathbf{P}_f(t \mid t-1)\} + E\left\{\sum_{t=0}^{T} \langle \hat{\boldsymbol{x}}(t \mid t-1), \mathbf{Q}_1(t)\hat{\boldsymbol{x}}(t \mid t-1)\rangle_{E_n}\right\}. \tag{10.134}$$

(The orthogonal projection theorem ensures the cross-terms are zero.) Returning to the full cost functional (10.121) obtains:

$$J_T = E\left\{\sum_{t=0}^{T-1} (\langle \hat{\boldsymbol{x}}(t \mid t-1), \mathbf{Q}_1(t)\hat{\boldsymbol{x}}(t \mid t-1)\rangle_{E_n} + \langle \boldsymbol{u}(t), \mathbf{R}_1(t)\boldsymbol{u}(t)\rangle_{E_m})\right.$$
$$\left. + \langle \hat{\boldsymbol{x}}(T \mid T-1), \mathbf{Q}_1(T)\hat{\boldsymbol{x}}(T \mid T-1)\rangle_{E_n} + \sum_{t=0}^{T} \operatorname{trace}\{\mathbf{Q}_1(t)\mathbf{P}_f(t \mid t-1)\}\right\}. \tag{10.135}$$

The last term in (10.135) is independent of the choice of $\{\boldsymbol{u}(t)\}$ and thus the optimization of (10.135) is subject only to the filter equation:

$$\hat{\boldsymbol{x}}(t+1 \mid t) = \mathbf{A}(t)\hat{\boldsymbol{x}}(t \mid t-1) + \mathbf{B}(t)\boldsymbol{u}(t) + \mathbf{K}_f(t)\boldsymbol{\nu}(t). \tag{10.136}$$

The innovations process $\{\boldsymbol{\nu}(t)\}$ is white, zero mean with the same statistical properties as $\{v(t)\}$. Clearly minimization of the cost functional (10.135) subject to (10.136) has exactly the structure as the complete state observations regulator problem of Theorem 10.13. Applying the results of that theorem yields an optimal control law:

$$u^0(t) = -\mathbf{K}_c(t)\hat{x}(t \mid t-1)$$

where $\mathbf{K}_c(t)$ is determined using equation (10.109) and (10.110), thus proving equations (10.124) to (10.127). Using (10.111) of Theorem 10.13 and equation (10.135) above yields the optimal cost expression (10.132). □

The stability of the solution can be inferred from equations (10.122), (10.124) and estimation error equation (10.133). It is easily shown that these form the composite system:

$$\begin{bmatrix} \tilde{x}(t+1 \mid t) \\ \hat{x}(t+1 \mid t) \end{bmatrix} = \begin{bmatrix} \mathbf{A}(t) - \mathbf{K}_f(t)\mathbf{C}(t) & \mathbf{O} \\ \mathbf{K}_f(t)C(t) & \mathbf{A}(t) - \mathbf{B}(t)\mathbf{K}_c(t) \end{bmatrix} \begin{bmatrix} \tilde{x}(t \mid t-1) \\ \hat{x}(t \mid t-1) \end{bmatrix}$$

$$+ \begin{bmatrix} \mathbf{D}(t) & -\mathbf{K}_f(t) \\ \mathbf{O} & \mathbf{K}_f(t) \end{bmatrix} \begin{bmatrix} \omega(t) \\ v(t) \end{bmatrix}$$

with initial condition $\begin{bmatrix} \tilde{x}(0 \mid 1-1) \\ \hat{x}(0 \mid -1) \end{bmatrix} = \begin{bmatrix} x(0) \\ \mathbf{o} \end{bmatrix}.$

Thus for time-invariant systems over an infinite time interval, where solution exist for the steady-state matrix Riccati difference equations, it is the eigenvalues of $\mathbf{A} - \mathbf{K}_f\mathbf{C}$ and $\mathbf{A} - \mathbf{B}\mathbf{K}_c$ which determine the stability of the solution. It has been seen from the filter and control problems that the systems represented by these matrices are asymptotically stable. This is yet one more result illustrating the 'separation principle' which applies in these stochastic optimal control problems.

10.3.4. Implications of the separation principle for time-varying systems

If the plant under consideration is time varying, the stochastic optimal controller includes time-varying gains within both the Kalman filter and the optimal control law. The Kalman filter gains are computed from a matrix Riccati difference equation evaluated over $[0, t]$ but the optimal control law involves the solution of a similar Riccati equation measured over $[t, T]$, where

$[0, T]$ represents the range of integration for the cost function:

$$J = E\left\{\sum_{t=0}^{T} (\langle x(t), \mathbf{Q}_1(t)x(t)\rangle_{E_n} + \langle u(t), \mathbf{R}_1(t)u(t)\rangle_{E_m}) + \langle x(T), \mathbf{Q}_1(T)x(T)\rangle_{E_n}\right\}.$$

Some systems will have parameters which vary in a predetermined manner with time and then the above results may readily be applied. Unfortunately, it is more likely that if the system is time varying, the variation of parameters will be unpredictable. The Kalman filter gain may, of course, be computed, if the system parameters are measurable but the control gain matrix cannot be solved since future values of the system matrices are involved.

Thus, for time-varying systems there is a basic dilemma concerning the control law calculation in conditions where the future parameter variations are completely unpredictable. One possibility is to use a Receding Horizon control philosophy with a short-time horizon assuming that the plant parameter variations can be approximated or predicted during the remaining optimization interval $(T - t)$.

10.4. EXTENSIONS AND ASSESSMENT OF TIME-DOMAIN LINEAR QUADRATIC GAUSSIAN CONTROLLERS

Having established the theoretical basis for the LQG optimal controller, it is appropriate to consider some of the practical aspects. Most of the comments to follow apply to both discrete and continuous time systems with minor modifications.

10.4.1. Reduced order LQG controllers

It is clearly desirable to reduce the complexity of controllers when this is possible without substantial loss of performance. The order of LQG controllers can be reduced in a number of 'special' situations. Friedland [16] showed that the order of the LQG controller is reduced substantially when either measurement noise ($\mathbf{R} \to \mathbf{O} \Rightarrow$ order $= n - r$), or the control weighting ($\mathbf{R}_1 \to \mathbf{O} \Rightarrow$ order $= n - m$), tends to zero. Some reduction in the order may also be achieved when either some or all of the measurements are noise free or corrupted by coloured noise (Bryson and Johansen [17], Fogel and Huang [18], Missaghie and Fairman [19]). The order of the Kalman filter may also be reduced by first reducing the order of the system model (Wilson and Mishra [20]).

10.4.2. State and state-estimate feedback

Most industrial systems have some groups of states which are accessible and can be accessed without significant measurement noise. For example, the input subsystems usually consist of actuators with electrical feedback signals of speed, voltage.

Substantial reductions in the orders of LQG controllers can sometimes be achieved when some states are directly accessible in a system (Grimble [21], [22]). Assume that the system can be partitioned, as in Fig. 10.4, into input (S_2) and output (S_1) subsystems and only the states of the input subsystem $\mathbf{x}_2$ are available for feedback, then the Kalman filter required has an order equal to that of the output subsystem. There are also advantages with respect to the robustness of such a design, since most of the non-linearities in systems are associated with the actuators and direct state feedback around these devices is therefore desirable. In fact, Grimble [23] has shown that full-state feedback robustness properties apply to perturbations at the input point X shown in Fig. 10.4.

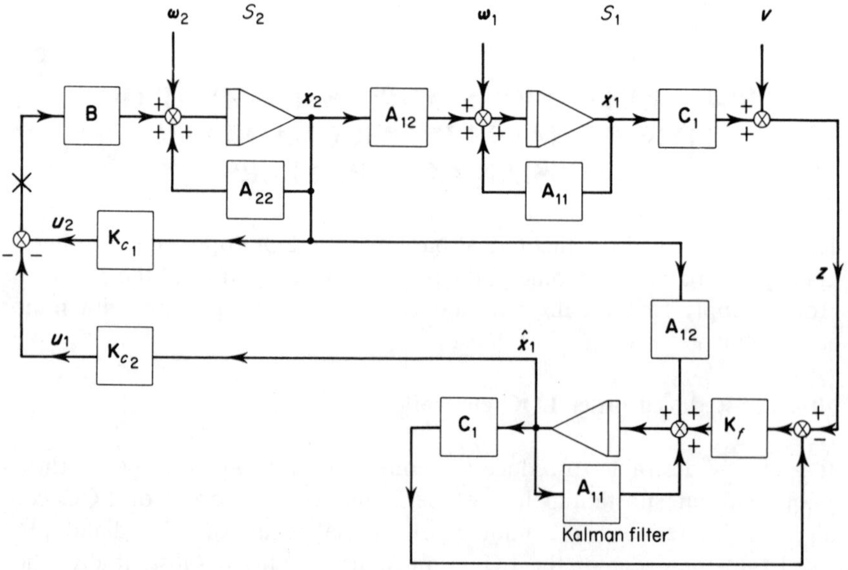

Fig. 10.4. Combined accessible state and state estimate feedback scheme

10.4.3. Time delays

If a continuous time plant includes output or input delay elements the LQP (Grimble [24]) or LQG (Grimble [25]) optimal controllers include a *Smith Predictor* type of structure. This accords with the classical solution to such

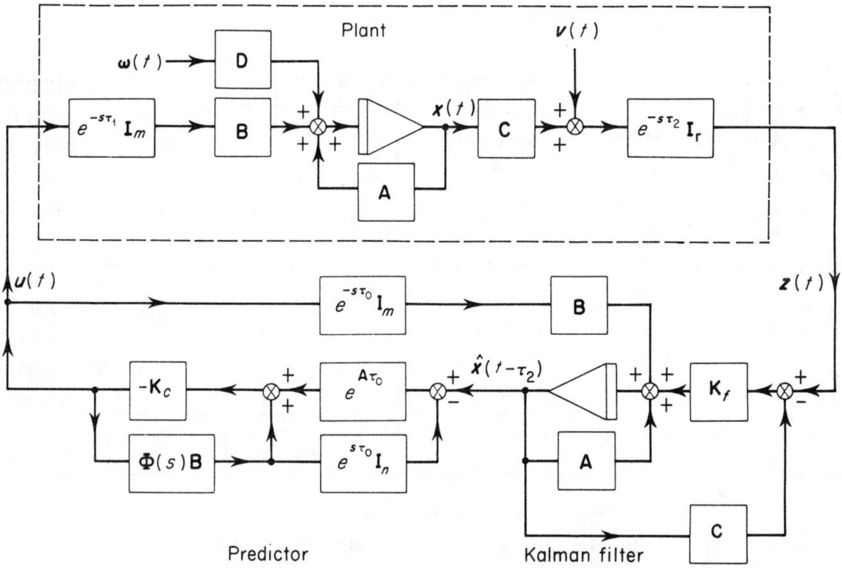

Fig. 10.5. Optimal regulator for a system with output and control delays ($\tau_0 = \tau_1 + \tau_2$)

problems but it is interesting to observe that this structure arises naturally in the optimal control problem solution. The modifications to the LQG Kalman filtering solution are illustrated in Fig. 10.5. This very practical structure is not so apparent in the solution of discrete-time problems where the delays may be implicit in the state description of the process (Lam [26]). The ship-positioning problem (see Chapter 13) provides a good example where measurements are delayed. In this case the sonar beacon signals are delayed by an amount depending upon the water depth.

10.4.4. Integral control action

Various methods of introducing integral action into the LQP problem solution are well established (Athans [27], Porter [28]) but the results for the stochastic LQG problem are not so well known (Grimble [29]). Define a dynamic weighting operator $(Ly)(t) \triangleq \int_0^t \mathbf{A}_2 y(\tau)\, d\tau$, and form a new stochastic cost functional as:

$$J_2 = \lim_{T\to\infty} \frac{1}{2T} E\left\{ \int_{-T}^{T} (\langle y(t), \mathbf{Q}_1 y(t)\rangle_{E_r} + \langle (Ly)(t), \mathbf{Q}_2 (Ly)(t)\rangle_{E_r} + \langle u^T(t), \mathbf{R}_1 u(t)\rangle_{E_m})\, dt \right\}$$

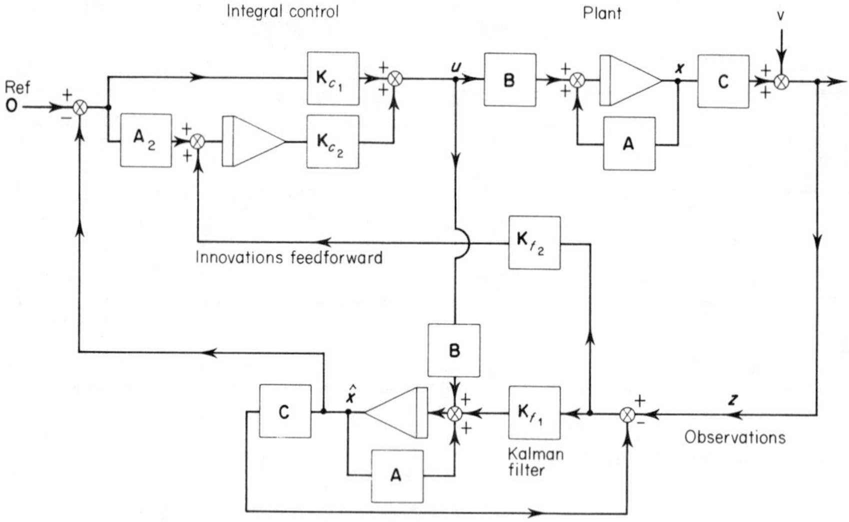

Fig. 10.6. Integral control for the LQG problem

where $\mathbf{Q}_2 \geq \mathbf{O}$ is a symmetric weighting matrix. The integral operator ensures that if the minimum cost is finite then the output must tend to zero in the steady state. The solution to the LQG problem with this criterion is shown in Fig. 10.6. The integral control gain matrix $\mathbf{K}_{c2}$ has two input channels: the first is the regulating error and the second involves an innovations feedforward term from the Kalman filter. Both terms can be justified from physical considerations.

10.4.5. Robustness and guaranteed stability margins for LQG regulators

By duality the Kalman filter has equivalent robustness properties to the state regulator of Section 2.6 (Chapter 2), when perturbations exist in the feedback loop of the filter. However, in a short important paper, Doyle [30], showed the surprising result that there are no guaranteed stability margins for LQG regulators. The previous robustness results do not apply at the input and output of the plant shown (points (1) and (2)) in Fig. 10.7 (Doyle and Stein [31]). The physical reason for this result is that, if say gain perturbations are present, the Kalman filter model no longer mirrors that of the system. That is, the input to the filter is not the same as that to the plant and the output from the plant is subjected to different gains to those seen by the filter outputs. This does not apply to the points (3) and (4) in Fig. 10.7 where stability margins are retained. Fortunately, there are steps which can be taken in the design process to improve the robustness of the LQG design, as discussed in the next section.

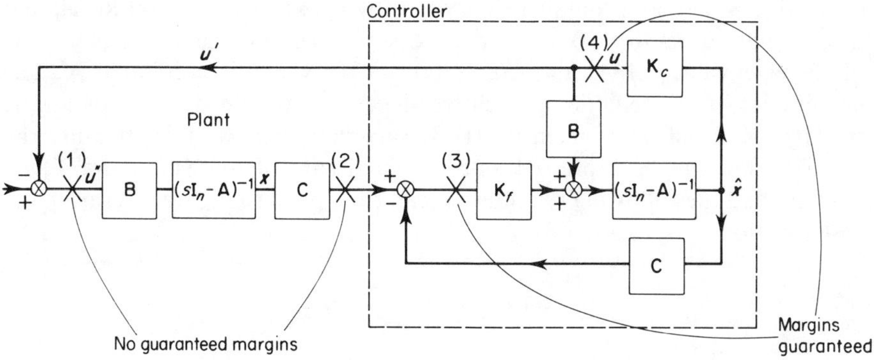

Fig. 10.7. The LQG controller and the points at which robustness results apply

10.4.6. Sensitivity and robust recovery procedures

The good robustness properties of the LQP state feedback controller can be recovered in the LQG case by using the 'Sensitivity Recovery' procedure due to Kwakernaak [32] and the 'Robust Recovery' procedure of Doyle and Stein [33]. In effect, full-state feedback-loop transfer properties can be recovered asymptotically if the system is minimum phase. This occurs at the expense of the optimal filtering properties since the noise intensity matrices are adjusted to achieve these desirable robustness properties.

The robust recovery procedure depends upon the observation that the transfer function between $\mathbf{u}''$ (with the loop broken at (1) in Fig. 10.7) and $\mathbf{u}'$ is in general different to the equivalent loop transfer for the LQP state regulator. However, if $\mathbf{Q}$ is replaced by $\mathbf{Q} = \mathbf{Q}_0 + q^2\mathbf{BVB}^T$ and $\mathbf{R} = \mathbf{R}_0$ ($\mathbf{Q}_0$, $\mathbf{R}_0$ are the nominal noise intensities and $\mathbf{V} = \mathbf{V}^T$ is any positive definite matrix) then as $q \to \infty$ the two transfers and hence robustness properties become identical for the LQG and LQP problems. Similar remarks apply to the sensitivity recovery procedure which can be applied at the loop breaking point (2).

Robust recovery using coloured noise

The above robust recovery procedure may be interpreted as introducing a fictitious white process noise source. In a generalization, Moore and Blight [34] have suggested the use of coloured noise to improve loop gain and phase margins. The additive fictitious input coloured noise reflects the uncertainty in the system model, particularly in the frequency bands where gain and phase margins are inadequate. The technique may be applied in some nonminimum phase situations.

Dynamic weightings elements

Safonov, Laub and Hartman [35] have shown that the LQG problem assigns frequency-dependent penalties on the sizes of the *singular values* of the sensitivity ($\mathbf{S}(j\omega)$) and complementary sensitivity ($\mathbf{T}(j\omega)$) matrices. Recall that the singular values ($\sigma_i(\mathbf{A})$) of the matrix $\mathbf{A}$ are the non-negative square roots of the eigenvalues of $\mathbf{A}^*\mathbf{A}$ where $\mathbf{A}^*(s) = \mathbf{A}^T(-s)$ is the adjoint of $\mathbf{A}$. The magnitudes of the penalties at each frequency is determined by the quadratic cost ($\mathbf{Q}_1, \mathbf{R}_1$) and noise intensity ($\mathbf{Q}, \mathbf{R}$) matrices. This follows because the cost-function can be expressed as a weighted sum of the $\mathbf{S}(j\omega)$ and $\mathbf{T}(j\omega)$ matrices and the following relationship can be applied to the trace terms in the cost function:

$$\operatorname{trace}\{\mathbf{A}\mathbf{A}^*\} = \sum_{i=1}^{\operatorname{rank}\mathbf{A}} \sigma_i^2(\mathbf{A}).$$

It follows that the feedback properties, including robustness, of the system can be tuned by use of the dynamic weighting elements ($\mathbf{Q}_1(s), \mathbf{R}_1(s)$) and frequency-dependent power spectrum matrices ($\mathbf{Q}(s), \mathbf{R}(s)$). This tuning process would normally require the use of computer-aided frequency response plots or simulation results.

Although the singular value approach is useful in detecting the near instability of a control system, it is sometimes unnecessarily conservative. This is due to the fact that some of the small perturbations which would theoretically destabilize the closed-loop system will never occur in the physical system. Nevertheless, they are still detected by a small singular value of the return-difference matrix. However, it provides one further means of assessing the performance of an LQG design which can be employed easily by using computer-aided design facilities.

10.4.7. Systems with random parameters

Another approach to the uncertainty problem is to represent the uncertain system parameters by random variables with known means and variances. This enables an equivalent model for a linear stochastic system to be defined which includes additional subsystems to represent the uncertainty (see Fig. 10.8). An input subsystem models the uncertainty in the plant dynamics and noise input matrices and the output subsystem represents the uncertainty in the output map. This enables a Generalized Wiener or Kalman filter to be defined (Grimble [36]) as shown in Fig. 10.8. The extra input subsystem ($\mathbf{A}_a, \mathbf{K}_a$) in the Kalman filter is clearly related to that which is introduced in the Robust Recovery Procedure when coloured noise is employed. An LQG-type controller for such a system uses the estimates $\hat{x}$ provided by the Generalized filter (Grimble [37]).

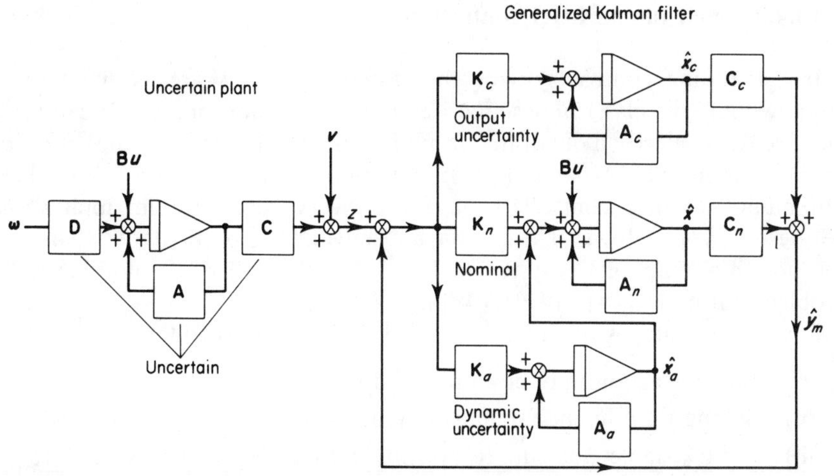

Fig. 10.8. Generalized Kalman filter for an uncertain system

10.4.8. Dual performance criterion

In an attempt to satisfy both robustness and performance requirements, a dual criterion (Grimble [38], Grimble and Johnson [45], Grimble [39]) has been defined which consists of both the usual LQG cost terms and additional sensitivity and complementary sensitivity weighting terms. The performance criterion for continuous systems has the form:

$$J_2 = \frac{1}{2\pi j}\ \text{trace}\left\{\int (\mathbf{S}^*\mathbf{Q}_2\mathbf{S}\boldsymbol{\Phi}_n + (\mathbf{I}_r - \mathbf{S})^*\mathbf{Q}_3(\mathbf{I}_r - \mathbf{S})\boldsymbol{\Phi}_n)\ ds\right\}$$

where $\boldsymbol{\Phi}_n$ is an appropriate noise spectrum. The solution to this problem is available in either state-space, Wiener transfer-function matrix or matrix-fraction-Diophantine equation forms. In the asymptotic cases as control weights are varied the cost can be reduced to either LQG or sensitivity (J_2) cost forms. The optimal controller therefore reduces to either a LQG controller or a robust optimal controller in the sense of Francis [40]. The relative size of the weighting matrix elements will determine the importance attached to the sensitivity, measured at particular points in the system.

A solution to this problem can also be obtained (Grimble [46]) by representing the system using the matrix fraction description due to Desoer *et al.* [41]. This solution has the advantage of a mathematical simplicity but the results are not so convenient for computation as the polynomial matrix approach algorithms (Grimble [39]).

10.4.9. Self-tuning LQG controllers

In self-tuning control systems the system dynamics are assumed to be unknown and the classical design criteria are not so appropriate. It can make more sense to specify a quadratic or minimum variance performance criterion to be minimized (Astrom [42]). Minimum Variance and Generalized Minimum Variance controllers have been very successful in applications (Hodgson [43]). However, these have disadvantages particularly when used for non-minimum phase systems. The LQG controllers do not have this problem (Grimble [44], [47], [48]).

Further advantages of the use of LQG designs are as follows:

(i) Integral action may be introduced easily.
(ii) Stochastic reference signals may be included.
(iii) Multivariable non-square systems with different delays in different loops can be controlled.

The LQG controller is a natural candidate to use with complex systems having many noise and disturbance inputs. The controller has better robustness and stability characteristics than single-stage control laws and is easier to design than say pole placement controllers. That is, it is usually easier to specify a performance criterion, for an unknown plant, than to choose a matrix which is physically reasonable having the required closed-loop pole characteristics. There is, of course, a greater computational load associated with LQG controllers. However, the rapid advances made with hardware and the useful properties of LQG controllers suggests that there will be an important role for this type of self-tuning system in the future (Grimble, Moir and Fung [49]). The LQG self-tuner is described in more detail in Chapter 12.

10.5. CONCLUSIONS

Since the 1971 special issue of the *IEEE Transaction on Automatic Control*, there has been a consistent growth in the theory and application of the LQP/LQG methodology. The optimal-control design procedures are versatile and standard results are available to describe and accommodate most engineering situations. One particular advantage is the *a priori* knowledge of certain performance characteristics for different plant and problem descriptions. If these minimum performance indices require improving, then as with any other design technique, the additional performance requirements must be achieved by an iterative design process. Overall, the LQG controllers are flexible and can be applied to a wide range of industrial situations, including those which require self-tuning or self-adaptive action.

The foregoing discussion demonstrates that LQG methods are supported by a range of useful results and practical procedures for design. The fact that the LQG controller is easily designed to compensate for noise and disturbances is a significant advantage over virtually all other multivariable design techniques.

10.6. REFERENCES

[1] Wonham, W. M. 1968. On the separation theorem of stochastic control, *SIAM J. Control*, **6**(2), 312–326.

[2] Davis, M. H. A. 1977. *Linear Estimation and Stochastic Control*, Chapman and Hall, London.

[3] Brooks, R. A. 1972. Linear stochastic control: An extended separation principle, *Journal of Mathematical Analysis and Applications*, **38**, 569–587.

[4] Balakrishnan, A. V. 1972. Stochastic control: A function space approach, *SIAM J. Control*, **10**(2), 285–297, May.

[5] Lindquist, A. 1973. On feedback control of linear stochastic systems, *SIAM J. Control*, **11**(2), 323–343, May.

[6] Bensoussan, A. and Viot, M. 1975. Optimal control of stochastic linear distributed parameter systems, *SIAM J. Control*, **13**(4), 904–926, July.

[7] Curtain, R. and Ichikawa, A. 1977. The separation principle for stochastic evolution equations, *SIAM J. Control and Optimization*, **15**(3), 367–385, May.

[8] Doob, J. L. 1953. *Stochastic processes*, John Wiley and Sons Inc., New York.

[9] Loeve, M. 1963. *Probability Theory*, Van Nostrand Reinhold Co., Ltd., New York.

[10] Naylor, A. W. and Sell, G. R. 1971. *Linear Operator Theory in Engineering and Science*, Holt, Rinehart and Winston Inc., New York.

[11] Taylor, A. E. 1958. *Introduction to Functional Analysis*, John Wiley and Sons Ltd., New York.

[12] Curtain, R. F. and Pritchard, A. J. 1977. *Functional Analysis in Modern Applied Mathematics*, Academic Press, London.

[13] Wonham, W. M. 1974. *Linear Multivariable Control: A Geometric Approach*, Springer-Verlag, Berlin.

[14] Kwakernaak, H. and Sivan, R. 1972. *Linear Optimal Control Systems*, Wiley-Interscience, New York.

[15] deRusso, P. M., Roy, R. J. and Close, C. M. 1967. *State Variables for Engineers*, John Wiley and Sons Ltd., New York.

[16] Friedland, B. 1971. Limiting forms of stochastic linear regulators, *Trans. ASME, Journal of Dynamic Systems, Measurements, and Control*, **93**, 134–141, September.

[17] Bryson, A. E. and Johansen, D. E. 1965. Linear filtering for time-varying systems using measurements containing coloured noise, *IEEE Trans. Autom. Control*, **AC-10**(1), 4–10.

[18] Fogel, E. and Huang, Y. F. 1980. Reduced order optimal state estimator for linear systems with partially noise-corrupted measurements, *IEEE Trans. Autom. Control*, **AC-25**(5), 994–996.

[19] Missaghie, M. M. and Fairman, F. W. 1978. Least squares observer design for continuous stochastic systems, *Int. J. System Science*, **9**(2), 137–146.

[20] Wilson, D. A. and Mishra, R. N. 1979. Design of low order estimators using reduced order models, *Int. J. Control*, **29**(3), 447–456.

[21] Grimble, M. J. 1980. Reduced order optimal controller for discrete-time stochastic systems, *IEE Proc.*, **127**(2), Pt. D, 55–63, March.

[22] Grimble, M. J. 1982. Structure of large stochastic optimal and suboptimal systems, *IEE Proc.*, **129**(5), Pt D, 167–176.

[23] Grimble, M. J. 1984. Robustness of combined state and state-estimate feedback control schemes, *IEEE Trans. on Auto. Control*, **AC-29**(7), 667–669, July.

[24] Grimble, M. J. 1980. The solution of finite-time optimal control problems with control time delays, *OCAM*, **1**, 263–277.

[25] Grimble, M. J. 1979. Solution of the stochastic optimal control problem in the *s*-domain for systems with time delay, *Proc. IEE*, **126**(7), 697–704, July.

[26] Lam, K. P. 1980. Implicit and explicit self-tuning controllers, Report 1334/80, 1980, Engineering Laboratory, University of Oxford, Oxford, UK.

[27] Athans, M. 1970. On the design of PID controllers using optimal linear regulator theory, *Automatica*, **7**, 643–647.

[28] Porter, B. 1971. Optimal control of multivariable linear systems incorporating integral feedback, *Electronic Letters*, **7**(8), 170–172, April.

[29] Grimble, M. J. 1979. The design of optimal stochastic regulating systems including integral action, *Proc. IEE*, **126**(9), 841–848, Sept.

[30] Doyle, J. C. 1978. Guaranteed margins for LQG regulators, *IEEE Trans. Auto. Control*, **AC-23**, 756–757.

[31] Doyle, J. C. and Stein, G. 1981. Multivariable feedback design: Concepts for a classical-modern synthesis, *IEEE Trans. Auto. Control*, **AC-26**(1), 4–16.

[32] Kwakernaak, H. 1969. Optimal low-sensitivity linear feedback systems, *Automatica*, **5**, 279–285.

[33] Doyle, J. C. and Stein, G. 1979. Robustness with observers, *IEEE Trans. Auto. Control*, **AC-24**(4), 607–611, Aug.

[34] Moore, B. C. and Blight, J. D. 1981. Performance and robustness design, Preprints 20th Control and Decision Conference pp. 1191–1200, San Diego, USA.

[35] Safonov, M. G., Laub, A. and Hartmann, G. L., Feedback properties of multivariable systems: The role and use of the return difference matrix, *IEEE Trans. Auto. Control*, **AC-26**(1), 47–65.

[36] Grimble, M. J. 1982. Generalized Wiener and Kalman filters for uncertain systems, Preprints 21st Control and Decision Conference, Orlando, Florida, Dec.

[37] Grimble, M. J. 1982. Optimal control of linear uncertain multivariable stochastic systems, *IEE Proc.* **129**(6), Pt. D, 263–270, Nov.

[38] Grimble, M. J. 1983. Robust LQG design of discrete systems using a dual criterion, Paper FA6:12.30, pp. 1196–1198, Preprints 22nd Control and Decision Conference.

[39] Grimble, M. J. 1985. LQG design of discrete systems using a dual criterion, *IEE Proc.*, **132**(2), 61–68, March.

[40] Francis, B. A. 1983. On the Wiener–Hopf approach to optimal feedback design, *Systems and Control Letters*, **2**, 197–201.

[41] Desoer, C. A., Liu, R. W., Murray, J. and Saeks, R. 1980. Feedback system design: The fractional representation approach to analysis and synthesis, *IEEE Trans. Auto. Control*, **AC-25**, 399–412, June.

[42] Astrom, K. J. 1979. Self-tuning regulators: Design principles and applications, Research Report LUTFD3/(TFRT-7177)/1-068/(1979), Dept. of Automatic Control, Lund Institute of Technology, Lund, Sweden.

[43] Hodgson, A. J. F. 1982. Problems of integrity in applications of adaptive

controllers, D.Phil. Thesis, Research Report 1436/82, 1982, Dept. of Engineering Science, University of Oxford, UK.

[44] Grimble, M. J. 1984. Implicit and explicit LQG self-tuning controllers, *Automatica*, **20**(5), 661–669.

[45] Grimble, M. J. and Johnson, M. A. 1984. Robustness and optimality: A dual performance index, Preprints IFAC World Conf., Budapest, July.

[46] Grimble, M. J. 1986. Dual criterion stochastic optimal control problem for robustness improvement, *IEEE Trans. Automatic Control*, **AC-31**(2), 181–185.

[47] Grimble, M. J. 1984. LQG multivariable controllers: Minimum variance interpretation for use in self-tuning systems, *Int. J. Control*, **40**(4), 831–842.

[48] Grimble, M. J. 1986. Self-tuners based on LQG theory, *Encyclopedia of Systems and Control*, ed. M. Singh, Pergamon Press, Oxford, UK.

[49] Grimble, M. J., Moir, T. J. and Fung, P. T. K. 1982. Comparison of WMV and LQG self-tuning controllers, IEEE Conf. on Applics. of Adaptive and Multivariable Control, Hull, UK.

CHAPTER 11

The Frequency Domain Analysis of the Stochastic Control Problem

11.1. INTRODUCTION

In the last decade there has been a revival of interest in the solution of stochastic linear quadratic optimal control problems described in the s-domain (Youla *et al.* [1], [2], Shaked [3], Grimble [4], [5], Kučera [6]). These techniques are mainly generalizations of the spectral factorization route.

A similar Wiener approach can be used to solve discrete-time stochastic optimal problems described in the z-domain. Early results were presented by Strejc [7] and more recent contributions have been made by Barrett [8], Gawthrop [9], Grimble [10] and Kučera [11]. If the control-weighting term in the linear quadratic cost functional and the measurement noise covariance matrix are set to zero, the general problem reduces to the so-called minimum variance control problem. Solutions for this class of problems, important in self-tuning applications, have been presented by Åström [12] and Peterka [13].

In most of the previous work, the systems have either deterministic or stochastic signals but not a mixture of both. Grimble [4], [10] has presented a novel two-stage controller-design procedure which permits the minimization of stochastic tracking errors and disturbances separately to the minimization of deterministic tracking errors and disturbance. An alternative approach to modelling random signals is to represent them by shape deterministic signals; for example, as a step function with random magnitude (Youla *et al.* [2], Grimble [14]). This particular approach has received little attention and is worthy of further research.

The main objective in this chapter is to give a unified presentation for the polynomial systems approach of Kučera [11] and the modern Wiener–Hopf approach of Youla *et al.* [1], [2]. These approaches are also related to the more classical techniques adopted by Shaked [3] and Grimble [10]. The analysis presented in this chapter is mainly based on the polynomial matrix approach of Kučera and is used to unify the various techniques. This approach has the potential for the calculation of the s- and z-domain solutions by straightforward computer-based algorithms. However, the method is currently restricted to infinite-time control problems and transport delays cannot be included in the continuous-time system descriptions. These restrictions do not apply to the non-polynomial solution procedures described in the later sections of this chapter.

The various relationships between the different system and descriptions and solution procedures are illustrated in Fig. 11.1. The arrows indicate where results from a particular procedure are either closely related or can be used to simplify the controller calculation in a different method. The matrix fractional approach of Desoer and colleagues [27] has been applied to the stochastic optimal control problem by Grimble [44]. However, this approach is not discussed in this text but an excellent presentation can be found in Vidyasagar [45].

The systems of interest in this chapter are assumed to be time invariant and the disturbances are (weakly) stationary stochastic processes with known rational spectral densities. The problem is to find a controller which minimizes a sum of weighted variances of the system input and output signals, subject to the constraint that the closed-loop system is stable. The system can be multivariable, non-square, open-loop unstable and/or non-minimum phase. The system can also include a transport delay and procedures are available to limit the effects of saturation. Discrete-time control problems will be considered in detail, but apart from the analysis of time delays, the approach for the continuous and discrete-time cases is very similar.

The plan of the chapter is as follows: Section 11.2 contains all the basic polynomial matrix analysis required in the chapter. The approach taken is to

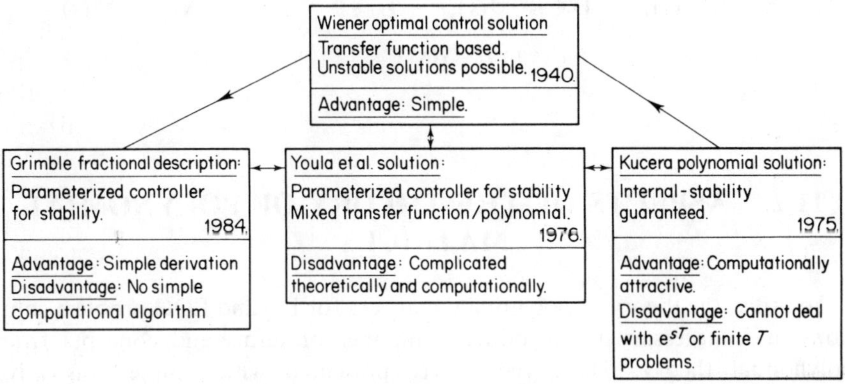

Fig. 11.1. Historical relationships between the different system representations

present the theory for scalar polynomials, generalize the results for polynomial matrices and concludes with material on Diophantine equations.

The application of the polynomial matrix theory to control system design is presented in Section 11.3. The topics investigated include the relationship between coprimeness and the concepts of controllability and observability, the specification of stochastic systems and the transformed versions of different species of quadratic cost functionals. A description of the class of stabilizing controllers, using the polynomial system concepts, closes the section.

Sections 11.4 and 11.5 contain a detailed analysis of the linear quadratic Gaussian (LQG) stochastic optimal control problem, as solved by the techniques of Kučera, Youla and the more classical gradient method. The analysis is performed for a discrete-time system description and summary theorems are presented for the continuous time-domain results using s-domain transfer-function matrices.

A brief review of analogous polynomial system results for the deterministic optimal control problem close the chapter.

Comments on notation

The set of polynomials in indeterminate s with real coefficients is denoted $\mathbb{P}(s)$. The set of ratios of polynomials of $\mathbb{P}(s)$ is denoted $\mathbb{R}(s)$. Thus if $a, b \in \mathbb{P}(s)$, then $r(s) = a(s)/b(s) \in \mathbb{R}(s)$.

The multivariable equivalent definitions are: Let $\mathbb{P}^{r\times m}(s)$ represent the set of polynomials in indeterminate s whose coefficient matrices satisfy $\mathbf{A}_i \in \mathbb{R}^{r\times m}$. Also let $\mathbb{R}^{r\times m}(s)$ denote the set of matrix fractions of the form:

$$\mathbf{G}(s) = \mathbf{N}(s)\mathbf{D}^{-1}(s) = \mathbf{D}_1^{-1}(s)\mathbf{N}_1(s)$$

where

$$\mathbf{N} \in \mathbb{P}^{r\times n}(s), \qquad \mathbf{D} \in \mathbb{P}^{n\times n}(s), \qquad \mathbf{D}_1 \in \mathbb{P}^{r\times r}(s), \qquad \mathbf{N}_1 \in \mathbb{P}^{r\times n}(s)$$

$$\text{and} \quad \mathbf{G}(s) \in \mathbb{R}^{r\times n}(s).$$

11.2. ASPECTS OF THE THEORY OF POLYNOMIAL MATRICES

To utilise the theory of polynomial matrices for LQ and LQG problem solutions, it is necessary to introduce a number of terms and concepts from polynomial theory. These results are presented with simple illustrative examples in this section.

11.2.1. Polynomial theory

The polynomial form of interest is given by $a(s)$ where the degree n is finite, the coefficients $a_k \in \mathbb{R}$ and s is an indeterminate variable, viz.:

$$a(s) \triangleq a_0 + a_1 s^1 + \cdots + a_{n-1} s^{n-1} + a_n s^n \in \mathbb{P}(s).$$

In the sequel, the indeterminate may be interpreted as the Laplace transform variable s in continuous time problems, the delay operator, or inverse of the z transform variable (z^{-1}) in discrete time problems.

DEFINITION 11.1. *Polynomial characteristics*

(i) If $a_n \neq 0$ then the degree of $a(s)$ is n; this is denoted by $\deg a = n$.

(ii) The leading coefficient of $a(s)$ is the coefficient associated with the highest power of the indeterminate s.

(iii) If the leading coefficient of polynomial $a(s)$ is unity then the polynomial is called monic. □

The potential to perform long division of polynomials is formalized as follows.

Theorem 11.1. ***Polynomial division.*** *Let a(s) and d(s) be two polynomials where d(s) ≠ 0. Then, there exists unique polynomials q(s) and r(s) such that*

(i) $a(s) = q(s)\ d(s) + r(s)$

and

(ii) either $r(s) = 0$ *or* $\deg r < \deg d$.

Proof (Hoffman and Kunze [15]). □

DEFINITION 11.2. *Polynomial divisors.* If from Theorem 11.1, $r(s) = 0$ then:

(i) Polynomial $a(s)$ is divisible (without remainder) by the polynomial $d(s)$.

(ii) Polynomial $d(s)$ is termed a divisor and $q(s)$ a quotient.

(iii) If $c(s)$ is a divisor of two polynomials $a(s)$ and $b(s)$ then $c(s)$ is termed a common divisor.

(iv) Every non-zero polynomial $a(s)$ and divisor $d(s)$ has a non-zero constant as a trivial common divisor. □

DEFINITION 11.3. *Greatest common divisor.* A polynomial $c(s)$ is the greatest common divisor of two polynomials $a(s)$ and $b(s)$ if:

(i) $c(s)$ is a common divisor of $a(s)$ and $b(s)$.
(ii) $c(s)$ is divisible by every common divisor of $a(s)$ and $b(s)$. □

Lemma 11.1. *The greatest common divisors of polynomials a(s) and b(s) are expressible in the form:*

$$c(s) = x(s)a(s) + y(s)b(s)$$

where x(s) and y(s) are polynomials.

Proof (Chen [16]). □

DEFINITION 11.4. *Coprime or relatively prime polynomials.* If the greatest common divisor of polynomials $a(s)$ and $b(s)$ is a non-zero constant then $a(s)$ and $b(s)$ are termed coprime or relatively prime polynomials. □

The greatest common divisor is therefore only unique up to a constant multiplier. A monic greatest common divisor is unique.

Theorem 11.2. ***Equivalent conditions for coprimeness.*** *The polynomials a(s) and b(s), with a(s) ≠ 0, are coprime if and only if one of the following equivalent conditions holds:*

(*i*) *For every* $s \in \mathbb{C}$, *the matrix*

$$\begin{bmatrix} a(s) \\ b(s) \end{bmatrix} \quad \textit{has rank } 1.$$

(*ii*) *There exist two polynomials* $x(s)$ *and* $y(s)$ *such that* $x(s)a(s) + y(s)b(s) = 1$.

(*iii*) *There exist no polynomials* $n(s)$ *and* $d(s)$ *with deg d < deg a such that:*

$$\frac{b(s)}{a(s)} = \frac{n(s)}{d(s)}$$

or equivalently

$$[-b(s) \;\; a(s)] \begin{bmatrix} d(s) \\ n(s) \end{bmatrix} = 0.$$

Proof (Chen [16]). □

DEFINITION 11.5. *Irreducible rational functions.* The rational function $b(s)/a(s)$ (where $a(s)$ and $b(s)$ are polynomials) is termed irreducible if $a(s)$ and $b(s)$ are coprime, otherwise $b(s)$ is reducible. □

Note, whereas polynomials of the indeterminate variable s with real coefficients are denoted by $p \in \mathbb{P}(s)$, real rational functions of the form $g(s) = b(s)/a(s)$ are denoted by $g \in \mathbb{R}(s)$.

Example 11.1. Consider polynomials:

$$a(s) = 6(s+1)(s+2)(s+3)$$

$$b(s) = (s+1)^2(s+2)(5s+4)$$

$$c(s) = s^2 + 4s + 2.$$

Thus, deg $a = 3$, deg $b = 4$ and deg $c = 2$. The leading coefficient of polynomials a, b, and c are 6, 5 and 1, respectively. Polynomial $c(s)$ is monic.

Consider first polynomial $a(s)$. If $d(s) = (s-3)$ then $a(s)$ may be written as:

$$a(s) = (6s^2 + 54s + 228)(s-3) + 720.$$

From Theorem 11.1, identify the quotient $(6s^2 + 54s + 228)$ and the remainder $r(s) = 720$, where deg $r <$ deg d. If $d(s) = (s+1)$ then $a(s)$ may be written:

$$a(s) = 6(s+2)(s+3) \cdot (s+1)$$

From Theorem 11.1 and Definition 11.2, identify the trivial remainder as $r(s) = 0$, hence $(s+1)$ is a divisor of $a(s)$ and $q(s) = 6(s+2)(s+3)$ the quotient.

Consider the polynomial $b(s)$, then if $d(s) = (s+1)^2$:

$$b(s) = (s+2)(5s+4) \cdot (s+1)^2.$$

From Theorem 11.1 and Definition 11.2, identify a trivial remainder $r(s) = 0$, quotient $q(s) = (s+2)(5s+4)$ and note that $(s+1)^2$ is a divisor of $b(s)$.

Thus, $(s+1)$ is a common divisor of both $a(s)$ and $b(s)$. The common divisors of $a(s)$ and $b(s)$ may be listed as $(s+1)$ and $(s+1)(s+2)$. From Definition 11.3 the greatest common divisor may be identified as $(s+1)(s+2)$.

Finally polynomials $a(s)$ and $c(s)$ form a coprime pair. To verify this assertion use conditions (i) from Theorem 11.2 namely:

$$\text{rank}\begin{bmatrix} a(s) \\ c(s) \end{bmatrix} = \text{rank}\begin{bmatrix} 6(s+1)(s+2)(s+3) \\ s^2 + 4s + 2 \end{bmatrix} = 1$$

for all the zeros of $a(s)$.

11.2.2. Polynomial matrices and matrix fractions

The theory for polynomial matrices follows that already presented for scalar polynomials. Polynomial matrices may be written using the indeterminate variable s as:

$$\mathbf{A}(s) = \mathbf{A}_0 + \mathbf{A}_1 s + \cdots + \mathbf{A}_{n-1}s^{n-1} + \mathbf{A}_n s^n \in \mathbb{P}^{r \times m}(s)$$

where $\mathbf{A}_i \in \mathbb{R}^{r \times m}, \qquad i = 0, 1, \ldots, n.$

Hence, a polynomial matrix is a matrix whose elements are polynomials in the indeterminate s, whilst a transfer-function matrix has elements which are rational functions of s.

DEFINITION 11.6. *Characteristics of polynomial matrices.* If $\mathbf{A}_n \neq \mathbf{O}$, then n is termed the degree of $\mathbf{A}(s)$ denoted deg $\mathbf{A} = n$.

If $r = m$ so that $\mathbf{A}$ is a square polynomial matrix, then

(a) $\det\{\mathbf{A}_n\} \neq 0$, implies $\mathbf{A}(s)$ is a proper polynomial matrix.

(b) $\det\{\mathbf{A}(s)\} =$ constant, independent of the indeterminate variable s, implies that the polynomial matrix $\mathbf{A}(s)$ is unimodular. □

Theorem 11.3. ***Inverse of the unimodular matrix.*** *A square polynomial matrix is unimodular if and only if its inverse is a polynomial matrix.*

Proof (Chen [16]). □

DEFINITION 11.7. *Left and right divisors.* Consider polynomial matrices $\mathbf{X}, \mathbf{Y}$ and $\mathbf{Z}$. If $\mathbf{X} = \mathbf{YZ}$ then

(i) $\mathbf{Y}$ is a left divisor of $\mathbf{X}$; $\mathbf{X}$ is a right multiple of $\mathbf{Y}$.

(ii) $\mathbf{Z}$ is a right divisor of $\mathbf{X}$; $\mathbf{X}$ is a left multiple of $\mathbf{Z}$. □

DEFINITION 11.8. *Common left divisor and greatest common left divisor.* Consider a set of polynomial matrices $\{\mathbf{X}_i(s)\}_{i=1}^{p}$ each having the same number of rows r and polynomial matrix $\mathbf{L} \in \mathbb{P}^{r\times r}(s)$

(i) If $\mathbf{L}(s)$ is a left divisor of each $\mathbf{X}_i(s)$, $i = 1, \ldots, p$ (namely $\mathbf{X}_i(s) = \mathbf{L}(s)\tilde{\mathbf{X}}_i(s)$, $i = 1, \ldots, p$) then $\mathbf{L}$ is termed a common left divisor of $\{\mathbf{X}_i(s)\}_{i=1}^{p}$.

(ii) If $\mathbf{L}(s)$ is a right multiple of every common left divisor, then it is termed the greatest common left divisor of the set $\{\mathbf{X}_i(s)\}_{i=1}^{p}$. □

DEFINITION 11.9. *Common right multiple and least common right multiple.* Consider a set of polynomial matrices $\{\mathbf{X}_i(s)\}_{i=1}^{p}$ each having the same number of rows r and a polynomial matrix $\mathbf{M} \in \mathbb{P}^{r\times r}(s)$.

(i) If $\mathbf{M}$ is a right multiple of each $\mathbf{X}_i(s), i = 1, \ldots p$ (namely $\mathbf{M}(s) = \mathbf{X}_i(s)\tilde{\mathbf{X}}_i(s)$, $i = 1, \ldots, p$) then $\mathbf{M}$ is termed a common right multiple of the set $\{\mathbf{X}_i(s)\}_{i=1}^{p}$.

(ii) If $\mathbf{M} \in \mathbb{P}^{r\times r}(s)$ is a left divisor of every common right multiple, then it is termed the least common right multiple of the set $\{\mathbf{X}_i(s)\}_{i=1}^{p}$. □

The definitions of common right divisor, greatest common right divisor, common left multiple and least common left multiple follow by analogy.

The two concepts of unimodularity and divisors are combined to enable the coprimeness of a polynomial matrix to be defined.

DEFINITION 11.10. *Coprime polynomial matrices.*

(i) Matrices $\mathbf{X} \in \mathbb{P}^{r\times m}(s)$ and $\mathbf{Y} \in \mathbb{P}^{r\times q}(s)$ are left coprime if their only common left divisors are unimodular matrices.

(ii) Matrices $\mathbf{M} \in \mathbb{P}^{l\times m}(s)$ and $\mathbf{N} \in \mathbb{P}^{r\times m}(s)$ are right coprime if their only common right divisors are unimodular. □

Two polynomial matrices of the same dimension may be left coprime but not necessarily right coprime and vice versa. Unimodular matrices may also be used to define an equivalence relationship.

DEFINITION 11.11. *Left- and right-equivalent polynomial matrices.* Consider polynomial matrices $\mathbf{X} \in \mathbb{P}^{r\times r}(s)$ and $\mathbf{Y} \in \mathbb{P}^{r\times m}(s)$ then

(i) If $\mathbf{X} = \mathbf{VY}$, where $\mathbf{V} \in \mathbb{P}^{r\times r}(s)$ is unimodular, then $\mathbf{X}$ and $\mathbf{Y}$ are termed right equivalent.

(ii) If $\mathbf{X} = \mathbf{YU}$, where $\mathbf{U} \in \mathbb{P}^{m\times m}(s)$ is unimodular, then $\mathbf{X}$ and $\mathbf{Y}$ are termed left equivalent.

(iii) Matrices $\mathbf{X}$ and $\mathbf{Y}$ are termed equivalent if $\mathbf{X} = \mathbf{VYU}$ where both $\mathbf{V}$ and $\mathbf{U}$ are unimodular matrices of compatible dimensions. □

The analogy with the scalar polynomial theory continues with a lemma for the greatest common right divisor.

Lemma 11.2. *Greatest common right divisor. A greatest common right divisor* $\mathbf{C}(s) \in \mathbb{P}^{m\times m}(s)$ *of the polynomial matrices* $\mathbf{A}_1(s) \in \mathbb{P}^{m\times m}(s)$ *and* $\mathbf{B}_1(s) \in \mathbb{P}^{r\times m}(s)$ *is expressible as:*

$$\mathbf{C}(s) = \mathbf{X}(s)\mathbf{A}_1(s) + \mathbf{Y}(s)\mathbf{B}_1(s) \tag{11.1}$$

where $\mathbf{X}(s) \in \mathbb{P}^{m\times m}(s)$ *and* $\mathbf{Y}(s) \in \mathbb{P}^{m\times r}(s)$.

Proof (Chen [16]). □

Theorem 11.4. *Equivalent conditions for coprime polynomial matrices. Let* $\mathbf{A}_1(s) \in \mathbb{P}^{m\times m}(s)$ *and* $\mathbf{B}_1(s) \in \mathbb{P}^{r\times m}(s)$*, then the following statements are equivalent:*

(i) Pair $\mathbf{A}_1, \mathbf{B}_1$ *are right coprime.*

(ii) The $(m + r)$ *rows of matrix* $\begin{bmatrix}\mathbf{A}_1(s)\\ \mathbf{B}_1(s)\end{bmatrix}$ *are right coprime.*

(iii) There exists a unimodular matrix $\mathbf{V}(s) \in \mathbb{P}^{(r+m)\times(r+m)}(s)$ *such that:*

$$\mathbf{V}(s)\begin{bmatrix}\mathbf{A}_1(s)\\ \mathbf{B}_1(s)\end{bmatrix} = \begin{bmatrix}\mathbf{I}_m\\ \mathbf{O}_{r\times m}\end{bmatrix} \tag{11.2}$$

(namely the Smith form is full rank).

(iv) There exist matrices $\mathbf{X}_1(s) \in \mathbb{P}^{m\times m}(s)$ *and* $\mathbf{Y}_1(s) \in \mathbb{P}^{m\times r}(s)$ *such that:*

$$\mathbf{X}_1(s)\mathbf{A}_1(s) + \mathbf{Y}_1(s)\mathbf{B}_1(s) = \mathbf{I}_m. \tag{11.3}$$

(This is the Bezout identity (Kailath [17]).)

(v)

$$Rank\begin{bmatrix}\mathbf{A}_1(s)\\ \mathbf{B}_1(s)\end{bmatrix} = m \text{ for all } s \in \mathbb{C}. \tag{11.4}$$

Proof (Antsaklis [18]). □

Example 11.2. *Conditions for coprimeness.* Let

$$\mathbf{A}_1(s) = \begin{bmatrix} s^2 & -1 \\ -s & s^2 \end{bmatrix} \quad \text{and} \quad \mathbf{B}_1(s) = \begin{bmatrix} s & -s \\ 0 & 1 \end{bmatrix}$$

(Wolovich [22]).

(a) Test for comprimeness of the rows of the composite matrix:

$$\begin{bmatrix}\mathbf{A}_1(s)\\ \hdashline \mathbf{B}_1(s)\end{bmatrix} = \begin{bmatrix} s^2 & -1 \\ -s & s^2 \\ \hdashline s & -s \\ 0 & 1 \end{bmatrix}.$$

Exhibit the following relationships:

$$[\,s^2 \quad -1\,] = [\,s \quad -1\,]\begin{bmatrix} s & 0 \\ 0 & 1 \end{bmatrix}$$

$$[\,-s \quad s^2\,] = [\,-1 \quad s^2\,]\begin{bmatrix} s & 0 \\ 0 & 1 \end{bmatrix}$$

$$[\,s \quad -s\,] = [\,1 \quad -s\,]\begin{bmatrix} s & 0 \\ 0 & 1 \end{bmatrix}$$

$$[\,0 \quad 1\,] = [\,0 \quad 1\,]\begin{bmatrix} s & 0 \\ 0 & 1 \end{bmatrix}.$$

Clearly, the rows have a common right divisor which is not unimodular, thus the rows are not right coprime and the pair $(\mathbf{A}_1(s), \mathbf{B}_1(s))$ is not right coprime.

(b) The Smith form (see Chapter 3) for $\begin{bmatrix} \mathbf{A}_1(s) \\ \mathbf{B}_1(s) \end{bmatrix}$ is given by:

$$\mathbf{V}(s)\begin{bmatrix} \mathbf{A}_1(s) \\ \mathbf{B}_1(s) \end{bmatrix} = \begin{bmatrix} 0 & 0 & 1 & s \\ 0 & 0 & 0 & 1 \\ 1 & 0 & -s & -s^2+1 \\ 0 & 1 & 1 & -s^2+s \end{bmatrix} \begin{bmatrix} s^2 & -1 \\ -s & s^2 \\ s & -s \\ 0 & 1 \end{bmatrix} = \begin{bmatrix} s & 0 \\ 0 & 1 \\ 0 & 0 \\ 0 & 0 \end{bmatrix}.$$

Hence, the invariants of $\begin{bmatrix} \mathbf{A}_1(s) \\ \mathbf{B}_1(s) \end{bmatrix}$ are s and 1 which does not conform with the requirement for m unity invariants as required in condition (iii) of Theorem 11.4.

(c) The rank property of this pair is:

$$\operatorname{rank}\begin{bmatrix} \mathbf{A}_1(s) \\ \mathbf{B}_1(s) \end{bmatrix} \neq m = 2 \text{ for all values of } s \in \mathbb{C}.$$

Since setting $s = 0$ yields:

$$\operatorname{rank}\begin{bmatrix} \mathbf{A}_1(0) \\ \mathbf{B}_1(0) \end{bmatrix} = \operatorname{rank}\begin{bmatrix} 0 & -1 \\ 0 & 0 \\ 0 & 0 \\ 0 & 1 \end{bmatrix} = 1 < m = 2.$$

The conclusion is that pair $(\mathbf{A}_1(s), \mathbf{B}_1(s))$ is not coprime.

Example 11.3. *Conditions for co-primeness.* Let

$$\mathbf{A}_1(s) = \begin{bmatrix} -4 & 3 \\ -3-5s & 3+4s \end{bmatrix} \quad \text{and} \quad \mathbf{B}_1(s) = \begin{bmatrix} 5+4s & -3-2s \\ 6 & -5 \end{bmatrix}.$$

(a) Test for right coprimeness of the rows of the composite matrix:

$$\begin{bmatrix} \mathbf{A}_1(s) \\ \hdashline \mathbf{B}_1(s) \end{bmatrix} = \begin{bmatrix} -4 & 3 \\ -3-5s & 3+4s \\ \hdashline 5+4s & -3-2s \\ 6 & -5 \end{bmatrix}.$$

Each of the above $m + r$ rows are right coprime and hence by Theorem 10.4, condition (ii) pair $(\mathbf{A}_1(s), \mathbf{B}_1(s))$ are right coprime.

(b) The unimodular matrix $\mathbf{V}$ may be obtained as:

$$\mathbf{V} = \begin{bmatrix} -2.5 & 0 & 0 & -1.5 \\ -3 & 0 & 0 & -2 \\ 2+4.5s & -1 & 1 & 1.5s \\ 1.5-0.5s & 1 & 0 & 0.5s+1.5 \end{bmatrix}$$

so that

$$\mathbf{V}(s)\begin{bmatrix} \mathbf{A}_1(s) \\ \mathbf{B}_1(s) \end{bmatrix} = \begin{bmatrix} \mathbf{I}_2 \\ \mathbf{O} \end{bmatrix}.$$

(c) If the unimodular matrix $\mathbf{V}$ is partitioned:

$$\mathbf{V} = \begin{bmatrix} \mathbf{X}_1(s) & \mathbf{Y}_1(s) \\ -\mathbf{B}_0(s) & \mathbf{A}_0(s) \end{bmatrix} \quad \text{then clearly}$$

$$\mathbf{V}(s)\begin{bmatrix} \mathbf{A}_1(s) \\ \mathbf{B}_1(s) \end{bmatrix} = \begin{bmatrix} \mathbf{X}_1(s) & \mathbf{Y}_1(s) \\ -\mathbf{B}_0(s) & \mathbf{A}_0(s) \end{bmatrix}\begin{bmatrix} \mathbf{A}_1(s) \\ \mathbf{B}_1(s) \end{bmatrix} = \begin{bmatrix} \mathbf{I}_2 \\ \mathbf{O}_2 \end{bmatrix}$$

hence

$$\mathbf{X}_1(s)\mathbf{A}_1(s) + \mathbf{Y}_1(s)\mathbf{B}_1(s) = \mathbf{I}_2.$$

Identify $$\mathbf{X}_1(s) = \begin{bmatrix} -2.5 & 0 \\ -3 & 0 \end{bmatrix} \quad \text{and} \quad \mathbf{Y}_1(s) = \begin{bmatrix} 0 & -1.5 \\ 0 & -2 \end{bmatrix}$$

as required for the Bezout identity.

(d) The rank condition is given by:

$$\operatorname{rank}\begin{bmatrix} \mathbf{A}_1(s) \\ \mathbf{B}_1(s) \end{bmatrix} = \operatorname{rank}\begin{bmatrix} -4 & 3 \\ -3-5s & 3+4s \\ 5+4s & -3-2s \\ 6 & -5 \end{bmatrix} = 2$$

for all values of $s \in \mathbb{C}$.

Uses of the matrix relationships for $\mathbf{V}^{-1}\mathbf{V}$ *and* $\mathbf{V}\mathbf{V}^{-1}$

The condition for coprimeness, using a unimodular matrix (equation (11.3), Theorem 11.4) can be exploited in a definition for the inverse of the matrix $\mathbf{V}$ discussed above. Partition the unimodular matrix $\mathbf{V}$ as:

$$\mathbf{V} = \begin{bmatrix} \mathbf{X}_1 & \mathbf{Y}_1 \\ -\mathbf{B}_0 & \mathbf{A}_0 \end{bmatrix} \tag{11.5}$$

where $\mathbf{V} \in \mathbb{P}^{(m+r)\times(m+r)}(s)$, $\mathbf{X}_1 \in \mathbb{P}^{m\times m}(s)$, $\mathbf{Y}_1 \in \mathbb{P}^{m\times r}(s)$, $\mathbf{B}_0 \in \mathbb{P}^{r\times m}(s)$ and $\mathbf{A}_0 \in \mathbb{P}^{r\times r}(s)$.

The following relationships hold:

(i) $\mathbf{V}$ can be used to demonstrate the right coprimeness of the pair $(\mathbf{A}_1, \mathbf{B}_1)$, viz.:

$$\begin{bmatrix} \mathbf{X}_1\mathbf{A}_1 + \mathbf{Y}_1\mathbf{B}_1 \\ -\mathbf{B}_0\mathbf{A}_1 + \mathbf{A}_0\mathbf{B}_1 \end{bmatrix} = \begin{bmatrix} \mathbf{I}_{m\times m} \\ \mathbf{O}_{r\times m} \end{bmatrix}. \tag{11.6}$$

(ii) $\mathbf{V}$ is unimodular, thus it has a unique inverse $\mathbf{V}^{-1}$ which is also unimodular. If the inverse is partitioned, viz.:

$$\mathbf{V}^{-1} = \begin{bmatrix} \hat{\mathbf{V}}_{11} & \hat{\mathbf{V}}_{12} \\ \hat{\mathbf{V}}_{21} & \hat{\mathbf{V}}_{22} \end{bmatrix} \tag{11.7}$$

then

$$\mathbf{V}\mathbf{V}^{-1} = \begin{bmatrix} \mathbf{X}_1 & \mathbf{Y}_1 \\ -\mathbf{B}_0 & \mathbf{A}_0 \end{bmatrix} \begin{bmatrix} \hat{\mathbf{V}}_{11} & \hat{\mathbf{V}}_{12} \\ \hat{\mathbf{V}}_{21} & \hat{\mathbf{V}}_{22} \end{bmatrix} = \begin{bmatrix} \mathbf{I}_{m\times m} & \mathbf{O} \\ \mathbf{O} & \mathbf{I}_{r\times r} \end{bmatrix}$$

from which it follows that:

$$\mathbf{X}_1\hat{\mathbf{V}}_{11} + \mathbf{Y}_1\hat{\mathbf{V}}_{21} = \mathbf{I}_{m\times m} \tag{11.8}$$

and

$$-\mathbf{B}_0\hat{\mathbf{V}}_{12} + \mathbf{A}_0\hat{\mathbf{V}}_{22} = \mathbf{I}_{r\times r}. \tag{11.9}$$

(iii) A comparison of (11.8) with (11.6) intimates that $\hat{\mathbf{V}}_{11}$ and $\hat{\mathbf{V}}_{21}$ may be identified as $\mathbf{A}_1$, and $\mathbf{B}_1$, respectively, so that (11.7) may be given the form:

$$\mathbf{V}^{-1} = \begin{bmatrix} \mathbf{A}_1 & -\mathbf{Y}_0 \\ \mathbf{B}_1 & \mathbf{X}_0 \end{bmatrix} \tag{11.10}$$

where

$$\mathbf{A}_1 \in \mathbb{P}^{m\times m}(s), \mathbf{Y}_0 \in \mathbb{P}^{m\times r}(s), \mathbf{B}_1 \in \mathbb{P}^{r\times m}(s) \quad \text{and} \quad \mathbf{X}_0 \in \mathbb{P}^{r\times r}(s).$$

(iv) Equation (11.9) is the Bezout identity for demonstrating that the pair $(\mathbf{A}_0, \mathbf{B}_0)$ are left coprime. In the notation of equation (11.10): $\mathbf{B}_0\mathbf{Y}_0 + \mathbf{A}_0\mathbf{X}_0 = \mathbf{I}_{r\times r}$.

To summarize, the four equation arisings from $\mathbf{V}\mathbf{V}^{-1} = \mathbf{I}_{(m+r)\times(m+r)}$ are:

$$\left.\begin{aligned} \mathbf{X}_1\mathbf{A}_1 + \mathbf{Y}_1\mathbf{B}_1 &= \mathbf{I}_{m\times m} && \text{11.11a} \\ \mathbf{B}_0\mathbf{Y}_0 + \mathbf{A}_0\mathbf{X}_0 &= \mathbf{I}_{r\times r} && \text{11.11b} \\ -\mathbf{X}_1\mathbf{Y}_0 + \mathbf{Y}_1\mathbf{X}_0 &= \mathbf{O}_{m\times r} && \text{11.11c} \\ -\mathbf{B}_0\mathbf{A}_1 + \mathbf{A}_0\mathbf{B}_1 &= \mathbf{O}_{r\times m} && \text{11.11d} \end{aligned}\right\} \tag{11.11}$$

Four similar equations arise from the relationship $\mathbf{V}^{-1}\mathbf{V} = \mathbf{I}_{(m+r)\times(m+r)}$

namely:

$$\left.\begin{aligned} &\mathbf{A}_1\mathbf{X}_1 + \mathbf{Y}_0\mathbf{B}_0 = \mathbf{I}_{m\times m} && 11.12\text{a} \\ &\mathbf{B}_1\mathbf{Y}_1 + \mathbf{X}_0\mathbf{A}_0 = \mathbf{I}_{r\times r} && 11.12\text{b} \\ &\mathbf{A}_1\mathbf{Y}_1 - \mathbf{Y}_0\mathbf{A}_0 = \mathbf{O}_{m\times r} && 11.12\text{c} \\ &\mathbf{B}_1\mathbf{X}_1 - \mathbf{X}_0\mathbf{B}_0 = \mathbf{O}_{r\times m}. && 11.12\text{d} \end{aligned}\right\} \qquad (11.12)$$

Equation (11.11d) above has a useful system implication when written in matrix fraction form. Hence, if the pair $(\mathbf{A}_1, \mathbf{B}_1)$ are right coprime, then there exist a left coprime pair $(\mathbf{A}_0, \mathbf{B}_0)$ such that:

$$\mathbf{A}_0^{-1}\mathbf{B}_0 = \mathbf{B}_1\mathbf{A}_1^{-1}. \qquad (11.13)$$

11.2.3. Polynomial Diophantine equations

The equations considered in this section are similar to those which are involved in the problem of obtaining an integer solution to an equation of the form $ax + by = c$ where a, b and c are given integers. The more general problem, where the triple $\{a, b, c\}$ represent known polynomials, is mathematically equivalent to the problem of solving Diophantine equations (as they are now called). Diophantus was a Greek mathematician living in Alexandria in the third century A.D. and there is some dispute whether he actually worked on integer equations of the above type (see the Preface to Vidyasagar [45]). However, the term 'Diophantine equation' is common in the self-tuning literature and will therefore be used here.

There are two types of polynomial matrix Diophantine equations: unilateral equations of the form $\mathbf{AX} + \mathbf{BY} = \mathbf{D}$ or $\mathbf{XA} + \mathbf{YB} = \mathbf{D}$ and bilateral equations which have the form $\mathbf{AX} + \mathbf{YB} = \mathbf{D}$. The conditions for the existence of solutions to unilateral and bilateral polynomial Diophantine equations are given in two theorems. These were first obtained by Roth [19] and were employed by Kučera [11] in one of the first important monographs on polynomial systems control theory.

Theorem 11.5. *Unilateral polynomial Diophantine equation. The unilateral matrix polynomial equation:*

$$\mathbf{AX} + \mathbf{BY} = \mathbf{D} \qquad (11.14)$$

has a solution if and only if the greatest common left divisor, $\mathbf{U}$ *of polynomial matrices* $\mathbf{A}$ *and* $\mathbf{B}$ *is a left divisor of* $\mathbf{D}$.

Proof.

(i) If: Let $\mathbf{U}$ be the greatest common left divisor then $\mathbf{A} = \mathbf{U}\mathbf{A}_0, \mathbf{B} = \mathbf{U}\mathbf{B}_0$ and $\mathbf{D} = \mathbf{U}\mathbf{D}_0$ and pair $(\mathbf{A}_0, \mathbf{B}_0)$ are left coprime, hence from (11.11b):

$$\mathbf{A}_0\mathbf{X}_0 + \mathbf{B}_0\mathbf{Y}_0 = \mathbf{I}_r.$$

Premultiply by $\mathbf{U}$ and postmultiply by $\mathbf{D}_0$ to obtain:

$$\mathbf{A}(\mathbf{X}_0\mathbf{D}_0) + \mathbf{B}(\mathbf{Y}_0\mathbf{D}_0) = \mathbf{D}$$

from which the pair $(\mathbf{X}_0\mathbf{D}_0, \mathbf{Y}_0\mathbf{D}_0)$ constitute a solution pair.

(ii) Only if: Let $\mathbf{X}_0, \mathbf{Y}_0$ be a solution pair and utilize $\mathbf{U}$ as the greatest common left divisor of $\mathbf{A}$ and $\mathbf{B}$, namely $\mathbf{A} = \mathbf{U}\mathbf{A}_0, \mathbf{B} = \mathbf{U}\mathbf{B}_0$ then in (11.14)

$$\mathbf{U}\mathbf{A}_0\mathbf{X}_0 + \mathbf{U}\mathbf{B}_0\mathbf{Y}_0 = \mathbf{U}(\mathbf{A}_0\mathbf{X}_0 + \mathbf{B}_0\mathbf{Y}_0) = \mathbf{D}$$

and hence $\mathbf{U}$ is a left divisor of $\mathbf{D}$. □

Theorem 11.6. ***Bilateral polynomial Diophantine equation.*** *The bilateral matrix polynomial equation:*

$$\mathbf{AX} + \mathbf{YB} = \mathbf{D} \tag{11.15}$$

has a solution if and only if the matrices:

$$\mathbf{L}_1 = \begin{bmatrix} \mathbf{A} & \mathbf{O} \\ \mathbf{O} & \mathbf{B} \end{bmatrix} \quad \textit{and} \quad \mathbf{L}_2 = \begin{bmatrix} \mathbf{A} & \mathbf{D} \\ \mathbf{O} & \mathbf{B} \end{bmatrix} \quad \textit{are equivalent,}$$

(*That is, there exist unimodular matrices* $\mathbf{U}_1, \mathbf{U}_2$ *such that* $\mathbf{L}_1 = \mathbf{U}_1\mathbf{L}_2\mathbf{U}_2$.)

Proof (Kučera [11]). □

Some useful results on this topic are to be found in Barnett [20].

Corollary 11.1. *The bilateral equation* (11.15) *has a solution if and only if matrices* $\mathbf{L}_1$, *and* $\mathbf{L}_2$ *have the same invariant polynomials.*

Proof (Barnett [20]). Equivalence of matrices $\mathbf{L}_1, \mathbf{L}_2$ occurs if and only if they have the same invariant polynomials. □

Theorem 11.7. ***Solution of the unilateral Diophantine equation.*** *Let* $[\mathbf{H}_0, \mathbf{G}_0]$ *denote a particular solution of the unilateral Diophantine equation:*

$$[\mathbf{H}, \mathbf{G}] \begin{bmatrix} \mathbf{A}_1 \\ \mathbf{B}_1 \end{bmatrix} = \mathbf{D} \in \mathbb{P}^{m \times m}(s).$$

The general solution becomes:

$$[\mathbf{H}, \mathbf{G}] = [\mathbf{H}_0, \mathbf{G}_0] + \mathbf{P}_0[-\mathbf{B}_0, \mathbf{A}_0]$$

where $[-\mathbf{B}_0, \mathbf{A}_0]$ *is a prime basis for the left kernel (nullspace) of*

$\begin{bmatrix} \mathbf{A}_1 \\ \mathbf{B}_1 \end{bmatrix}$ and $\mathbf{P}_0 \in \mathbb{P}^{m \times r}(s)$ *is an arbitrary polynomial matrix.*

Proof. That the general solution satisfies the Diophantine equation may easily be confirmed by direct substitution. □

11.3. POLYNOMIAL REPRESENTATIONS FOR STOCHASTIC SYSTEMS

It is now common practice to represent coloured-noise disturbances or measurement-noise signals by the output of a transfer function driven by a white-noise input. The transfer-function might be obtained from an experimental procedure, using for example, a variant of the extended least-squares identification algorithm. Alternatively, the transfer-function might be fitted to an empirical disturbance spectrum which is based on the physics of the particular problem. An example of the latter arises in the ship-positioning problem (Chapter 13) where a standard sea-wave spectrum might be used to model sea conditions. This type of spectrum gives average wave conditions as determined by an international team of experts. The model will not, of course, accurately replicate any particular sea conditions, even for the agreed wind strength and sea state, but the sea-state model is often adequate for the design of wave filters.

The representation of a transfer function, as a ratio of polynomials is an obvious but important concept. The extension of the idea to represent a transfer-function matrix using polynomial matrices is straightforward. There are two advantages of the polynomial systems approach:

(i) Identification algorithms are usually based upon the estimation of parameters in a polynomial (matrix) model.

(ii) It is difficult to construct computer algorithms to solve optimization problems based on transfer-function matrix system descriptions (although Program CC developed by Peter Thompson, Systems Technology Inc. uses this approach). The equivalent polynomial equations have established numerical solution procedures available.

One disadvantage of the polynomial system approach which arises in the solution of continuous-time optimal control problems for a system containing time delays is that the relationship with time-domain operators is not so straightforward. The transfer-function approach, however, can be directly related to its time-domain operator origins which is sometimes an advantage for problems of the above type. However, for discrete systems, where the time delay is naturally incorporated into the polynomial model description the methodology has much to commend it, particularly in areas like multivariable self-tuning control (Chapter 12).

11.3.1. ARMAX models and the polynomial representation

In many fields of study, a general linear stochastic system is characterized by an input–output difference equation. Thus, it is assumed that the current output $y(t) \in \mathbb{R}^r$ can be described by a linear combination of past outputs $\{y(i) \in \mathbb{R}^r; i = t-1, t-2, \ldots, t-n_a\}$, past inputs $\{u(i) \in \mathbb{R}^m; i = t-k, t-k-1, \ldots, t-k-n_b\}$ and a white-noise disturbance sequence $\{\xi(i) \in \mathbb{R}^l; i = t, t-1, \ldots, t-n_c\}$, viz.:

$$\mathbf{A}_0 y(t) + \mathbf{A}_1 y(t-1) + \cdots \mathbf{A}_{n_a} y(t-n_a)$$

$$= \mathbf{B}_0 u(t-k) + \mathbf{B}_1 u(t-k-1) + \cdots + \mathbf{B}_{n_b} u(t-k-n_b)$$

$$+ \mathbf{C}_0 \xi(t) + \mathbf{C}_1 \xi(t-1) + \cdots + \mathbf{C}_{n_c} \xi(t-n_c) \tag{11.16}$$

where delay of order $k \geq 1$ ensures that the system is physically realizable and the coefficient matrices are normally assumed to be time invariant with $\mathbf{A}_0 \in \mathbb{R}^{r \times r}$ full rank.

If, for the moment, the delay operator is denoted by:

$$q^{-1} u(t) \triangleq u(t-1)$$

the process model (11.16) can be written succinctly in polynomial form as:

$$\mathbf{A}(q^{-1}) y(t) = \mathbf{B}(q^{-1}) u(t) + \mathbf{C}(q^{-1}) \xi(t) \tag{11.17}$$

where

$$\mathbf{A}(q^{-1}) = \mathbf{A}_0 + \mathbf{A}_1 q^{-1} + \cdots + \mathbf{A}_{n_a} q^{-n_a} \in \mathbb{P}^{r \times r}(q^{-1})$$

$$\mathbf{B}(q^{-1}) = q^{-k}(\mathbf{B}_0 + \mathbf{B}_1 q^{-1} + \cdots + \mathbf{B}_{n_b} q^{-n_b}) \in \mathbb{P}^{r \times m}(q^{-1})$$

and

$$\mathbf{C}(q^{-1}) = \mathbf{C}_0 + \mathbf{C}_1 q^{-1} + \cdots + \mathbf{C}_{n_c} q^{-n_c} \in \mathbb{P}^{r \times l}(q^{-1}).$$

This is known as the ARMAX model (*A*uto*R*egressive *M*oving *A*verage with e*X*ogenous variable). The term $\mathbf{A}(q^{-1})y(t)$ is a linear regression of $y(t)$ on its own past values (autoregressive), the term $\mathbf{C}(q^{-1})\xi(t)$ defines a moving average of the noise inputs and the input variable is known (in the econometric literature) as an exogenous variable. The operator equation (11.17) has the solution:

$$y(t) = \mathbf{W}(q^{-1})u(t) + \mathbf{W}_0(q^{-1})\xi(t) \tag{11.18}$$

where

$$\mathbf{W}(q^{-1}) = \mathbf{A}^{-1}(q^{-1})\mathbf{B}(q^{-1}) \in \mathbb{R}^{r\times m}(q^{-1}) \tag{11.19}$$

and

$$\mathbf{W}_0(q^{-1}) = \mathbf{A}^{-1}(q^{-1})\mathbf{C}(q^{-1}) \in \mathbb{R}^{r\times l}(q^{-1}). \tag{11.20}$$

Consider now a transform analysis for the system described by equation (11.16). Begin by introducing the definitions of the bilateral Laplace and z transforms.

Bilateral Laplace and z transforms

The bilateral Laplace transform is defined by:

$$\mathcal{L}_2\{f(t)\} \triangleq f(s) = \int_{-\infty}^{\infty} e^{-st} f(t)\,dt, \qquad c_1 < \mathcal{Re}(s) < c_2, s \in \mathbb{C}$$

and the inversion formula by:

$$\mathcal{L}_2^{-1}\{f(s)\} \triangleq f(t) = \frac{1}{2\pi j}\int_{c-j\infty}^{c+j\infty} e^{st} f(s)\,ds, \qquad c_1 < c < c_2.$$

The bilateral z transform is defined as:

$$\mathcal{Z}_2\{f(t)\} = f(z^{-1}) = \sum_{t=-\infty}^{\infty} f(t)z^{-t}, \qquad R_1 < |z| < R_2, z \in \mathbb{C}$$

and the inversion formula by:

$$\mathcal{Z}_2^{-1}\{f(z^{-1})\} \triangleq f(t) = \frac{1}{2\pi j}\oint_{\Gamma} z^{t-1} f(z^{-1})\,dz$$

where Γ is a circular contour in the z-plane lying in the region of convergence given by the annulus $R_1 < |z| < R_2$ and t is a discrete-time index lying in the integers $t = 0, \pm 1, \pm 2, \dots$. The real scalars c_1, c_2, R_1, R_2 specify the regions of convergence of the transforms.

ARMAX model: A z-transform analysis.

Lemma 11.3. *Let* $\{x(t); x(t) \in \mathbb{R}^n; t = 0 \pm 1, \pm 2, \dots\}$ *have the bilateral z-transform* $\mathcal{Z}_2\{x(t)\} = \mathbf{x}(z^{-1})$, *then for* $\mathbf{A} \in \mathbb{R}^{n \times n}$ *and k, an integer,*

$$\mathcal{Z}_2\{\mathbf{A}x(t-k)\} = \mathbf{A}\mathbf{z}^{-k}\mathbf{x}(z^{-1})$$

where $R_1 < |z| < R_2$ *and* $z \in \mathbb{C}$.

Applying this result to the ARMAX model equation (11.16) yields:

$$\mathbf{A}(z^{-1})\mathbf{y}(z^{-1}) = \mathbf{B}(z^{-1})\mathbf{u}(z^{-1}) + \mathbf{C}(z^{-1})\boldsymbol{\xi}(z^{-1}) \tag{11.21}$$

or equivalently:

$$\mathbf{y}(z^{-1}) = \mathbf{W}(z^{-1})\mathbf{u}(z^{-1}) + \mathbf{W}_0(z^{-1})\boldsymbol{\xi}(z^{-1}) \tag{11.22}$$

where

$$\mathbf{A}(z^{-1}) = \mathbf{A}_0 + \mathbf{A}_1 z^{-1} + \cdots + \mathbf{A}_{n_a} z^{-n_a}$$

$$\mathbf{B}(z^{-1}) = z^{-k}(\mathbf{B}_0 + \mathbf{B}_1 z^{-1} + \cdots + \mathbf{B}_{n_b} z^{-n_b})$$

$$\mathbf{C}(z^{-1}) = \mathbf{C}_0 + \mathbf{C}_1 z^{-1} + \cdots + \mathbf{C}_{n_c} z^{-n_c}$$

$$\mathbf{W}(z^{-1}) = \mathbf{A}^{-1}(z^{-1})\mathbf{B}(z^{-1}), \mathbf{W}_0(z^{-1}) = \mathbf{A}^{-1}(z^{-1})\mathbf{C}(z^{-1})$$

with

$$\mathbf{y}(z^{-1}) = \mathcal{Z}_2\{y(t); t = 0, \pm 1, \pm 2, \dots\},$$

$$\mathbf{u}(z^{-1}) = \mathcal{Z}_2\{u(t);\ t = 0, \pm 1, \pm 2, \dots\}$$

$$\boldsymbol{\xi}(z^{-1}) = \mathcal{Z}_2\{\xi(t);\ t = 0, \pm 1, \pm 2, \dots\},$$

$$\text{and} \qquad R_1 < |z| < R_2,\ z \in \mathbb{C}.$$

Comparing equations (11.17) and (11.21), similarly comparing equations (11.18) and (11.22) reveals that the polynomial forms of the quantities $\mathbf{A}(\cdot)$, $\mathbf{B}(\cdot)$, $\mathbf{C}(\cdot)$, $\mathbf{W}(\cdot)$ and $\mathbf{W}_0(\cdot)$ have been preserved in both approaches. This is quite important in certain areas, for example, in the self-tuning literature. In that particular topic it is usual to define the system matrices in terms of the delay operator q^{-1}, namely $\mathbf{W}(q^{-1}) = \mathbf{A}^{-1}(q^{-1})\mathbf{B}(q^{-1})$, yet as the above analysis shows, the same coefficient matrices are obtained when writing the z-transform equivalent. However, mathematically the delay operator q^{-1}, and the complex number z^{-1} are quite different objects even though their effects when viewed from the time domain have certain similarities. In this text, just one symbol z^{-1} is utilized to represent both the delay operator and the z-transform complex number. This has the advantage that the ARMAX models in Chapter 12 on self-tuning are of the same form as those used in the z-domain solution to the LQG problem presented here. It is also this reason which motivates the writing of z-transforms as functions of z^{-1}. This enables the z-domain LQG problem solution to be employed directly in Chapter 12 without having to change the form of the polynomial matrices in the controller expression. There should be no confusion in the use of z^{-1} for both the delay operator and the complex number associated with the z-transform since the interpretation should be clear from the context of the expressions involved. In the remainder of this chapter, a transform analysis is presented and z^{-1} will denote the complex number associated with the z transform.

11.3.2. State-space and polynomial system concepts

The usual state-space description for either discrete or continuous systems is readily given a polynomial systems representation. For present discussions, the emphasis is on a square (equal numbers of outputs and inputs) discrete state-space description. Note that non-square systems may be analysed in a similar fashion. Consider, then the state-space discrete system, tacitly assuming $r = m \ll n$:

$$x(t+1) = \tilde{\mathbf{A}}x(t) + \tilde{\mathbf{B}}u(t) + \tilde{\mathbf{D}}\xi(t) \tag{11.23}$$

$$y(t) = \tilde{\mathbf{C}}x(t) \tag{11.24}$$

where

$$x(t) \in \mathbb{R}^n,\ u(t) \in \mathbb{R}^m,\ y(t) \in \mathbb{R}^m,\ \xi(t) \in \mathbb{R}^l \quad \text{and}$$

$\xi(t)$ represents a white-noise source modelling the process noise and the time index t is an integer. Using the z-transform, equations (11.23) and (11.24) become:

$$\mathbf{y}(z^{-1}) = \mathbf{W}(z^{-1})\mathbf{u}(z^{-1}) + \mathbf{W}_0(z^{-1})\xi(z^{-1}) \tag{11.25}$$

where

$$\mathbf{W}(z^{-1}) = \tilde{\mathbf{C}}(\mathbf{I}_n - z^{-1}\tilde{\mathbf{A}})^{-1}z^{-1}\tilde{\mathbf{B}} \tag{11.26}$$

and

$$\mathbf{W}_0(z^{-1}) = \tilde{\mathbf{C}}(\mathbf{I}_n - \mathbf{z}^{-1}\tilde{\mathbf{A}})^{-1}z^{-1}\tilde{\mathbf{D}}. \tag{11.27}$$

Two aspects are pursued in this section, namely how (11.26) is written in matrix fraction form and the relationship between controllability and left coprimeness. Introduce for $\tilde{\mathbf{C}}(\mathbf{I}_n - z^{-1}\tilde{\mathbf{A}})^{-1} \in \mathbb{R}^{m\times n}(z^{-1})$, the Smith–McMillan form as:

$$\tilde{\mathbf{C}}(\mathbf{I}_n - z^{-1}\tilde{\mathbf{A}})^{-1} = \mathbf{M}(z^{-1})[\mathbf{\Gamma}(z^{-1}) \,\vdots\, \mathbf{O}]\mathbf{N}(z^{-1}) \tag{11.28}$$

where

$$\mathbf{M}(z^{-1}) \in \mathbb{P}^{m\times m}(z^{-1}) \text{ and } \mathbf{N}(z^{-1}) \in \mathbb{P}^{n\times n}(z^{-1})$$

are unimodular,

$$\mathbf{\Gamma}(z^{-1}) = \text{diag}\{\gamma_1(z^{-1}), \ldots, \gamma_m(z^{-1})\} \in \mathbb{R}^{m\times m}(z^{-1})$$

and

$$\{\gamma_i(z^{-1}) = n_i(z^{-1})/d_i(z^{-1});\ \gamma_i \in \mathbb{R}(z^{-1}); n_i,\ d_i \in \mathbb{P}(z^{-1});\ i = 1, \ldots, m\}.$$

It is assumed at this stage that the pair $(\tilde{\mathbf{A}}, \tilde{\mathbf{C}})$ is completely observable in order not to complicate the discussion unnecessarily.

Partition $\mathbf{\Gamma}(z^{-1})$ as:

$$\mathbf{\Gamma}(z^{-1}) = [\mathbf{\Gamma}_1(z^{-1})]^{-1}\mathbf{\Gamma}_2(z^{-1}) \tag{11.29}$$

where

$$\mathbf{\Gamma}_1(z^{-1}) = \text{diag}\{d_1(z^{-1}), \ldots, d_m(z^{-1})\} \in \mathbb{P}^{m\times m}(z^{-1})$$

and

$$\mathbf{\Gamma}_2(z^{-1}) = \text{diag}\{n_1(z^{-1}), \ldots, n_m(z^{-1})\} \in \mathbb{P}^{m\times m}(z^{-1}).$$

Hence, using equation (11.29), equation (11.28) becomes:

$$\tilde{\mathbf{C}}(\mathbf{I}_n - z^{-1}\tilde{\mathbf{A}})^{-1} = \mathbf{M}(z^{-1})[\mathbf{\Gamma}_1(z^{-1})]^{-1}[\mathbf{\Gamma}_2(z^{-1}) \,\vdots\, \mathbf{O}]\mathbf{N}(z^{-1}). \tag{11.30}$$

Substituting (11.30) into (11.25) yields:

$$\mathbf{y}(z^{-1}) = \mathbf{A}^{-1}(z^{-1})(\mathbf{B}(z^{-1})\mathbf{u}(z^{-1}) + \mathbf{C}(z^{-1})\boldsymbol{\xi}(z^{-1})) \tag{11.31}$$

where

$$\mathbf{A}(z^{-1}) = \mathbf{\Gamma}_1(z^{-1})\mathbf{M}^{-1}(z^{-1}) \in \mathbb{P}^{m\times m}(z^{-1}) \tag{11.32}$$

$$\mathbf{B}(z^{-1}) = z^{-1}[\mathbf{\Gamma}_2(z^{-1}) \,\vdots\, \mathbf{O}]\mathbf{N}(z^{-1})\tilde{\mathbf{B}} \in \mathbb{P}^{m\times m}(z^{-1}) \tag{11.33}$$

and

$$\mathbf{C}(z^{-1}) = z^{-1}[\mathbf{\Gamma}_2(z^{-1}) \,\vdots\, \mathbf{O}]\mathbf{N}(z^{-1})\tilde{\mathbf{D}} \in \mathbb{P}^{m\times l}(z^{-1}). \tag{11.34}$$

Thus the discrete state-space system description given by equations (11.23) and (11.24) has a set of (non-unique) polynomial representations given by equations (11.31) to (11.34). Although this is a slightly artificial route to a polynomial system representation it can be used to expose the relationship between left co-primeness and controllability.

Polynomial matrix $\mathbf{A}(z^{-1})$ has all its invariants on the diagonal of $\mathbf{\Gamma}_1$. For the Smith form of $\mathbf{B}(z^{-1})$(assumed to be of full normal rank) obtain:

$$\mathbf{B}(z^{-1}) = z^{-1}\mathbf{M}_1(z^{-1})\tilde{\mathbf{\Gamma}}_2(z^{-1})\mathbf{N}_1(z^{-1}) \tag{11.35}$$

where

$$\mathbf{M}_1(z^{-1}), \tilde{\mathbf{\Gamma}}_2(z^{-1}),\ \mathbf{N}_1(z^{-1}) \in \mathbb{P}^{m\times m}(z^{-1})$$

with $\mathbf{M}_1$ and $\mathbf{N}_1$ unimodular and $\tilde{\mathbf{\Gamma}}_2(z^{-1}) = \mathrm{diag}\{\tilde{n}_1(z^{-1}), \ldots, \tilde{n}_m(z^{-1})\}$.

Suppose that there exists an uncontrollable mode in the system then such a mode will cancel in the transfer-function matrix $\tilde{\mathbf{C}}(\mathbf{I}_n - z^{-1}\tilde{\mathbf{A}})^{-1}z^{-1}\tilde{\mathbf{B}}$. Hence, using equations (11.32) and (11.35) yields:

$$\tilde{\mathbf{C}}(\mathbf{I}_n - z^{-1}\tilde{\mathbf{A}})^{-1}z^{-1}\tilde{\mathbf{B}} = \mathbf{M}(z^{-1})\mathbf{\Gamma}_1^{-1}(z^{-1})z^{-1}\mathbf{M}_1(z^{-1})\tilde{\mathbf{\Gamma}}_2(z^{-1})\mathbf{N}_1(z^{-1}). \tag{11.36}$$

Note that $\mathbf{\Gamma}_1^{-1}(z^{-1})\mathbf{M}_1(z^{-1})\tilde{\mathbf{\Gamma}}_2(z^{-1}) \in \mathbb{R}^{m\times m}(z^{-1})$ comprises two diagonal matrices of full normal rank ($\mathbf{\Gamma}_1^{-1}(z^{-1})$ and $\tilde{\mathbf{\Gamma}}_2(z^{-1})$) interspaced by a unimodular matrix $\mathbf{M}_1(z^{-1}) \in \mathbb{P}^{m\times m}(z^{-1})$, so that the Smith–McMillan form is readily obtained as:

$$\mathbf{\Gamma}_1^{-1}(z^{-1})\mathbf{M}_1(z^{-1})\tilde{\mathbf{\Gamma}}_2(z^{-1}) = \tilde{\mathbf{M}}_1(z^{-1})\mathbf{\Gamma}_1^{-1}(z^{-1})\tilde{\mathbf{\Gamma}}_2(z^{-1})\tilde{\mathbf{N}}_1(z^{-1}). \tag{11.37}$$

Hence, equation (11.36) becomes:

$$\tilde{\mathbf{C}}(\mathbf{I}_n - z^{-1}\tilde{\mathbf{A}})^{-1}z^{-1}\tilde{\mathbf{B}} = \mathbf{M}_2(z^{-1})\mathbf{\Gamma}_1^{-1}(z^{-1})z^{-1}\tilde{\mathbf{\Gamma}}_2(z^{-1})\mathbf{N}_2(z^{-1})$$

where

$$\mathbf{M}_2(z^{-1}) \triangleq \mathbf{M}(z^{-1})\tilde{\mathbf{M}}_1(z^{-1}) \in \mathbb{P}^{m\times m}(z^{-1}),$$
$$\mathbf{N}_2(z^{-1}) \triangleq \tilde{\mathbf{N}}_1(z^{-1})\mathbf{N}_1(z^{-1}) \in \mathbb{P}^{m\times m}(z^{-1}),$$

with $\mathbf{M}_2$ and $\mathbf{N}_2$ both unimodular. The uncontrollable modes disappear from the transfer function (11.36) by cancellation of factors between $\boldsymbol{\Gamma}_1^{-1}(z^{-1})$ and $\tilde{\boldsymbol{\Gamma}}_2(z^{-1})$ to give:

$$\begin{aligned}\tilde{\mathbf{C}}(\mathbf{I}_n - z^{-1}\tilde{\mathbf{A}})^{-1}z^{-1}\tilde{\mathbf{B}} &= [\mathbf{A}(z^{-1})]^{-1}\mathbf{B}(z^{-1})\\ &= \mathbf{M}_2(z^{-1})\boldsymbol{\Gamma}_3^{-1}(z^{-1})z^{-1}\boldsymbol{\Gamma}_4(z^{-1})\mathbf{N}_2(z^{-1})\\ &= [\mathbf{A}_0(z^{-1})]^{-1}\mathbf{B}_0(z^{-1})\end{aligned}$$

where

$$\mathbf{A}_0(z^{-1}) \triangleq \boldsymbol{\Gamma}_3(z^{-1})\mathbf{M}_2^{-1}(z^{-1}) \in \mathbb{P}^{m\times m}(z^{-1}) \tag{11.38}$$

$$\mathbf{B}_0(z^{-1}) \triangleq z^{-1}\boldsymbol{\Gamma}_4(z^{-1})\mathbf{N}_2(z^{-1}) \in \mathbb{P}^{m\times m}(z^{-1}) \tag{11.39}$$

and the pair $\boldsymbol{\Gamma}_3(z^{-1}), \boldsymbol{\Gamma}_4(z^{-1})$ are obtained from pair $\boldsymbol{\Gamma}_1(z^{-1})$, $\tilde{\boldsymbol{\Gamma}}_2(z^{-1})$ after the cancellation of common factors.

The following remarks are pertinent:

1. The assumption that a system representation is left coprime does not guarantee that the underlying physical system is controllable. This has been demonstrated by the analysis above where pair $(\mathbf{A}_0, \mathbf{B}_0)$ is left coprime (by construction) yet the system pair $(\mathbf{A}, \mathbf{B})$ possessed an uncontrollable mode. Left coprimeness of the system representation does however ensure that no cancellation will take place in forming the transfer-function matrix $\mathbf{A}^{-1}(z^{-1})\mathbf{B}(z^{-1})$. Similar considerations apply to the transfer-function matrix between the noise process $\boldsymbol{\xi}$ and the output y.
2. In the case of polynomial systems derived from the discrete state-space system (11.23), (11.24), the system is completely controllable if and only if the pair $(\mathbf{I}_n - z^{-1}\tilde{\mathbf{A}}, z^{-1}\tilde{\mathbf{B}})$ is left coprime. Similarly, complete observability is equivalent to the right coprimeness of the pair $(z^{-1}\tilde{\mathbf{C}}, \mathbf{I}_n - z^{-1}\tilde{\mathbf{A}})$.

   For continuous state-space systems the conditions are:

   Pair $(\tilde{\mathbf{C}}, \tilde{\mathbf{A}})$ is completely controllable if and only if the pair $(\tilde{\mathbf{C}}, s\mathbf{I}_n - \tilde{\mathbf{A}})$ is left coprime.

Pair $(\tilde{\mathbf{A}}, \tilde{\mathbf{C}})$ is completely observable if and only if the pair $(s\mathbf{I} - \tilde{\mathbf{A}}, \tilde{\mathbf{C}})$ is right coprime (Blomberg and Ylinen [47], Wolovich [22]).

The importance of the relationship between coprimeness and controllability/observability concepts arises in the need to give solvability conditions for the LQG problem. In the state-space formulation the usual assumption is that of a stabilizable-detectable system. Thus, in the polynomial system formulation, the solvability lemma includes the clause: the system has no unstable hidden modes. Further formal discussion of the solvability discussions is given in Section 11.4.2.

11.3.3. Stochastic system models for the LQG problem

The stochastic system of interest is displayed in matrix fraction form in Fig. 11.2. Only minor variations occur for the discrete- and continuous-time cases (the analyses are closely analogous), however, the discrete system description in the z-domain is pursued here:

System output equations

$$\mathbf{y}(z^{-1}) = \mathbf{A}^{-1}(z^{-1})(\mathbf{B}(z^{-1})\mathbf{u}(z^{-1}) + \mathbf{C}(z^{-1})\boldsymbol{\xi}(z^{-1})). \tag{11.40}$$

Observation process

$$\mathbf{z}_0(z^{-1}) = \mathbf{y}(z^{-1}) + \mathbf{v}(z^{-1}). \tag{11.41}$$

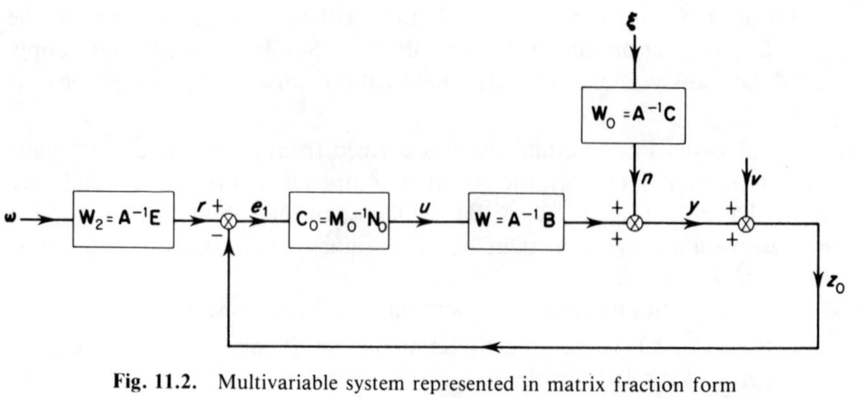

Fig. 11.2. Multivariable system represented in matrix fraction form

Controller input

$$\mathbf{e}_1(z^{-1}) = \mathbf{r}(z^{-1}) - \mathbf{z}_0(z^{-1}). \tag{11.42}$$

Reference generation process

$$\mathbf{r}(z^{-1}) = \mathbf{A}^{-1}(z^{-1})\mathbf{E}(z^{-1})\boldsymbol{\omega}(z^{-1}). \tag{11.43}$$

Tracking error

$$\mathbf{e}(z^{-1}) = \mathbf{r}(z^{-1}) - \mathbf{y}(z^{-1}). \tag{11.44}$$

The transfer-function matrices in the above system are represented by polynomial matrix fractions. Any system transfer-function matrix $\mathbf{W}(z^{-1})$ can be written using a pair of left coprime polynomial matrices $\mathbf{A}_0, \mathbf{B}_0$ or a right coprime pair $(\mathbf{A}_1, \mathbf{B}_1)$, viz.:

$$\mathbf{W}(z^{-1}) = \mathbf{A}_0^{-1}(z^{-1})\mathbf{B}_0(z^{-1}) = \mathbf{B}_1(z^{-1})\mathbf{A}_1^{-1}(z^{-1}). \tag{11.45}$$

Similarly, if $\mathbf{W}_0(z^{-1})$ and $\mathbf{W}_2(z^{-1})$ represent the disturbance and the reference generation subsystems respectively, then the total system may be represented in terms of any left co-prime matrix fraction decomposition:

$$[\mathbf{W}(z^{-1}), \mathbf{W}_0(z^{-1}), \mathbf{W}_2(z^{-1})] = [\mathbf{A}(z^{-1})]^{-1}[\mathbf{B}(z^{-1}), \mathbf{C}(z^{-1}), \mathbf{E}(z^{-1})]. \tag{11.46}$$

For notational simplicity, the dependence on z^{-1} is usually omitted, hence the above matrix becomes:

$$[\mathbf{W}, \mathbf{W}_0, \mathbf{W}_2] = \mathbf{A}^{-1}[\mathbf{B}, \mathbf{C}, \mathbf{E}]. \tag{11.47}$$

Note, that assuming the pair $(\mathbf{A}, [\mathbf{B}, \mathbf{C}, \mathbf{E}])$ is left co-prime does not imply that the individual pairs $(\mathbf{A}, \mathbf{B})$, $(\mathbf{A}, \mathbf{C})$ and $(\mathbf{A}, \mathbf{E})$ are necessarily left coprime.

The system in (11.40) to (11.44) may be multivariable and non-square. In time-domain notation, the outputs to be controlled are $y(t) \in \mathbb{R}^r$ and the applied control signals are $u(t) \in \mathbb{R}^m$, where the time index t is an integer.

Partial state

If the disturbance $n(t) = \mathbf{W}_0(z^{-1})\xi(t) = \mathbf{A}^{-1}(z^{-1})\mathbf{C}(z^{-1})\xi(t)$ is zeroed then the system equation (11.40) becomes:

$$y(t) = \mathbf{A}^{-1}(z^{-1})\mathbf{B}(z^{-1})u(t).$$

In terms of a right coprime pair: $y(t) = \mathbf{B}_1(z^{-1})\mathbf{A}_1^{-1}(z^{-1})\mathbf{u}(t)$, namely

$$\left.\begin{aligned} &\mathbf{A}_1(z^{-1})\boldsymbol{\zeta}(t) = \mathbf{u}(t) \\ \text{and} \\ &y(t) = \mathbf{B}_1(z^{-1})\boldsymbol{\zeta}(t). \end{aligned}\right\} \tag{11.48}$$

The quantity $\boldsymbol{\zeta}(t)$ is known as a partial state. The partial state does not usually represent a physically measurable variable, however, this description does have the potential as a physical model.

System model

The discrete-time system description is assumed to have at least a single-step delay in all signal paths. This realizes the constraint that a physical system will not usually respond to current signal inputs instantaneously. Hence,

$$\mathbf{B}(z^{-1}) = z^{-k}(\mathbf{B}_0 + \mathbf{B}_1 z^{-1} + \mathbf{B}_2 z^{-2} + \cdots + \mathbf{B}_{n_b} z^{-n_b}), \qquad k \geq 1$$

and $\mathbf{B}(0) = \mathbf{O} \in \mathbb{R}^{r\times m}$; also $\mathbf{A}(0) \in \mathbb{R}^{r\times r}$ is assumed to have full rank. Steady-state or infinite-time control problems are analysed in the sequel, hence, the system initial conditions can be neglected. Essentially, the initial transient effects are assumed to have decayed to zero.

The controller $\mathbf{C}_0(z^{-1})$ may be represented using either stable transfer-function matrices, $\mathbf{M}_0, \mathbf{N}_0$ as $\mathbf{C}_0(z^{-1}) = \mathbf{M}_0^{-1}(z^{-1})\mathbf{N}_0(z^{-1})$;

$$\mathbf{M}_0(z^{-1}) \in \mathbb{R}^{m\times m}(z^{-1}), \mathbf{N}_0(z^{-1}) \in \mathbb{R}^{m\times r}(z^{-1}),$$

or using polynomial matrices $\mathbf{H}_0, \mathbf{G}_0$ as

$$\mathbf{C}_0(z^{-1}) = \mathbf{H}_0^{-1}(z^{-1})\mathbf{G}_0(z^{-1}),$$

where

$$\mathbf{H}_0(z^{-1}) \in \mathbb{P}^{m\times m}(z^{-1}) \quad \text{and} \quad \mathbf{G}_0(z^{-1}) \in \mathbb{P}^{m\times r}(z^{-1}).$$

The controller is assumed to be implemented in minimal order form so that unstable hidden modes are removed. This procedure of minimal order realization is acceptable in systems with known accurate models but would be difficult to achieve in self-tuning systems.

The measurement noise v, process noise $\boldsymbol{\xi}$ and reference subsystem input noise $\boldsymbol{\omega}$ are assumed to be mutually independent, zero-mean, stationary white-noise sequences with covariances $\mathbf{R} > \mathbf{O}, \mathbf{Q}_0 \geq \mathbf{O}$ and $\mathbf{Q}_2 \geq \mathbf{O}$, respectively. The optimal control cost-functional weighting matrices are $\mathbf{Q}_1 \geq \mathbf{O}$ and $\mathbf{R}_1 > \mathbf{O}$ for the output and control terms, respectively.

Spectral factorization

The following spectral factors may be defined for use later in the chapter (see equations (11.127) and (11.128)).

Filter spectral factor

$$\mathbf{Y}_f\mathbf{Y}_f^* = \mathbf{W}_0\mathbf{Q}_0\mathbf{W}_0^* + \mathbf{W}_2\mathbf{Q}_2\mathbf{W}_2^* + \mathbf{R} \tag{11.49}$$

$$= \mathbf{A}^{-1}(\mathbf{C}\mathbf{Q}_0\mathbf{C}^* + \mathbf{E}\mathbf{Q}_2\mathbf{E}^* + \mathbf{A}\mathbf{R}\mathbf{A}^*)(\mathbf{A}^*)^{-1}.$$

Hence

$$\mathbf{Y}_f = \mathbf{A}^{-1}\mathbf{D}$$

where $\mathbf{D}\mathbf{D}^* = \mathbf{C}\mathbf{Q}_0\mathbf{C}^* + \mathbf{E}\mathbf{Q}_2\mathbf{E}^* + \mathbf{A}\mathbf{R}\mathbf{A}^* \in \mathbb{P}^{r\times r}(z^{-1}), \mathbf{D} \in \mathbb{P}^{r\times r}(z^{-1})$

and $\mathbf{D}$ is a strictly Hurwitz polynomial matrix, so that $\mathbf{D}^{-1}$ is asymptotically stable.

Control spectral factor

$$\mathbf{Y}_c^*\mathbf{Y}_c = \mathbf{W}^*\mathbf{Q}_1\mathbf{W} + \mathbf{R}_1 \tag{11.50}$$

$$= (\mathbf{A}_1^*)^{-1}(\mathbf{B}_1^*\mathbf{Q}_1\mathbf{B}_1 + \mathbf{A}_1^*\mathbf{R}_1\mathbf{A}_1)\mathbf{A}_1^{-1}.$$

Hence

$$\mathbf{Y}_c = \mathbf{D}_1\mathbf{A}_1^{-1}$$

where

$$\mathbf{D}_1^*\mathbf{D}_1 = \mathbf{B}_1^*\mathbf{Q}_1\mathbf{B}_1 + \mathbf{A}_1^*\mathbf{R}_1\mathbf{A}_1 \in \mathbb{P}^{m\times m}(z^{-1}), \mathbf{D}_1 \in \mathbb{P}^{m\times m}(z^{-1})$$

and $\mathbf{D}_1$ is also a strictly Hurwitz polynomial matrix.

Theorem 11.8. *A general polynomial matrix expression of the form:* $\mathbf{A}\mathbf{R}\mathbf{A}^* + \mathbf{C}\mathbf{Q}_0\mathbf{C}^* + \mathbf{E}\mathbf{Q}_2\mathbf{E}^*$ *has a left spectral factor* $\mathbf{D}$ *such that* $\mathbf{D}\mathbf{D}^* = \mathbf{A}\mathbf{R}\mathbf{A}^* + \mathbf{C}\mathbf{Q}_0\mathbf{C}^* + \mathbf{E}\mathbf{Q}_2\mathbf{E}^*$ *if and only if rank* $[\mathbf{A}\mathbf{R}, \mathbf{C}\mathbf{Q}_0, \mathbf{E}\mathbf{Q}_2] = r$.

A general polynomial matrix expression of the form $\mathbf{A}_1^*\mathbf{R}_1\mathbf{A}_1 + \mathbf{B}_1^*\mathbf{Q}_1\mathbf{B}_1$ *has* *a right spectral factor* $\mathbf{D}_1$ *such that* $\mathbf{D}_1^*\mathbf{D}_1 = \mathbf{A}_1^*\mathbf{R}_1\mathbf{A}_1 + \mathbf{B}_1^*\mathbf{Q}_1\mathbf{B}_1$ *if and only if*

$$\text{rank}\begin{bmatrix}\mathbf{R}_1\mathbf{A}_1\\ \mathbf{Q}_1\mathbf{B}_1\end{bmatrix} = m.$$

Proof (Kučera, 1979[11]). □

Corollary 10.2. *The assumptions made for the system matrices, the covariances and the optimal cost-functional always ensure the existence of a filter spectral factor (left spectral factor* $\mathbf{D}$*) and a control spectral factor (right spectral factor* $\mathbf{D}_1$*).*

Proof:

(i) *Filter case.* It is assumed that $\mathbf{A}(z^{-1}) \in \mathbb{P}^{r\times r}(z^{-1})$ is of full normal rank, hence rank $\{\mathbf{A}(z^{-1})\} = r$. The covariance $\mathbf{R} \in \mathbb{R}^{r\times r}$ is assumed positive definite giving:

$$\text{rank } [\mathbf{AR}] = \text{rank}[\mathbf{AR}, \mathbf{CQ}_0, \mathbf{EQ}_2] = r.$$

(ii) *Control case.* It is assumed that $\mathbf{A}_1(z^{-1}) \in \mathbb{P}^{m\times m}(z^{-1})$ is also of full normal rank, hence rank $[\mathbf{A}_1(z^{-1})] = m$. The control weighting matrix $\mathbf{R}_1$ is assumed to be positive definite giving:

$$\text{rank } [\mathbf{R}_1\mathbf{A}_1] = \text{rank}\begin{bmatrix}\mathbf{R}_1\mathbf{A}_1\\ \mathbf{Q}_1\mathbf{B}_1\end{bmatrix} = m.$$ □

The spectral factors are unique up to real orthogonal multiplier (Youla [21]).

Relationship to the generalized spectral factor

Recall that in Chapter 2 a stable spectral factor and a generalized spectral factor were defined and the latter was shown to be particularly important in the analysis for an open-loop unstable system. Using the polynomial system representation, the spectral factors obtained ($\mathbf{Y}_c = \mathbf{D}_1\mathbf{A}_1^{-1}$ and $\mathbf{Y}_f = \mathbf{A}^{-1}\mathbf{D}$) are identical to the generalized spectral factor of Shaked [3]. That is, the spectral factor pole polynomial is found to be the same as the open-loop system pole polynomial. This confirms the previous observation noted in Chapter 2 that the stable spectral factor is only appropriate to the analysis of stable open-loop systems as used in the earlier Wiener optimal control solution prevalent in the literature.

To conclude this section, it should be noted that there are several useful introductory texts on polynomial systems and polynomial matrix theory. Among the most notable are those of Kučera [11], Kailath [17], Wolovich

[22], Barnett [20], [23] and finally Gohberg, Lancaster and Rodman [24]. More advanced presentations can be found in Barnett [23] and Blomberg and Ylinen [47].

11.3.4. Signal analysis and power spectra

Some of the well-known results from signal analysis are reviewed in this section. The emphasis is on the properties and results of direct relevance to the solution of the LQG problem. In particular, Parseval's Theorem, the linear system relationships for power spectra and the definitions for covariance functions are pursued.

The section opens with various forms of Parseval's Theorem.

Lemma 11.4. *Convolution of two scalar continuous-time signals. Let* $x, y \in L_2(-\infty, \infty)$ *then*

$$\int_{-\infty}^{\infty} x(t)y(t)\,dt = \frac{1}{2\pi j}\int_{-j\infty}^{j\infty} x^*(s)y(s)\,ds = \frac{1}{2\pi j}\int_{-j\infty}^{j\infty} x(s)y^*(s)\,ds \quad (11.51)$$

$$= \frac{1}{2\pi}\int_{-\infty}^{\infty} x^*(j\omega)y(j\omega)\,d\omega = \frac{1}{2\pi}\int_{-\infty}^{\infty} x(j\omega)y^*(j\omega)\,d\omega \quad (11.52)$$

where

$$x^*(s) \triangleq x(-s) \quad \text{and} \quad y^*(s) \triangleq y(-s).$$

Proof. The condition $x, y \in L_2(-\infty, \infty)$ ensures that the left-hand side exists (Schwartz inequality) and that the line $s = j\omega(-\infty < \omega < \infty)$ lies within the region of convergence of both bilateral transforms. Thus,

$$\int_{-\infty}^{\infty} x(t)y(t)\,dt = \int_{-\infty}^{\infty} x(t)\left\{\frac{1}{2\pi j}\int_{-j\infty}^{j\infty} e^{st}y(s)\,ds\right\}dt$$

$$= \frac{1}{2\pi j}\int_{-j\infty}^{j\infty}\left\{\int_{-\infty}^{\infty} e^{st}x(t)\,dt\right\}y(s)\,ds$$

$$= \frac{1}{2\pi j}\int_{-j\infty}^{j\infty} x(-s)y(s)\,ds \triangleq \frac{1}{2\pi j}\int_{-j\infty}^{j\infty} x^*(s)y(s)\,ds.$$

Introduce $s = j\omega(-\infty < \omega < \infty)$ then $ds = jd\omega$ and

$$\frac{1}{2\pi j}\int_{-j\infty}^{j\infty} x(-s)y(s)\,ds = \frac{1}{2\pi}\int_{-\infty}^{\infty} x(-j\omega)y(j\omega)\,d\omega$$

$$= \frac{1}{2\pi}\int_{-\infty}^{\infty} x^{*}(j\omega)y(j\omega)\,d\omega$$

where

$$x^{*}(j\omega) \triangleq x(-j\omega).$$

The other pair of results follow similarly. □

Lemma 11.5. ***Parseval's theorem for continuous-time vector signals.*** *Let* $\boldsymbol{x} \in L_2^n(-\infty, \infty)$ *then,*

$$\int_{-\infty}^{\infty} \langle \boldsymbol{x}(t), \boldsymbol{x}(t)\rangle_{E_n}\,dt = \frac{1}{2\pi j}\int_{-j\infty}^{j\infty} \mathbf{x}^{*}(s)\mathbf{x}(s)\,ds \qquad (11.53)$$

$$= \frac{1}{2\pi}\int_{-\infty}^{\infty} \mathbf{x}^{*}(j\omega)\mathbf{x}(j\omega)\,d\omega \qquad (11.54)$$

where $\mathbf{x}(s)$ *denotes the bilateral Laplace transform of* $\boldsymbol{x}(t)$ *and* $\mathbf{x}^{*}(s) \triangleq \mathbf{x}^{T}(-s)$.

Proof. The condition $\boldsymbol{x} \in L_2^n(-\infty, \infty)$ ensures the line $s = j\omega$ lies within the region of convergence of the bilateral Laplace transform. Thus:

$$\int_{-\infty}^{\infty} \langle \boldsymbol{x}(t), \boldsymbol{x}(t)\rangle_{E_n}\,dt = \int_{-\infty}^{\infty} \boldsymbol{x}^{T}(t)\boldsymbol{x}(t)\,dt$$

$$= \int_{-\infty}^{\infty} \boldsymbol{x}^{T}(t)\left\{\frac{1}{2\pi j}\int_{-j\infty}^{j\infty} e^{st}\mathbf{x}(s)\,ds\right\}dt$$

$$= \frac{1}{2\pi j}\int_{-j\infty}^{j\infty}\left\{\int_{\infty}^{\infty} \boldsymbol{x}^{T}(t)\,e^{st}\,dt\right\}\mathbf{x}(s)\,ds$$

$$= \frac{1}{2\pi j}\int_{-j\infty}^{j\infty} \mathbf{x}^{T}(-s)\mathbf{x}(s)\,ds$$

$$= \frac{1}{2\pi j}\int_{-j\infty}^{j\infty} \mathbf{x}^{*}(s)\mathbf{x}(s)\,ds; \qquad \mathbf{x}^{*}(s) \triangleq \mathbf{x}^{T}(-s).$$

Using a change of variable $s = j\omega$ with $-\infty < \omega < \infty$, and $ds = jd\omega$ readily yields (11.54). □

Lemma 11.6. ***Convolution result for two scalar discrete-time signals*** *Let x and $y \in l_2(-\infty, \infty)$ then,*

$$\sum_{t=-\infty}^{\infty} x(t)y(t) = \frac{1}{2\pi j} \oint_{|z|=1} x^*(z^{-1})y(z^{-1}) \frac{dz}{z}$$

$$= \frac{1}{2\pi j} \oint_{|z|=1} x(z^{-1})y^*(z^{-1}) \frac{dz}{z} \qquad (11.55)$$

where

$$x^*(z) = x(z^{-1}) \quad \text{and} \quad y^*(z) = y(z^{-1}).$$

Proof. The condition $x, y \in l_2(-\infty, \infty)$ ensures that the left-hand side exists and that $|z| = 1$ lies within the region of convergence of both bilateral z-transforms. Thus,

$$\sum_{t=-\infty}^{\infty} x(t)y(t) = \sum_{t=-\infty}^{\infty} x(t)\left\{\frac{1}{2\pi j} \oint_{|z|=1} z^{t-1}y(z^{-1})\, dz\right\}$$

$$= \frac{1}{2\pi j} \oint_{|z|=1} \left\{\sum_{t=-\infty}^{\infty} x(t)z^t\right\} y(z^{-1}) \frac{dz}{z}$$

$$= \frac{1}{2\pi j} \oint_{|z|=1} x(z)y(z^{-1}) \frac{dz}{z} \triangleq \frac{1}{2\pi j} \oint_{|z|=1} x^*(z^{-1})y(z^{-1}) \frac{dz}{z}$$

where $x^*(z^{-1}) \triangleq x(z)$. The remaining result can be demonstrated similarly. □

Lemma 11.7. ***Parseval's theorem for discrete-time vector signals.*** *Let $\boldsymbol{x} \in l_2^n(-\infty, \infty)$ then,*

$$\sum_{t=-\infty}^{\infty} \langle \boldsymbol{x}(t), \boldsymbol{x}(t)\rangle_{E_n} = \frac{1}{2\pi j} \oint_{|z|=1} \mathbf{x}^*(z^{-1})\mathbf{x}(z^{-1}) \frac{dz}{z} \qquad (11.56)$$

where $\mathbf{x}^*(z^{-1}) \triangleq \mathbf{x}^T(z)$, and $\mathbf{x}(z^{-1})$ is the bilateral z-transform of $\boldsymbol{x}$.

Proof. The condition that $\boldsymbol{x} \in l_2^n(-\infty, \infty)$ ensures that the contour $|z| = 1$

lies in the region of convergence. Thus,

$$\sum_{t=-\infty}^{\infty} \langle x(t), x(t)\rangle_{E_n} = \sum_{t=-\infty}^{\infty} x^T(t)x(t)$$

$$= \sum_{t=-\infty}^{\infty} x^T(t)\left\{\frac{1}{2\pi j}\oint_{|z|=1} z^{t-1}\mathbf{x}(z^{-1})\, dz\right\}$$

$$= \frac{1}{2\pi j}\oint_{|z|=1}\left\{\sum_{t=-\infty}^{\infty} x^T(t)z^t\right\}\mathbf{x}(z^{-1})\,\frac{dz}{z}$$

$$= \frac{1}{2\pi j}\oint_{|z|=1} \mathbf{x}^T(z)\mathbf{x}(z^{-1})\,\frac{dz}{z}$$

giving

$$\sum_{t=-\infty}^{\infty} \langle x(t), x(t)\rangle_{E_n} = \frac{1}{2\pi j}\oint_{|z|=1} \mathbf{x}^*(z^{-1})\mathbf{x}(z^{-1})\,\frac{dz}{z}$$

where $\mathbf{x}^*(z^{-1}) \triangleq \mathbf{x}^T(z)$, and $\mathbf{x}(z^{-1})$ is the bilateral z-transform. □

The various forms of Parseval's theorem are important because of the role they play in the derivation of the results for power spectra to follow.

Power density spectra

The definition of the continuous-time power density spectrum is first introduced. Consider a truncation of a deterministic scalar signal $x: \mathbb{R} \to \mathbb{R}$, defined as:

$$x_T(t) = \begin{cases} x(t) & -T < t < T \\ 0 & \text{elsewhere.} \end{cases}$$

Noting that $x_T \in L_2(-\infty, \infty)$, the average power is defined as:

$$P = \lim_{T\to\infty} \frac{1}{2T}\int_{-T}^{T} x^2(t)\, dt.$$

From Lemma 11.4:

$$P = \lim_{T\to\infty} \frac{1}{2T}\left(\frac{1}{2\pi}\int_{-\infty}^{\infty} x_T(j\omega)x_T^*(j\omega)\, d\omega\right)$$

$$= \frac{1}{2\pi}\int_{-\infty}^{\infty}\left\{\lim_{T\to\infty}\frac{x_T(j\omega)x_T^*(j\omega)}{2T}\right\} d\omega.$$

When this limit exists, the power density spectrum of the signal $x\colon \mathbb{R} \to \mathbb{R}$ is defined by:

$$\Phi_{xx}(j\omega) = \lim_{T\to\infty} \frac{1}{2T}\, x_T(j\omega)x_T^*(j\omega)$$

where

$$x_T^*(j\omega) \triangleq x_T(-j\omega), \qquad \Phi_{xx}^*(j\omega) = \Phi_{xx}(j\omega) \geq 0 \quad \text{and} \quad \Phi_{xx}(j\omega) \in \mathbb{R}.$$

The power density spectrum matrix is analogously defined for a signal vector $\boldsymbol{x}\colon \mathbb{R} \to \mathbb{R}^n$ as:

$$\boldsymbol{\Phi}_{xx}(j\omega) = \lim_{T\to\infty} \frac{1}{2T}\, \mathbf{x}_T(j\omega)\mathbf{x}_T^*(j\omega) \tag{11.57}$$

where $\mathbf{x}_T^*(j\omega) \triangleq \mathbf{x}_T^T(-j\omega)$ and $\boldsymbol{x}_T \in L_2^n(-\infty, \infty)$ is a vector of truncated signals. Note that $\boldsymbol{\Phi}_{xx}^*(j\omega) = \boldsymbol{\Phi}_{xx}(j\omega) \geq \mathbf{O}$, $\boldsymbol{\Phi}_{xx}(j\omega) \in \mathbb{R}^{n\times n}$, $-\infty < \omega < \infty$, and that the power density matrix has elements giving power and cross-power density spectra for the elements of the signal vector $\boldsymbol{x}$.

The power-density spectrum matrix for a random process is defined as the ensemble average of the power density spectra of all the sample functions, viz.:

$$\boldsymbol{\Phi}_{xx}(j\omega) = \lim_{T\to\infty} \frac{1}{2T}\, E\{\mathbf{x}_T(j\omega)\mathbf{x}_T^*(j\omega)\}, \tag{11.58}$$

where $-\infty < \omega < \infty$.

For the case of discrete signals introduce the truncation for the vector signals process $\boldsymbol{x}\colon \mathbb{R} \to \mathbb{R}^n$

$$\boldsymbol{x}_K(t) = \begin{cases} \boldsymbol{x}(t) & -K \leq t \leq K \\ 0 & \text{elsewhere} \end{cases}$$

Noting that $\boldsymbol{x}_K \in l_2^n(-\infty, \infty)$, a matrix of average power terms is defined as:

$$\mathbf{P} = \lim_{K\to\infty} \frac{1}{2K+1} \left\{ \sum_{t=-\infty}^{\infty} \boldsymbol{x}_K(t)\boldsymbol{x}_K^T(t) \right\} \in \mathbb{R}^{n\times n}.$$

From Lemma 11.7:

$$\mathbf{P} = \lim_{K\to\infty} \frac{1}{2K+1} \left\{ \frac{1}{2\pi j} \oint_{|z|=1} \mathbf{x}_K(z^{-1})\mathbf{x}_K^*(z^{-1})\, \frac{dz}{z} \right\} \in \mathbb{R}^{n\times n}$$

$$= \frac{1}{2\pi j} \oint_{|z|=1} \left\{ \lim_{K\to\infty} \frac{1}{2K+1}\, \mathbf{x}_K(z^{-1})\mathbf{x}_K^*(z^{-1}) \right\} \frac{dz}{z}.$$

The power-density spectral matrix is thus defined as:

$$\mathbf{\Phi}_{xx}(z^{-1}) = \lim_{K\to\infty}\left\{\frac{1}{2K+1}\,\mathbf{x}_K(z^{-1})\mathbf{x}_K^*(z^{-1})\right\} \tag{11.59}$$

where $\mathbf{x}_K^*(z^{-1}) \triangleq \mathbf{x}_K^T(z)$ and $\mathbf{x}_K(z^{-1})$ is the bilateral z-transform of the signal $x \in l_2^n(-\infty, \infty)$. Note that the function dependence of $\mathbf{\Phi}_{xx}$ is defined in terms of z^{-1} but the domain of $\mathbf{\Phi}_{xx}$ is $z \in \mathbb{C}$ on a suitable region of convergence. For $|z| = 1$ the following relations hold:

$$\mathbf{\Phi}_{xx}^*(z^{-1}) = \mathbf{\Phi}_{xx}(z^{-1}) \geq \mathbf{O}, \qquad \mathbf{\Phi}_{xx}(z^{-1}) \in \mathbb{R}^{n\times n}.$$

A discrete-time random process has a power-density spectrum matrix defined analogously to equation (11.59):

$$\mathbf{\Phi}_{xx}(z^{-1}) = \lim_{K\to\infty}\left\{\frac{1}{2K+1}\,E\{\mathbf{x}_K(z^{-1})\mathbf{x}_K^*(z^{-1})\}\right\}. \tag{11.60}$$

Autocorrelation matrices

The autocorrelation matrix is defined for a continuous-time signal process as the ensemble average:

$$\mathbf{R}_{xx}(t, \tau) = E\{x(t)x^T(\tau)\}$$

and for a wide sense stationary process as:

$$\mathbf{R}_{xx}(\tau) = E\{x(t+\tau)x^T(t)\}. \tag{11.61}$$

The relationship between the autocorrelation matrix and the spectral density matrix can be retrieved as follows. From equation (11.58):

$$\mathbf{\Phi}_{xx}(j\omega) = \lim_{T\to\infty}\frac{1}{2T}\,E\{\mathbf{x}_T(j\omega)\mathbf{x}_T^*(j\omega)\}$$

where $\mathbf{x}_T(j\omega)$ is the bilateral Laplace transform of the truncated signal $x_T \in L_2^n(-\infty, \infty)$, evaluated for $s = j\omega$. Thus,

$$\mathbf{\Phi}_{xx}(j\omega) = \lim_{T\to\infty}\frac{1}{2T}\,E\left\{\int_{-T}^{T} x(t_1)\,e^{-j\omega t_1}\,dt_1 \int_{-T}^{T} x^T(t_2)\,e^{j\omega t_2}\,dt_2\right\}$$

$$= \lim_{T\to\infty}\frac{1}{2T}\int_{-T}^{T}\int_{-T}^{T} E\{x(t_1)x^T(t_2)\}\,e^{-j\omega(t_1-t_2)}\,dt_1 dt_2. \tag{11.62}$$

Wiener–Khintchine pair

Using equation (11.61) for a wide sense stationary process and some simple integral manipulation (defining $t_1 = t + \tau$ and $t_2 = t$ (Assefi, [25])) equation (11.62) becomes:

$$\mathbf{\Phi}_{xx}(j\omega) = \int_{-\infty}^{\infty} \mathbf{R}_{xx}(\tau)\ e^{-j\omega\tau}\ d\tau. \tag{11.63}$$

This is the Fourier transform of the autocorrelation matrix. Thus using the Fourier transform inversion formula:

$$\mathbf{R}_{xx}(\tau) = \frac{1}{2\pi} \int_{-\infty}^{\infty} \mathbf{\Phi}_{xx}(j\omega)\ e^{j\omega\tau}\ d\omega. \tag{11.64}$$

Equations (11.63) and (11.64) form the Wiener–Khintchine pair. Clearly, a bilateral Laplace transform pair can also be obtained:

$$\mathbf{\Phi}_{xx}(s) = \int_{-\infty}^{\infty} \mathbf{R}_{xx}(\tau)\ e^{-s\tau}\ d\tau, \qquad \alpha < \mathscr{Re}(s) < \beta$$

and

$$\mathbf{R}_{xx}(\tau) = \frac{1}{2\pi j} \int_{c-j\infty}^{c+j\infty} \mathbf{\Phi}_{xx}(s)\ e^{s\tau}\ ds, \qquad \alpha < c < \beta$$

where the scalars α and β determine the region of convergence.

Mean square matrix

The mean-square value matrix $\mathbf{P}_{xx}(t) = E\{\boldsymbol{x}(t)\boldsymbol{x}^T(t)\}$ can be evaluated from the spectral density matrix:

$$\mathbf{P}_{xx}(t) = \mathbf{R}_{xx}(0) = \frac{1}{2\pi \mathrm{j}} \int_{-j\infty}^{j\infty} \mathbf{\Phi}_{xx}(s)\ ds \tag{11.65}$$

$$= \frac{1}{2\pi} \int_{-\infty}^{\infty} \mathbf{\Phi}_{xx}(j\omega)\ d\omega. \tag{11.66}$$

Discrete processes

A similar relationship between the correlation matrix and the power density exists for a discrete random process. As before, the correlation matrix is defined (for a wide sense stationary process) as:

$$\mathbf{R}_{xx}(\tau) = E\{\boldsymbol{x}(t+\tau)\boldsymbol{x}^T(t)\}$$

where the underlying time index set is integer. From equation (11.60):

$$\Phi_{xx}(z^{-1}) = \lim_{K\to\infty}\left\{\frac{1}{2K+1}\,E\{\mathbf{x}_K(z^{-1})\mathbf{x}_K^*(z^{-1})\}\right\}$$

where $\mathbf{x}_K(z^{-1})$ is the bilateral z-transform of the truncated signal $x_K \in l_2^n(-\infty,\infty)$. Hence, substituting for $\mathbf{x}_K(z^{-1})$ obtains:

$$\Phi_{xx}(z^{-1}) = \lim_{K\to\infty}\left\{\frac{1}{2K+1}\,E\left\{\sum_{t_1=-K}^{K} x(t_1)z^{-t_1}\sum_{t_2=-K}^{K} x^T(t_2)z^{t_2}\right\}\right\}$$

$$= \lim_{K\to\infty}\frac{1}{2K+1}\sum_{t_1=-K}^{K}\sum_{t_2=-K}^{K} E\{x(t_1)x^T(t_2)\}z^{-(t_1-t_2)}$$

$$= \lim_{K\to\infty}\frac{1}{2K+1}\sum_{t_1=-K}^{K}\sum_{t_2=-K}^{K} \mathbf{R}_{xx}(t_1-t_2)z^{-(t_1-t_2)}.$$

Thus,

$$\Phi_{xx}(z^{-1}) = \sum_{t=-\infty}^{\infty}\mathbf{R}_{xx}(t)z^{-t} = \mathcal{Z}_2\{\mathbf{R}_{xx}(t)\}. \tag{11.67}$$

(This follows from a simplification of the double summation and the use of the limit.) Using the inversion formula for the z-transform yields:

$$\mathbf{R}_{xx}(t) = \frac{1}{2\pi j}\oint_{|z|=1} z^{t-1}\Phi_{xx}(z^{-1})\,dz. \tag{11.68}$$

The mean square value matrix $\mathbf{P}_{xx}(t)$ can be evaluated from the spectral density matrix as:

$$\mathbf{P}_{xx}(t) = E\{x(t)x^T(t)\} = \mathbf{R}_{xx}(0) = \frac{1}{2\pi j}\oint_{|z|=1}\Phi_{xx}(z^{-1})\,\frac{dz}{z}. \tag{11.69}$$

Equations (11.66) and (11.69) have an important role in subsequent sections in the construction of cost functionals.

Filtered spectral density relationships

Several cases arise and these are itemized for convenience.

Lemma 11.8. ***Continuous time process.*** *If the system outputs and inputs are related by the transform equation* $\mathbf{y}(s) = \mathbf{W}(s)\mathbf{u}(s)$ *then the power-density* *spectra are related by:*

$$\mathbf{\Phi}_{yy}(j\omega) = \mathbf{W}(j\omega)\mathbf{\Phi}_{uu}(\omega)\mathbf{W}^*(j\omega) \tag{11.70}$$

where $\mathbf{\Phi}_{yy}(j\omega)$ and $\mathbf{\Phi}_{uu}(j\omega)$ are the output and input power-density spectra, respectively.

Lemma 11.9. ***Output mean-square relation.*** *The mean-square output is given by:*

$$E\{y(t)y^T(t)\} = \mathbf{R}_{yy}(0) = \frac{1}{2\pi j}\int_{-j\infty}^{j\infty} \mathbf{W}(s)\mathbf{\Phi}_{uu}(s)\mathbf{W}^*(s)\, ds \tag{11.71}$$

$$= \frac{1}{2\pi}\int_{-\infty}^{\infty} \mathbf{W}(j\omega)\mathbf{\Phi}_{uu}(j\omega)\mathbf{W}^*(j\omega)\, d\omega. \tag{11.72}$$

Similar results hold for discrete processes:

Lemma 11.10. ***Discrete time processes.*** *If the system outputs and inputs are related by the transform equation $\mathbf{y}(z^{-1}) = \mathbf{W}(z^{-1})\mathbf{u}(z^{-1})$ then the power-density spectra are related by:*

$$\mathbf{\Phi}_{yy}(z^{-1}) = \mathbf{W}(z^{-1})\mathbf{\Phi}_{uu}(z^{-1})\mathbf{W}^*(z^{-1}) \tag{11.73}$$

where $\mathbf{\Phi}_{yy}(z^{-1})$ and $\mathbf{\Phi}_{uu}(z^{-1})$ are the output and input power-density spectra, respectively.

Lemma 11.11. ***Output mean-square relation.*** *The mean-square output matrix is given by:*

$$E\{y(t)y^T(t)\} = \mathbf{R}_{yy}(0) \tag{11.74}$$

$$= \frac{1}{2\pi j}\oint_{|z|=1} \mathbf{W}(z^{-1})\mathbf{\Phi}_{uu}(z^{-1})\mathbf{W}^*(z^{-1})\,\frac{dz}{z}. \tag{11.75}$$

11.3.5. The class of stabilizing controllers

An important recent development in systems theory was the parameterization of the class of stabilizing controllers for the system configuration shown in Fig. 11.2. This idea was first published in the literature by Kučera [26] and by Youla, Jabr and Bongiorno [2]. Subsequent developments were published by Desoer, Liu, Murray and Saeks [27], and the concept has been utilized in the studies of optimal robustness (Zames and Francis [28], Kwakernaak [29]).

In this section, the Youla parameterization is obtained and then related to the work of Kučera. Using the dimensions implied by the signals in Fig. 11.2, it is useful to summarize the various polynomial matrices and matrix fractions to be used in this and subsequent sections. Note that the treatment presented in this section is applicable to both continuous and discrete time-systems. Consequently, to avoid specifying the indeterminate s or z^{-1}, the classes of matrices and matrix fractions are denoted using the $(\cdot)$ convention.

System polynomial matrices

$$\mathbf{A}_0 \in \mathbb{P}^{r\times r}(\cdot), \qquad \mathbf{B}_0 \in \mathbb{P}^{r\times m}(\cdot), \qquad \mathbf{A}_1 \in \mathbb{P}^{m\times m}(\cdot), \qquad \mathbf{B}_1 \in \mathbb{P}^{r\times m}(\cdot)$$

Controller polynomial matrices

$$\mathbf{H}_0 \in \mathbb{P}^{m\times m}(\cdot), \qquad \mathbf{G}_0 \in \mathbb{P}^{m\times r}(\cdot), \qquad \mathbf{H}_1 \in \mathbb{P}^{r\times r}(\cdot), \qquad \mathbf{G}_1 \in \mathbb{P}^{m\times r}(\cdot)$$

Related polynomial matrices

$$\mathbf{D}_0 \in \mathbb{P}^{m\times m}(\cdot), \qquad \mathbf{P}_0 \in \mathbb{P}^{m\times r}(\cdot), \qquad \mathbf{X}_0 \in \mathbb{P}^{r\times r}(\cdot), \qquad \mathbf{Y}_0 \in \mathbb{P}^{m\times r}(\cdot)$$

$$\mathbf{D}_1 \in \mathbb{P}^{r\times r}(\cdot), \qquad \mathbf{P}_1 \in \mathbb{P}^{m\times r}(\cdot), \qquad \mathbf{X}_1 \in \mathbb{P}^{m\times m}(\cdot), \qquad \mathbf{Y}_1 \in \mathbb{P}^{m\times r}(\cdot)$$

Controller matrix fractions

$$\mathbf{C}_0 \in \mathbb{R}^{m\times r}(\cdot), \qquad \mathbf{M}_0 \in \mathbb{R}^{m\times m}(\cdot), \qquad \mathbf{N}_0 \in \mathbb{R}^{m\times r}(\cdot), \qquad \mathbf{K} \in \mathbb{R}^{m\times r}(\cdot)$$

$$\mathbf{M}_1 \in \mathbb{R}^{r\times r}(\cdot), \qquad \mathbf{N}_1 \in \mathbb{R}^{m\times r}(\cdot)$$

The system equations may be listed from Fig. 11.2 as:

$$\mathbf{W} = \mathbf{B}_1\mathbf{A}_1^{-1} = \mathbf{A}_0^{-1}\mathbf{B}_0 \in \mathbb{R}^{r\times m}(\cdot)$$

The plant is assumed to be represented in either a left or right coprime matrix fraction form. Thus, stable modes which might cancel when forming $\mathbf{W} = \mathbf{A}^{-1}\mathbf{B}$ will not appear in the characteristic polynomial of the modelled closed-loop system. Since such modes are stable and represent uncontrollable and/or unobservable modes in the physical system they can be neglected from the present analysis. However, if these modes are lightly damped oscillatory modes, then some action to change the basic physical system might be required to avoid unacceptable internal oscillations within the system. Of course, feedback action alone cannot affect such modes hence the need for the above assumption.

Let z_p be the partial state of the system, then:

$$\mathbf{A}_1 z_p = u \tag{11.76}$$

and

$$y = \mathbf{B}_1 z_p$$

where $n = \mathbf{o}$ and $v = \mathbf{o}$. Using the controller equation:

$$u = \mathbf{H}_0^{-1}\mathbf{G}_0 e_1$$

and hence

$$u = \mathbf{H}_0^{-1}\mathbf{G}_0(r - \mathbf{B}_1\mathbf{A}_1^{-1}u) \tag{11.77}$$

or using the partial state equation (11.76),

$$(\mathbf{H}_0\mathbf{A}_1 + \mathbf{G}_0\mathbf{B}_1)z_p = \mathbf{G}_0 r. \tag{11.78}$$

The stability of the system (11.77) depends on the polynomial matrix $(\mathbf{H}_0\mathbf{A}_1 + \mathbf{G}_0\mathbf{B}_1)$; if this is non-Hurwitz and there is no cancellation of unstable modes with $\mathbf{G}_0$, the partial state:

$$z_p = (\mathbf{H}_0\mathbf{A}_1 + \mathbf{G}_0\mathbf{B}_1)^{-1}\mathbf{G}_0 r$$

will include these modes. If there is a cancellation of unstable modes in the transfer $\mathbf{B}_1(\mathbf{H}_0\mathbf{A}_1 + \mathbf{G}_0\mathbf{B}_1)^{-1}\mathbf{G}_0$ (viz., $y = \mathbf{B}_1(\mathbf{H}_0\mathbf{A}_1 + \mathbf{G}_0\mathbf{B}_1)^{-1}\mathbf{G}_0 r$) then the system has unstable uncontrollable and/or uncontrollable–unobservable modes so that the closed-loop system is unstable.

The internal stability of optimal control designs has posed a problem in the past since controllers, in some cases, may try to cancel unstable plant poles or zeros. The resulting system is, of course, unstable. To avoid this situation, a method is required for constraining the optimal solution. One approach is to use a control-law structure chosen from the class of stabilizing controllers.

Lemma 11.12. ***The class of stabilizing controllers.*** *Let $\mathbf{D}_0$ be an arbitrary strictly Hurwitz $(m \times m)$ polynomial matrix then the class of stabilizing controllers is generated by $\mathbf{C}_0 = \mathbf{H}_0^{-1}\mathbf{G}_0$ where:*

$$[\mathbf{H}_0 \;\; \mathbf{G}_0] = [\mathbf{D}_0 \;\; \mathbf{P}_0]\begin{bmatrix} \mathbf{X}_1 & \mathbf{Y}_1 \\ -\mathbf{B}_0 & \mathbf{A}_0 \end{bmatrix}. \tag{11.79}$$

The polynomial matrices $\mathbf{X}_1$, and $\mathbf{Y}_1$ satisfy the Bezout identity $\mathbf{X}_1\mathbf{A}_1 + \mathbf{Y}_1\mathbf{B}_1 = \mathbf{I}_m$ and the $(m \times r)$ polynomial matrix $\mathbf{P}_0$ is arbitrary.

Proof. The Bezout identity for the right coprime pair $(\mathbf{A}_1, \mathbf{B}_1)$ is (11.11a):

$$\mathbf{X}_1\mathbf{A}_1 + \mathbf{Y}_1\mathbf{B}_1 = \mathbf{I}_m.$$

Premultiplying by $\mathbf{D}_0$ yields the unilateral Diophantine equation:

$$\tilde{\mathbf{H}}_0\mathbf{A}_1 + \tilde{\mathbf{G}}_0\mathbf{B}_1 = \mathbf{D}_0 \tag{11.80}$$

where $\tilde{\mathbf{H}}_0 = \mathbf{D}_0\mathbf{X}_1$ and $\tilde{\mathbf{G}}_0 = \mathbf{D}_0\mathbf{Y}_1$ is a particular solution. From Theorem 11.7, the general solution is given by:

$$[\mathbf{H}_0 \quad \mathbf{G}_0] = [\tilde{\mathbf{H}}_0 \quad \tilde{\mathbf{G}}_0] + \mathbf{P}_0[-\mathbf{B}_0 \quad \mathbf{A}_0]$$

$$= \mathbf{D}_0[\mathbf{X}_1 \quad \mathbf{Y}_1] + \mathbf{P}_0[-\mathbf{B}_0 \quad \mathbf{A}_0]$$

$$[\mathbf{H}_0 \quad \mathbf{G}_0] = [\mathbf{D}_0 \quad \mathbf{P}_0]\begin{bmatrix} \mathbf{X}_1 & \mathbf{Y}_1 \\ -\mathbf{B}_0 & \mathbf{A}_0 \end{bmatrix}.$$

Using the pair $[\mathbf{H}_0 \quad \mathbf{G}_0]$ to define the controller ensures that $\mathbf{H}_0\mathbf{A}_1 + \mathbf{G}_0\mathbf{B}_1 = \mathbf{D}_0$ where $\mathbf{D}_0$ is a strictly Hurwitz polynomial matrix. The characteristic polynomial for the closed-loop system $\rho_c = \det(\mathbf{D}_0)$ is therefore strictly Hurwitz and the system is asymptotically stable. □

This parametric relationship for the class of all stabilizing controllers was first given by Antsaklis [18]. The pair $\mathbf{H}_0, \mathbf{G}_0$ are right coprime if and only if $\mathbf{D}_0$ and $\mathbf{P}_0$ are right coprime. This follows from (11.79), recall that matrix

$$\begin{bmatrix} \mathbf{X}_1 & \mathbf{Y}_1 \\ -\mathbf{B}_0 & \mathbf{A}_0 \end{bmatrix} \text{ is unimodular,}$$

then (11.79) may be rewritten as:

$$[\mathbf{D}_0 \quad \mathbf{P}_0] = [\mathbf{H}_0 \quad \mathbf{G}_0]\begin{bmatrix} \mathbf{A}_1 & -\mathbf{Y}_0 \\ \mathbf{B}_1 & \mathbf{X}_0 \end{bmatrix}. \tag{11.81}$$

Hence, $[\mathbf{D}_0, \mathbf{P}_0]$ right co-prime implies $[\mathbf{H}_0 \quad \mathbf{G}_0]$ right coprime (Theorem 11.4). The controller $\mathbf{C}_0 = \mathbf{H}_0^{-1}\mathbf{G}_0$ is in left-matrix fraction form and an equivalent right-matrix fraction form can also be derived.

Lemma 11.13. *Let $\mathbf{D}_1$ be an arbitrary strictly Hurwitz $(r \times r)$ polynomial matrix then the class of stabilizing right-matrix fraction controllers $\mathbf{C}_0 = \mathbf{G}_1\mathbf{H}_1^{-1}$ is generated by:*

$$\begin{bmatrix} \mathbf{G}_1 \\ \mathbf{H}_1 \end{bmatrix} = \begin{bmatrix} \mathbf{A}_1 & -\mathbf{Y}_0 \\ -\mathbf{B}_1 & -\mathbf{X}_0 \end{bmatrix}\begin{bmatrix} \mathbf{P}_1 \\ -\mathbf{D}_1 \end{bmatrix} \tag{11.82}$$

where the polynomial matrices $\mathbf{X}_0$, *and* $\mathbf{Y}_0$ *satisfy the Bezout identity* $\mathbf{B}_0\mathbf{Y}_0 + \mathbf{A}_0\mathbf{X}_0 = \mathbf{I}_r$ *and the polynomial matrix* $\mathbf{P}_1$ *is arbitrary.*

Proof. Let $[\mathbf{P}_1^T \quad -\mathbf{D}_1^T]$ satisfy

$$[\mathbf{D}_0 \quad \mathbf{P}_0]\begin{bmatrix} \mathbf{P}_1 \\ -\mathbf{D}_1 \end{bmatrix} = \mathbf{O}$$

(where the arbitrary choice of $\mathbf{D}_0, \mathbf{P}_0$ ensures $\mathbf{D}_1, \mathbf{P}_1$ may also be chosen aribitarily). Using (11.81) yields:

$$[\mathbf{H}_0 \quad \mathbf{G}_0]\begin{bmatrix} \mathbf{A}_1 & -\mathbf{Y}_0 \\ \mathbf{B}_1 & \mathbf{X}_0 \end{bmatrix}\begin{bmatrix} \mathbf{P}_1 \\ -\mathbf{D}_1 \end{bmatrix} = \mathbf{O}$$

Define

$$\begin{bmatrix} \mathbf{G}_1 \\ -\mathbf{H}_1 \end{bmatrix} = \begin{bmatrix} \mathbf{A}_1 & -\mathbf{Y}_0 \\ \mathbf{B}_1 & \mathbf{X}_0 \end{bmatrix}\begin{bmatrix} \mathbf{P}_1 \\ -\mathbf{D}_1 \end{bmatrix}$$

then

$$[\mathbf{H}_0 \quad \mathbf{G}_0]\begin{bmatrix} \mathbf{G}_1 \\ -\mathbf{H}_1 \end{bmatrix} = \mathbf{O}$$

and $\mathbf{C}_0 = \mathbf{H}_0^{-1}\mathbf{G}_0 = \mathbf{G}_1\mathbf{H}_1^{-1}$ where the pair $\mathbf{G}_1, \mathbf{H}_1$ are given parametrically by (11.82). From (11.82), using the unimodular inverse:

$$\begin{bmatrix} \mathbf{P}_1 \\ -\mathbf{D}_1 \end{bmatrix} = \begin{bmatrix} \mathbf{X}_1 & \mathbf{Y}_1 \\ -\mathbf{B}_0 & \mathbf{A}_0 \end{bmatrix}\begin{bmatrix} \mathbf{G}_1 \\ -\mathbf{H}_1 \end{bmatrix}$$

so that:

$$\mathbf{A}_0\mathbf{H}_1 + \mathbf{B}_0\mathbf{G}_1 = \mathbf{D}_1$$

Hence, the controller and the closed-loop system is asymptotically stable if and only if $\mathbf{D}_1$ is strictly Hurwitz. □

The two Lemma 11.12 and 11.13 may be reformulated to give the Youla form for the stabilizing controller viz.:

Lemma 11.14. ***Youla form of stabilizing controller.*** *The left-matrix fraction* $\mathbf{C}_0 = \mathbf{H}_0^{-1}\mathbf{G}_0$ *is a stabilizing controller if and only if*

$$\mathbf{C}_0 = \mathbf{H}_0^{-1}\mathbf{G}_0 = \mathbf{M}_0^{-1}\mathbf{N}_0 \tag{11.83}$$

where the rational matrices $\mathbf{M}_0$ *and* $\mathbf{N}_0$ *satisfy:*

$$\mathbf{M}_0 = (\mathbf{X}_1 - \mathbf{K}\mathbf{B}_0) \quad \textit{and} \quad \mathbf{N}_0 = (\mathbf{Y}_1 + \mathbf{K}\mathbf{A}_0) \tag{11.84}$$

for some stable rational gain matrix $\mathbf{K}$, *where* $\det(\mathbf{X}_1 - \mathbf{K}\mathbf{B}_0) \neq 0$.

Alternatively, the right-matrix fraction is a stabilizing controller if and only if

$$\mathbf{C}_1 = \mathbf{G}_1\mathbf{H}_1^{-1} = \mathbf{N}_1\mathbf{M}_1^{-1} \tag{11.85}$$

where the rational matrices $\mathbf{M}_1$ *and* $\mathbf{N}_1$ *satisfy:*

$$\mathbf{M}_1 = \mathbf{X}_0 - \mathbf{B}_1\mathbf{K} \quad \text{and} \quad \mathbf{N}_1 = \mathbf{Y}_0 + \mathbf{A}_1\mathbf{K} \tag{11.86}$$

with $\det(\mathbf{X}_0 - \mathbf{B}_1\mathbf{K}) \neq 0$.

Proof. Let $\mathbf{K} \triangleq \mathbf{D}_0^{-1}\mathbf{P}_0 = \mathbf{P}_1\mathbf{D}_1^{-1}$ then (11.83) and (11.85) follow directly from (11.79) and (11.82), respectively. □

For given system matrices, the only unknown in the controller expressions (11.83) and (11.84) is the rational gain matrix $\mathbf{K}$. This may be chosen by the usual Wiener optimization argument (Youla *et al.* [2]). Note, that whilst $\mathbf{H}_0, \mathbf{G}_0, \mathbf{H}_1$ and $\mathbf{G}_1$ are polynomial matrices; $\mathbf{M}_0, \mathbf{N}_0, \mathbf{M}_1$ and $\mathbf{N}_1$ are rational matrices.

Stability lemma

The lemma which follows is required in the solution of the LQG optimal control problem. It demonstrates that the closed-loop system stabilizes the closed-loop system.

Lemma 11.15. ***Stability lemma.*** *The closed-loop system with both open-loop system and controller free of unstable hidden modes is asymptotically stable if and only if the controller may be expressed in the form:*

$$\mathbf{C}_0 = \mathbf{M}_0^{-1}\mathbf{N}_0$$

where $\mathbf{M}_0$ *and* $\mathbf{N}_0$ *are asymptotically stable rational matrices satisfying:*

$$\mathbf{M}_0\mathbf{A}_1 + \mathbf{N}_0\mathbf{B}_1 = \mathbf{I}_m \tag{11.87}$$

Equivalently, asymptotic stability follows if and only if

$$\mathbf{C}_0 = \mathbf{N}_1\mathbf{M}_1^{-1}$$

where $\mathbf{M}_1$, *and* $\mathbf{N}_1$ *are asymptotically stable rational matrices satisfying:*

$$\mathbf{A}_0\mathbf{M}_1 + \mathbf{B}_0\mathbf{N}_1 = \mathbf{I}_r \tag{11.88}$$

Proof. From (11.84) $\mathbf{M}_0 = \mathbf{X}_1 - \mathbf{K}\mathbf{B}_0$, $\mathbf{N}_0 = \mathbf{Y}_1 + \mathbf{K}\mathbf{A}_0$. Using $\mathbf{K} = \mathbf{D}_0^{-1}\mathbf{P}_0$ yields:

$$\mathbf{D}_0\mathbf{M}_0 = \mathbf{D}_0\mathbf{X}_1 - \mathbf{P}_0\mathbf{B}_0 = \mathbf{H}_0 \quad \text{hence } \mathbf{M}_0 = \mathbf{D}_0^{-1}\mathbf{H}_0$$

and similarly

$$\mathbf{D}_0\mathbf{N}_0 = \mathbf{D}_0\mathbf{Y}_1 + \mathbf{P}_0\mathbf{A}_0 = \mathbf{G}_0 \quad \text{hence } \mathbf{N}_0 = \mathbf{D}_0^{-1}\mathbf{G}_0.$$

Finally,

$$\mathbf{M}_0\mathbf{A}_1 + \mathbf{N}_0\mathbf{B}_1 = \mathbf{D}_0^{-1}(\mathbf{H}_0\mathbf{A}_1 + \mathbf{G}_0\mathbf{B}_1) = \mathbf{I}_m$$

to prove (11.87). The proof of (11.88) is analogous. □

The proofs of the above lemmas establishes the link between the quantities used in the Kučera optimization theories [11], namely $\mathbf{H}_0, \mathbf{G}_0, \mathbf{H}_1, \mathbf{G}_1$ (polynomial matrices) and $\mathbf{M}_0, \mathbf{N}_0, \mathbf{M}_1, \mathbf{N}_1$ (rational matrices) [11] and the gain **K** which features in the Youla, Bongiorno and Jabr minimization procedure [2]. In summary:

$$\mathbf{M}_0 = \mathbf{X}_1 - \mathbf{K}\mathbf{B}_0 = \mathbf{D}_0^{-1}\mathbf{H}_0 \tag{11.89}$$

$$\mathbf{N}_0 = \mathbf{Y}_1 + \mathbf{K}\mathbf{A}_0 = \mathbf{D}_0^{-1}\mathbf{G}_0 \tag{11.90}$$

$$\mathbf{M}_1 = \mathbf{X}_0 - \mathbf{B}_1\mathbf{K} = \mathbf{H}_1\mathbf{D}_1^{-1} \tag{11.91}$$

$$\mathbf{N}_1 = \mathbf{Y}_0 + \mathbf{A}_1\mathbf{K} = \mathbf{G}_1\mathbf{D}_1^{-1}. \tag{11.92}$$

Lemma 11.16. *Useful identities. The rational matrices* $\mathbf{M}_0, \mathbf{N}_0, \mathbf{M}_1, \mathbf{N}_1$ *defined above satisfy:*

$$\left.\begin{aligned} \mathbf{N}_1\mathbf{A}_0 &= \mathbf{A}_1\mathbf{N}_0 \\ \mathbf{M}_1\mathbf{B}_0 &= \mathbf{B}_1\mathbf{M}_0 \\ \mathbf{M}_1\mathbf{A}_0 + \mathbf{B}_1\mathbf{N}_0 &= \mathbf{I}_r \\ \mathbf{A}_1\mathbf{M}_0 + \mathbf{N}_1\mathbf{B}_0 &= \mathbf{I}_m. \end{aligned}\right\} \tag{11.93}$$

Proof. Use the identities

$$\mathbf{W} = \mathbf{A}_0^{-1}\mathbf{B}_0 = \mathbf{B}_1\mathbf{A}_1^{-1},\ \mathbf{C}_0 = \mathbf{M}_0^{-1}\mathbf{N}_0 = \mathbf{N}_1\mathbf{M}_1^{-1}, \mathbf{A}_0\mathbf{M}_1 + \mathbf{B}_0\mathbf{N}_1 = \mathbf{I}_r$$

and

$$\mathbf{M}_0\mathbf{A}_1 + \mathbf{N}_0\mathbf{B}_1 = \mathbf{I}_m.$$

The proof is in four parts:

(i) $\mathbf{C}_0(\mathbf{I}_r + \mathbf{W}\mathbf{C}_0)^{-1} = (\mathbf{I}_m + \mathbf{C}_0\mathbf{W})^{-1}\mathbf{C}_0$

$$\mathbf{N}_1\mathbf{M}_1^{-1}(\mathbf{I}_r + \mathbf{A}_0^{-1}\mathbf{B}_0\mathbf{N}_1\mathbf{M}_1^{-1})^{-1} = (\mathbf{I}_m + \mathbf{M}_0^{-1}\mathbf{N}_0\mathbf{B}_1\mathbf{A}_1^{-1})^{-1}\mathbf{M}_0^{-1}\mathbf{N}_0$$

$$\mathbf{N}_1(\mathbf{A}_0\mathbf{M}_1 + \mathbf{B}_0\mathbf{N}_1)^{-1}\mathbf{A}_0 = \mathbf{A}_1(\mathbf{M}_0\mathbf{A}_1 + \mathbf{N}_0\mathbf{B}_1)^{-1}\mathbf{N}_0$$

$$\mathbf{N}_1\mathbf{A}_0 = \mathbf{A}_1\mathbf{N}_0. \tag{11.94}$$

(ii) $\mathbf{W}(\mathbf{I}_m + \mathbf{C}_0\mathbf{W})^{-1} = (\mathbf{I}_r + \mathbf{W}\mathbf{C}_0)^{-1}\mathbf{W}$

$$\mathbf{B}_1\mathbf{A}_1^{-1}(\mathbf{I}_m + \mathbf{M}_0^{-1}\mathbf{N}_0\mathbf{B}_1\mathbf{A}_1^{-1})^{-1} = (\mathbf{I}_r + \mathbf{A}_0^{-1}\mathbf{B}_0\mathbf{N}_1\mathbf{M}_1^{-1})^{-1}\mathbf{A}_0^{-1}\mathbf{B}_0$$

$$\mathbf{B}_1(\mathbf{M}_0\mathbf{A}_1 + \mathbf{N}_0\mathbf{B}_1)^{-1}\mathbf{M}_0 = \mathbf{M}_1(\mathbf{A}_0\mathbf{M}_1 + \mathbf{B}_0\mathbf{N}_1)^{-1}\mathbf{B}_0$$

$$\mathbf{B}_1\mathbf{M}_0 = \mathbf{M}_1\mathbf{B}_0.$$

(iii) $\mathbf{A}_0\mathbf{M}_1 + \mathbf{B}_0\mathbf{N}_1 = \mathbf{I}_r$

$\mathbf{A}_0\mathbf{M}_1\mathbf{A}_0 + \mathbf{B}_0\mathbf{N}_1\mathbf{A}_0 = \mathbf{A}_0$ (use 11.94)

$\mathbf{A}_0\mathbf{M}_1\mathbf{A}_0 + \mathbf{B}_0\mathbf{A}_1\mathbf{N}_0 = \mathbf{A}_0$ (use $\mathbf{A}_0\mathbf{B}_1 = \mathbf{B}_0\mathbf{A}_1$)

$\mathbf{A}_0\mathbf{M}_1\mathbf{A}_0 + \mathbf{A}_0\mathbf{B}_1\mathbf{N}_0 = \mathbf{A}_0$

$$\mathbf{M}_1\mathbf{A}_0 + \mathbf{B}_1\mathbf{N}_0 = \mathbf{I}_r.$$

(iv) $\mathbf{M}_0\mathbf{A}_1 + \mathbf{N}_0\mathbf{B}_1 = \mathbf{I}_m$

$\mathbf{A}_1\mathbf{M}_0\mathbf{A}_1 + \mathbf{A}_1\mathbf{N}_0\mathbf{B}_1 = \mathbf{A}_1$ (use 11.94)

$\mathbf{A}_1\mathbf{M}_0\mathbf{A}_1 + \mathbf{N}_1\mathbf{A}_0\mathbf{B}_1 = \mathbf{A}_1$ (use $\mathbf{A}_0\mathbf{B}_1 = \mathbf{B}_0\mathbf{A}_1$)

$\mathbf{A}_1\mathbf{M}_0\mathbf{A}_1 + \mathbf{N}_1\mathbf{B}_0\mathbf{A}_1 = \mathbf{A}_1$

$$\mathbf{A}_1\mathbf{M}_0 + \mathbf{N}_1\mathbf{B}_0 = \mathbf{I}_m.$$

□

Sensitivity matrices

Two sensitivity matrices which are useful in assessing the robustness of the system can be defined. The control-sensitivity matrix (inverse of the control node return difference matrix):

$$\mathbf{S}_u \triangleq (\mathbf{I}_m + \mathbf{C}_0\mathbf{W})^{-1} \in \mathbb{R}^{m\times m}(\cdot).$$

Output sensitivity matrix (inverse of the output node return difference matrix):

$$\mathbf{S}_y \triangleq (\mathbf{I}_r + \mathbf{W}\mathbf{C}_0)^{-1} \in \mathbb{R}^{r\times r}(\cdot)$$

Using the above relationships:

$$\begin{aligned}\mathbf{S}_u &= (\mathbf{I}_m + \mathbf{M}_0^{-1}\mathbf{N}_0\mathbf{B}_1\mathbf{A}_1^{-1})^{-1}\\ &= \mathbf{A}_1\mathbf{M}_0 = \mathbf{I}_m - \mathbf{N}_1\mathbf{B}_0 \qquad (11.95)\end{aligned}$$

and

$$\begin{aligned}\mathbf{S}_y &= (\mathbf{I}_r + \mathbf{A}_0^{-1}\mathbf{B}_0\mathbf{N}_1\mathbf{M}_1^{-1})^{-1}\\ &= \mathbf{M}_1\mathbf{A}_0 = \mathbf{I}_r - \mathbf{B}_1\mathbf{N}_0. \qquad (11.96)\end{aligned}$$

In classical control designs the norm of $\mathbf{S}(j\omega)$ is kept small over a frequency range of interest to limit the effect of parameter variations and disturbances. The selection of the optimal-control and noise-weighting matrices will affect the sensitivity in optimal designs. The selection of weighting matrices to achieve good sensitivity and robustness properties has been considered by Safanov, Laub and Hartmann [30] and in recent publications by Grimble [31], [33] and Grimble and Johnson [32]. The complementary sensitivity matrices also enter the optimal robustness analyses and these are defined as:

Control complementary sensitivity matrix:

$$\mathbf{T}_u = \mathbf{I}_m - \mathbf{S}_u = \mathbf{N}_1\mathbf{B}_0 \in \mathbb{R}^{m\times m}(\cdot). \qquad (11.97)$$

Output complementary sensitivity matrix:

$$\mathbf{T}_y = \mathbf{I}_r - \mathbf{S}_y = \mathbf{B}_1\mathbf{N}_0 \in \mathbb{R}^{r\times r}(\cdot). \qquad (11.98)$$

This completes the background for the application of polynomial matrices to system theory and prepares the way for the polynomial system solution of the LQG optimal control problem. The results on the class of stabilizing controllers applies to both optimal and non-optimal systems.

11.4. THE LQG STOCHASTIC OPTIMAL CONTROL PROBLEM

The LQG stochastic optimal control problem and its solution is presented in this section. The polynomial system concepts introduced in earlier sections are utilized in the derivations. A detailed analysis of the problem is given in terms of the z-domain system description and the section closes with summary theorems for the s-domain system results.

The basic problem is to determine the closed-loop controller $\mathbf{C}_0(z^{-1})$ which minimizes a cost-functional of the steady-state mean-square error and control variations. This optimum must be achieved whilst minimizing the effects of process and measurement noise. In fact, this is the most important advantage of the LQG design approach over the multivariable frequency-domain methodologies. That is, the balance between disturbance and measurement noise rejection and system performance can be achieved in a relatively straightforward manner. Once the noise statistics and the system have been specified, the filtering characteristics of the LQG controller have been determined (in state-space terms, the Kalman filter is specified) which leaves the designer free to concentrate on achieving the desired robustness and performance objectives.

The section commences with the problem definition, and assumptions. This is followed by the details of the proof which naturally falls into a number of areas including:

(i) The manipulations involved in the completion of squares within the cost functional.
(ii) The introduction of $\mathbf{G}, \mathbf{F}$ and $\mathbf{H}, \mathbf{F}$ Diophantine equations.
(iii) Stability considerations.
(iv) A proof that the Diophantine equations utilized have a unique minimum degree solution.

11.4.1. System, controller, and cost functional

The matrix fraction form given in Fig. 11.2 is utilized so that the following system equations obtain.

System output equation

$$y = \mathbf{A}^{-1}(z^{-1})(\mathbf{B}(z^{-1})u + \mathbf{C}(z^{-1})\xi) \tag{11.99}$$

where

$$\mathbf{A}^{-1}\mathbf{B} = \mathbf{A}_0^{-1}\mathbf{B}_0 = \mathbf{B}_1\mathbf{A}_1^{-1}. \tag{11.100}$$

Observation process

$$z_0 = y + v. \tag{11.101}$$

Controller input

$$e_1 = r - z_0 \tag{11.102}$$

Reference generation process

$$r = \mathbf{A}^{-1}(z^{-1})\mathbf{E}(z^{-1})\omega. \tag{11.103}$$

Tracking error

$$e = r - y. \tag{11.104}$$

Controller equation

$$u = \mathbf{C}_0(z^{-1})e_1 \tag{11.105}$$

where

$$\mathbf{C}_0(z^{-1}) = \mathbf{M}_0^{-1}(z^{-1})\mathbf{N}_0(z^{-1}) \tag{11.106}$$

and for asymptotic stability (Lemma 11.15):

$$\mathbf{M}_0\mathbf{A}_1 + \mathbf{N}_0\mathbf{B}_1 = \mathbf{I}_m. \tag{11.107}$$

The transfer function between the reference input r and the control input u is denoted by $\mathbf{G}_c(z^{-1})$ and obtains as:

$$\begin{aligned}\mathbf{G}_c &= (\mathbf{I}_m + \mathbf{C}_0\mathbf{W})^{-1}\mathbf{C}_0 \\ &= (\mathbf{I}_m + \mathbf{M}_0^{-1}\mathbf{N}_0\mathbf{B}_1\mathbf{A}_1^{-1})^{-1}\mathbf{M}_0^{-1}\mathbf{N}_0 = \mathbf{A}_1\mathbf{N}_0.\end{aligned} \tag{11.108}$$

The control, system output and tracking error signals become:

$$u = \mathbf{A}_1\mathbf{N}_0(r - v - n) \tag{11.109}$$

$$y = \mathbf{W}\mathbf{A}_1\mathbf{N}_0(r - v) + (\mathbf{I}_r - \mathbf{W}\mathbf{A}_1\mathbf{N}_0)n \tag{11.110}$$

$$e = (\mathbf{I}_r - \mathbf{W}\mathbf{A}_1\mathbf{N}_0)(r - n) + \mathbf{W}\mathbf{A}_1\mathbf{N}_0 v. \tag{11.111}$$

LQG quadratic cost functional

In time-domain terms the cost functional to be minimized is quadratic in the tracking error and control terms:

$$J = E\{\boldsymbol{e}^T(t)\mathbf{Q}_1\boldsymbol{e}(t) + \boldsymbol{u}^T(t)\mathbf{R}_1\boldsymbol{u}(t)\} \tag{11.112}$$

where $E\{\cdot\}$ denotes the unconditioned expectation operator. The cost functional (11.112) represents a steady-state cost criterion so that the system is assumed to have been in operation since $t_0 = -\infty$ (or in practical terms sufficiently early for initial condition transients to have decayed to zero).

In the usual steady-state LQG problem $\mathbf{Q}_1 \geq \mathbf{O}$ and $\mathbf{R}_1 > \mathbf{O}$ are symmetric constant weighting matrices. However, the following solution procedure can readily be modified to accommodate dynamic cost weightings:

$$\mathbf{Q}_1 = \mathbf{Q}_c^*\mathbf{Q}_c; \qquad \mathbf{Q}_c \triangleq \mathbf{Q}_c(z^{-1})$$

$$\mathbf{R}_1 = \mathbf{R}_c^*\mathbf{R}_c; \qquad \mathbf{R}_c \triangleq \mathbf{R}_c(z^{-1})$$

with $\mathbf{Q}_c(0)$ and $\mathbf{R}_c(0)$ both full rank (see Grimble [46]).

The following equivalence obtains:

$$\begin{aligned} J &= E\{\boldsymbol{e}^T(t)\mathbf{Q}_1\boldsymbol{e}(t) + \boldsymbol{u}^T(t)\mathbf{R}_1\boldsymbol{u}(t)\} \\ &= E\{\text{trace}(\mathbf{Q}_1\ \boldsymbol{e}(t)\boldsymbol{e}^T(t)) + \text{trace}(\mathbf{R}_1\boldsymbol{u}(t)\boldsymbol{u}^T(t))\} \\ &= \text{trace}(E\{\mathbf{Q}_1\boldsymbol{e}(t)\boldsymbol{e}^T(t)\}) + \text{trace}(E\{\mathbf{R}_1\boldsymbol{u}(t)\boldsymbol{u}^T(t)\}). \end{aligned}$$

Using the autocorrelation–power spectrum result of Section 11.3.4, namely equation (11.69) obtains:

$$J = \frac{1}{2\pi j}\oint_{|z|=1} \text{trace}(\mathbf{Q}_1\mathbf{\Phi}_{ee}(z^{-1})) + \text{trace}(\mathbf{R}_1\mathbf{\Phi}_{uu}(z^{-1}))\ \frac{dz}{z} \tag{11.113}$$

where $\mathbf{\Phi}_{ee}$ and $\mathbf{\Phi}_{uu}$ denote the rational spectral densities corresponding to the error and control signals, respectively. The result of Lemma 11.11, and equations (11.109) to (11.111) can be used to derive expressions for $\mathbf{\Phi}_{ee}(z^{-1})$ and $\mathbf{\Phi}_{uu}(z^{-1})$ to simplify the cost functional (11.113). Hence, the following relations obtain:

$$\mathbf{\Phi}_{ee} = (\mathbf{I}_r - \mathbf{W}\mathbf{A}_1\mathbf{N}_0)(\mathbf{\Phi}_{rr} + \mathbf{\Phi}_{nn})(\mathbf{I}_r - \mathbf{W}\mathbf{A}_1\mathbf{N}_0)^* + \mathbf{W}\mathbf{A}_1\mathbf{N}_0\mathbf{\Phi}_{vv}\mathbf{N}_0^*\mathbf{A}_1^*\mathbf{W}^* \tag{11.114}$$

$$\mathbf{\Phi}_{uu} = \mathbf{A}_1\mathbf{N}_0(\mathbf{\Phi}_{rr} + \mathbf{\Phi}_{vv} + \mathbf{\Phi}_{nn})\mathbf{N}_0^*\mathbf{A}_1^*. \tag{11.115}$$

From Fig. 11.2, the following spectral densities may be defined:

$$\mathbf{\Phi}_{rr} = \mathbf{W}_2\mathbf{Q}_2\mathbf{W}_2^* \quad \text{where } \mathbf{W}_2 = \mathbf{A}^{-1}\mathbf{E}, \qquad \mathbf{Q}_2 = E\{\omega(t)\omega^T(t)\}$$

$$\mathbf{\Phi}_{nn} = \mathbf{W}_0\mathbf{Q}_0\mathbf{W}_0^* \quad \text{where } \mathbf{W}_0 = \mathbf{A}^{-1}\mathbf{C}, \qquad \mathbf{Q}_0 = E\{\xi(t)\xi^T(t)\}$$

$$\mathbf{\Phi}_{vv} = \mathbf{R} = \mathbf{E}\{v(t)v^T(t)\}$$

and

$$\mathbf{\Phi}_0 \triangleq \mathbf{\Phi}_{rr} + \mathbf{\Phi}_{nn}. \tag{11.116}$$

11.4.2. Solvability lemma and solution theorem for LQG controller

The LQG stochastic optimal control problem is that of determining a controller $\mathbf{C}_0(z^{-1})$ to minimize the cost index (11.112), or in equivalent transfer function form (11.113), subject to the system equations (11.109), (11.110) and (11.111) or equivalent power-density spectra (11.114), (11.115) and (11.116). The existence of a solution is guaranteed by the following lemma.

Lemma 11.17. *Solvability conditions for the* **LQG** *problem.* *The* LQG *optimization problem is solvable if and only if:*

(i) *The system is free of unstable hidden modes.*

(ii) *The greatest common left divisor* $\mathbf{U}_0$ *of the system polynomials* $\mathbf{A}, \mathbf{B}$ *is a strictly Hurwitz polynomial matrix.*

(iii) *The matrices* $\mathbf{M}_0$ *and* $\mathbf{N}_0$ *are strictly stable transfer functions satisfying the conditions of the stability Lemma* 11.15.

Proof. Condition (i) must be satisfied since a system with unstable hidden modes cannot be stabilized by feedback action. Condition (ii) is necessary if the cost functional (11.112, 11.113) is to remain finite, since the greatest common divisor $\mathbf{U}_0$ of the pair $(\mathbf{A}, \mathbf{B})$ represents the modes of the disturbance subsystem which are not present in the plant. Clearly, such modes must be stable if the tracking error and control variances are to remain finite (see Problem 11.8.1). Condition (iii) is necessary and sufficient for closed-loop stability. Furthermore, the cost functional (11.112, 11.113) is finite if and only if the transfer-function matrices $\mathbf{M}_0$, and $\mathbf{N}_0$ represent asymptotically stable systems. That is if (ii) is satisfied then:

$$\mathbf{M}_1\mathbf{A}_0 + \mathbf{B}_1\mathbf{N}_0 = \mathbf{I}_r, \qquad \mathbf{N}_1\mathbf{A}_0 = \mathbf{A}_1\mathbf{N}_0.$$

Thus in the transfer functions of (11.109) to (11.111):

$$\mathbf{A}_1\mathbf{N}_0\mathbf{A}^{-1}\mathbf{C} = \mathbf{N}_1\mathbf{U}_0^{-1}\mathbf{C}$$

$$\mathbf{A}_1\mathbf{N}_0\mathbf{A}^{-1}\mathbf{E} = \mathbf{N}_1\mathbf{U}_0^{-1}\mathbf{E}$$

$$(\mathbf{I}_r - \mathbf{A}^{-1}\mathbf{B}\mathbf{A}_1\mathbf{N}_0)\mathbf{A}^{-1}\mathbf{E} = \mathbf{M}_1\mathbf{U}_0^{-1}\mathbf{E}$$

$$(\mathbf{I}_r - \mathbf{A}^{-1}\mathbf{B}\mathbf{A}_1\mathbf{N}_0)\mathbf{A}^{-1}\mathbf{C} = \mathbf{M}_1\mathbf{U}_0^{-1}\mathbf{C}$$

$$\mathbf{A}^{-1}\mathbf{B}\mathbf{A}_1\mathbf{N}_0 = \mathbf{B}_1\mathbf{N}_0$$

and these are all strictly stable ensuring the finiteness of the cost functional defined by equation (11.113) with the expressions (11.114) and (11.115). □

The main theorem may now be presented. It will be seen that the LQG controller calculation involves the solution of two Diophantine equations and two spectral factorizations.

Theorem 11.9. ***Optimal closed-loop controller solution.*** *Assume that the* LQG *stochastic optimal control problem satisfies the solvability conditions of Lemma 11.17, and that the optimal controller* $\mathbf{C}_0(z^{-1})$ *is realized without unstable hidden modes. The optimal controller is then given by;*

$$\mathbf{C}_0(z^{-1}) = \mathbf{H}_0^{-1}(z^{-1})\mathbf{G}_0(z^{-1}) \tag{11.117}$$

where $\mathbf{H}_0$, $\mathbf{G}_0$ *are obtained from any left coprime matrix fraction representation for* $\mathbf{H}\mathbf{D}_3^{-1}$ *and* $\mathbf{G}\mathbf{D}_2^{-1}$:

$$[\mathbf{H}\mathbf{D}_3^{-1}, \mathbf{G}\mathbf{D}_2^{-1}] = \mathbf{D}_f^{-1}[\mathbf{H}_0 \quad \mathbf{G}_0] \tag{11.118}$$

and the polynomial matrices **H**, **G**, *and* **F** *are obtained as the minimum degree solution with respect to* **F** *of the Diophantine equations:*

$$\bar{\mathbf{D}}_1\mathbf{G} + \mathbf{F}\mathbf{A}_2 = \bar{\mathbf{B}}_1\mathbf{Q}_1\mathbf{D}_2 \tag{11.119}$$

$$\bar{\mathbf{D}}_1\mathbf{H} - \mathbf{F}\mathbf{B}_3 = \bar{\mathbf{A}}_1\mathbf{R}_1\mathbf{D}_3 \tag{11.120}$$

where

$$\bar{\mathbf{D}}_1 \triangleq z^{-g}\mathbf{D}_1^*, \bar{\mathbf{B}}_1 \triangleq z^{-g}\mathbf{B}_1^* \quad \textit{and} \quad \bar{\mathbf{A}}_1 \triangleq z^{-g}\mathbf{A}_1^*.$$

The scalar $g \triangleq \max(n_{d_1}, n_{b_1}, n_{a_1})$.
The right coprime matrix fraction decomposition of $\mathbf{W}(z^{-1})$ *is calculated*

using:

$$\mathbf{B}_1\mathbf{A}_1^{-1} = \mathbf{W} = \mathbf{A}^{-1}\mathbf{B} \tag{11.121}$$

and the right coprime matrix fraction representations of $\mathbf{D}^{-1}\mathbf{B}$ *and* $\mathbf{D}^{-1}\mathbf{A}$ *are defined via:*

$$\mathbf{D}^{-1}\mathbf{A} = \mathbf{A}_2\mathbf{D}_2^{-1} \tag{11.122}$$

$$\mathbf{D}^{-1}\mathbf{B} = \mathbf{B}_2\mathbf{D}_3^{-1} \tag{11.123}$$

The stable (strictly Hurwitz) spectral factors $\mathbf{D}$ *and* $\mathbf{D}_1$ *are calculated from the equations:*

$$\mathbf{D}\mathbf{D}^* = \mathbf{C}\mathbf{Q}_0\mathbf{C}^* + \mathbf{E}\mathbf{Q}_2\mathbf{E}^* + \mathbf{A}\mathbf{R}\mathbf{A}^* \tag{11.124}$$

$$\mathbf{D}_1^*\mathbf{D}_1 = \mathbf{B}_1^*\mathbf{Q}_1\mathbf{B}_1 + \mathbf{A}_1^*\mathbf{R}_1\mathbf{A}_1. \tag{11.125}$$

The proof of the optimal controller theorem

The various aspects of the proof are discussed in some detail in subsequent sections.

Expansion of the cost functional

The first step in the proof involves an expansion of the cost functional (11.113) using equations (11.114) to (11.116), viz.:

$$J = \frac{1}{2\pi j}\oint_{|z|=1} \{\mathrm{trace}(\mathbf{Q}_1\mathbf{\Phi}_{ee}) + \mathrm{trace}(\mathbf{R}_1\mathbf{\Phi}_{uu})\}\frac{dz}{z}$$

$$= \frac{1}{2\pi j}\oint_{|z|=1} \{\mathrm{trace}(\mathbf{Q}_1(\mathbf{I}_r - \mathbf{W}\mathbf{A}_1\mathbf{N}_0)(\mathbf{\Phi}_{rr} + \mathbf{\Phi}_{nn})(\mathbf{I}_r - \mathbf{W}\mathbf{A}_1\mathbf{N}_0)^*$$

$$+ \mathbf{Q}_1\mathbf{W}\mathbf{A}_1\mathbf{N}_0\mathbf{\Phi}_{vv}\mathbf{N}_0^*\mathbf{A}_1^*\mathbf{W}^*)$$

$$+ \mathrm{trace}(\mathbf{R}_1\mathbf{A}_1\mathbf{N}_0(\mathbf{\Phi}_{rr} + \mathbf{\Phi}_{vv} + \mathbf{\Phi}_{nn})\mathbf{N}_0^*\mathbf{A}_1^*)\}\frac{dz}{z}.$$

Use the trace identity, trace $(\mathbf{XY}) = \mathrm{trace}(\mathbf{YX})$ and recall $\mathbf{\Phi}_0 \triangleq \mathbf{\Phi}_{rr} + \mathbf{\Phi}_{nn}$ to obtain:

$$J = \frac{1}{2\pi j} \oint_{|z|=1} \{\text{trace}((\mathbf{N}_0^* \mathbf{A}_1^* \mathbf{W}^* \mathbf{Q}_1 \mathbf{W} \mathbf{A}_1 \mathbf{N}_0 + \mathbf{Q}_1 - \mathbf{N}_0^* \mathbf{A}_1^* \mathbf{W}^* \mathbf{Q}_1 - \mathbf{Q}_1 \mathbf{W} \mathbf{A}_1 \mathbf{N}_0) \mathbf{\Phi}_0)$$

$$+ \text{trace}(\mathbf{N}_0^* \mathbf{A}_1^* \mathbf{R}_1 \mathbf{A}_1 \mathbf{N}_0 (\mathbf{\Phi}_0 + \mathbf{\Phi}_{vv})) + \text{trace}(\mathbf{N}_0^* \mathbf{A}_1^* \mathbf{W}^* \mathbf{Q}_1 \mathbf{W} \mathbf{A}_1 \mathbf{N}_0 \mathbf{\Phi}_{vv})\} \frac{dz}{z}$$

$$= \frac{1}{2\pi j} \oint_{|z|=1} \{\text{trace}(\mathbf{N}_0^* \mathbf{A}_1^* (\mathbf{W}^* \mathbf{Q}_1 \mathbf{W} + \mathbf{R}_1) \mathbf{A}_1 \mathbf{N}_0 (\mathbf{\Phi}_0 + \mathbf{\Phi}_{vv}))$$

$$+ \text{trace}((\mathbf{Q}_1 - \mathbf{N}_0^* \mathbf{A}_1^* \mathbf{W}^* \mathbf{Q}_1 - \mathbf{Q}_1 \mathbf{W} \mathbf{A}_1 \mathbf{N}_0) \mathbf{\Phi}_0)\} \frac{dz}{z}. \tag{11.126}$$

This is simplified further using the filter and control spectral factors. Recall equation (11.49):

$$\mathbf{Y}_f \mathbf{Y}_f^* = \mathbf{W}_0 \mathbf{Q}_0 \mathbf{W}_0^* + \mathbf{W}_2 \mathbf{Q}_2 \mathbf{W}_2^* + \mathbf{R} = \mathbf{A}^{-1} \mathbf{D} \mathbf{D}^* \mathbf{A}^{*-1} = \mathbf{\Phi}_0 + \mathbf{\Phi}_{vv} \tag{11.127}$$

where

$$\mathbf{D}\mathbf{D}^* = \mathbf{C}\mathbf{Q}_0\mathbf{C}^* + \mathbf{E}\mathbf{Q}_2\mathbf{E}^* + \mathbf{A}\mathbf{R}\mathbf{A}^*$$

and equation (11.50):

$$\mathbf{Y}_c^* \mathbf{Y}_c = \mathbf{W}^* \mathbf{Q}_1 \mathbf{W} + \mathbf{R}_1 = (\mathbf{A}_1^*)^{-1} \mathbf{D}_1^* \mathbf{D}_1 \mathbf{A}_1^{-1}$$

where

$$\mathbf{D}_1^* \mathbf{D}_1 = \mathbf{B}_1^* \mathbf{Q}_1 \mathbf{B}_1 + \mathbf{A}_1^* \mathbf{R}_1 \mathbf{A}_1. \tag{11.128}$$

(These relations give equations (11.124) and (11.125) of the theorem.) Substituting into equation (11.126) yields:

$$J = \frac{1}{2\pi j} \oint_{|z|=1} J_N(z^{-1}) \frac{dz}{z} \tag{11.129}$$

where

$$J_N(z^{-1}) = \text{trace}(\mathbf{N}_0^* \mathbf{D}_1^* \mathbf{D}_1 \mathbf{N}_0 \mathbf{A}^{-1} \mathbf{D} \mathbf{D}^* \mathbf{A}^{*-1})$$

$$+ \text{trace}((\mathbf{Q}_1 - \mathbf{N}_0^* \mathbf{B}_1^* \mathbf{Q}_1 - \mathbf{Q}_1 \mathbf{B}_1 \mathbf{N}_0) \mathbf{\Phi}_0). \tag{11.130}$$

Completing squares

Consider the identity,

$$(\mathbf{D}^*\mathbf{A}^{*-1}\mathbf{N}_0^*\mathbf{D}_1^* - \mathbf{D}^{-1}\mathbf{A}\boldsymbol{\Phi}_0\mathbf{Q}_1\mathbf{B}_1\mathbf{D}_1^{-1})(\mathbf{D}_1\mathbf{N}_0\mathbf{A}^{-1}\mathbf{D} - \mathbf{D}_1^{*-1}\mathbf{B}_1^*\mathbf{Q}_1\boldsymbol{\Phi}_0\mathbf{A}^*\mathbf{D}^{*-1})$$

$$= \mathbf{D}^*\mathbf{A}^{*-1}\mathbf{N}_0^*\mathbf{D}_1^*\mathbf{D}_1\mathbf{N}_0\mathbf{A}^{-1}\mathbf{D} - \mathbf{D}^*\mathbf{A}^{*-1}\mathbf{N}_0^*\mathbf{D}_1^*\mathbf{D}_1^{*-1}\mathbf{B}_1^*\mathbf{Q}_1\boldsymbol{\Phi}_0\mathbf{A}^*\mathbf{D}^{*-1}$$

$$- \mathbf{D}^{-1}\mathbf{A}\boldsymbol{\Phi}_0\mathbf{Q}_1\mathbf{B}_1\mathbf{D}_1^{-1}\mathbf{D}_1\mathbf{N}_0\mathbf{A}^{-1}\mathbf{D}$$

$$+ \mathbf{D}^{-1}\mathbf{A}\boldsymbol{\Phi}_0\mathbf{Q}_1\mathbf{B}_1\mathbf{D}_1^{-1}\mathbf{D}_1^{*-1}\mathbf{B}_1^*\mathbf{Q}_1\boldsymbol{\Phi}_0\mathbf{A}^*\mathbf{D}^{*-1}.$$

Noting that

$$\operatorname{trace}(\mathbf{D}^*\mathbf{A}^{*-1}\mathbf{N}_0^*\mathbf{D}_1^*\mathbf{D}_1^{*-1}\mathbf{B}_1^*\mathbf{Q}_1\boldsymbol{\Phi}_0\mathbf{A}^*\mathbf{D}^{*-1}) = \operatorname{trace}(\mathbf{N}_0^*\mathbf{B}_1^*\mathbf{Q}_1\boldsymbol{\Phi}_0)$$

$$\operatorname{trace}(\mathbf{D}^{-1}\mathbf{A}\boldsymbol{\Phi}_0\mathbf{Q}_1\mathbf{B}_1\mathbf{D}_1^{-1}\mathbf{D}_1\mathbf{N}_0\mathbf{A}^{-1}\mathbf{D}) = \operatorname{trace}(\mathbf{Q}_1\mathbf{B}_1\mathbf{N}_0\boldsymbol{\Phi}_0).$$

Thus (11.130) becomes:

$$J_N(z^{-1}) = \operatorname{trace}((\mathbf{D}^*\mathbf{A}^{*-1}\mathbf{N}_0^*\mathbf{D}_1^* - \mathbf{D}^{-1}\mathbf{A}\boldsymbol{\Phi}_0\mathbf{Q}_1\mathbf{B}_1\mathbf{D}_1^{-1})$$

$$\times (\mathbf{D}_1\mathbf{N}_0\mathbf{A}^{-1}\mathbf{D} - \mathbf{D}_1^{*-1}\mathbf{B}_1^*\mathbf{Q}_1\boldsymbol{\Phi}_0\mathbf{A}^*\mathbf{D}^{*-1}))$$

$$+ \operatorname{trace}(\mathbf{Q}_1(\mathbf{I}_r - \mathbf{B}_1\mathbf{D}_1^{-1}\mathbf{D}_1^{*-1}\mathbf{B}_1^*\mathbf{Q}_1\boldsymbol{\Phi}_0\mathbf{A}^*\mathbf{D}^{*-1}\mathbf{D}^{-1}\mathbf{A})\boldsymbol{\Phi}_0).$$

This expression may be written in the form:

$$J_N(z^{-1}) = \operatorname{trace}(\mathbf{S}_1^*\mathbf{S}_1 + \mathbf{S}_2) \qquad (11.131)$$

where

$$\mathbf{S}_1 = \mathbf{D}_1\mathbf{N}_0\mathbf{A}^{-1}\mathbf{D} - \mathbf{D}_1^{*-1}\mathbf{B}_1^*\mathbf{Q}_1\boldsymbol{\Phi}_0\mathbf{A}^*\mathbf{D}^{*-1} \qquad (11.132)$$

$$\mathbf{S}_2 = \mathbf{Q}_1(\mathbf{I}_r - \mathbf{B}_1\mathbf{D}_1^{-1}\mathbf{D}_1^{*-1}\mathbf{B}_1^*\mathbf{Q}_1\boldsymbol{\Phi}_0\mathbf{A}^*\mathbf{D}^{*-1}\mathbf{D}^{-1}\mathbf{A})\boldsymbol{\Phi}_0. \qquad (11.133)$$

Note that $\mathbf{S}_2$ is wholly independent of controller parameters and is therefore unaffected by any subsequent optimization of the cost functional.

Use equation (11.127) in (11.132) to obtain;

$$\mathbf{S}_1 = \mathbf{D}_1\mathbf{N}_0\mathbf{A}^{-1}\mathbf{D} - \mathbf{D}_1^{*-1}\mathbf{B}_1^*\mathbf{Q}_1\boldsymbol{\Phi}_0\mathbf{A}^*\mathbf{D}^{*-1}$$

$$= \mathbf{D}_1\mathbf{N}_0\mathbf{A}^{-1}\mathbf{D} - \mathbf{D}_1^{*-1}\mathbf{B}_1^*\mathbf{Q}_1(\mathbf{A}^{-1}\mathbf{D}\mathbf{D}^*\mathbf{A}^{*-1} - \mathbf{R})\mathbf{A}^*\mathbf{D}^{*-1}$$

$$\mathbf{S}_1 = \mathbf{D}_1\mathbf{N}_0\mathbf{A}^{-1}\mathbf{D} - \mathbf{D}_1^{*-1}\mathbf{B}_1^*\mathbf{Q}_1\mathbf{A}^{-1}\mathbf{D} + \mathbf{D}_1^{*-1}\mathbf{B}_1^*\mathbf{Q}_1\mathbf{R}\mathbf{A}^*\mathbf{D}^{*-1}. \qquad (11.134)$$

The second term of $\mathbf{S}_1$ contains the transforms of non-causal or unstable terms and must be partitioned using the first of the Diophantine equations.

G, F *Diophantine equation*

The partition of the term $\mathbf{D}_1^{*-1}\mathbf{B}_1^*\mathbf{Q}_1\mathbf{A}^{-1}\mathbf{D}$ is obtained using a Diophantine equation in the polynomial matrix pair $(\mathbf{F}, \mathbf{G})$. Write $\mathbf{A}^{-1}\mathbf{D}$ as the right co-prime factorization:

$$\mathbf{A}^{-1}\mathbf{D} = \mathbf{D}_2\mathbf{A}_2^{-1} \tag{11.135}$$

and introduce the Diophantine equation:

$$\bar{\mathbf{D}}_1\mathbf{G} + \mathbf{F}\mathbf{A}_2 = \bar{\mathbf{B}}_1\mathbf{Q}_1\mathbf{D}_2 \tag{11.136}$$

where $\bar{\mathbf{D}}_1 = \mathbf{D}_1^* z^{-g}$, $\bar{\mathbf{B}}_1 = \mathbf{B}_1^* z^{-g}$ and g is chosen to ensure $\bar{\mathbf{D}}_1$ and $\bar{\mathbf{B}}_1$ are polynomials in z^{-1}. (At this juncture $g \triangleq \max(n_{d_1}, n_{b_1})$.)

Straightforward manipulation of (11.136) yields:

$$\mathbf{D}_1^{*-1}\mathbf{B}_1^*\mathbf{Q}_1\mathbf{A}^{-1}\mathbf{D} = \mathbf{G}\mathbf{A}_2^{-1} + z^g\mathbf{D}_1^{*-1}\mathbf{F}. \tag{11.137}$$

(This step is equivalent to the use of a partial fraction expansion in the Wiener–Hopf solution of optimal control problems discussed in Chapter 2.) Substitution of (11.137) into (11.134) obtains:

$$\mathbf{S}_1 = \mathbf{D}_1\mathbf{N}_0\mathbf{A}^{-1}\mathbf{D} - \mathbf{G}\mathbf{A}_2^{-1} + \mathbf{D}_1^{*-1}\mathbf{B}_1^*\mathbf{Q}_1\mathbf{R}\mathbf{A}^*\mathbf{D}^{*-1} - z^g\mathbf{D}_1^{*-1}\mathbf{F} = \mathbf{W}_s + \mathbf{S}_3 \tag{11.138}$$

where (using 11.135):

$$\mathbf{W}_s = (\mathbf{D}_1\mathbf{N}_0\mathbf{D}_2 - \mathbf{G})\mathbf{A}_2^{-1} \tag{11.139}$$

$$\mathbf{S}_3 = \mathbf{D}_1^{*-1}(\mathbf{B}_1^*\mathbf{Q}_1\mathbf{R}\mathbf{A}^*\mathbf{D}^{*-1} - z^g\mathbf{F}). \tag{11.140}$$

The above formulae (11.139) and (11.140) are used in the minimization process to determine the optimal transfer function $\mathbf{N}_0$. Combining equations (11.129), (11.131) and (11.138) the cost-functional to be minimized has the form:

$$J = \frac{1}{2\pi j}\oint_{|z|=1} \operatorname{trace}(\mathbf{W}_s^*\mathbf{W}_s + \mathbf{W}_s^*\mathbf{S}_3 + \mathbf{S}_3^*\mathbf{W}_s + \mathbf{S}_3^*\mathbf{S}_3 + \mathbf{S}_2)\,\frac{dz}{z}. \tag{11.141}$$

H, F *Diophantine equation*

The motivation for the use of the **H, F** Diophantine equation is more difficult to explain. However, it is intrinsic to the stability characteristics of the pair $\mathbf{M}_0, \mathbf{N}_0$ and to the optimality of $\mathbf{M}_0$ and is justified in the sequel.

Write a coprime factorization for $\mathbf{D}^{-1}\mathbf{B}$ as:

$$\mathbf{D}^{-1}\mathbf{B} = \mathbf{B}_3\mathbf{D}_3^{-1} \tag{11.142}$$

and introduce a second Diophantine equation in terms of polynomial matrices **H** and **F**:

$$\bar{\mathbf{D}}_1\mathbf{H} - \mathbf{F}\mathbf{B}_3 = \bar{\mathbf{A}}_1\mathbf{R}_1\mathbf{D}_3 \tag{11.143}$$

where $\bar{\mathbf{D}}_1 = \mathbf{D}_1^* z^{-g}$ and $\bar{\mathbf{A}}_1 = \mathbf{A}_1^* z^{-g}$; index g chosen to ensure $\bar{\mathbf{D}}_1$ and $\bar{\mathbf{A}}_1$ are polynomials in z^{-1}.

Lemma 10.18. ***Relationship between* G *and* H**

$$\mathbf{D}_1^*\mathbf{H}\mathbf{D}_3^{-1}\mathbf{A}_1 + \mathbf{D}_1^*\mathbf{G}\mathbf{D}_2^{-1}\mathbf{B}_1 = \mathbf{D}_1^*\mathbf{D}_1 \tag{11.144}$$

equivalently

$$\mathbf{D}_1^{-1}\mathbf{H}\mathbf{D}_3^{-1}\mathbf{A}_1 + \mathbf{D}_1^{-1}\mathbf{G}\mathbf{D}_2^{-1}\mathbf{B}_1 = \mathbf{I}_m. \tag{11.145}$$

Proof. Rearrange equation (11.136) as:

$$\mathbf{D}_1^*\mathbf{G} + z^g\mathbf{F}\mathbf{A}_2 = \mathbf{B}_1^*\mathbf{Q}_1\mathbf{D}_2$$

Post multiply by $\mathbf{D}_2^{-1}\mathbf{B}_1$ hence:

$$\mathbf{D}_1^*\mathbf{G}\mathbf{D}_2^{-1}\mathbf{B}_1 + z^g\mathbf{F}\mathbf{A}_2\mathbf{D}_2^{-1}\mathbf{B}_1 = \mathbf{B}_1^*\mathbf{Q}_1\mathbf{B}_1.$$

Use (11.135) to obtain:

$$\mathbf{D}_1^*\mathbf{G}\mathbf{D}_2^{-1}\mathbf{B}_1 + z^g\mathbf{F}\mathbf{D}^{-1}\mathbf{A}\mathbf{B}_1 = \mathbf{B}_1^*\mathbf{Q}_1\mathbf{B}_1. \tag{11.146}$$

Rearrange equation (11.143) as:

$$\mathbf{D}_1^*\mathbf{H} - z^g\mathbf{F}\mathbf{B}_3 = \mathbf{A}_1^*\mathbf{R}_1\mathbf{D}_3.$$

Postmultiply by $\mathbf{D}_3^{-1}\mathbf{A}_1$ to obtain:

$$\mathbf{D}_1^*\mathbf{H}\mathbf{D}_3^{-1}\mathbf{A}_1 - \mathbf{z}^g\mathbf{F}\mathbf{B}_3\mathbf{D}_3^{-1}\mathbf{A}_1 = \mathbf{A}_1^*\mathbf{R}_1\mathbf{A}_1.$$

Use (11.142) to simplify:

$$\mathbf{D}_1^*\mathbf{H}\mathbf{D}_3^{-1}\mathbf{A}_1 - z^g\mathbf{F}\mathbf{D}^{-1}\mathbf{B}\mathbf{A}_1 = \mathbf{A}_1^*\mathbf{R}_1\mathbf{A}_1. \tag{11.147}$$

Add equations (11.146) and (11.147) to obtain:

$$\mathbf{D}_1^*\mathbf{H}\mathbf{D}_3^{-1}\mathbf{A}_1 + \mathbf{D}_1^*\mathbf{G}\mathbf{D}_2^{-1}\mathbf{B}_1 + z^g\mathbf{F}\mathbf{D}^{-1}(\mathbf{A}\mathbf{B}_1 - \mathbf{B}\mathbf{A}_1) = \mathbf{B}_1^*\mathbf{Q}_1\mathbf{B}_1 + \mathbf{A}_1^*\mathbf{R}_1\mathbf{A}_1.$$

Use (11.121) and (11.128) to obtain (11.144).

Premultiply (11.144) by $\mathbf{D}_1^{-1}\mathbf{D}_1^{*-1}$ to obtain (11.145). □

It is useful, particularly in a stability analysis and when simplifying the controller, to utilize a left coprime matrix fraction representation for pair $(\mathbf{H}\mathbf{D}_3^{-1}, \mathbf{G}\mathbf{D}_2^{-1})$. Recall the equation (11.145) as:

$$\mathbf{H}\mathbf{D}_3^{-1}\mathbf{A}_1 + \mathbf{G}\mathbf{D}_2^{-1}\mathbf{B}_1 = \mathbf{D}_1. \tag{11.148}$$

Let any left coprime matrix fraction representation for $\mathbf{H}\mathbf{D}_3^{-1}, \mathbf{G}\mathbf{D}_2^{-1}$ be defined as:

$$[\mathbf{H}\mathbf{D}_3^{-1} \quad \mathbf{G}\mathbf{D}_2^{-1}] = \mathbf{D}_f^{-1}[\mathbf{H}_0 \quad \mathbf{G}_0]. \tag{11.149}$$

Using (11.149) in (11.148) gives:

$$\mathbf{H}_0\mathbf{A}_1 + \mathbf{G}_0\mathbf{B}_1 = \mathbf{D}_f\mathbf{D}_1. \tag{11.150}$$

Demonstration that $\mathbf{W}_s$ *has no poles outside the unit-circle*

To be able to simplify the expanded cost functional (11.141) it is necessary to demonstrate that $\mathbf{W}_s$ is analytic outside the unit-circle of the z-plane. This is achieved by showing $\mathbf{W}_s$ to be a stable transfer-function matrix. The candidate pair $\mathbf{M}_0, \mathbf{N}_0$ which are being optimized are assumed asymptotically stable and to satisfy $\mathbf{M}_0\mathbf{A}_1 + \mathbf{N}_0\mathbf{B}_1 = \mathbf{I}_m$. An alternative representation for $\mathbf{C}_0(z^{-1}) = \mathbf{N}_1\mathbf{M}_1^{-1}$, where pair $\mathbf{M}_1, \mathbf{N}_1$ are also asymptotically stable may be used. From Lemma 11.15, equation (11.88) gives:

$$\mathbf{A}_0\mathbf{M}_1 + \mathbf{B}_0\mathbf{N}_1 = \mathbf{I}_r.$$

Let the greatest common left divisor of $(\mathbf{A}, \mathbf{B})$ again be denoted by $\mathbf{U}_0$ so that $\mathbf{A} = \mathbf{U}_0\mathbf{A}_0, \mathbf{B} = \mathbf{U}_0\mathbf{B}_0$. By the solvability lemma $\mathbf{U}_0$ is strictly Hurwitz. Premultiply the above equation by $\mathbf{D}^{-1}\mathbf{U}_0$ to obtain:

$$\mathbf{D}^{-1}\mathbf{A}\mathbf{M}_1 + \mathbf{D}^{-1}\mathbf{B}\mathbf{N}_1 = \mathbf{D}^{-1}\mathbf{U}_0.$$

Using (11.135) and (11.142) the above equation becomes:

$$(\mathbf{A}_2\mathbf{D}_2^{-1}\mathbf{M}_1 + \mathbf{B}_3\mathbf{D}_3^{-1}\mathbf{N}_1)\mathbf{U}_0^{-1}\mathbf{D} = \mathbf{I}_r. \tag{11.151}$$

Substitute relation (11.151) in the expression for (11.139), viz.:

$$\begin{aligned}\mathbf{W}_s &= (\mathbf{D}_1\mathbf{N}_0\mathbf{D}_2 - \mathbf{G})\mathbf{A}_2^{-1}\\ &= \mathbf{D}_1\mathbf{N}_0\mathbf{D}_2\mathbf{A}_2^{-1} - \mathbf{G}\mathbf{A}_2^{-1}\\ &= \mathbf{D}_1\mathbf{N}_0\mathbf{A}^{-1}\mathbf{D} - \mathbf{G}\mathbf{A}_2^{-1}(\mathbf{A}_2\mathbf{D}_2^{-1}\mathbf{M}_1 + \mathbf{B}_3\mathbf{D}_3^{-1}\mathbf{N}_1)\mathbf{U}_0^{-1}\mathbf{D}.\end{aligned} \tag{11.152}$$

From Lemma 11.16, equation (11.93).

$$\mathbf{N}_1\mathbf{A}_0 = \mathbf{A}_1\mathbf{N}_0$$

hence

$$\mathbf{N}_1\mathbf{U}_0^{-1}\mathbf{U}_0\mathbf{A}_0 = \mathbf{N}_1\mathbf{U}_0^{-1}\mathbf{A} = \mathbf{A}_1\mathbf{N}_0$$

and

$$\mathbf{N}_0\mathbf{A}^{-1} = \mathbf{A}_1^{-1}\mathbf{N}_1\mathbf{U}_0^{-1}. \tag{11.153}$$

Using (11.153) in (11.152) yields:

$$\begin{aligned}\mathbf{W}_s &= (\mathbf{D}_1\mathbf{A}_1^{-1}\mathbf{N}_1 - \mathbf{G}\mathbf{D}_2^{-1}\mathbf{M}_1 - \mathbf{G}\mathbf{A}_2^{-1}\mathbf{B}_3\mathbf{D}_3^{-1}\mathbf{N}_1)\mathbf{U}_0^{-1}\mathbf{D}\\ &= (\mathbf{D}_1 - \mathbf{G}\mathbf{A}_2^{-1}\mathbf{B}_3\mathbf{D}_3^{-1}\mathbf{A}_1)\mathbf{A}_1^{-1}\mathbf{N}_1\mathbf{U}_0^{-1}\mathbf{D} - \mathbf{G}\mathbf{D}_2^{-1}\mathbf{M}_1\mathbf{U}_0^{-1}\mathbf{D}.\end{aligned} \tag{11.154}$$

Consider then,

$$\begin{aligned}\mathbf{G}\mathbf{A}_2^{-1}\mathbf{B}_3\mathbf{D}_3^{-1}\mathbf{A}_1 &= \mathbf{G}\mathbf{A}_2^{-1}\mathbf{D}^{-1}\mathbf{B}\mathbf{A}_1\\ &= \mathbf{G}\mathbf{D}_2^{-1}\mathbf{A}^{-1}\mathbf{B}\mathbf{A}_1 = \mathbf{G}\mathbf{D}_2^{-1}\mathbf{B}_1.\end{aligned}$$

Returning to equation (11.154):

$$\mathbf{W}_s = (\mathbf{D}_1 - \mathbf{G}\mathbf{D}_2^{-1}\mathbf{B}_1)\mathbf{A}_1^{-1}\mathbf{N}_1\mathbf{U}_0^{-1}\mathbf{D} - \mathbf{G}\mathbf{D}_2^{-1}\mathbf{M}_1\mathbf{U}_0^{-1}\mathbf{D}.$$

The second Diophantine equation in $\mathbf{H}, \mathbf{F}$ enabled equation (11.148) to be derived and using (11.148) the above equation for $\mathbf{W}_s$ can be simplified as:

$$\mathbf{W}_s = (\mathbf{H}\mathbf{D}_3^{-1}\mathbf{N}_1 - \mathbf{G}\mathbf{D}_2^{-1}\mathbf{M}_1)\mathbf{U}_0^{-1}\mathbf{D}. \tag{11.155}$$

This expression for $\mathbf{W}_s$ contains only stable matrices hence all its singularities lie within the unit circle $|z| = 1$.

Minimization of the cost-functional

To consider the minimization of the cost functional recall the above equations (11.141), (11.139), (11.140) and (11.133). These are summarized below to permit the minimization process to be discussed easily.

$$J = \frac{1}{2\pi j} \oint_{|z|=1} \text{trace}(\mathbf{W}_s^*\mathbf{W}_s + \mathbf{W}_s^*\mathbf{S}_3 + \mathbf{S}_3^*\mathbf{W}_s + \mathbf{S}_3^*\mathbf{S}_3 + \mathbf{S}_2) \frac{dz}{z} \quad (11.141)$$

where

$$\mathbf{W}_s = (\mathbf{D}_1\mathbf{N}_0\mathbf{D}_2 - \mathbf{G})\mathbf{A}_2^{-1} \quad (11.139)$$

$$\mathbf{S}_3 = \mathbf{D}_1^{*-1}(\mathbf{B}_1^*\mathbf{Q}_1\mathbf{R}\mathbf{A}^*\mathbf{D}^{*-1} - z^g\mathbf{F}) \quad (11.140)$$

$$\mathbf{S}_2 = \mathbf{Q}_1(\mathbf{I}_r - \mathbf{B}_1\mathbf{D}_1^{-1}\mathbf{D}_1^{*-1}\mathbf{B}_1^*\mathbf{Q}_1\mathbf{\Phi}_0\mathbf{A}^*\mathbf{D}^{*-1}\mathbf{D}^{-1}\mathbf{A})\mathbf{\Phi}_0. \quad (11.133)$$

Note first that $\mathbf{S}_2$ is independent of the controller parameters thus this term does not enter the optimization process, similar comments apply to the term $\mathbf{S}_3^*\mathbf{S}_3$. It has been demonstrated above (equation (11.155)) that $\mathbf{W}_s$ has singularities only within the unit-circle in the z plane. From (11.140) $\mathbf{S}_3$ only has singularities outside the unit circle $|z| = 1$, hence by Cauchy's residue theorem:

$$\frac{1}{2\pi j} \oint_{|z|=1} \text{trace}(\mathbf{W}_s^*\mathbf{S}_3) \frac{dz}{z} = 0.$$

Some simple manipulation of contour integrals yields the result that:

$$\frac{1}{2\pi j} \oint_{|z|=1} \text{trace}(\mathbf{W}_s^*\mathbf{S}_3) \frac{ds}{z} = \frac{1}{2\pi j} \oint_{|z|=1} \text{trace}(\mathbf{S}_3^*\mathbf{W}_s) \frac{dz}{z} = 0$$

and the residues of the second integral sum exactly to zero.
The remaining term to be optimized is just that involving $\mathbf{W}_s^*\mathbf{W}_s$. Using the properties of Lemma 11.11:

$$\frac{1}{2\pi j} \oint_{|z|=1} \mathbf{W}_s^*(z^{-1})\mathbf{W}_s(z^{-1}) \frac{dz}{z} \geq \mathbf{O}.$$

Hence, this term has a zero minimum value if (and only if):

$$\mathbf{W}_s^{\circ}(z^{-1}) = (\mathbf{D}_1\mathbf{N}_0^{\circ}\mathbf{D}_2 - \mathbf{G})\mathbf{A}_2^{-1} = \mathbf{O} \quad (11.156)$$

or

$$\mathbf{N}_0^0 = \mathbf{D}_1^{-1}\mathbf{G}\mathbf{D}_2^{-1}. \tag{11.157}$$

From the stability result (11.107):

$$\begin{aligned}\mathbf{M}_0^0 &= (\mathbf{I}_m - \mathbf{N}_0^0\mathbf{B}_1)\mathbf{A}_1^{-1}\\ &= (\mathbf{I}_m - \mathbf{D}_1^{-1}\mathbf{G}\mathbf{D}_2^{-1}\mathbf{B}_1)\mathbf{A}_1^{-1}\\ &= \mathbf{D}_1^{-1}(\mathbf{D}_1 - \mathbf{G}\mathbf{D}_2^{-1}\mathbf{B}_1)\mathbf{A}_1^{-1}.\end{aligned}$$

Use (11.148) to obtain

$$\mathbf{M}_0^0 = \mathbf{D}_1^{-1}\mathbf{H}\mathbf{D}_3^{-1}. \tag{11.158}$$

Note immediately from (11.148)

$$\mathbf{M}_0^0\mathbf{A}_1 + \mathbf{N}_0^0\mathbf{B}_1 = \mathbf{I}_m. \tag{11.159}$$

If the $\mathbf{H}_0, \mathbf{G}_0$ factorization is used the following is an alternative form for the controller equations:

$$\mathbf{C}_0(z^{-1}) = (\mathbf{M}_0^0)^{-1}\mathbf{N}_0^0 = \mathbf{D}_3\mathbf{H}^{-1}\mathbf{G}\mathbf{D}_2^{-1} = \mathbf{H}_0^{-1}\mathbf{G}_0. \tag{11.160}$$

Return-difference matrix for optimal controller
The return-difference matrix becomes:

$$\begin{aligned}(\mathbf{I}_m + \mathbf{C}_0\mathbf{W}) &= (\mathbf{I}_m + \mathbf{H}_0^{-1}\mathbf{G}_0\mathbf{B}_1\mathbf{A}_1^{-1})\\ &= \mathbf{H}_0^{-1}(\mathbf{H}_0\mathbf{A}_1 + \mathbf{G}_0\mathbf{B}_1)\mathbf{A}_1^{-1}.\end{aligned}$$

Using equation (11.150) yields,

$$(\mathbf{I}_m + \mathbf{C}_0\mathbf{W}) = \mathbf{H}_0^{-1}\mathbf{D}_f\mathbf{D}_1\mathbf{A}_1^{-1}.$$

The return-difference matrix then gives:

$$\frac{\rho_c(z^{-1})}{\rho_0(z^{-1})} = \frac{\text{closed-loop characteristic polynomial}}{\text{open-loop characteristic polynomial}} = \frac{\det(\mathbf{D}_f\mathbf{D}_1)}{\det(\mathbf{H}_0\mathbf{A}_1)}.$$

Thus, the degree of stability of the closed-loop system may be readily determined by factoring the polynomial $\det(\mathbf{D}_f\mathbf{D}_1)$. Recall also that the zeros of

$\det(\mathbf{D}_f)$ are included within the set of zeros of $\det(\mathbf{D})$ and since both $\mathbf{D}_1$ and $\mathbf{D}$ are strictly Hurwitz spectral factors, the closed-loop system is guaranteed to be asymptotically stable.

Optimal cost value

Clearly substituting $\mathbf{W}_s^o = \mathbf{O}$ into equation (11.141) yields an optimal cost value:

$$J^0 = \frac{1}{2\pi j} \oint_{|z|=1} \operatorname{trace}(\mathbf{S}_3^* \mathbf{S}_3 + \mathbf{S}_2) \frac{dz}{z} \tag{11.161}$$

where

$$\left.\begin{aligned} \mathbf{S}_3 &= \mathbf{D}_1^{*-1}(\mathbf{B}_1^*\mathbf{Q}_1\mathbf{R}\mathbf{A}^*\mathbf{D}^{*-1} - z^g\mathbf{F}) \\ \mathbf{S}_2 &= \mathbf{Q}_1(\mathbf{I}_r - \mathbf{B}_1\mathbf{D}_1^{-1}\mathbf{D}_1^{*-1}\mathbf{B}_1^*\mathbf{Q}_1\mathbf{\Phi}_0\mathbf{A}^*\mathbf{D}^{*-1}\mathbf{D}^{-1}\mathbf{A})\mathbf{\Phi}_0. \end{aligned}\right\} \tag{11.162}$$

The demonstration of a unique minimum degree solution for the pair of Diophantine equations follows in the next section.

11.4.3. The controller Diophantine equations: solvability condition

The final part of the proof comprises the conditions necessary to solve the pair of Diophantine equations and the exhibition of a solution. To calculate the LQG controller $\mathbf{C}_0(z^{-1})$ the following Diophantine equations require a solution:

$$\bar{\mathbf{D}}_1\mathbf{G} + \mathbf{F}\mathbf{A}_2 = \bar{\mathbf{B}}_1\mathbf{Q}_1\mathbf{D}_2 \tag{11.136}$$

$$\bar{\mathbf{D}}_1\mathbf{H} - \mathbf{F}\mathbf{B}_3 = \bar{\mathbf{A}}_1\mathbf{R}_1\mathbf{D}_3 \tag{11.143}$$

where

$$\mathbf{A}^{-1}\mathbf{D} = \mathbf{D}_2\mathbf{A}_2^{-1} \tag{11.135}$$

$$\mathbf{D}^{-1}\mathbf{B} = \mathbf{B}_3\mathbf{D}_3^{-1} \tag{11.142}$$

with

$$\bar{\mathbf{D}}_1 = \mathbf{D}_1^* z^{-g}, \bar{\mathbf{B}}_1 = \mathbf{B}_1^* z^{-g}, \bar{\mathbf{A}}_1 = \mathbf{A}_1^* z^{-g} \quad \text{and} \, g \triangleq \max(n_{d_1}, n_{b_1}, n_{a_1}).$$

These may be written more concisely in matrix form as:

$$\bar{\mathbf{D}}_1[\mathbf{G}, \mathbf{H}] + \mathbf{F}[\mathbf{A}_2, -\mathbf{B}_3] = [\bar{\mathbf{B}}_1\mathbf{Q}_1\mathbf{D}_2, \bar{\mathbf{A}}_1\mathbf{R}_1\mathbf{D}_3]. \tag{11.163}$$

To show that this equation is solvable, first denote the greatest common left divisor of $\mathbf{A}_2$ and $\mathbf{B}_3$ by $\mathbf{U}_2$ and let

$$\mathbf{P} = \begin{bmatrix} \mathbf{P}_{11} & \mathbf{P}_{12} \\ \mathbf{P}_{21} & \mathbf{P}_{22} \end{bmatrix}$$

be a $(r+m)$ square unimodular matrix such that $[\mathbf{A}_2, -\mathbf{B}_3]\mathbf{P} = [\mathbf{O}, \mathbf{U}_2]$, (this equation follows from condition (iii) of Theorem 11.4) and

$$\mathbf{P}_{11} \in \mathbb{P}^{r\times m}(z^{-1}), \qquad \mathbf{P}_{12} \in \mathbb{P}^{r\times r}(z^{-1}),\ \mathbf{P}_{21} \in \mathbb{P}^{m\times m}(z^{-1}), \qquad \mathbf{P}_{22} \in \mathbb{P}^{m\times r}(z^{-1})).$$

Multiplying out obtain $\mathbf{A}_2\mathbf{P}_{11} = \mathbf{B}_3\mathbf{P}_{21}$ so that

$$\mathbf{B}\mathbf{D}_3\mathbf{P}_{21} = \mathbf{D}\mathbf{B}_3\mathbf{P}_{21} = \mathbf{D}\mathbf{A}_2\mathbf{P}_{11} = \mathbf{A}\mathbf{D}_2\mathbf{P}_{11}$$

or

$$\mathbf{B}(\mathbf{D}_3\mathbf{P}_{21}) = \mathbf{A}(\mathbf{D}_2\mathbf{P}_{11}).$$

Recall that $\mathbf{B}\mathbf{A}_1 = \mathbf{A}\mathbf{B}_1$ where $\mathbf{A}_1$ and $\mathbf{B}_1$ are right coprime. Thus, there exists a polynomial matrix $\mathbf{U}_3$ such that

$$\mathbf{D}_2\mathbf{P}_{11} = \mathbf{B}_1\mathbf{U}_3 \text{ and } \mathbf{D}_3\mathbf{P}_{21} = \mathbf{A}_1\mathbf{U}_3.$$

Postmultiplying the Diophantine equation by the $\mathbf{P}$ matrix gives:

$$\bar{\mathbf{D}}_1[\mathbf{G}, \mathbf{H}]\mathbf{P} + \mathbf{F}[\mathbf{O}, \mathbf{U}_2] = [\bar{\mathbf{B}}_1\mathbf{Q}_1\mathbf{D}_2, \bar{\mathbf{A}}_1\mathbf{R}_1\mathbf{D}_3]\mathbf{P}.$$

Let $[\mathbf{G}', \mathbf{H}'] = [\mathbf{G}, \mathbf{H}]\mathbf{P}$ then the resulting two equations become:

$$\bar{\mathbf{D}}_1\mathbf{G}' = \bar{\mathbf{B}}_1\mathbf{Q}_1\mathbf{D}_2\mathbf{P}_{11} + \bar{\mathbf{A}}_1\mathbf{R}_1\mathbf{D}_3\mathbf{P}_{21} \tag{11.164}$$

$$\bar{\mathbf{D}}_1\mathbf{H}' + \mathbf{F}\mathbf{U}_2 = \bar{\mathbf{B}}_1\mathbf{Q}_1\mathbf{D}_2\mathbf{P}_{12} + \bar{\mathbf{A}}_1\mathbf{R}_1\mathbf{D}_3\mathbf{P}_{22}. \tag{11.165}$$

The first of these equations may be simplified by substituting for $\mathbf{P}_{11}$ and $\mathbf{P}_{21}$ to obtain:

$$\bar{\mathbf{D}}_1\mathbf{G}' = (\bar{\mathbf{B}}_1\mathbf{Q}_1\mathbf{B}_1 + \bar{\mathbf{A}}_1\mathbf{R}_1\mathbf{A}_1)\mathbf{U}_3$$

or

$$\mathbf{G}' = \mathbf{D}_1\mathbf{U}_3.$$

Attention may now return to the second equation. Note from the definitions

$\mathbf{D}^{-1}\mathbf{A} = \mathbf{A}_2\mathbf{D}_2^{-1}$ and $\mathbf{D}^{-1}\mathbf{B} = \mathbf{B}_3\mathbf{D}_3^{-1}$ and both $\mathbf{A}_2, \mathbf{D}_2$ and $\mathbf{B}_3, \mathbf{D}_3$ are right coprime. Also let $\mathbf{U}_2$ denote the greatest common left divisor of $\mathbf{A}_2$ and $\mathbf{B}_3$ and obtain:

$$\mathbf{D}^{-1}\mathbf{U}_0\mathbf{A}_0 = \mathbf{U}_2\tilde{\mathbf{A}}_2\mathbf{D}_2^{-1} \quad \text{and} \quad \mathbf{D}^{-1}\mathbf{U}_0\mathbf{B}_0 = \mathbf{U}_2\tilde{\mathbf{B}}_3\mathbf{D}_3^{-1}$$

where $\mathbf{A}_2 = \mathbf{U}_2\tilde{\mathbf{A}}_2$ and $\mathbf{B}_3 = \mathbf{U}_2\tilde{\mathbf{B}}_3$. Clearly, $\det(\mathbf{U}_2)$ is contained as a factor of $\det(\mathbf{U}_0)$. Now $\mathbf{U}_0$ is by the definition of the system, stable and hence $\det(\mathbf{U}_2)$ and $\det(\bar{\mathbf{D}}_1)$ are coprime. The matrices:

$$\begin{bmatrix} \bar{\mathbf{D}}_1 & \mathbf{O} \\ \mathbf{O} & \mathbf{U}_2 \end{bmatrix} \quad \text{and} \quad \begin{bmatrix} \bar{\mathbf{D}}_1 & \bar{\mathbf{B}}_1\mathbf{Q}_1\mathbf{D}_2\mathbf{P}_{12} + \bar{\mathbf{A}}_1\mathbf{R}_1\mathbf{D}_3\mathbf{P}_{22} \\ \mathbf{O} & \mathbf{U}_2 \end{bmatrix}$$

therefore have the same invariant polynomials of $1, 1, \ldots, 1, \det(\bar{\mathbf{D}}_1)\det(\mathbf{U}_2)$, (Recall the definition of invariant polynomials given in the description of the Smith form in Chapter 3. That is, if d_j is the greatest common divisor of all jth order minors of $\mathbf{L}_1$ ($j = 1, 2, \ldots$, rank $\mathbf{L}_1$), with d_j taken to be monic, then d_j is divisible by d_{j-1} and $\gamma_j = d_j/d_{j-1}$ represent the invariant factors or polynomials of $\mathbf{L}_1$.) It follows from Theorem 11.6 that equation (11.165) is solvable.

The general solution of the equation (11.136) and (11.143) is required and may now be derived. First, the kernel of the equations (11.164) and (11.165) is found. The solution of the equations:

$$\bar{\mathbf{D}}_1\mathbf{G}' = \mathbf{O}$$

$$\bar{\mathbf{D}}_1\mathbf{H}' + \mathbf{F}\mathbf{U}_2 = \mathbf{O}$$

is clearly $\mathbf{G}' = \mathbf{O}$, $\mathbf{H}' = \mathbf{L}\mathbf{U}_2$ and $\mathbf{F} = -\bar{\mathbf{D}}_1\mathbf{L}$ where $\mathbf{L} \in \mathbb{P}^{m \times r}(z^{-1})$ is an arbitrary polynomial matrix. Note that this gives:

$$[\mathbf{G}, \mathbf{H}]\mathbf{P} = [\mathbf{G}', \mathbf{H}'] = [\mathbf{O}, \mathbf{L}\mathbf{U}_2] = [\mathbf{L}\mathbf{A}_2, -\mathbf{L}\mathbf{B}_3]\mathbf{P}$$

and recall that $\mathbf{P}$ is unimodular. Denote a particular solution of the equations (11.136) and (11.143) by $\mathbf{G}_1, \mathbf{H}_1$ and $\mathbf{F}_1$, then the general solution follows as:

$$\mathbf{G} = \mathbf{G}_1 + \mathbf{L}\mathbf{A}_2$$

$$\mathbf{H} = \mathbf{H}_1 - \mathbf{L}\mathbf{B}_3 \tag{11.166}$$

$$\mathbf{F} = \mathbf{F}_1 - \bar{\mathbf{D}}_1\mathbf{L}.$$

Note that $\bar{\mathbf{D}}_1$ is of the form $z^{-g}(\mathbf{D}_{10} + \mathbf{D}_{11}z + \cdots + \mathbf{D}_{1n}z^n)$ where since $\mathbf{D}_1$ is the control spectral factor $\mathbf{D}_{10}$ is known to be full rank. The polynomial matrix

$\bar{\mathbf{D}}_1$ is therefore regular or proper (that is the leading coefficient matrix is non-singular). Now the particular solution satisfies $\mathbf{F}_1 = \bar{\mathbf{D}}_1\mathbf{L} + \mathbf{F}$ or $\bar{\mathbf{D}}_1^{-1}\mathbf{F}_1 = \mathbf{L} + \bar{\mathbf{D}}_1^{-1}\mathbf{F}$ and because $\bar{\mathbf{D}}_1$ is proper the minimum degree solution with respect to $\mathbf{F}$ can be found by use of a left division algorithm (Kučera [11]). Write,

$$\mathbf{F}_1 = \bar{\mathbf{D}}_1\mathbf{U}_1 + \mathbf{V}_1$$

where the $\deg(\mathbf{V}_1) < \deg(\bar{\mathbf{D}}_1)$, then clearly the minimum degree solution is given by setting $\mathbf{L} = \mathbf{U}_1$:

$$\mathbf{G} = \mathbf{G}_1 + \mathbf{U}_1\mathbf{A}_2 \tag{11.167}$$

$$\mathbf{H} = \mathbf{H}_1 - \mathbf{U}_1\mathbf{B}_3 \tag{11.168}$$

$$\mathbf{F} = \mathbf{V}_1. \tag{11.169}$$

The $\mathbf{U}_1$ and $\mathbf{V}_1$ are determined uniquely by the division algorithm. Also, note that $\deg(\mathbf{F}) = n_f < \deg(\bar{\mathbf{D}}_1)$. The uniqueness of the controller $\mathbf{C}_0(z^{-1})$ follows from that of $\mathbf{G}$, $\mathbf{H}$ and $\mathbf{F}$. The controller depends upon the inverse of the $\mathbf{H}$ polynomial matrix and this is always causal because of the assumption that the system contains at least a single-step delay. This may be demonstrated using (11.157) and (11.158):

$$\mathbf{M}_0\mathbf{A}_1 + \mathbf{N}_0\mathbf{B}_1 = \mathbf{D}_1^{-1}\mathbf{H}\mathbf{D}_3^{-1}\mathbf{A}_1 + \mathbf{D}_1^{-1}\mathbf{G}\mathbf{D}_2^{-1}\mathbf{B}_1 = \mathbf{I}_m$$

by noting that $\mathbf{D}_1(0)$, $\mathbf{D}_3(0)$ and $\mathbf{A}_1(0)$ are full rank by definition and hence, $\mathbf{H}(0)$ is full rank. The above results may now be summarized in the following lemma.

Lemma 11.19. *Diophantine equation solution.* *Assume that* $\mathbf{U}_0$ *the greatest common divisor of* $\mathbf{A}$ *and* $\mathbf{B}$ *is strictly stable and let* $\mathbf{G}_1, \mathbf{H}_1$ *and* $\mathbf{F}_1$ *denote a particular solution of the equations.*

$$\bar{\mathbf{D}}_1\mathbf{G} + \mathbf{F}\mathbf{A}_2 = \bar{\mathbf{B}}_1\mathbf{Q}_1\mathbf{D}_2 \tag{11.136}$$

$$\bar{\mathbf{D}}_1\mathbf{H} - \mathbf{F}\mathbf{B}_3 = \bar{\mathbf{A}}_1\mathbf{R}_1\mathbf{D}_3. \tag{11.143}$$

Then a minimum degree solution with respect to $\mathbf{F}$ *exists and is unique and is given by:*

$$\mathbf{G} = \mathbf{G}_1 + \mathbf{U}_1\mathbf{A}_2 \tag{11.167}$$

$$\mathbf{H} = \mathbf{H}_1 - \mathbf{U}_1\mathbf{B}_3 \tag{11.168}$$

$$\mathbf{F} = \mathbf{V}_1 \tag{11.169}$$

where $\mathbf{U}_1$ *and* $\mathbf{V}_1$ *are determined by the division algorithms:*

$$\mathbf{F}_1 = \bar{\mathbf{D}}_1\mathbf{U}_1 + \mathbf{V}_1 \ \textit{and}\ \deg(\mathbf{V}_1) < \deg(\bar{\mathbf{D}}_1).$$

Other algorithms for solving matrix Diophantine equations have been described by several authors including Chang and Pearson [34] and Feinstein and Bar-Ness [35].

11.4.4. LQG stochastic optimal control: an example

To illustrate the steps in the calculation of the LQG optimal controller an example is presented.

Example 11.4. Consider the following multivariable system and noise descriptions.

$$\mathbf{A}(z^{-1})y(t) = \mathbf{B}(z^{-1})u(t) + \mathbf{C}(z^{-1})\xi(t)$$

where

$$\mathbf{A}(z^{-1}) = \mathbf{I}_2 + z^{-1}\mathbf{A}_1$$

$$\mathbf{B}(z^{-1}) = z^{-1}\mathbf{B}_0 + z^{-2}\mathbf{B}_1 = z^{-1}(1 - 2z^{-1})\begin{bmatrix} 4 & 2 \\ 1 & 4 \end{bmatrix}$$

$$\mathbf{C}(z^{-1}) = \mathbf{I}_2 + z^{-1}\mathbf{C}_1 = c\mathbf{I}_2 \quad \text{where } c \triangleq 1 + 0.5z^{-1}$$

and

$$\mathbf{A}_1 = \begin{bmatrix} -3 & 0 \\ 0 & -3 \end{bmatrix}, \qquad \mathbf{C}_1 = \begin{bmatrix} 0.5 & 0 \\ 0 & 0.5 \end{bmatrix},$$

$$\mathbf{B}_0 = \begin{bmatrix} 4 & 2 \\ 1 & 4 \end{bmatrix}, \qquad \mathbf{B}_1 = \begin{bmatrix} -8 & -4 \\ -2 & -8 \end{bmatrix}.$$

The system is open-loop unstable and non-minimum phase. Assume that the reference in Fig. 11.2 is set to zero and the measurement noise is zero. Also, let the process noise covariance $\mathbf{Q} = \mathbf{I}_2$. Thus,

$$\mathbf{W}(z^{-1}) = \mathbf{A}(z^{-1})^{-1}\mathbf{B}(z^{-1}) = \mathbf{B}_1(z^{-1})\mathbf{A}_1(z^{-1})^{-1}$$

where

$$\mathbf{A}_1(z^{-1}) = (1 - 3z^{-1})\mathbf{I}_2, \qquad \mathbf{B}_1(z^{-1}) = z^{-1}(1 - 2z^{-1})\begin{bmatrix} 4 & 2 \\ 1 & 4 \end{bmatrix},$$

and hence

$$\mathbf{A}_1(z^{-1}) = \mathbf{A}(z^{-1}) \text{ and } \mathbf{B}_1(z^{-1}) = \mathbf{B}(z^{-1}).$$

The optimal control weighting matrices are chosen as

$$\mathbf{Q}_1 = \mathbf{I}_2 \quad \text{and} \quad \mathbf{R}_1 = \begin{bmatrix} 17 & 12 \\ 12 & 20 \end{bmatrix}.$$

Solution. The spectral factor (11.124) follows as $\mathbf{D}(z^{-1}) = \mathbf{C}(z^{-1}) = c\mathbf{I}_2$ and from (11.125):

$$\mathbf{D}_1^*\mathbf{D}_1 = (1-2z)(1-2z^{-1})\begin{bmatrix} 17 & 12 \\ 12 & 20 \end{bmatrix} + (1-3z)(1-3z^{-1})\begin{bmatrix} 17 & 12 \\ 12 & 20 \end{bmatrix}$$

$$= 13.09(1-0.382z)(1-0.382z^{-1})\begin{bmatrix} 17 & 12 \\ 12 & 20 \end{bmatrix}$$

or

$$\mathbf{D}_1(z^{-1}) = 3.618(1-0.382z^{-1})\begin{bmatrix} 4 & 2 \\ 1 & 4 \end{bmatrix} = d_1(z^{-1})\begin{bmatrix} 4 & 2 \\ 1 & 4 \end{bmatrix}$$

for this problem $\mathbf{D}_2 = \mathbf{D}, \mathbf{A}_2 = \mathbf{A}, \mathbf{D}_3 = \mathbf{D}$ and $\mathbf{B}_3 = \mathbf{B}$. The Diophantine equations (11.136) and (11.143) are given as:

$$\bar{\mathbf{D}}_1\mathbf{G} + \mathbf{F}\mathbf{A} = \bar{\mathbf{B}}\mathbf{D}$$

$$\bar{\mathbf{D}}_1\mathbf{H} - \mathbf{F}\mathbf{B} = \bar{\mathbf{A}}\mathbf{R}_1\mathbf{D}.$$

These equations may be written as:

$$\bar{\mathbf{D}}_1[\mathbf{G},\mathbf{H}] + \mathbf{F}[\mathbf{A}, -\mathbf{B}] = [\bar{\mathbf{B}}, \bar{\mathbf{A}}\mathbf{R}_1]\mathbf{D}.$$

Note that $\mathbf{D}_1$ and $\mathbf{B}$ may be written in the form $\mathbf{D}_1(z^{-1}) = d_1(z^{-1})\mathbf{M}$ and

$$\mathbf{B}(z^{-1}) = b(z^{-1})\mathbf{M} \quad \text{where } \mathbf{M} \triangleq \begin{bmatrix} 4 & 2 \\ 1 & 4 \end{bmatrix}.$$

Thus, if $\mathbf{A}$ is written as $\mathbf{A} = a\mathbf{I}_2$, then

$$\bar{d}_1\mathbf{M}^T[\mathbf{G},\mathbf{H}] + \mathbf{F}[a\mathbf{I}_2, -b\mathbf{M}] = [b\mathbf{M}^T, \bar{a}\mathbf{M}^T\mathbf{M}]\mathbf{D}.$$

This equation is clearly satisfied if $\mathbf{F} = f\mathbf{M}^T, \mathbf{H} = h\mathbf{M}, \mathbf{G} = g\mathbf{I}_2$ where f, h and g are scalar polynomials satisfying:

$$\bar{d}_1[g,h] + f[a, -b] = [\bar{b}, \bar{a}]c. \tag{11.170}$$

Equation (11.170) becomes:

$$3.618(1-0.382z)z^{-2}[g,h]+f[(1-3z^{-1}),\ -z^{-1}(1-2z^{-1})]$$

$$=[z^{-1}(1-2z),z^{-2}(1-3z)](1+0.5\,z^{-1})$$

$$[(3.618z^{-2}-1.382z^{-1})g,(3.618z^{-1}-1.382)h]$$

$$+f[1-3z^{-1},\ -1+2z^{-1}]=[-2+0.5z^{-2},\ -3-0.5z^{-1}+0.5z^{-2}].$$

Since deg $\bar{\mathbf{D}}_1=2>$ deg $\mathbf{F}$ this implies $\mathbf{F}=f_0+f_1z^{-1}$. The above equation may be solved by obtaining the minimum degree solution with respect to $\mathbf{F}$ using the algorithms in Kučera [11]. Alternatively, by equating coefficients of z^0, z^{-1} and z^{-2} it follows that $\mathbf{G}=g_0$ and $\mathbf{H}=h_0+h_1z^{-1}$ and hence, a solution is required to the equations:

$$[-1.382z^{-1}g_0+3.618z^{-2}g_0,\ -1.382h_0+(3.618h_0-1.382h_1)z^{-1}+3.618h_1z^{-2}]$$

$$+[f_0+(-3f_0+f_1)z^{-1}-3f_1z^{-2},\ -f_0+(2f_0-f_1)z^{-1}+2f_1z^{-2}]$$

$$=[-2+0.5z^{-2},\ -3-0.5z^{-1}+0.5z^{-2}]$$

giving

$$[0,\ -1.382h_0]+[f_0,\ -f_0]=[-2,\ -3]$$

$$[-1.382g_0,(3.618h_0-1.382h_1)]+[(-3f_0+f_1),2f_0-f_1]=[0,\ -0.5]$$

$$[3.618g_0,3.618h_1]+[-3f_1,2f_1]=[0.5,0.5].$$

From the first of these equations:

$$f_0=-2 \quad \text{and} \quad h_0=3.618.$$

The remaining equations may be written in the form:

$$\mathbf{X}\boldsymbol{\theta}=\mathbf{y}$$

where $\boldsymbol{\theta}=[g_0,h_1,f_1]^T$ or,

$$\begin{bmatrix}-1.382 & 0 & 1\\ 0 & -1.382 & -1\\ 3.618 & 0 & -3\\ 0 & 3.618 & 2\end{bmatrix}\begin{bmatrix}g_0\\ h_1\\ f_1\end{bmatrix}=\begin{bmatrix}-6\\ -9.59\\ 0.5\\ 0.5\end{bmatrix}.$$

Now the rank $[\mathbf{X}]=3$ and rank$[\mathbf{X},\mathbf{y}]=3$ and hence, the equations have a unique solution. The matrix $\mathbf{X}$ has a left inverse and since the equations are known to have a solution, $\mathbf{X}\boldsymbol{\theta}=\mathbf{y}$ has precisely one solution (Noble [36]), given by $\boldsymbol{\theta}=(\mathbf{X}^T\mathbf{X})^{-1}\mathbf{X}^T\mathbf{y}$ $=[33.15,\ -21.87,39.82]^T$. Hence

$$[\mathbf{H}c^{-1},\mathbf{G}c^{-1}]=[\mathbf{M}hc^{-1},\mathbf{I}_2gc^{-1}]$$

$$=c^{-1}[\mathbf{H}_0,\mathbf{G}_0]$$

Thus, $\mathbf{H}_0 = \mathbf{M}(3.618 - 21.87z^{-1})$ and $\mathbf{G}_0 = \mathbf{I}_2(33.15)$ and the optimal controller follows from (11.117):

$$\mathbf{C}_0(z^{-1}) = \frac{1}{1.53 - 9.24z^{-1}} \begin{bmatrix} 4 & -2 \\ -1 & 4 \end{bmatrix}.$$

The stability of the closed-loop system can be verified from the return-difference matrix relationship:

$$\mathbf{I}_2 + \mathbf{W}\mathbf{C}_0 = \mathbf{I}_2 + \mathbf{I}_2 \frac{14z^{-1}(1 - 2z^{-1})}{(1 - 3z^{-1})(1.53 - 9.24z^{-1})}$$

$$\det(\mathbf{I}_2 + \mathbf{W}\mathbf{C}_0) = \left(\frac{(2.66 - z^{-1})(2.05 + z^{-1})0.28}{(1 - 3z^{-1})(1.53 - 9.24z^{-1})}\right)^2.$$

The closed-loop system characteristic equation gives the poles at $z = 1/2.66$ and $z = -1/2.05$ and thus the system is stable as required.

11.4.5. Minimum-variance control: a special LQG controller

The term minimum variance is usually reserved for LQG controllers which have been derived with the control costing set to zero, namely $\mathbf{R}_1 = \mathbf{O}$. In such cases, the two Diophantine equations in the controller calculation are replaced by a single Diophantine equation. The complexity of the solution procedure is therefore reduced which is of considerable benefit in adaptive control applications.

Setting $\mathbf{R}_1 = \mathbf{O}$, the two Diophantine equations (11.119) and (11.120) become:

$$\bar{\mathbf{D}}_1\mathbf{G} + \mathbf{F}\mathbf{A}_2 = \bar{\mathbf{B}}_1\mathbf{Q}_1\mathbf{D}_2$$

and

$$\bar{\mathbf{D}}_1\mathbf{H} = \mathbf{F}\mathbf{B}_3.$$

Combining these equations, noting that $g = n_{d_1} = n_{b_1}$,

$$\bar{\mathbf{D}}_1\mathbf{G}\mathbf{A}_2^{-1}\mathbf{B}_3 + \bar{\mathbf{D}}_1\mathbf{H} = \bar{\mathbf{B}}_1\mathbf{Q}_1\mathbf{D}_2\mathbf{A}_2^{-1}\mathbf{B}_3. \qquad (11.171)$$

Define

$$\mathbf{A}_3 \text{ and } \mathbf{B}_4 \text{ via } \mathbf{A}_2^{-1}\mathbf{B}_3 = \mathbf{B}_4\mathbf{A}_3^{-1} \qquad (11.172)$$

then equation (11.171) becomes:

$$\mathbf{H}\mathbf{A}_3 + \mathbf{G}\mathbf{B}_4 = \bar{\mathbf{D}}_1^{-1}\mathbf{B}_1\mathbf{Q}_1\mathbf{D}_2\mathbf{B}_4. \qquad (11.173)$$

From equation (11.125), $\mathbf{R}_1 = \mathbf{O}$ yields:

$$\mathbf{D}_1^*\mathbf{D}_1 = \mathbf{B}_1^*\mathbf{Q}_1\mathbf{B}_1$$

hence

$$\mathbf{D}_1 = \bar{\mathbf{D}}_1^{-1}\bar{\mathbf{B}}_1\mathbf{Q}_1\mathbf{B}_1.$$

Postmultiply by $\mathbf{A}_1^{-1}$ to obtain:

$$\mathbf{D}_1\mathbf{A}_1^{-1} = \bar{\mathbf{D}}_1^{-1}\bar{\mathbf{B}}_1\mathbf{Q}_1\mathbf{B}_1\mathbf{A}_1^{-1} = \bar{\mathbf{D}}_1^{-1}\bar{\mathbf{B}}_1\mathbf{Q}_1\mathbf{A}^{-1}\mathbf{B}.$$

Postmultiply by $\mathbf{D}_3$ to obtain

$$\mathbf{D}_1\mathbf{A}_1^{-1}\mathbf{D}_3 = \bar{\mathbf{D}}_1^{-1}\bar{\mathbf{B}}_1\mathbf{Q}_1\mathbf{A}^{-1}\mathbf{B}\mathbf{D}_3 = \bar{\mathbf{D}}_1^{-1}\bar{\mathbf{B}}_1\mathbf{Q}_1\mathbf{A}^{-1}\mathbf{D}\mathbf{B}_3$$

$$\mathbf{D}_1\mathbf{A}_1^{-1}\mathbf{D}_3 = \bar{\mathbf{D}}_1^{-1}\bar{\mathbf{B}}_1\mathbf{Q}_1\mathbf{D}_2\mathbf{A}_2^{-1}\mathbf{B}_3. \tag{11.174}$$

Using (11.172) to replace $\mathbf{A}_2^{-1}\mathbf{B}_3$ in (11.173) and (11.174) obtains:

$$\mathbf{H}\mathbf{A}_3 + \mathbf{G}\mathbf{B}_4 = \mathbf{D}_1\mathbf{A}_1^{-1}\mathbf{D}_3\mathbf{A}_3 \tag{11.175}$$

Define

$$\mathbf{D}_5 = \mathbf{A}_1^{-1}\mathbf{D}_3\mathbf{A}_3 \tag{11.176}$$

Note that $\mathbf{D}_5$ is polynomial. That is,

$$\mathbf{B}_1\mathbf{D}_5 = \mathbf{B}_1\mathbf{A}_1^{-1}\mathbf{D}_3\mathbf{A}_3 = \mathbf{A}^{-1}\mathbf{B}\mathbf{D}_3\mathbf{A}_3 = \mathbf{A}^{-1}\mathbf{D}\mathbf{B}_3\mathbf{A}_3 = \mathbf{D}_2\mathbf{A}_2^{-1}\mathbf{A}_2\mathbf{B}_4$$

or

$$\mathbf{B}_1\mathbf{D}_5 = \mathbf{B}_1\mathbf{A}_1^{-1}\mathbf{D}_3\mathbf{A}_3 = \mathbf{D}_2\mathbf{B}_4.$$

Since $\mathbf{D}_2\mathbf{B}_4$ is a polynomial matrix and the pair $\mathbf{A}_1, \mathbf{B}_1$ are right coprime it follows that $\mathbf{D}_5 = \mathbf{A}_1^{-1}\mathbf{D}_3\mathbf{A}_3$ must be a polynomial matrix as noted.

Thus, together (11.175) and (11.176) give the single Diophantine equation to be solved as:

$$\mathbf{H}\mathbf{A}_3 + \mathbf{G}\mathbf{B}_4 = \mathbf{D}_1\mathbf{D}_5. \tag{11.177}$$

This equation is similar to equation (11.150) which results from combining the two Diophantine equations of the general LQG problem. For minimum variance controllers only a single equation must be solved (viz 11.177) and the original equations (11.119) and (11.120) are automatically satisfied. In the general LQG problem it is necessary to solve both (11.119) and (11.120) before calculating (11.150).

Several of the self-tuning regulators and controllers employ a single Diophantine equation controller calculation of the above form (see for example, Wellstead *et al.* [37]). Clearly, such methods are closely related to the minimum variance control philosophy.

11.4.6. The LQG stochastic optimal controller: *s*-domain theorem

To conclude this section on the LQG stochastic optimal controller, the *s*-domain theorem is given (without proof): this corresponds to the continuous time version of the LQG problem. The system configuration of Fig. 11.2 is retained and the basic system equations are:

System output equation

$$\mathbf{y}(s) = \mathbf{A}^{-1}(s)(\mathbf{B}(s)\mathbf{u}(s) + \mathbf{C}(s)\boldsymbol{\xi}(s))$$

where

$$\mathbf{A}^{-1}\mathbf{B} = \mathbf{A}_0^{-1}\mathbf{B}_0 = \mathbf{B}_1\mathbf{A}_1^{-1}.$$

Observation process

$$\mathbf{z}_0(s) = \mathbf{y}(s) + \mathbf{v}(s).$$

Controller input

$$\mathbf{e}_1(s) = \mathbf{r}(s) - \mathbf{z}_0(s).$$

Reference generation process

$$\mathbf{r}(s) = \mathbf{A}^{-1}(s)\mathbf{E}(s)\boldsymbol{\omega}(s).$$

Tracking error

$$\mathbf{e}(s) = \mathbf{r}(s) - \mathbf{y}(s).$$

Controller equation

$$\mathbf{u}(s) = \mathbf{C}_0(s)\mathbf{e}_1(s)$$

where

$$\mathbf{C}_0(s) = \mathbf{M}_0^{-1}(s)\mathbf{N}_0(s)$$

and for asymptotic stability

$$\mathbf{M}_0(s)\mathbf{A}_1(s) + \mathbf{N}_0(s)\mathbf{B}_1(s) = \mathbf{I}_m.$$

Cost functional (time domain)

$$J = E\{\boldsymbol{e}^T(t)\mathbf{Q}_1\boldsymbol{e}(t) + \boldsymbol{u}^T(t)\mathbf{R}_1\boldsymbol{u}(t)\}.$$

Cost functional (s-domain)

$$J = \frac{1}{2\pi j}\int_{-j\infty}^{j\infty} \text{trace}(\mathbf{Q}_1\mathbf{\Phi}_{ee}(s) + \mathbf{R}_1\mathbf{\Phi}_{uu}(s))\, ds.$$

Spectral densities and covariance matrices

$$\mathbf{\Phi}_{rr} = \mathbf{W}_2\mathbf{Q}_2\mathbf{W}_2^*, \quad \mathbf{W}_2(s) = \mathbf{A}^{-1}\mathbf{E}, \quad E\{\boldsymbol{\omega}(t)\boldsymbol{\omega}^T(\tau)\} = \mathbf{Q}_2\delta(t-\tau)$$

$$\mathbf{\Phi}_{nn} = \mathbf{W}_0\mathbf{Q}_0\mathbf{W}_0^*, \quad \mathbf{W}_0 = \mathbf{A}^{-1}\mathbf{C}, \quad E\{\boldsymbol{\xi}(t)\boldsymbol{\xi}^T(\tau)\} = \mathbf{Q}_0\,\delta(t-\tau)$$

$$E\{\boldsymbol{v}(t)\boldsymbol{v}^T(\tau)\} = \mathbf{R}\,\delta(t-\tau) \quad \text{and} \quad \mathbf{\Phi}_0 \triangleq \mathbf{\Phi}_{rr} + \mathbf{\Phi}_{nn}.$$

The problem remains one of selecting a stable transfer-function matrix pair $\mathbf{M}_0$, $\mathbf{N}_0$ to minimize the cost index above. The solvability conditions are those of Lemma 11.17 where the polynomial matrices are polynomials of s instead of z^{-1}. The solution theorem which summarizes the results follows as.

Theorem 11.10. ***LQG optimal closed-loop controller solution.*** *Assume that the* LQG *stochastic optimal control problem satisfies the solvability conditions of Lemma* 11.17 *and that the optimal controller* $\mathbf{C}_0(s)$ *is realized without unstable hidden modes. The controller is given as:*

$$\mathbf{C}_0(s) = \mathbf{H}_0^{-1}(s)\mathbf{G}_0(s)$$

where $\mathbf{H}_0$, $\mathbf{G}_0$ *are obtained from any left coprime matrix fraction representation for* $\mathbf{HD}_3^{-1}$ and $\mathbf{GD}_2^{-1}$:

$$[\mathbf{HD}_3^{-1} \quad \mathbf{GD}_2^{-1}] = \mathbf{D}_f^{-1}[\mathbf{H}_0 \quad \mathbf{G}_0].$$

The polynomial matrices $\mathbf{H}$, $\mathbf{G}$ *and* $\mathbf{F}$ *are obtained as the minimum degree solution with respect to* $\mathbf{F}$ *of the Diophantine equations:*

$$\mathbf{D}_1^*\mathbf{G} + \mathbf{F}\mathbf{A}_2 = \mathbf{B}_1^*\mathbf{Q}_1\mathbf{D}_2$$

$$\mathbf{D}_1^*\mathbf{H} - \mathbf{F}\mathbf{B}_3 = \mathbf{A}_1^*\mathbf{R}_1\mathbf{D}_3$$

where deg $\mathbf{F}(\text{row}_i) <$ deg $\mathbf{D}_1(\text{row}_i)$; $i = 1, \ldots, m$.

The right coprime matrix fraction decomposition of $\mathbf{W}(s)$ *is calculated as*

$$\mathbf{W} = \mathbf{A}^{-1}\mathbf{B} = \mathbf{B}_1\mathbf{A}_1^{-1}$$

and the right coprime matrix fraction representations of $\mathbf{D}^{-1}\mathbf{B}$ *and* $\mathbf{D}^{-1}\mathbf{A}$ *are defined as:*

$$\mathbf{D}^{-1}\mathbf{A} = \mathbf{A}_2\mathbf{D}_2^{-1}$$

$$\mathbf{D}^{-1}\mathbf{B} = \mathbf{B}_3\mathbf{D}_3^{-1}.$$

The stable spectral factors $\mathbf{D}$ *and* $\mathbf{D}_1$ *are calculated from the equations:*

Filter spectral factor

$$\mathbf{D}\mathbf{D}^* = \mathbf{C}\mathbf{Q}_0\mathbf{C}^* + \mathbf{E}\mathbf{Q}_2\mathbf{E}^* + \mathbf{A}\mathbf{R}\mathbf{A}^*.$$

Controller spectral factor

$$\mathbf{D}_1^*\mathbf{D}_1 = \mathbf{B}_1^*\mathbf{Q}_1\mathbf{B}_1 + \mathbf{A}_1^*\mathbf{R}_1\mathbf{A}_1.$$

11.5. THE WIENER AND GRADIENT SOLUTIONS TO THE LQG STOCHASTIC OPTIMAL CONTROL PROBLEM

There is a close relationship between the matrix fraction approach described in previous sections and the older Wiener–Hopf solutions to the LQG stochastic optimal control problem. The main difference is that the partial fraction expansion procedures of the Wiener–Hopf solutions are replaced by polynomial matrix Diophantine equations. In this section, the relationship between the two approaches is explored; further demonstrations of the connections are given in Section 7.4.

Stability, however, is treated differently by the two procedures and in some problems, the older Wiener methods can yield an unstable closed-loop design. This problem was described briefly in Chapter 2 earlier.

Particular topics discussed in this section include the Youla parameterization for the optimal controller, the gradient solution to the LQG stochastic optimal control problem and a brief résumé of the results for the deterministic LQ optimal control problem.

11.5.1. The Youla parameterization and the modern Wiener–Hopf solution procedures

The published work by Youla, Jabr and Bongiorno [2] treated the stochastic optimal control problem in the continuous-time domain; solutions being given in terms of s-domain transfer functions. The discussion presented here is for the discrete-time domain and is somewhat simpler than the Youla *et al.* solution procedure.

A stabilizing control law is derived using the so-called Youla parameterization for the controller. This was described previously in Section 11.3.5 (Lemma 11.14). The optimization proceeds by minimizing the same cost functional ((11.112) or (11.113)) with respect to all $(m \times r)$ real rational matrices $\mathbf{K}(z^{-1})$ which are analytic in $|z| \geq 1$ (that is, matrices which are asymptotically stable) and which satisfy the condition $\det(\mathbf{X}_1 - \mathbf{K}\mathbf{B}_0) \neq 0$ or alternatively $\det(\mathbf{X}_0 - \mathbf{B}_1\mathbf{K}) \neq 0$ (see Lemma 11.14).

From Lemma 11.14, the controller is given by

$$\mathbf{C}_0(z^{-1}) = (\mathbf{X}_1 - \mathbf{K}\mathbf{B}_0)^{-1}(\mathbf{Y}_1 + \mathbf{K}\mathbf{A}_0) \tag{11.178}$$

where $\mathbf{X}_1, \mathbf{Y}_1$ satisfy

$$\mathbf{X}_1\mathbf{A}_1 + \mathbf{Y}_1\mathbf{B}_1 = \mathbf{I}_m. \tag{11.179}$$

Recalling also equation (11.84):

$$\mathbf{M}_0 = \mathbf{X}_1 - \mathbf{K}\mathbf{B}_0 \tag{11.180}$$

$$\mathbf{N}_0 = \mathbf{Y}_1 + \mathbf{K}\mathbf{A}_0 \tag{11.181}$$

From Section 11.4.2, the spectral factors are used to reduce the cost functional to the form

$$J = \frac{1}{2\pi j} \oint_{|z|=1} J_N(z^{-1}) \frac{dz}{z} \tag{11.129}$$

where $J_N(z^{-1}) = \mathrm{trace}(\mathbf{S}_1^*\mathbf{S}_1 + \mathbf{S}_2)$ (11.131)

$$\mathbf{S}_1 = \mathbf{D}_1\mathbf{N}_0\mathbf{A}^{-1}\mathbf{D} - \mathbf{D}_1^{*-1}\mathbf{B}_1^*\mathbf{Q}_1\mathbf{\Phi}_0\mathbf{A}^*\mathbf{D}^{*-1} \tag{11.132}$$

$$\mathbf{S}_2 = \mathbf{Q}_1(\mathbf{I}_r - \mathbf{B}_1\mathbf{D}_1^{-1}\mathbf{D}_1^{*-1}\mathbf{B}_1^*\mathbf{Q}_1\mathbf{\Phi}_0\mathbf{A}^*\mathbf{D}^{*-1}\mathbf{D}^{-1}\mathbf{A})\mathbf{\Phi}_0. \tag{11.133}$$

In the expression for $\mathbf{S}_1$, the term in $\mathbf{N}_0$ is expanded using (11.181) to give:

$$\mathbf{D}_1\mathbf{N}_0\mathbf{A}^{-1}\mathbf{D} = \mathbf{D}_1(\mathbf{Y}_1 + \mathbf{K}\mathbf{A}_0)\mathbf{A}^{-1}\mathbf{D} = \mathbf{D}_1\mathbf{Y}_1\mathbf{A}^{-1}\mathbf{D} + \mathbf{D}_1\mathbf{K}\mathbf{A}_0\mathbf{A}_0^{-1}\mathbf{U}_0^{-1}\mathbf{D}$$

$$\mathbf{D}_1\mathbf{N}_0\mathbf{A}^{-1}\mathbf{D} = \mathbf{D}_1\mathbf{Y}_1\mathbf{A}^{-1}\mathbf{D} + \mathbf{D}_1\mathbf{K}\mathbf{U}_0^{-1}\mathbf{D}. \tag{11.182}$$

Incorporating (11.182) into (11.132), gives for S_1:

$$\mathbf{S}_1 = \mathbf{D}_1\mathbf{Y}_1\mathbf{A}^{-1}\mathbf{D} + \mathbf{D}_1\mathbf{K}\mathbf{U}_0^{-1}\mathbf{D} - \mathbf{D}_1^{*-1}\mathbf{B}_1^*\mathbf{Q}_1\mathbf{\Phi}_0\mathbf{A}^*\mathbf{D}^{*-1}$$

$$= \mathbf{D}_1(\mathbf{Y}_1\mathbf{A}_0^{-1} + \mathbf{K})\mathbf{U}_0^{-1}\mathbf{D} - \mathbf{D}_1^{*-1}\mathbf{B}_1^*\mathbf{Q}_1\mathbf{\Phi}_0\mathbf{A}^*\mathbf{D}^{*-1} \tag{11.183}$$

Introduce the discrete equivalent of the Youla brace notation:

$$f(z^{-1}) \triangleq \{f(z^{-1})\}_- + \{f(z^{-1})\}_+ \quad \text{where } \{f(z^{-1})\}_+$$

represents the terms in $f(z^{-1})$ with poles strictly inside the unit circle in the z-plane. This is the traditional interpretation of these terms which differs from the causality arguments given in Chapter 2. The term $\{f(z^{-1})\}_-$ includes those terms of $f(z^{-1})$ with poles lying outside or on the unit circle ($|z| \geq 1$) of the z-plane. Thus, the expression for $\mathbf{S}_1$ (11.183) may be written using the brace notation as:

$$\mathbf{S}_1 = \{\mathbf{D}_1(\mathbf{Y}_1\mathbf{A}_0^{-1} + \mathbf{K})\mathbf{U}_0^{-1}\mathbf{D}\}_+ - \{\mathbf{D}_1^{*-1}\mathbf{B}_1^*\mathbf{Q}_1\mathbf{\Phi}_0\mathbf{A}^*\mathbf{D}^{*-1}\}_+$$

$$+ \{\mathbf{D}_1(\mathbf{Y}_1\mathbf{A}_0^{-1} + \mathbf{K})\mathbf{U}_0^{-1}\mathbf{D}\}_- - \{\mathbf{D}_1^{*-1}\mathbf{B}_1^*\mathbf{Q}_1\mathbf{\Phi}_0\mathbf{A}^*\mathbf{D}^{*-1}\}_-$$

$$= [\mathbf{D}_1(\mathbf{Y}_1\mathbf{A}_0^{-1} + \mathbf{K})\mathbf{U}_0^{-1}\mathbf{D} - \{\mathbf{D}_1(\mathbf{Y}_1\mathbf{A}_0^{-1} + \mathbf{K})\mathbf{U}_0^{-1}\mathbf{D}\}_-$$

$$- \{\mathbf{D}_1^{*-1}\mathbf{B}_1^*\mathbf{Q}_1\mathbf{\Phi}_0\mathbf{A}^*\mathbf{D}^{*-1}\}_+]$$

$$+ \{\mathbf{D}_1\mathbf{Y}_1\mathbf{A}_0^{-1}\mathbf{U}_0^{-1}\mathbf{D}\}_- - \{\mathbf{D}_1^{*-1}\mathbf{B}_1^*\mathbf{Q}_1\mathbf{\Phi}_0\mathbf{A}^*\mathbf{D}^{*-1}\}_-$$

where $\{\mathbf{D}_1\mathbf{K}\mathbf{U}_0^{-1}\mathbf{D}\}_- = \mathbf{O}$ since $\mathbf{U}_0$ is strictly Hurwitz.

Thus

$$\mathbf{S}_1 = \mathbf{W}_{s1} + \mathbf{S}_{31}$$

where

$$\mathbf{W}_{s1} \triangleq [\mathbf{D}_1(\mathbf{Y}_1\mathbf{A}_0^{-1} + \mathbf{K})\mathbf{U}_0^{-1}\mathbf{D} - \{\mathbf{D}_1\mathbf{Y}_1\mathbf{A}^{-1}\mathbf{D}\}_-$$

$$- \{\mathbf{D}_1^{*-1}\mathbf{B}_1^*\mathbf{Q}_1\mathbf{\Phi}_0\mathbf{A}^*\mathbf{D}^{*-1}\}_+] \tag{11.184}$$

$$\mathbf{S}_{31} \triangleq \{\mathbf{D}_1\mathbf{Y}_1\mathbf{A}^{-1}\mathbf{D}\}_- - \{\mathbf{D}_1^{*-1}\mathbf{B}_1^*\mathbf{Q}_1\mathbf{\Phi}_0\mathbf{A}^*\mathbf{D}^{*-1}\}_-. \tag{11.185}$$

The term $\mathbf{S}_{31}$ is analytic on $|z| = 1$. The assumption that the cost functional, J is finite with equation (11.131) implies that $\mathbf{S}_1$ is analytic on $|z| = 1$. However, $\mathbf{S}_{31} = \mathbf{S}_1 - \mathbf{W}_{s1}$ where $\mathbf{W}_{s1}$ is analytic on $|z| = 1$ and hence $\mathbf{S}_{31}$ possesses the desired analyticity.

Thus, the term $\mathbf{W}_{s1}$ (equation (11.184)) is analytic in the region $|z| \geq 1$, whilst the term $\mathbf{S}_{31}$ is analytic in the region $|z| \leq 1$. Following arguments analogous to those presented in Section 11.4.2 for the minimization of the cost functional, the condition of optimality is $\mathbf{W}_{s1}^0 = \mathbf{O}$, so that (11.184) can be rearranged for the optimal gain as:

$$\mathbf{K}^0(z^{-1}) = \mathbf{D}_1^{-1}[\{\mathbf{D}_1^{*-1}\mathbf{B}_1^*\mathbf{Q}_1\mathbf{\Phi}_0\mathbf{A}^*\mathbf{D}^{*-1}\}_+$$

$$+ \{\mathbf{D}_1\mathbf{Y}_1\mathbf{A}^{-1}\mathbf{D}\}_-]\mathbf{D}^{-1}\mathbf{U}_0 - \mathbf{Y}_1\mathbf{A}^{-1}\mathbf{U}_0$$

$$= \mathbf{D}_1^{-1}[\{\mathbf{D}_1^{*-1}\mathbf{B}_1^*\mathbf{Q}_1\mathbf{\Phi}_0\mathbf{A}^*\mathbf{D}^{*-1} - \mathbf{D}_1\mathbf{Y}_1\mathbf{A}^{-1}\mathbf{D}\}_+]\mathbf{D}^{-1}\mathbf{U}_0. \tag{11.186}$$

This is the first derivation recorded for the optimal controller of a discrete system, analogous to the equivalent s-domain results presented by Youla and co-workers [2]. The expression for $\mathbf{K}^0(z^{-1})$ may be simplified as follows. Using equation (11.127) to simplify the term within braces:

$$\{\mathbf{D}_1^{*-1}\mathbf{B}_1^*\mathbf{Q}_1\mathbf{\Phi}_0\mathbf{A}^*\mathbf{D}^{*-1} - \mathbf{D}_1\mathbf{Y}_1\mathbf{A}^{-1}\mathbf{D}\}_+$$

$$= \{\mathbf{D}_1^{*-1}\mathbf{B}_1^*\mathbf{Q}_1(\mathbf{A}^{-1}\mathbf{D}\mathbf{D}^*\mathbf{A}^{*-1} - \mathbf{R})\mathbf{A}^*\mathbf{D}^{*-1} - \mathbf{D}_1\mathbf{Y}_1\mathbf{A}^{-1}\mathbf{D}\}_+$$

$$= \{\mathbf{D}_1^{*-1}\mathbf{B}_1^*\mathbf{Q}_1\mathbf{D}_2\mathbf{A}_2^{-1} - \mathbf{D}_1^{*-1}\mathbf{B}_1^*\mathbf{Q}_1\mathbf{R}\mathbf{A}^*\mathbf{D}^{*-1} - \mathbf{D}_1\mathbf{Y}_1\mathbf{A}^{-1}\mathbf{D}\}_+.$$

Note from (11.90), (11.92) and (11.93) that $\mathbf{A}_1\mathbf{Y}_1 = \mathbf{Y}_0\mathbf{A}_0$ and from (11.12b) that $\mathbf{B}_1\mathbf{Y}_1 = \mathbf{I}_r - \mathbf{X}_0\mathbf{A}_0$. Using these results with equations (11.137) and (11.148) and noting $\mathbf{\Phi}_0 = \mathbf{A}^{-1}\mathbf{D}\mathbf{D}^*\mathbf{A}^{*-1} - \mathbf{R}$ yields:

$$\{\mathbf{D}_1^{*-1}\mathbf{B}_1^*\mathbf{Q}_1\mathbf{\Phi}_0\mathbf{A}^*\mathbf{D}^{*-1} - \mathbf{D}_1\mathbf{Y}_1\mathbf{A}^{-1}\mathbf{D}\}_+$$

$$= \{\mathbf{G}\mathbf{A}_2^{-1} + \bar{\mathbf{D}}_1^{-1}\mathbf{F} - \mathbf{D}_1^{*-1}\mathbf{B}_1^*\mathbf{Q}_1\mathbf{R}\mathbf{A}^*\mathbf{D}^{*-1}$$

$$-(\mathbf{H}\mathbf{D}_3^{-1}\mathbf{A}_1+\mathbf{G}\mathbf{D}_2^{-1}\mathbf{B}_1)\mathbf{Y}_1\mathbf{A}^{-1}\mathbf{D}\}_+$$

$$=\{\mathbf{G}\mathbf{A}_2^{-1}-\mathbf{H}\mathbf{D}_3^{-1}\mathbf{Y}_0\mathbf{U}_0^{-1}\mathbf{D}-\mathbf{G}\mathbf{D}_2^{-1}(\mathbf{I}_r-\mathbf{X}_0\mathbf{A}_0)\mathbf{A}^{-1}\mathbf{D}\}_+$$

$$=-\mathbf{H}\mathbf{D}_3^{-1}\mathbf{Y}_0\mathbf{U}_0^{-1}\mathbf{D}+\mathbf{G}\mathbf{D}_2^{-1}\mathbf{X}_0\mathbf{U}_0^{-1}\mathbf{D} \tag{11.187}$$

and hence,

$$\mathbf{K}^0(z^{-1})=\mathbf{D}_1^{-1}(-\mathbf{H}\mathbf{D}_3^{-1}\mathbf{Y}_0+\mathbf{G}\mathbf{D}_2^{-1}\mathbf{X}_0). \tag{11.188}$$

The expression for the optimal controller $\mathbf{C}_0(z^{-1})$ may similarly be simplified using (11.148):

$$\mathbf{X}_1-\mathbf{K}^0\mathbf{B}_0=\mathbf{X}_1-\mathbf{D}_1^{-1}(-\mathbf{H}\mathbf{D}_3^{-1}(\mathbf{I}_m-\mathbf{A}_1\mathbf{X}_1)+\mathbf{G}\mathbf{D}_2^{-1}\mathbf{B}_1\mathbf{X}_1)$$

$$=\mathbf{D}_1^{-1}\mathbf{H}\mathbf{D}_3^{-1}+(\mathbf{I}_m-\mathbf{D}_1^{-1}\mathbf{H}\mathbf{D}_3^{-1}\mathbf{A}_1-\mathbf{D}_1^{-1}\mathbf{G}\mathbf{D}_2^{-1}\mathbf{B}_1)\mathbf{X}_1$$

$$=\mathbf{D}_1^{-1}\mathbf{H}\mathbf{D}_3^{-1} \tag{11.189}$$

$$\mathbf{Y}_1+\mathbf{K}^0\mathbf{A}_0=\mathbf{Y}_1+\mathbf{D}_1^{-1}(-\mathbf{H}\mathbf{D}_3^{-1}\mathbf{A}_1\mathbf{Y}_1+\mathbf{G}\mathbf{D}_2^{-1}(\mathbf{I}_r-\mathbf{B}_1\mathbf{Y}_1))$$

$$=\mathbf{D}_1^{-1}\mathbf{G}\mathbf{D}_2^{-1}+(\mathbf{I}_r-\mathbf{D}_1^{-1}\mathbf{H}\mathbf{D}_3^{-1}\mathbf{A}_1-\mathbf{D}_1^{-1}\mathbf{G}\mathbf{D}_2^{-1}\mathbf{B}_1)\mathbf{Y}_1$$

$$=\mathbf{D}_1^{-1}\mathbf{G}\mathbf{D}_2^{-1} \tag{11.190}$$

and (11.178) gives:

$$\mathbf{C}_0(z^{-1})=\mathbf{D}_3\mathbf{H}^{-1}\mathbf{G}\mathbf{D}_2^{-1}. \tag{11.191}$$

This is identical to the expression for the optimal controller (11.160) derived via the Kučera type of analysis developed in the previous section.

Transfer-function matrix between reference and control-input

Recall from equation (11.108) that the transfer function between reference and control input is given by:

$$\mathbf{G}_c(z^{-1})=(\mathbf{I}_m+\mathbf{C}_0\mathbf{W})^{-1}\mathbf{C}_0=\mathbf{A}_1\mathbf{N}_0.$$

Substituting for $\mathbf{N}_0$, equation (11.181), yields:

$$\mathbf{G}_c(z^{-1})=\mathbf{A}_1\mathbf{K}^0\mathbf{A}_0+\mathbf{A}_1\mathbf{Y}_1$$

and using (11.186) gives:

$$\mathbf{G}_c(z^{-1}) = \mathbf{A}_1\mathbf{D}_1^{-1}[\{\mathbf{D}_1^{*-1}\mathbf{B}_1^*\mathbf{Q}_1\mathbf{\Phi}_0\mathbf{A}^*\mathbf{D}^{*-1}\}_+ + \{\mathbf{D}_1\mathbf{Y}_1\mathbf{A}^{-1}\mathbf{D}\}_-]\mathbf{D}^{-1}\mathbf{A}. \tag{11.192}$$

This expression is the discrete equivalent to that presented in Youla *et al.* [2]. Using equations (11.148) and (11.12b), the last term in (11.192) may be simplified, namely:

$$\mathbf{D}_1\mathbf{Y}_1\mathbf{A}^{-1}\mathbf{D} = (\mathbf{H}\mathbf{D}_3^{-1}\mathbf{A}_1 + \mathbf{G}\mathbf{D}_2^{-1}\mathbf{B}_1)\mathbf{Y}_1\mathbf{A}^{-1}\mathbf{D}$$

$$= \mathbf{H}\mathbf{D}_3^{-1}\mathbf{Y}_0\mathbf{U}_0^{-1}\mathbf{D} + \mathbf{G}\mathbf{D}_2^{-1}(\mathbf{I}_r - \mathbf{X}_0\mathbf{A}_0)\mathbf{A}^{-1}\mathbf{D}$$

thus

$$\{\mathbf{D}_1\mathbf{Y}_1\mathbf{A}^{-1}\mathbf{D}\}_+ = (\mathbf{H}\mathbf{D}_3^{-1}\mathbf{Y}_0 - \mathbf{G}\mathbf{D}_2^{-1}\mathbf{X}_0)\mathbf{U}_0^{-1}\mathbf{D} + \{\mathbf{G}\mathbf{A}_2^{-1}\}_+$$

and

$$\{\mathbf{D}_1\mathbf{Y}_1\mathbf{A}^{-1}\mathbf{D}\}_- = \{\mathbf{G}\mathbf{A}_2^{-1}\}_-.$$

Note also that (see the derivation of (11.188)):

$$\{\mathbf{D}_1^{*-1}\mathbf{B}_1^*\mathbf{Q}_1\mathbf{\Phi}_0\mathbf{A}^*\mathbf{D}^{*-1}\}_+ = \{\mathbf{G}\mathbf{A}_2^{-1}\}_+$$

hence substituting in (11.192) yields:

$$\mathbf{G}_c(z^{-1}) = \mathbf{A}_1\mathbf{D}_1^{-1}(\{\mathbf{G}\mathbf{A}_2^{-1}\}_+ + \{\mathbf{G}\mathbf{A}_2^{-1}\}_-)\mathbf{D}^{-1}\mathbf{A} = \mathbf{A}_1\mathbf{D}_1^{-1}\mathbf{G}\mathbf{D}_2^{-1}.$$

Recall $\mathbf{N}_0 = \mathbf{D}_1^{-1}\mathbf{G}\mathbf{D}_2^{-1}$, so that $\mathbf{G}_c(z^{-1}) = \mathbf{A}_1\mathbf{N}_0$ corresponding to equation (11.108). The procedure to be followed to solve the LQG stochastic optimal control problem is encapsulated in the following theorem.

Theorem 11.11. *Discrete optimal controller solution via the Youla parameterization.* *The optimal controller for the* LQG *stochastic optimal control problem, as defined in Theorem 11.9, can be calculated by the Youla parameterization as follows.*

For system matrices $\mathbf{A}, \mathbf{B}$ *determine coprime pairs:*

$$\mathbf{A}^{-1}\mathbf{B} = \mathbf{A}_0^{-1}\mathbf{B}_0 = \mathbf{B}_1\mathbf{A}_1^{-1}. \tag{11.193}$$

Note that $\mathbf{A} = \mathbf{U}_0\mathbf{A}_0$, $\mathbf{B} = \mathbf{U}_0\mathbf{B}_0$. (11.194)

Determine stable spectral factors $\mathbf{D}$ *and* $\mathbf{D}_1$ *from the equations:*

$$\mathbf{DD}^* = \mathbf{CQ}_0\mathbf{C}^* + \mathbf{EQ}_2\mathbf{E}^* + \mathbf{ARA}^* \tag{11.195}$$

$$\mathbf{D}_1^*\mathbf{D}_1 = \mathbf{B}_1^*\mathbf{Q}_1\mathbf{B}_1 + \mathbf{A}_1^*\mathbf{RA}_1. \tag{11.196}$$

Determine polynomial matrices $\mathbf{X}_1, \mathbf{Y}_1$ *satisfying*

$$\mathbf{X}_1\mathbf{A}_1 + \mathbf{Y}_1\mathbf{B}_1 = \mathbf{I}_m. \tag{11.197}$$

Calculate

$$\mathbf{\Phi}_0 = \mathbf{\Phi}_{rr} + \mathbf{\Phi}_{nn} \tag{11.198}$$

$$= \mathbf{A}^{-1}(\mathbf{CQ}_0\mathbf{C}^* + \mathbf{EQ}_2\mathbf{E}^*)\mathbf{A}^{-1}. \tag{11.199}$$

Let $\mathbf{L} = \mathbf{B}_1^*\mathbf{Q}_1\mathbf{\Phi}_0\mathbf{A}^*$. (11.200)

The optimal parameterized gain $\mathbf{K}^0(z^{-1})$ *is given by the following equivalent expressions:*

$$\mathbf{K}^0(z^{-1}) = \mathbf{D}_1^{-1}(\{\mathbf{D}_1^{*-1}\mathbf{LD}^{*-1}\}_+ + \{\mathbf{D}_1\mathbf{Y}_1\mathbf{A}^{-1}\mathbf{D}\}_-)\mathbf{D}^{-1}\mathbf{U}_0 - \mathbf{Y}_1\mathbf{A}^{-1}\mathbf{U}_0 \tag{11.201}$$

$$= \mathbf{D}_1^{-1}(\{\mathbf{D}_1^{*-1}\mathbf{LD}^{*-1} - \mathbf{D}_1\mathbf{Y}_1\mathbf{A}^{-1}\mathbf{D}\}_+)\mathbf{D}^{-1}\mathbf{U}_0. \tag{11.202}$$

The optimal controller is given by:

$$\mathbf{C}_0(z^{-1}) = (\mathbf{X}_1 - \mathbf{K}^0\mathbf{B}_0)^{-1}(\mathbf{Y}_1 + \mathbf{K}^0\mathbf{A}_0) \tag{11.203}$$

and the optimal reference to control input transfer-function matrix is:

$$\mathbf{G}_c(z^{-1}) = \mathbf{A}_1\mathbf{D}_1^{-1}(\{\mathbf{D}_1^{*-1}\mathbf{LD}^{*-1}\}_+ + \{\mathbf{D}_1\mathbf{Y}_1\mathbf{A}^{-1}\mathbf{D}\}_-)\mathbf{D}^{-1}\mathbf{A}. \tag{11.204}$$

Proof. This follows from the discussion presented above, and in particular collating equations: (11.178), (11.186) and (11.192). □

This theorem applies generically to the s-domain problem description provided the correct interpretation of the $\{\cdot\}_+, \{\cdot\}_-$ braces is adopted.

11.5.2. The variational solution to the LQG stochastic optimal control problem

Throughout this text has been an underlying philosophy of posing the control design as an optimization problem in Hilbert space and using the gradient conditions to obtain a solution. In the early part of this chapter, there was a digression to introduce polynomial system concepts. Subsequently, the important contributions made by Kučera and Youla and co-workers were pursued. These are essentially methods for solving the optimation problem by transfer-function matrix techniques. This final section presents an investigation of the problem using the time-domain gradient and causality concepts pursued previously (Chapter 2 for example). The solutions by Kučera, Youla *et al.* and the gradient methods are equivalent. However, each approach offers different insights. For example, the Kučera and Youla approaches have *a priori* constraints on stability, whilst the gradient techniques apply to both infinite- and finite-time interval problems.

For the time-domain approach, the system description and cost criterion are the same as those presented at the beginning of the chapter. The cost-functional to be minimized may be written in the form:

$$J(u) = \lim_{N \to \infty} \frac{1}{2N+1} E\left\{ \sum_{t=-N}^{N} \langle (Q_c e)(t), (Q_c e)(t) \rangle_{E_r} \right.$$

$$\left. + \langle (R_c u)(t), (R_c u)(t) \rangle_{E_m} \right\} \qquad (11.205)$$

where $Q_c(0) \geq \mathbf{O}$ and $R_c(0) > \mathbf{O}$. The operators Q_c and R_c are included for greater generality and are assumed to represent proper stable dynamical systems. The operator R_c can be selected to minimize the effect of system saturation. This operator can be used to couple the control inputs to the system-states which are particularly sensitive to either large changes or saturation.

Condition for optimality

System equations (11.109) to (11.111) are recalled, however, here the operators are to be interpreted as being in the time domain where the underlying signal spaces are $l_2^m[-N, N]$ and $l_2^r[-N, N]$ viz.:

$$u(t) \triangleq (G_c(r - v - n))(t) \qquad (11.206)$$

$$y(t) \triangleq (WG_c(r - v))(t) + ((I_r - WG_c)n)(t) \qquad (11.207)$$

$$e(t) \triangleq ((I_r - WG_c)(r - n))(t) + (WG_c v)(t). \qquad (11.208)$$

The necessary and sufficient conditions for optimality which prescribe the controller C_0 are obtained next. The cost functional (11.205) may be written:

$$J = \lim_{N \to \infty} \frac{1}{2N+1} (J_N(u)) \qquad (11.209)$$

where $J_N(u)$ is defined in terms of the $l_2^r[-N, N], l_2^m[-N, N]$ Hilbert space inner products (the $l_2^r[-N, N]$ inner product: $\langle x, y\rangle_{H_r} \triangleq \sum_{t=-T}^{T} \mathbf{x}(t)^T\mathbf{y}(t)$) as:

$$J_N(u) = E\{\langle Q_c e, Q_c e\rangle_{H_r} + \langle R_c u, R_c u\rangle_{H_m}\}.$$

Using (11.208), and (11.206) yields:

$$\begin{aligned} J_N &= E\{\langle e, Q_c^* Q_c e\rangle_{H_r} + \langle u, R_c^* R_c u\rangle_{H_m}\} \\ &= E\{\langle ((I_r - WG_c)(r-n) + WG_c v), Q_1((I_r - WG_c)(r-n) + WG_c v)\rangle_{H_r} \\ &\quad + \langle G_c(r-v-n), R_1 G_c(r-v-n)\rangle_{H_m}\} \end{aligned}$$

where

$$Q_1 \triangleq Q_c^* Q_c \quad \text{and} \quad R_1 \triangleq R_c^* R_c.$$

Thus,

$$\begin{aligned} J_N = {} & E\{\langle (r-n), (I_r - WG_c)Q_1(I_r - WG_c)(r-n)\rangle_{H_r}\} \\ & + E\{\langle v, G_c^* W^* Q_1 W G_c v\rangle_{H_m}\} \\ & + E\{\langle (r-n-v), G_c^* R_1 G_c(r-n-v)\rangle_{Hm}\} \\ & + E\{\langle (r-n), (I_r - WG_c)^* Q_1 W G_c v\rangle_{H_r} \\ & + \langle v, G_c^* W^* Q_1(I_r - WG_c)(r-n)\rangle_{H_m}\}. \end{aligned} \tag{11.210}$$

This expression can be simplified by using the trace identity and the definition of the correlation function, namely:

$$E\{\langle x, Ax\rangle_{H_n}\} = \sum_{t=-N}^{N} \operatorname{trace}((A\phi_{xx})(t)) \tag{11.211}$$

where

$$\phi_{xx}(t) \triangleq E\{x(i+t)x^T(i)\} \in \mathbb{R}^{n\times n}.$$

Introduce the following correlation matrices:

$$\begin{aligned} \phi_{rr}(t) &= E\{r(i+t)r^T(i)\} \\ \phi_{nn}(t) &= E\{n(i+t)n^T(i)\} \\ \phi_{vv}(t) &= E\{v(i+t)v^T(i)\} \end{aligned} \tag{11.212}$$

and

$$\phi_{cc}(t) \triangleq \phi_{rr}(t) + \phi_{nn}(t) + \phi_{vv}(t)$$

$$\phi_0(t) \triangleq \phi_{rr}(t) + \phi_{nn}(t)$$

Recall also the various independence properties between the noise sources and the stochastic reference; so that use of (11.211) and (11.212) in (11.210) yields:

$$\begin{aligned}
J_N = & \sum_{t=-N}^{N} \{\text{trace}(((\boldsymbol{I}_r - \boldsymbol{W}\boldsymbol{G}_c)^* \boldsymbol{Q}_1 (\boldsymbol{I}_r - \boldsymbol{W}\boldsymbol{G}_c)\phi_0)(t)) \\
& + \text{trace}((\boldsymbol{G}_c^* \boldsymbol{W}^* \boldsymbol{Q}_1 \boldsymbol{W}\boldsymbol{G}_c \phi_{vv})(t)) + \text{trace}((\boldsymbol{G}_c^* \boldsymbol{R}_1 \boldsymbol{G}_c \phi_{cc})(t))\} \\
= & \sum_{t=-N}^{N} \{\text{trace}((\boldsymbol{G}_c^* (\boldsymbol{W}^* \boldsymbol{Q}_1 \boldsymbol{W} + \boldsymbol{R}_1)\boldsymbol{G}_c \phi_{cc})(t)) \\
& - \text{trace}((\boldsymbol{G}_c^* \boldsymbol{W}^* \boldsymbol{Q}_1 \phi_0)(t)) - \text{trace}((\boldsymbol{Q}_1 \boldsymbol{W}\boldsymbol{G}_c \phi_0)(t)) \\
& + \text{trace}((\boldsymbol{Q}_1 \phi_0)(t))\}
\end{aligned}$$

$$\begin{aligned}
J_N = & \sum_{t=-N}^{N} \{\text{trace}(((\boldsymbol{W}^* \boldsymbol{Q}_1 \boldsymbol{W} + \boldsymbol{R}_1)\boldsymbol{G}_c \phi_{cc} \boldsymbol{G}_c^*)(t)) \\
& - 2\ \text{trace}((\boldsymbol{W}^* \boldsymbol{Q}_1 \phi_0 \boldsymbol{G}_c^*)(t)) + \text{trace}((\boldsymbol{Q}_1 \phi_0)(t)). \qquad (11.213)
\end{aligned}$$

The discrete-time operator $\boldsymbol{G}_c$ in the cost functional (11.213) represents the closed-loop system transfer between input reference signal $\boldsymbol{r}$ and control input $\boldsymbol{u}$. The time-domain optimization procedure follows the usual variational argument. Let $\boldsymbol{G}_c^0$ represent the optimal transfer and $\varepsilon\tilde{\boldsymbol{G}}_c$ some physically realizable variation of this operator. Replacing $\boldsymbol{G}_c$ by $\boldsymbol{G}_c^0 + \varepsilon\tilde{\boldsymbol{G}}_c$, the infinite-time cost functional (11.209) and (11.213) becomes:

$$\begin{aligned}
J(\boldsymbol{u}) = & J^0 + \varepsilon^2 \lim_{N\to\infty} \frac{1}{2N+1} \sum_{t=-N}^{N} \text{trace}(((\boldsymbol{W}^* \boldsymbol{Q}_1 \boldsymbol{W} + \boldsymbol{R}_1)\tilde{\boldsymbol{G}}_c \phi_{cc} \tilde{\boldsymbol{G}}_c^*)(t)) \\
& + 2\varepsilon \lim_{N\to\infty} \frac{1}{2N+1} \sum_{t=-N}^{N} \text{trace}((((\boldsymbol{W}^* \boldsymbol{Q}_1 \boldsymbol{W} + \boldsymbol{R}_1)\boldsymbol{G}_c^0 \phi_{cc} \\
& - \boldsymbol{W}^* \boldsymbol{Q}_1 \phi_0)\tilde{\boldsymbol{G}}_c^*)(t)).
\end{aligned}$$

A necessary condition for optimality is given by:

$$\left.\frac{\partial J(\boldsymbol{u})}{\partial \varepsilon}\right|_{\varepsilon=0} = 0 \qquad (11.214)$$

which yields:

$$\lim_{N\to\infty} \frac{1}{(2N+1)} \sum_{t=-N}^{N} \text{trace}((((\boldsymbol{W}^*\boldsymbol{Q}_1\boldsymbol{W}+\boldsymbol{R}_1)\boldsymbol{G}_c^0\boldsymbol{\phi}_{cc} - \boldsymbol{W}^*\boldsymbol{Q}_1\boldsymbol{\phi}_0)\tilde{\boldsymbol{G}}_c^*)(t)) = 0. \tag{11.215}$$

For a given index t, let the matrix $X(n-t)$ be defined as:

$$X(n-t) = ((\boldsymbol{W}^*\boldsymbol{Q}_1\boldsymbol{W}+\boldsymbol{R}_1)\boldsymbol{G}_c^0\boldsymbol{\phi}_{cc} - \boldsymbol{W}^*\boldsymbol{Q}_1\boldsymbol{\phi}_0)(n-t)$$

where $n \in [t, N]$ and n is the index of summation for the adjoint operator. The condition of optimality (11.215) becomes:

$$(X\tilde{\boldsymbol{G}}_c^*)(t) = \mathbf{o} \text{ for all indices } t \in (-\infty, \infty).$$

Since $\tilde{\boldsymbol{G}}_c^*$ is an arbitrary adjoint, the term $X(n-t)$ must be zero for all integers $[t, \infty)$. The necessary condition (11.215) may be expressed as:

$$X(t) = ((\boldsymbol{W}^*\boldsymbol{Q}_1\boldsymbol{W}+\boldsymbol{R}_1)\boldsymbol{G}_c^0\boldsymbol{\phi}_{cc} - \boldsymbol{W}^*\boldsymbol{Q}_1\boldsymbol{\phi}_0)(t) = \mathbf{O} \quad \text{for all integers } t \geq 0. \tag{11.216}$$

This also yields a sufficient condition since $\partial^2 J/\partial\varepsilon^2 > 0$.

Solution for the optimal closed-loop controller

The solution for the optimal controller $\mathbf{C}_0^0(z^{-1})$ is obtained below. This is derived by transforming the optimality condition (11.216) into the z-domain and then solving for $\mathbf{C}_0^0(z^{-1})$ from the expression obtained for $\mathbf{G}_c^0(z^{-1})$. This approach has the advantage that the relationship between the time-domain and transform techniques is established. If the classical Wiener approach is followed, this can lead to an unstable closed-loop solution when the system is open-loop unstable. This problem was avoided in the work of Kučera [11] and Youla *et al.* [2] by the use of the controller parameterization. The difficulty caused by a direct application of the Wiener approach is that physically realizable transforms are not simply those transforms having poles within the unit circle of the z-plane. These terms were denoted by the brace notation $\{\cdot\}_+$ in the previous Youla *et al.* analysis. This becomes apparent when the causality of the time-domain operators and their transforms is taken into consideration. The modification required to apply the Wiener technique involves a correct causality interpretation for the brace notation $\{\cdot\}_+$.

Since the adjoint operators in (11.216) are non-causal, the bilateral z-transform is used, although the actual solution will only involve single sided z-transforms. The use of the z-transform facilitates a direct notational transfer between expressions in the delay operator and the resulting z-transform. In the

z-domain (11.216) becomes:

$$\mathbf{X}(z^{-1}) = (\mathbf{W}^*(z^{-1})\mathbf{Q}_1(z^{-1})\mathbf{W}(z^{-1}) + \mathbf{R}_1(z^{-1}))\mathbf{G}_c^0(z^{-1})\mathbf{\Phi}_{cc}(z^{-1})$$
$$- \mathbf{W}^*(z^{-1})\mathbf{Q}_1(z^{-1})\mathbf{\Phi}_0(z^{-1}) \quad (11.217)$$

where $\mathbf{X}(z^{-1})$ is the z-transform of a discrete-time function which is zero for all time indices $t \geq 0$. Introduce the generalized spectral factors:

$$\mathbf{Y}_c^*(z^{-1})\mathbf{Y}_c(z^{-1}) = \mathbf{W}^*(z^{-1})\mathbf{Q}_1(z^{-1})\mathbf{W}(z^{-1}) + \mathbf{R}_1(z^{-1}) \quad (11.218)$$

$$\mathbf{Y}_f(z^{-1})\mathbf{Y}_f^*(z^{-1}) = \mathbf{\Phi}_{rr}(z^{-1}) + \mathbf{\Phi}_{nn}(z^{-1}) + \mathbf{\Phi}_{vv}(z^{-1}). \quad (11.219)$$

Note also the definition for $\mathbf{\Phi}_0(z^{-1})$, namely:

$$\mathbf{\Phi}_0(z^{-1}) = \mathbf{\Phi}_{rr}(z^{-1}) + \mathbf{\Phi}_{nn}(z^{-1}). \quad (11.220)$$

Using (11.218) and (11.219) gives:

$$\mathbf{X}(z^{-1}) = \mathbf{Y}_c^*(z^{-1})\mathbf{Y}_c(z^{-1})\mathbf{G}_c^0(z^{-1})\mathbf{Y}_f(z^{-1})\mathbf{Y}_f^*(z^{-1})$$
$$- \mathbf{W}^*(z^{-1})\mathbf{Q}_1(z^{-1})\mathbf{\Phi}_0(z^{-1}).$$

This rearranges as:

$$\mathbf{Y}_c^{*-1}(z^{-1})\mathbf{X}(z^{-1})\mathbf{Y}_f^{*-1}(z^{-1}) = \mathbf{Y}_c(z^{-1})\mathbf{G}_c^0(z^{-1})\mathbf{Y}_f(z^{-1})$$
$$- \mathbf{Y}_c^{*-1}(z^{-1})\mathbf{W}^*(z^{-1})\mathbf{Q}_1(z^{-1})\mathbf{\Phi}_0(z^{-1})\mathbf{Y}_f^{*-1}(z^{-1}). \quad (11.221)$$

The left side of this equation represents a matrix function $\mathbf{X}(t)$ (zero for $t \geq 0$) which is input to a non-causal system, thus

$$\{\mathbf{Y}_c^*(z^{-1})\mathbf{X}(z^{-1})\mathbf{Y}_f^*(z^{-1})\}_+ = \mathbf{O}.$$

where the brace $\{\cdot\}_+$ denotes the transform of the time function over the positive ($t \geq 0$) time interval. Equating the positive-time transforms of the right-hand side of (11.221), after rearrangement, yields:

$$\mathbf{G}_c^0(z^{-1}) = \mathbf{Y}_c^{-1}(z^{-1})\{\mathbf{Y}_c^{*-1}(z^{-1})\mathbf{W}^*(z^{-1})\mathbf{Q}_1(z^{-1})\mathbf{\Phi}_0(z^{-1})\mathbf{Y}_f^{*-1}(z^{-1})\}_+\mathbf{Y}_f^{-1}(z^{-1})$$
$$(11.222)$$

where $\mathbf{\Phi}_0(z^{-1})$ is given by (11.220).

The optimal closed-loop controller may now be calculated as:

$$\mathbf{C}_0^0(z^{-1}) = \mathbf{G}_c^0(z^{-1})(\mathbf{I}_m - \mathbf{W}(z^{-1})\mathbf{G}_c^0(z^{-1}))^{-1}$$

$$= (\mathbf{I}_r - \mathbf{G}_c^0(z^{-1})\mathbf{W}(z^{-1}))^{-1}\mathbf{G}_c^0(z^{-1}). \tag{11.223}$$

An example to illustrate the solution procedure follows.

Example 11.5. *Stochastic regulator problem.* Consider the problem of determining the optimal controller for a unity feedback system. The zero-mean, white-noise signals ξ and v are assumed to have covariances $Q = 1, R = r^2$, respectively. The stochastic reference r is assumed to be zero. The weighting matrices in the cost functional are defined as $Q_1 = 1$ and $R_1 = r_1^2$. The system transfer functions are defined as:

$$W(z^{-1}) = A^{-1}(z^{-1})\tilde{B}(z)z^{-k} \tag{11.224}$$

with

$$A(z^{-1}) = 1 - 1.7z^{-1} + 0.7z^{-2}$$

$$\tilde{B}(z^{-1}) = (1 + 0.5z^{-1})$$

and

$$W_0(z^{-1}) = A^{-1}(z^{-1})C(z^{-1}) \tag{11.225}$$

where

$$C(z^{-1}) = 1 + 1.5z^{-1} + 0.9z^{-2}.$$

It is required to evaluate the optimal controller for a delay of $k = 1$, and $k = 2$. The minimum-variance controller (obtained by setting $r = r_1 = 0$) is also required.

Solution. Use (11.212), (11.218) and (11.219) to calculate the spectral factors

$$Y_c^*(z^{-1})Y_c(z^{-1}) = W^*(z^{-1})Q_1W(z^{-1}) + R_1$$

$$= \frac{(1 + 0.5z)(1 + 0.5z^{-1})}{A(z)A(z^{-1})} + r_1^2$$

$$= \frac{0.7r_1^2(z^2 + z^{-2}) + (0.5 - 2.89r_1^2)(z + z^{-1}) + 1.25 + 4.38r_1^2}{(1 - 1.7z + 0.7z^2)(1 - 1.7z^{-1} + 0.7z^{-2})}.$$

The numerator and denominator polynomials are parahermitian and as a consequence may be expressed as polynomials in $(z + z^{-1})$. The generalized spectral factor (Shaked [3]) can be given the form:

$$Y_c(z^{-1}) = \frac{D_1(z^{-1})}{A(z^{-1})} = \frac{\alpha_1 + \beta_1 z^{-1} + \gamma_1 z^{-2}}{A(z^{-1})}. \tag{11.226}$$

Similarly,

$$\begin{aligned}
Y_f(z^{-1})Y_f^*(z^{-1}) &= \Phi_{rr}(z^{-1}) + \Phi_{nn}(z^{-1}) + \Phi_{vv}(z^{-1}) \\
&= W_0(z^{-1})QW_0^*(z^{-1}) + R \\
&= \frac{(1+1.5z+0.9z^2)(1+1.5z^{-1}+0.9z^{-2})}{A(z)A(z^{-1})} + R_1^2 \\
&= \frac{(0.9+0.7r^2)(z^2+z^{-2}) + (2.85-2.89r^2)(z+z^{-1}) + 4.06 + 4.38r^2}{(1-1.7z^{-1}+0.7z^{-2})(1-1.7z+0.7z^2)}
\end{aligned}$$

and

$$Y_f(z^{-1}) = \frac{D(z^{-1}}{A(z^{-1})} = \frac{(\alpha+\beta z^{-1}+\gamma z^{-2})}{A(z^{-1})}. \tag{11.227}$$

The closed-loop transfer-matrix $G_c(z^{-1})$ is given by

$$G_c(z^{-1}) = Y_c^{-1}(z^{-1})\{Y_c^{*-1}(z^{-1})W^*(z^{-1})Q_1(z^{-1})\Phi_0(z^{-1})Y_f^{*-1}(z^{-1})\}_+ Y_f^{-1}(z^{-1}). \tag{11.228}$$

Note

$$\Phi_0(z^{-1}) = \Phi_{rr}(z^{-1}) + \Phi_{nn}(z^{-1}) = \Phi_{nn}(z^{-1}) \qquad (r=0). \tag{11.229}$$

Substituting from (11.224), to (11.227) and (11.229) in (11.228) gives:

$$\begin{aligned}
G_c^0(z^{-1}) &= \frac{A(z^{-1})}{D_1(z^{-1})}\left\{\frac{A(z)}{D_1(z)}\cdot\frac{\tilde{B}(z)z^k}{A(z)}\cdot\frac{C(z^{-1})}{A(z^{-1})}\cdot\frac{C(z)}{A(z)}\cdot\frac{A(z)}{D(z)}\right\}_+ \frac{A(z^{-1})}{D(z^{-1})} \\
&= \frac{A(z^{-1})}{D_1(z^{-1})}\left\{z^k\,\frac{C(z^{-1})\tilde{B}(z)C(z)}{A(z^{-1})D_1(z)D(z)}\right\}_+ \frac{A(z^{-1})}{D(z^{-1})}
\end{aligned}$$

and the transfer for $C_0^0(z^{-1})$ follows from (11.223).

The minimum-variance controller is obtained by setting $r_1 = r = 0$. In this case $D_1(z^{-1}) = \tilde{B}(z^{-1})$ and $D(z^{-1}) = C(z^{-1})$, hence:

$$G_c^0(z^{-1}) = \frac{A^2(z^{-1})}{D_1(z^{-1})D(z^{-1})}\left\{\frac{z^kC(z^{-1})}{A(z^{-1})}\right\}_+$$

Consider first the situation when the transport delay is $k=1$. The realizable transform is given as:

$$\begin{aligned}
\left\{z^k\,\frac{C(z^{-1})}{A(z^{-1})}\right\}_+ &= \left\{z\,\frac{(1+1.5z^{-1}+0.9z^{-2})}{A(z^{-1})}\right\}_+ \\
&= \left\{\frac{z(1+3.2z^{-1}+0.2z^{-2})}{A(z^{-1})}\right\}_+ \\
&= (3.2+0.2z^{-1})/A(z^{-1})
\end{aligned}$$

$$G_c(z^{-1}) = \frac{A(z^{-1})(3.2+0.2z^{-1})}{D_1(z^{-1})D(z^{-1})}$$

but

$$1 - W(z^{-1})G_c(z^{-1}) = 1 - \frac{(3.2 + 0.2z^{-1})z^{-1}}{D(z^{-1})} = \frac{A(z^{-1})}{D(z^{-1})}.$$

The closed-loop minimum-variance controller follows from equation (11.223) as:

$$C_0(z^{-1}) = \frac{3.2 + 0.2z^{-1}}{1 + 0.5z^{-1}}.$$

For the case of a two-step delay ($k = 2$):

$$\left\{z^k \frac{C(z^{-1})}{A(z^{-1})}\right\}_+ = \left\{z^2 \frac{(1 + 1.5z^{-1} + 0.9z^{-2})}{A(z^{-1})}\right\}_+$$

$$= \left\{z^2 + 3.2z + \frac{5.64 - 2.24z^{-1}}{A(z^{-1})}\right\}_+$$

$$G_c(z^{-1}) = \frac{A(z^{-1})(5.64 - 2.24z^{-1})}{D_1(z^{-1})D(z^{-1})}$$

but

$$1 - W(z^{-1})G_c(z^{-1}) = 1 - \frac{(5.64 - 2.24z^{-1})}{D(z^{-1})}z^{-2} = \frac{A(z^{-1})(1 + 3.2z^{-1})}{D(z^{-1})}$$

hence

$$C_0(z^{-1}) = \frac{5.64 - 2.24z^{-1}}{(1 + 3.2z^{-1})(1 + 0.5z^{-1})}$$

$$= \frac{5.64 - 2.24z^{-1}}{1 + 3.7z^{-1} + 1.6z^{-2}}$$

This is identical to the minimum-variance controller derived by Åström [12], Åström and Wittenmark [38].

Equivalence of the matrix fraction and gradient solutions

The question now arises as to whether the solution derived by gradient method is essentially the same as that derived by Kučera [11] and by Youla *et al.* [2]. To investigate this relationship the matrix fraction description will again be employed.

Recall from the definition of the spectral factors (11.127) and (11.128):

$$\mathbf{Y}_c^*\mathbf{Y}_c = \mathbf{A}_1^{*-1}\mathbf{D}_1^*\mathbf{D}_1\mathbf{A}_1^{-1}, \qquad \mathbf{Y}_c = \mathbf{D}_1\mathbf{A}_1^{-1}$$

$$\mathbf{Y}_f\mathbf{Y}_f^* = \mathbf{A}^{-1}\mathbf{D}\mathbf{D}^*\mathbf{A}^{*-1}, \qquad \mathbf{Y}_f = \mathbf{A}^{-1}\mathbf{D} = \mathbf{D}_2\mathbf{A}_2^{-1}.$$

It is interesting that the definition of a generalized spectral factor, proposed by Shaked [3], arises naturally from the connection with polynomial matrices. Thus, the observation by Shaked that the spectral factor should have the same pole-polynomial as the system is a natural consequence of the above identification.

In terms of matrix fractions, the realizable transform becomes:

$$\{\mathbf{Y}_c^{*-1}(z^{-1})\mathbf{W}^*(z^{-1})\mathbf{Q}_1(z^{-1})\mathbf{\Phi}_0(z^{-1})\mathbf{Y}_f^*(z^{-1})^{-1}\}_+ = \{\mathbf{D}_1^{*-1}\mathbf{B}_1^*\mathbf{Q}_1\mathbf{\Phi}_0\mathbf{A}^*\mathbf{D}^{*-1}\}_+. \tag{11.230}$$

Substituting for $\mathbf{\Phi}_0$ and using the Diophantine equation (10.136), the above equation (11.230) may be simplified:

$$\begin{aligned}\{\mathbf{D}_1^{*-1}\mathbf{B}_1^*\mathbf{Q}_1\mathbf{\Phi}_0\mathbf{A}^*\mathbf{D}^{*-1}\}_+ &= \{\mathbf{D}_1^{*-1}\mathbf{B}_1^*\mathbf{Q}_1\mathbf{D}_2\mathbf{A}_2^{-1} - \mathbf{D}_1^{*-1}\mathbf{B}_1^*\mathbf{Q}_1\mathbf{R}\mathbf{A}^*\mathbf{D}^{*-1}\}_+ \\ &= \{\mathbf{G}\mathbf{A}_2^{-1} + \bar{\mathbf{D}}_1^{-1}\mathbf{F}\}_+ = \{\mathbf{G}\mathbf{A}_2^{-1}\}_+.\end{aligned} \tag{11.231}$$

Interpreting the brace $\{\cdot\}_+$ as representing the causal (rather than stable) terms $\{\mathbf{G}\mathbf{A}_2^{-1}\}_+ = \mathbf{G}\mathbf{A}_2^{-1}$, hence (11.222),

$$\begin{aligned}\mathbf{G}_c^{\mathrm{o}} &= \mathbf{Y}_c^{-1}(z^{-1})\{\mathbf{D}_1^{*-1}\mathbf{B}_1^*\mathbf{Q}_1\mathbf{\Phi}_0\mathbf{A}^*\mathbf{D}^{*-1}\}_+\mathbf{Y}_f^{-1}(z^{-1}) \\ &= \mathbf{A}_1\mathbf{D}_1^{-1}\mathbf{G}\mathbf{A}_2^{-1}\mathbf{A}_2\mathbf{D}_2^{-1} = \mathbf{A}_1\mathbf{D}_1^{-1}\mathbf{G}\mathbf{D}_2^{-1} \\ \mathbf{G}_c^{\mathrm{o}} &= \mathbf{A}_1\mathbf{N}_0^{\mathrm{o}}.\end{aligned} \tag{11.232}$$

This is clearly identical to the optimal transfer derived by both the Kučera and Youla approaches. It is straightforward to determine the optimal controller expression as:

$$\mathbf{C}_0^{\mathrm{o}}(z^{-1}) = \mathbf{D}_3\mathbf{H}^{-1}\mathbf{G}\mathbf{D}_2^{-1}. \tag{11.233}$$

Related work

If the system is time varying and non-stationary, the open-loop minimum-variance control problem can be solved by a time-domain analysis which is similar to the operator based solution described in Section 11.5.2 (Syed and Meditch [39], [40]). A feedback solution to this problem was also described by Syed and Meditch [41].

Whitbeck [42] has recently discussed a direct approach to the solution of the classical Wiener–Hopf control problem for linear systems and stationary noise. The paper is interesting since it uses the realistic case of a continuous-time system model and cost function coupled with a discrete-time control law and a hold device.

11.6. THE POLYNOMIAL SOLUTION OF THE DETERMINISTIC OPTIMAL LQ CONTROL PROBLEM

Although this chapter has been concerned with the LQG stochastic optimal control problem, the deterministic optimal LQ control problem of Chapter 2 is briefly reconsidered. The results are given for the case where the system is represented in the matrix fraction form introduced earlier. An example is also presented to demonstrate how the polynomial system approach resolves the problem of an unstable open-loop system.

11.6.1. Deterministic control problem and solution

Consider for simplicity a regulator problem for the feedback configuration of Fig. 11.2, where the reference $\boldsymbol{r} = \mathbf{o}$. Assume also that $\mathbf{W}_0$ is square and minimum phase and that the driving signal $\boldsymbol{\xi}$ is an impulse function. The cost function becomes:

$$J = \int_0^\infty (\langle y^T(t)\mathbf{Q}_1 y(t)\rangle_{E_r} + \langle u(t), \mathbf{R}_1 u(t)\rangle_{E_m})\, dt$$

where $\mathbf{Q}_1 \geq \mathbf{O}$ and $\mathbf{R}_1 > \mathbf{O}$ are constant symmetric matrices.

Algorithm 11.1. *Deterministic output feedback* LQ *controller*

1. Calculate a left coprime system matrix decomposition:

$$[\mathbf{W} \quad \mathbf{W}_0] = \mathbf{A}^{-1}[\mathbf{B} \quad \mathbf{C}]$$

and a right coprime system matrix decomposition:

$$\mathbf{W} = \mathbf{B}_1\mathbf{A}_1^{-1}$$

and right coprime factorizations of $\mathbf{C}^{-1}\mathbf{B} = \mathbf{B}_3\mathbf{D}_3^{-1}$ and $\mathbf{C}^{-1}\mathbf{A} = \mathbf{A}_2\mathbf{C}_2^{-1}$.

2. Obtain the stable spectral factor $\mathbf{D}_1$ using:

$$\mathbf{B}_1^*\mathbf{Q}_1\mathbf{B}_1 + \mathbf{A}_1^*\mathbf{R}_1\mathbf{A}_1 = \mathbf{D}_1^*\mathbf{D}_1.$$

3. Calculate any polynomial matrix solution $\mathbf{G}_1, \mathbf{H}_1$ and $\mathbf{F}_1$ of the bilateral equations:

$$\mathbf{D}_1^*\mathbf{G} + \mathbf{F}\mathbf{A}_2 = \mathbf{B}_1^*\mathbf{Q}_1\mathbf{D}_2$$

$$\mathbf{D}_1^*\mathbf{H} - \mathbf{F}\mathbf{B}_3 = \mathbf{A}_1^*\mathbf{R}_1\mathbf{D}_3.$$

4. Use a single-sided left division algorithm to calculate the polynomial matrix quotient $\mathbf{U}_1$ and $\mathbf{V}_1$:

$$\mathbf{F}_1 = \mathbf{D}_1\mathbf{U}_1 + \mathbf{V}_1$$

satisfying $\deg(\mathbf{V}_1(\text{row}_i)) < \deg(\mathbf{D}_1(\text{row}_i))$, $i = 1, 2, \ldots m$.

5. Find

$$\mathbf{G} = \mathbf{G}_1 + \mathbf{U}_1\mathbf{A}_2$$

$$\mathbf{H} = \mathbf{H}_1 - \mathbf{U}_1\mathbf{B}_3$$

$$\mathbf{F} = \mathbf{V}_1.$$

6. Obtain any left coprime representation for $\mathbf{H}\mathbf{D}_3^{-1}$ and $\mathbf{G}\mathbf{D}_2^{-1}$ using:

$$[\mathbf{H}\mathbf{D}_3^{-1} \quad \mathbf{G}\mathbf{D}_2^{-1}] = \mathbf{D}_f^{-1}[\mathbf{H}_0 \quad \mathbf{G}_0].$$

7. The optimal controller is realized without unstable hidden modes and is given as:

$$\mathbf{C}_0 = \mathbf{H}_0^{-1}\mathbf{G}_0. \qquad \square$$

The similarity with the results for the stochastic case is obvious. The proof is given in Grimble [43].

Example 11.6. *Unstable open-loop system.* Consider the following scalar system:

$$\dot{x} = ax + u, \qquad a > 0$$

$$y = x \qquad x_0 = x(0).$$

and assume that an output feedback regulator is required to minimize the cost function:

$$J = \int_0^\infty (qx^2(t) + ru^2(t))\, dt, \qquad q \geq 0, r > 0.$$

Solution.

$$W = W_0 = 1/(s - a), \qquad A = A_2 = s - a, \qquad D_2 = D_3 = D_0 = 1,$$

$$B_3 = B_2 = B_1 = B, \qquad A_1 = A \text{ and } D$$

satisfies:

$$B_1^* Q B_1 + A_1^* R A_1 = D^* D$$

or

$$q + (-s^2 + a^2)r = D^*D \Rightarrow D = \sqrt{r}(s + b)$$

where $rb^2 = a^2r + q$. The Diophantine equations in Step 3 become:

$$\sqrt{r}(-s + b)G + F(s - a) = q$$

$$\sqrt{r}(-s + b)H - F = (-s - a)r.$$

Equating coefficients assuming $G = g$, $F = f$, $H = h$ (scalars):

$$\begin{array}{ll} -\sqrt{r}g + f = 0, & \sqrt{r}bg - af = q \\ = \sqrt{r}h = -r, & \sqrt{r}bh - f = -ar \end{array} \Rightarrow \begin{cases} g = \sqrt{r}(b + a) \\ h = \sqrt{r} \\ f = r(b + a). \end{cases}$$

The feedback controller immediately follows as:

$$u^0(t) = -(g/h)x(t) = -(b + a)x(t)$$

and the system is identical to that obtained by the Riccati equation solution of Section 2.4. The closed-loop system is therefore known to be asymptotically stable.

11.7. CONCLUSIONS

The design methods presented in this chapter are applicable to linear, constant, discrete- or continuous-time systems where the spectral densities of the additive stochastic disturbances are modelled by rational transfer functions. The system must be stabilizable and detectable but can be non-square, unstable and/or non-minimum phase. The system can contain transport delays. The effects of plant saturation can be limited by careful choice of the dynamical operators which occur in the performance criterion.

There are several practical arguments in favour of the transfer-function approaches to the steady-state stochastic control problem in comparison with the more usual time-domain approaches. The transfer-function methods avoid difficulties with singular noise or control weighting matrices, furthermore, systems having coloured-noise inputs need not be treated as a special case. The time-domain Riccati equation approach is commonly believed to be more suitable for numerical calculations on a digitial computer. However, a full comparison of the computational problems associated with the two approaches is required now that algorithms for polynomial matrix manipulations and calculations are readily available.

Stylistically, the Kučera and Youla publications are very different, as are their respective techniques of analysis. However, the problem description is essentially the same and so it is not surprising that the solution procedures produce equivalent results. A demonstration that this is the case has not appeared before.

In the gradient solution technique no *a priori* constraint is placed on stability. It is therefore rather more remarkable that the solution by this route corresponds to those of Kučera and Youla. To achieve this result, it was necessary to reconsider the definition of the spectral factors and to note the causal interpretation of the brace $\{\cdot\}_+$ term rather than use the more usual definitions based on stability concepts. The text by Kučera is recommended for an excellent tutorial exposition of the polynomial systems approach [11].

11.8. PROBLEMS

1. Consider the model for a system shown in Fig. 11.3. Show that the plant with

$$W = A^{-1}B = \frac{s+1}{(s+1)(s-1)}$$

and disturbance

$$W_0 = A^{-1}C = \frac{1}{(s+1)(s-1)}$$

can be stabilized and the cost

$$J = E\{y^2(t) + u^2(t)\}$$

remain finite (note that $U_0 = (s+1)$). Show that the system in Fig. 11.4 can be stabilized but that a finite cost J cannot be achieved (note that $U_0 = (s-1)$).

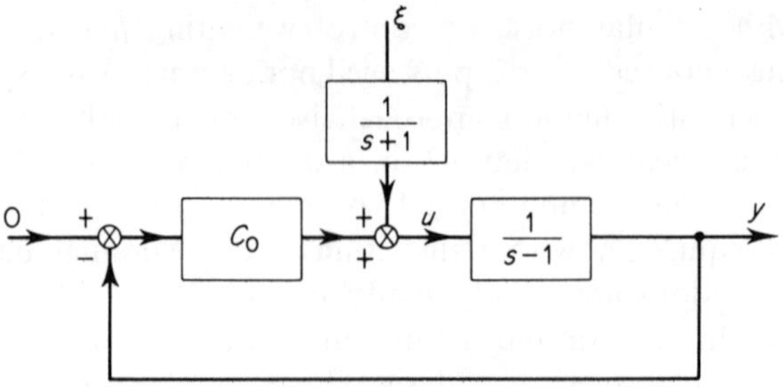

Fig. 11.3. System model for problem 11.8.1

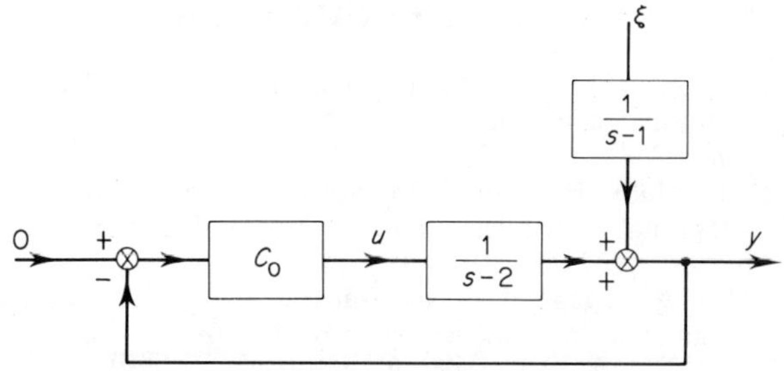

Fig. 11.4. Second system model for Problem 11.8.1

2. Verify the steps of the analysis used in the text to expose the relationship between left coprimeness and controllability. Equations (11.31) to (11.34) show

$$\mathbf{y}(z^{-1}) = \mathbf{A}^{-1}(z^{-1})(\mathbf{B}(z^{-1})\mathbf{u}(z^{-1}) + \mathbf{C}(z^{-1})\boldsymbol{\xi}(z^{-1}))$$

where

$$\mathbf{A} = \boldsymbol{\Gamma}_1(z^{-1})\mathbf{M}^{-1}(z^{-1}) \in \mathbb{P}^{m\times m}(z^{-1})$$

and

$$\mathbf{B} = z^{-1}[\boldsymbol{\Gamma}_2(z^{-1})\ \mathbf{O}]\mathbf{N}(z^{-1})\tilde{\mathbf{B}} \in \mathbb{P}^{m\times m}(z^{-1})$$

$$\mathbf{C} = z^{-1}[\boldsymbol{\Gamma}_2(z^{-1})\ \mathbf{O}]\mathbf{N}(z^{-1})\tilde{\mathbf{C}} \in \mathbb{P}^{m\times l}(z^{-1}).$$

Demonstrate that $\mathbf{A}(z^{-1})$ has all its invariants on the diagonal of $\boldsymbol{\Gamma}_1(z^{-1})$. Using the definitions for $\tilde{\boldsymbol{\Gamma}}_1$ and $\tilde{\boldsymbol{\Gamma}}_2$ of equations (11.36) prove equation (11.36) that the Smith–McMillan form for $\mathbf{H}(z^{-1})$ is obtained as:

$$\mathbf{H}(z^{-1}) \triangleq \boldsymbol{\Gamma}_1^{-1}(z^{-1})\mathbf{M}_1(z^{-1})\tilde{\boldsymbol{\Gamma}}_2(z^{-1}) = \tilde{\mathbf{M}}_1(z^{-1})\tilde{\boldsymbol{\Gamma}}_1^{-1}(z^{-1})\boldsymbol{\Gamma}_2(z^{-1})\tilde{\mathbf{N}}_1(z^{-1})$$

where $\tilde{\mathbf{M}}_1$, and $\tilde{\mathbf{N}}_1$ are unimodular matrices.

3. A system is described in the discrete equation form:

$$A(z^{-1})y(t) = z^{-k}B(z^{-1})u(t) + C(z^{-1})\xi(t)$$

where $\xi(t)$ is zero mean white noise of unit covariance. The time delay $k > 1$ and $B(0) = b_0 > 0$. Derive an expression for the optimal feedback control law which will minimize the output variance:

$$J = E\{y^2(t+k)\}.$$

11.9. REFERENCES

[1] Youla D. C., Bongiorno J. J. and Jabr H. A. 1976. Modern Wiener–Hopf design of optimal controllers—Part 1: The single input–output case, *IEEE Trans. Auto. Control,* **AC-21**(1), 3–13, February.

[2] Youla D. C., Jabr, H. A. and Bongiorno J. J. 1976. Modern Wiener–Hopf design of optimal controllers—Part 2: The multivariable case, *IEEE Trans Auto. Control,* **AC-21**(3), 319–338, June.

[3] Shaked, U. 1976, A general transfer function approach to the steady-state linear quadratic Gaussian stochastic control problem, *Int. J. Control,* **24**(6), 771–800.

[4] Grimble, M. J. 1978. The design of stochastic optimal feedback control systems, *Proc. IEE,* **125**(11), 1275–1284, Nov.

[5] Grimble, M. J. 1979. Solution of the stochastic optimal control problem in the s-domain for systems with time delay, *Proc. IEE,* **126**(7), 697–704, July.

[6] Kučera, V. 1980. Stochastic multivariable control: A polynomial equation approach, *IEEE Trans. Auto. Control,.* **AC-25**(5) 913–919, Oct.

[7] Strejc, V. 1966. The physical realizability of an optimal v-parameter discrete linear control system determined in Wiener's sense, Preprints IFAC Conference (8B1–8B6), June.

[8] Barrett, J. F. 1976. Discrete time stochastic linear quadratic regulation by spectral factorization, Report CAMS/76/2, Dept of Engineering, Cambridge University, Cambridge, UK.

[9] Gawthrop, P. J. 1976. Frequency-domain solution of the optimum steady-state regulator problem, Report No. 1145/76, Department of Engineering Science, Oxford University, Oxford, UK.

[10] Grimble, M. J. 1979. Solution of the discrete-time stochastic optimal control problem in the z-domain, *Int. J. System Science,* **10**(12), 1369–1390.

[11] Kučera V. 1979. *Discrete-Linear Control: The Polynomial Equation Approach,* John Wiley and Sons Ltd., Chichester.

[12] Astrom, K. J. 1970. *Introduction to Stochastic Control Theory,* Academic Press, New York, USA.

[13] Peterka, V. 1972. On steady-state minimum variance control strategy, *Kybernetika,* **8**(3), 219–232.

[14] Grimble, M. J. 1981. Design of closed-loop optimal controllers for systems with shape deterministic inputs, *Trans. Inst. Measurement and Control,* **3**(2), 79–88, April–June.

[15] Hoffman K. and Kunze R. 1971. *Linear Algebra*, 2nd Edition, Prentice-Hall, Inc., Englewood Cliffs, New Jersey.

[16] Chen, C.-T. 1984. *Linear System Theory and Design,* Holt, Rinehart and Winston, New York.

[17] Kailath, T. 1980. *Linear Systems,* Prentice-Hall Inc., Englewood Cliffs, New Jersey.

[18] Antsaklis, P. 1979. Some relations satisfied by prime polynomials matrices and their role in linear multivariable system theory, *IEEE Trans. Auto. Control,* **AC-24**(4), 611–616, Aug.

[19] Roth, W. E. 1952. The equation $AX - YB = C$ and $AX - XB = C$ in matrices, *Proc. Amer. Math. Soc.,* **3**, 392–396.

[20] Barnett, S. 1971. *Matrices in Control Theory,* Van Nostrand–Reinhold, London.

[21] Youla, D. C. 1961. On the factorization of rational matrices, *IRE. Trans.* **IT-7**, 172–189.

[22] Wolovich, W. A. 1974. *Linear Multivariable Systems,* Springer-Verlag, New York.

[23] Barnett, S. 1984. *Polynomials and Linear Control Systems,* Marcel Dekker Inc., New York.

[24] Gohberg, I., Lancaster, P. and Rodman, L. 1982. *Matrix Polynomials,* Academic Press, New York.

[25] Assefi T. 1979. *Stochastic Processes and Estimation with Applications,* John Wiley and Sons Ltd., New York.

[26] Kučera, V. 1974. Closed-loop stability of discrete linear single-variable systems, *Kybernetika,* **10**(2), 146–171.

[27] Desoer, C. A., Liu, R. N., Murray, J. and Saeks, R. 1980. Feedback system design: The fractional representation approach to analysis and synthesis, *IEEE Trans. Auto. Control,* **AC-25**(3), 399–412.

[28] Zames, G. and Francis, B. A. 1981. A new approach to classical frequency methods: feedback and minimax sensitivity, CDC Conference, San Diego, California, USA.

[29] Kwakernaak, H. 1982. Optimal robustness of linear feedback systems, Memorandum No. 395, Department of Applied Mathematics, Twente University of Technology, Enschede, Netherlands.

[30] Safonov, M. G., Laub, A. J. and Hartmann, G. L. 1981. Feedback properties of multivariable systems: The role and use of the return-difference matrix, *IEEE Trans. Auto Control,* **AC-26**(1), 47–65.

[31] Grimble, M. J. 1983. Robust LQG design of discrete systems using a dual criterion, Paper FA6:12.30, pp.1196–1198, Preprints 22nd Control and Decision Conference, USA.

[32] Grimble, M. J. and Johnson M. A. 1984. Robustness and optimality: A dual performance index, *Proc. IFAC 9th Triennial World Congress, Optimal Control Systems II,* pp. 323–328, Budapest, Hungary.

[33] Grimble, M. J. 1985. LQG design of discrete systems using a dual criterion, *IEE Proc.* **132** (2) Part D, 6–68, March.

[34] Chang, B. C. and Pearson, J. B. 1982. Algorithms for the solution of polynomial equations arising in multivariable control theory, Technical Report No. 8208, Electrical Engineering Dept., Rice University, Houston, Texas, USA.

[35] Feinstein, J. and Bar-Ness, Y. 1982 The solution of the matrix polynomial equation: $A(s)X(s)+B(s)Y(s)=C(s)$, American Control Conference, pp. 316–321, Arlington, Virginia.

[36] Noble, B. 1969. *Applied Linear Algebra,* Prentice-Hall, New Jersey, USA.

[37] Wellstead, P. E., Prager, P. and Zanker, P. 1979. Pole assignment self-tuning regulator, *Proc. IEE,* **126**(8), 781–787.

[38] Astrom, K. J. and Wittenmark, B. 1973. On self-tuning regulators, *Automatica,* **9**, 185–199.

[39] Syed, V. H. and Meditch, J. S. 1978. Minimum variance controllers for feedback stochastic systems, Preprints JACC, pp. 649–660, Philadelphia, Oct.

[40] Syed, V. H. and Meditch J. S. 1978. The optimal tracking problem for non-stationary stochastic processes, The Seventh World Congress of the IFAC, pp. 2049–2056, Helsinki, Finland.

[41] Meditch, J. S. and Syed, V. H. 1977. On the design of semifree feedback stochastic systems: Solution via Volterra kernels, *Preprints JACC,* **2**, 1640–1646.

[42] Whitbeck, R. F. 1981. Direct Wiener–Hopf solutions of filters/observers and optimal coupler problem. *J. Guidance and Control,* **4**(3) 329–336.

[43] Grimble, M. J. 1983. Polynomial matrix solution of the optimal deterministic continuous time servomechanism problem, Research Report ICU/30/Nov 1983, Industrial Control Unit, University of Strathcylde, Glasgow, Scotland, UK.

[44] Grimble, M. J. 1986. Dual criterion stochastic optimal control problem for robustness improvement, *IEEE Trans. Auto. Control,* **AC-31**(2), 181–185, Feb.

[45] Vidyasagar, M. 1985. *Control Systems Synthesis: A Factorization Approach,* MIT Press, Massachusetts, USA.

[46] Grimble, M. J. 1986. Controllers for LQG self-tuning applications with coloured measurement noise and dynamic costing, *Proc IEE,* **133**(1), Part D. 19–29, January.

[47] Blomberg, H. and Ylinen, R. 1983. *Algebraic Theory for Multivariable Linear Systems,* Academic Press, London, England.

CHAPTER 12

Optimal Self-tuning Control Systems

12.1. INTRODUCTION

To use the techniques of the previous chapters, it was assumed that the plant-model parameters were known with a suitable and sufficient degree of accuracy. For some systems, however, the plant parameters may be varying or unknown and even the plant-model structure may be subject to some uncertainty. The self-tuning philosophy was developed to accommodate these relaxed conditions of system description. It is a scheme which combines the on-line identification of plant parameters with the computation of appropriate control action. In the explicit algorithms these two functions are separated, namely the identification algorithm provides estimates of plant parameters and these are input to a controller calculation. The more subtle implicit algorithms unite these two functions and identify the parameters of the controllers directly.

As one of the most recent and exciting innovations to derive from the philosophy of optimal control theory, it is perhaps appropriate that the theory sections of this book should conclude with a chapter on self-tuning controllers and their design.

12.1.1. A historical perspective

With some considerable foresight, R. E. Kalman proposed a self-tuning philosophy in the late fifties (Kalman [1]). The use of a discrete (z-transfer function) system representation was assumed along with the suggestion that system coefficients should be treated as unknowns. A least-squares approach was proposed for the calculation of the unknown coefficients and an optimal controller for the control computation. A rather primitive self-optimizing controller was built by Kalman but these early attempts had to await the arrival of powerful modern computers before useful progress was made.

This maturation in computer hardware accompanied by an increased sophistication of systems theory took only twelve years, for in 1970, Peterka published the first treatment of a modern self-tuning regulator (Peterka [2]). The plant description took the form of an autoregressive moving average model (Åström [3]):

$$\mathbf{A}(z^{-1})y(t) = z^{-k}\mathbf{B}(z^{-1})u(t) + \mathbf{C}(z^{-1})\xi(t)$$

and a minimum-variance control law was derived. The parameters were estimated via a least-squares identification algorithm. At the time, the method attracted little attention and the paper itself left several questions, concerned with the practical application of the technique, unanswered. The discrete-system description used by Peterka was adopted by other researchers since it is particularly appropriate for the identification problem. It is only recently that continuous time plant models have been employed in self-tuning systems (Gawthrop [4]).

Following the work of Peterka, the self-tuning regulator and the self-tuning controller were proposed and these are now the most popular optimal self-tuning systems. Åström and Wittenmark at the Lund Institute of Technology, Sweden, further developed the self-tuning regulator from that of Peterka and the report of their work (Åström and Wittenmark [5]) has become a standard reference on self-tuning, provoking much of the subsequent development of the subject.

Clarke and Gawthrop [6] later produced a self-tuning controller, utilizing a generalized minimum-variance control law, previously proposed by Hastings-James [7]. This particular control law was obtained from a quadratic type of performance index which included a control term. This was an important innovation since it provided greater flexibility and widened the range of system responses which could be achieved.

The first multivariable self-tuning regulator was devised by Borisson [8] using the minimum variance control strategy. The self-tuning controller of Clarke and Gawthrop was more recently given a multivariable format by Koivo [9].

The popularity of the self-tuning schemes is undoubtedly due to the simplicity of the concept, the ease of implementation and their success in a number of diverse industrial applications. A typical example is provided by the control of a slab-reheating furnace in the steel industry (Reeve [10]). Such furnaces reheat slabs from ambient to a temperature at which they become sufficiently pliable for further processing in a rolling-mill (Gray *et al.* [11]). The furnace parameters vary in sympathy with the changing operating conditions and environment (as exemplified by a change of steel type, variations in the calorific value of the fuel or the long term deterioration of the furnace insulation). The self-tuning philosophy is appropriate to this situation and the scheme is able to continuously monitor changes in the system parameters over long operating periods. The Lackenby Steelworks (British Steel Corporation) have implemented a number of such control schemes on their reheat furnaces.

Other applications include self-tuning control for the dynamic positioning of oil drill ships (Fung and Grimble [12]), control of a catalytic reactor (Harris *et al.* [13]), control of a nuclear reactor (Allidina and Hughes [14]) and control of a distillation column (Dahlquist, [15]).

12.1.2. The general principles of optimal adaptive control

Optimal adaptive control schemes comprise both the on-line identification function and the subsequent control action. The system input signal has therefore the dual role of providing data for system parameter estimation and providing control action, hence the terminology 'dual control' . Following Jacobs and Patchell [16], the optimal control signal may be categorized as:

1. Certainty equivalent control: the control which would be used if the estimated parameters were the actual ones.
2. Caution: the component which recognizes that the error in the parameter estimates may cause excessive excursions in the control signal. This component accommodates uncertainties in the parameter estimates and is a function of the estimated parameters and their accuracy.
3. Probing: this component of the control signal is used to inject an optimal test signal into the process to improve the estimates of uncertain parameters.

The self-tuning philosophy approximates the optimal control signal by the first component. Hence, the self-tuner may be classified as a certainty equivalent control law since any uncertainty in the parameter estimates is disregarded in the computation of the controller, (Clarke *et al.* [17]).

Controllers which utilize the interaction between identification and control are termed 'dual controllers' (Sternby [18]). Thus far, only a rudimentary theoretical development of dual controllers is available which comprises of analytical solutions for simple systems only; consequently this subject is pursued no further here.

A control problem is said to be separable (Jacobs [19]) if the optimal control law can be written in the form:

$$\boldsymbol{u}(t) = \text{function}(\hat{\boldsymbol{x}}(t \mid t))$$

where the estimate $\hat{\boldsymbol{x}}(t \mid t)$ is (often) the conditional mean of the state vector $\boldsymbol{x}(t)$. A separable stochastic control problem is also certainty equivalent if the function of the control law above is identical to that obtained by solving the equivalent deterministic optimal control problem. In the discussion of this chapter, there is no need to distinguish between these two concepts. (Jacobs [19] contains a useful discussion of these topics.)

The theory and practice of self-tuning regulators falls naturally into a description of the various forms of controllers and a study of appropriate parameter estimation techniques. Unification of these two topics defines self-tuning regulators and controllers. Many combinations of controller design methods and estimation procedures are possible, including, for example,

Controller design methods

(i) Mininum variance control [28].
(ii) LQG design [54], [58].
(iii) Pole placement methods [43].
(iv) Phase and gain margin adjustment procedures.

Parameter estimation methods

(i) Stochastic approximation.
(ii) Recursive least-squares.
(iii) Extended least-squares.
(iv) Multistage least-squares.
(v) Instrumental variables [55].
(vi) Recursive maximum likelihood methods.

The method of implementing the system identification stage can be used to classify two types of self-tuning schemes. An implicit identification method identifies the *controller* parameters directly as shown in Fig. 12.1. This obviates the need to identify the plant parameters and perform a second

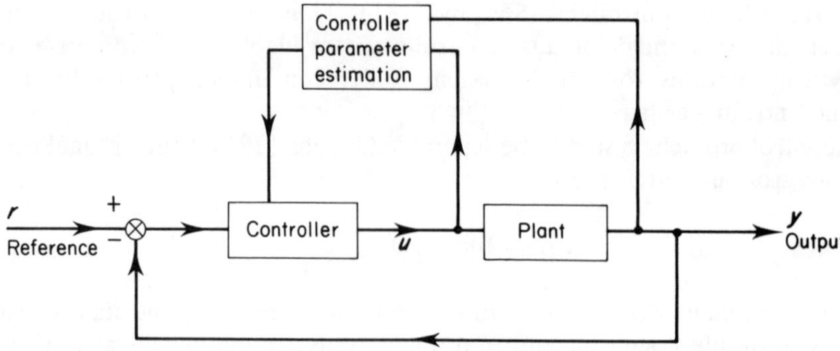

Fig. 12.1. Implicit self-tuning control system (indirect identification)

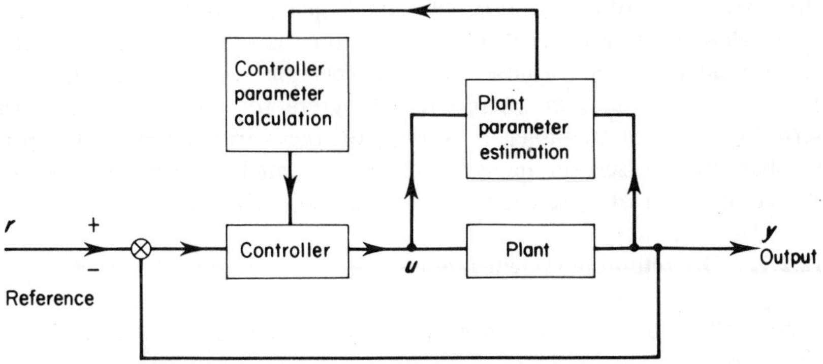

Fig. 12.2. Explict self-tuning control system (direct identification)

calculation to evaluate the control action. A scheme which separates out the identification and control calculation is known as an explicit self-tuning scheme. This is illustrated in Fig. 12.2.

Implicit self-tuning schemes are sometimes thought to be more robust than their explicit counterparts but normally require the estimation of more parameters (Åström and Wittenmark [5]). The separation of the control and estimation calculations which characterize the explicit schemes, has the advantage of providing a structure which is easier to validate and debug at the programming stage.

The notation of this chapter differs slightly from that in preceding chapters since the literature on self-tuning control now has accepted symbols. For example, A is used in the transfer-function description of the plant conflicting with its previous role as the state-space system matrix. Causal transfer functions are written for the discrete domain using polynomials in z^{-1}.

12.2. OPTIMAL CONTROL LAWS FOR SINGLE-INPUT/SINGLE-OUTPUT SYSTEMS

This introduction to the optimal control laws used by self-tuning system commences with a discrete-plant description continues with the specification of the cost criterion and concludes with a derivation of some of the control laws.

The following derivation of the control laws is limited to various optimal controllers obtained under the assumption of known plant parameters. The system is assumed to be in the steady state so that initial conditions may be neglected and steady-state performance criteria can be used. Initially, the discussion is for single-input/single-output systems, which is appropriate since multivariable self-tuners are often direct generalizations of the single-input/single-output cases. A discrete-time model is used for either situation, multivariable or single variable. This reflects the fact that self-tuners are normally implemented on micro/minicomputers for which a discrete-time description is most suitable. A survey of the various types of optimal controllers used in self-tuning systems was presented in Grimble *et al.* [20] and a comparison of different schemes is given in Kurz *et al.* [21].

12.2.1. Discrete-time system models

A controlled autoregressive moving average model is assumed to describe the plant process:

$$\sum_{i=0}^{n_a} a_i y(t-i) = \sum_{i=0}^{n_b} b_i u(t-k-i) + \sum_{i=0}^{n_c} c_i \xi(t-i). \tag{12.1a}$$

In z-domain polynomial form an equivalent representation is:

$$A(z^{-1})y(t) = B(z^{-1})u(t) + C(z^{-1})\xi(t) \tag{12.1b}$$

where z is the forward shift operator $z^k y(t) \triangleq y(t+k)$ and t is the sampling instant. The time delay is assumed to be an integer number of sample intervals $k \geq 1$. This definition of k implies b_0 is non-zero and the polynomials $A(\cdot)$, $B(\cdot)$ and $C(\cdot)$ have the form:

$$\left.\begin{aligned} A(z^{-1}) &= 1 + a_1 z^{-1} + \cdots + a_{n_a} z^{-n_a} \\ B(z^{-1}) &= (b_0 + b_1 z^{-1} + \cdots + b_{n_k} z^{-n_k}) z^{-k} \\ C(z^{-1}) &= 1 + c_1 z^{-1} + \cdots + c_{n_c} z^{-n_c} \end{aligned}\right\} \quad (b_0 \neq 0) \tag{12.2}$$

where $n_k \triangleq n_b - k$, and the two coefficients a_0 and c_0 are normalized to unity. It is clearly possible to divide the polynomial coefficients of (11.1a) by a_0 if $a_0 \neq 1$ so that $A(0) = 1$ and similarly the coefficient c_0 can be taken as unity by adjusting the amplitude of the modelled disturbance process $\{\xi(t)\}$. (The argument of these polynomials (z^{-1}) is often omitted to simplify the notation.)

The plant output is $y(t)$ and the control is denoted by $u(t)$, while the disturbance $\xi(t)$ is a weakly stationary sequence of uncorrelated random variables with zero mean. An upper bound on the polynomial orders n_a, n_b and n_c is assumed known, as is the order of the delay, integer k. In the absence of *a priori* knowledge of delay order k may be set to unity, and the actual presence of a higher-order delay will reveal itself in the estimation step where near zero values for $b_0, b_1, \ldots$ will be obtained. Hence, the delay k can be identified during self-tuning by the same process used for estimating the plant polynomial coefficients.

Rewriting (12.1b) in transfer-function form obtains:

$$y(t) = W(z^{-1})u(t) + W_0(z^{-1})\xi(t)$$

where

$$W(z^{-1}) \triangleq B(z^{-1})/A(z^{-1}) \text{ and } W_0(z^{-1}) \triangleq C(z^{-1})/A(z^{-1}).$$

If the plant $W(z^{-1})$ is non-minimum phase then the numerator polynomial $B(z^{-1})$ is factored as $B(z^{-1}) = B_1(z^{-1})B_2(z^{-1})$ where $B_1(z^{-1})$ has all its zeros strictly inside the unit circle ($B_2(z^{-1})$ includes all the delay and non-minimum phase terms) and the polynomial $B_2(\cdot)$ has the form:

$$B_2(z^{-1}) = (b_{20} + b_{21}z^{-1} + \cdots + b_{2n_0}z^{-n_0})z^{-k} \tag{12.3}$$

($n_0 \triangleq n_2 - k$). The orders of the polynomials $B_1(\cdot)$ and $B_2(\cdot)$ are denoted by n_1 and n_2, respectively. Polynomial $B_2(\cdot)$ includes the delay k and is assumed to be normalized so that the lowest-order term is unity; if the order of the delay k is known, b_{20} is set to unity. Finally, the reciprocal polynomial corresponding to $B_2(z^{-1})$ is defined as:

$$\tilde{B}_2(z^{-1}) = z^{-n_2}B_2(z). \tag{12.4}$$

The feedback control system is shown in Fig. 12.3(a). There is a stochastic reference signal $r(t)$ and a deterministic reference signal $w(t)$. The feedback and feedforward controllers to be determined are denoted by $C_0(z^{-1})$ and

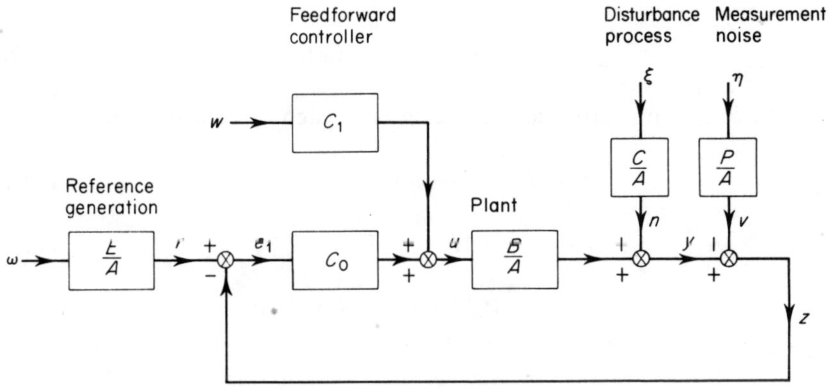

Fig. 12.3(a). Feedback control system

$C_1(z^{-1})$, respectively. The driving noise sources $\{\eta(t)\}$ and $\{\omega(t)\}$ are taken to be weakly stationary sequences of uncorrelated random variables with zero means. The stochastic signals $\eta(t)$, $\xi(t)$ and $\omega(t)$ are mutually independent with respective variances R, Q and Q_2. The polynomials $C(z^{-1})$ and $E(z^{-1})$ of Fig. 12.3(a) may (without loss of generality) be assumed to have no zeros outside or on the unit circle in the z-domain. The Representation and Spectral Factorization theorems may be used to justify this assertion (Åström [3]). Note that in most of the systems of interest $P(z^{-1}) = A(z^{-1})$ and the measurement noise $v(t)$ is white noise.

The polynomial $A(z^{-1})$ represents the least common denominator of the transfer functions for the plant, reference and disturbance subsystems: which by assumption does not contain unstable hidden modes. The tracking error signals and controller inputs are defined as:

$$e_0 = r - y \quad \text{(stochastic error)} \tag{12.5}$$

$$e_1 = r - z \quad \text{(controller input).} \tag{12.6}$$

The reference and disturbance subsystems may be denoted (as obtained previously) by:

$$W_2(z^{-1}) = E(z^{-1})/A(z^{-1}) \tag{12.7}$$

$$W_0(z^{-1}) = C(z^{-1})/A(z^{-1}). \tag{12.8}$$

Let the transfer function $M_0(z^{-1})$ be defined as:

$$M_0 = C_0/(1 + C_0 W)$$

where the argument z^{-1} has been omitted for brevity. Note further that

$$1 - WM_0 = 1/(1 + C_0 W). \tag{12.9}$$

The control, plant output and tracking error signals then become:

$$u = M_0(r - v - n) \tag{12.10}$$

$$y = (1 - WM_0)n + WM_0(r - v) \tag{12.11}$$

$$e_0 = (1 - WM_0)(r - n) + WM_0 v \tag{12.12}$$

where

$$n \triangleq W_0 \xi. \tag{12.13}$$

The controller C_0 can be represented in terms of the polynomials C_n and C_d as $C_0 = C_n/C_d$.

Generalized spectral factors in the s-domain have been introduced in Chapter 2 and in this analysis their discrete analogue is required. These are defined here for use later in the chapter. The optimal control performance criterion weighting terms will be denoted by Q_1 and R_1.

The generalized spectral factors Y, Y_1 are then defined using:

$$YY^* = (EQ_2E^* + CQC^* + PRP^*)/(AA^*) \tag{12.14}$$

$$Y_1^*Y_1 = (B^*Q_1B + A^*R_1A)/(A^*A) \tag{12.15}$$

where $Y^*(z^{-1}) = Y(z)$. The strictly Hurwitz polynomials (polynomials with zeros within the unit disc) D and D_1 satisfy:

$$YY^* = DD^*/(AA^*) \Rightarrow Y = D/A \tag{12.16}$$

$$Y_1^*Y_1 = D_1^*D_1/(A^*A) \Rightarrow Y_1 = D_1/A. \tag{12.17}$$

With the plant model determined and the preliminary definitions for the spectral factors given, the selection of a performance criterion may commence.

12.2.2. Optimal performance criteria

The minimum-variance controller which is used in the self-tuning regulator of Åström and Wittenmark [5] minimizes the criteria:

$$J = E\{y^2(t)\}$$

$$= \lim_{N\to\infty} E\left\{\frac{1}{N}\sum_{i=1}^{N} y^2(i)\right\} \quad \text{(invoking ergodicity)}$$

where the expectation operator is unconditional and $y(i)$ represents the system output. Borisson [8] has shown that the minimum-variance controller minimizes both of the above criteria.

The generalized minimum-variance controller which forms the basis of the self-tuning controller of Clarke and Gawthrop [6] utilizes a cost functional:

$$J_1 = E\{\phi^2(t+k) \mid t\}$$

where the expectation at time $t + k$ is conditional on all input/output data acquired up to time t. The conditional expectation is, with some abuse of notation, denoted by $E\{\,\cdot \mid t\}$. The function $\phi(t)$ represents the generalized system output:

$$\phi(t+k) = P_1(z^{-1})y(t+k) + Q_1(z^{-1})u(t) - R_1(z^{-1})w(t)$$

where $w(t)$ is the set point signal which is known at time t. The transfer functions P_1, Q_1 and R_1 are chosen to achieve a desired performance from the system. An equivalent alternative formulation for the criterion is:

$$J_2 = E\left\{\sum_{i=t}^{t+k} \phi^2(i) \mid t\right\}$$

since $\phi(i)$ is not affected by $u(t)$ for $i < t$ and the criterion is not used to find $u(i)$ for $i > t$. The time over which the cost-function is measured is clearly finite and this gives rise to the term 'single-stage cost functional'.

There was initially some confusion about the significance of the cost function employed in the self-tuning controller. This was clarified by MacGregor and Tidwell [22] who showed that the cost-function J_1 was not the same as that used in LQG optimal control problems. The differences arise from the use of a conditional rather than an unconditional expectation operator.

The first types of controller to be considered in the following are based on cost functions involving the unconditional expectation, hence these include LQG optimal control laws and the minimum-variance controller. A second group of control laws are derived using the conditional expectation and these include the *generalized minimum-variance controller* (Clarke and Hastings-James [56]) and the relatively new *weighted minimum-variance controller* (Grimble [23]).

12.2.3. Linear-quadratic-Gaussian optimal control laws

The LQG optimal control laws will now be obtained in a suitable form for implementation in self-tuning control systems. Assume that the reference signal is stochastic (so that $w = 0$) and that the measurement noise is white ($P = A$). Consider the system (12.1b) and the feedback configuration shown in Fig. 12.3(a).

Theorem 12.1. ***Optimal LQG controller.*** *Assume that the performance criterion to be minimized has the form:*

$$\mathrm{J} = \mathrm{E}\{Q_1\, e_0^2(t) + R_1 u^2(t)\} \tag{12.18}$$

and that the system has reached the steady-state operating condition (that is $t_0 \to -\infty$). The optimal controller is given as:

$$C_0 = G_0/H_0. \tag{12.19}$$

The following coupled Diophantine equations, in terms of the polynomials G, H and F, provide the required particular solution G_0, H_0 with minimal degree with respect to F:

$$\bar{D}_1 G + FA = \bar{B} Q_1 D \tag{12.20}$$

$$\bar{D}_1 H - FB = \bar{A} R_1 D$$

where

$$g \triangleq \max(n_{d_1}, n_b, n_a) \quad \text{and} \quad \bar{D}_1 \triangleq z^{-g} D_1^*, \qquad \bar{B} \triangleq z^{-g} B^* \quad \text{and} \quad \bar{A} \triangleq z^{-g} A^*.$$

The equations can be combined to obtain the implied equation:

$$AH + BG = D_1 D \tag{12.21}$$

and the closed-loop characteristic polynomial $\rho_c = D_1 D$.

Proof. The above theorem might be constructed directly from the results of Theorem 11.9 or from the Wiener solution (equations 11.222–223) of the LQG problem (Grimble [24]) as follows:

$$C_0 = M_0 (I - W M_0)^{-1} \tag{12.22}$$

where

$$M_0 \triangleq Y_1^{-1} \{ (Y_1^*)^{-1} W^* Q_1 \Phi_0 (Y^*)^{-1} \}_+ Y^{-1}$$

and

$$(Y_1^*)^{-1} \triangleq Y_1^T(z)^{-1}, \; W^* \triangleq W^T(z).$$

Also,

$$\Phi_0 \triangleq \Phi_{rr} + W_0 \Phi_{\xi\xi} W_0^*.$$

Substituting for the plant transfer functions defined in Section 12.2.1 and from (12.20) note that,

$$\left\{ \frac{W^* Q_1 \Phi_0}{Y_1^* Y^*} \right\}_+ = \left\{ Q_1 \left(\frac{DB^*}{AD_1^*} - \frac{RA^*B^*}{D_1^* D^*} \right) \right\}_+$$

$$= \left\{ \frac{F}{\bar{D}_1} + \frac{G}{A} - \frac{Q_1 R \bar{A} \bar{B}}{\bar{D}_1 \bar{D}} \right\}_+ = \frac{G}{A}.$$

The Diophantine equation (12.20) has therefore enabled the partial fraction expansion stage to be simplified when calculating the causal transform $\{\cdot\}_+$. Hence,

$$M_0 = \frac{GA}{D_1 D}. \tag{12.23}$$

Substituting (12.23) into (12.22) and using (12.21) gives (12.19) as required. □

The proof that the Diophantine equations have a unique minimal degree solution is given in Kučera [25] and in Section 11.2.3. The following corollary may be obtained which simplifies the computation of the controller:

Corollary 12.1. ***Optimal LQG controller.*** *Assume that the polynomials in the above Diophantine equations are relatively prime and that* $n_a > n_d - k$ *and* $n_b > n_d$. *In this case, the controller* $C_0 = G_0/H_0$ *can be obtained from the unique solution of the single Diophantine equation* (12.21) *with* $n_g \triangleq n_a - 1$ *and* $n_h \triangleq n_b - 1$.

Proof (Grimble [54]). □

(Recall that two polynomials are relatively prime if they have no common factor.) The above corollary intimates that only one Diophantine equation needs to be solved (rather than both (12.20) and (12.21)) if the plant polynomials satisfy the relatively prime condition.

Special cases

Minimum-variance control laws have been widely used in self-tuning systems. These are obtained below as special cases of the LQG control laws (Åström and Wittenmark [27]). Consider a regulating system ($Q_2 = 0$) where the measurement noise is zero ($R = 0$) and assume that the cost function (12.18) includes only output weighting ($Q_1 = 1$, $R_1 = 0$).

Corollary 12.2. ***Minimum-variance controller (minimum-phase plant).*** *If the plant is assumed to be minimum phase the optimal minimum-variance controller follows from* (12.19) *where H and G are given by the solution of the Diophantine equation:*

$$AH + z^{-k}B_1 G = B_1 C \tag{12.24}$$

and

$$n_{g_0} \triangleq n_a - 1.$$

Corollary 12.3. *Minimum-variance controller (non-minimum-phase plant). If the plant is non-minimum phase, then the Diophantine equation to be solved in calculating the optimal controller, becomes:*

$$AH + B_1B_2G = \tilde{B}_1B_2z^kC$$

where

$$n_{g_0} \triangleq n_a - 1.$$

Proof. Corollaries 12.2 and 12.3 follow directly from Corollary 12.1 by recalling $B = B_2B_1$ and that B_2 includes both the non-minimum phase and the delay (z^{-k}) terms. □

There is no need to have two separate controllers for minimum- and non-minimum-phase plants since they are special cases of the more general LQG controller. However, in most of the self-tuning literature, the minimum-variance controller employed is that for a minimum-phase plant description. This particular controller results in an unstable closed-loop system when the plant is non-minimum phase and thus it is important to distinguish between the minimum-variance controller derived for a minimum-phase plant and that derived for a non-minimum-phase plant.

Measurement noise

The plant configuration of Fig. 12.3(a) is rather complicated, consequently (neglecting temporarily the set point signal $w(t)$) it might be thought expedient to combine all the noise sources to obtain the equivalent system of Fig. 12.3(b). The system equation then becomes:

$$Ae_1 = D\varepsilon - Bu$$

where D is defined in (12.16).

Most of the self-tuning control schemes are based on plants of the form shown in Fig. 12.3(b) with only the single stochastic input. It is of interest to consider whether there is any advantage in using the simplified system of Fig. 12.3(b) by minimizing the variance of e_1,

$$J_1 \triangleq E\{Q_1\, e_1^2(t) + R_1u^2(t)\}$$

rather than the variance of the tracking error e_0,

$$J \triangleq E\{Q_1\, e_0^2(t) + R_1u^2(t)\}.$$

The reference ($r(\cdot)$) and the disturbance ($n(\cdot)$) signals of LQG theory may

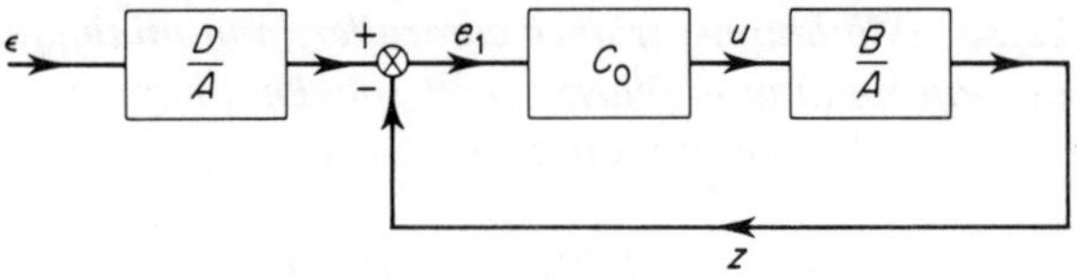

Fig. 12.3(b). Innovations form of the plant model

certainly be combined to form one effective stochastic signal. However, the LQG controller to minimize the weighted variance of e_1 is not necessarily identical to the one which minimizes the weighted variance of e_0. Thus, it is not always desirable to consider the plant of Fig. 12.3(b) and derive the corresponding controller when the measurement noise v is significant and control of the tracking error e_0 is of importance.

To demonstrate this difference, consider the two problems and the expression for the respective LQG controllers:

(a) Plant of Fig. 12.3(b) and minimize $E\{Q_1 e_1^2(t) + R_1 u^2(t)\}$.

This situation is equivalent to the case of zero measurement noise but with a disturbance of $(D/A)\varepsilon$. Thus $YY^* = \Phi_0 = (DD^*)/(AA^*)$ and from (12.22):

$$M_0 = Y_1^{-1}\{(Y_1^*)^{-1}W^*Q_1 Y\}_+ Y^{-1}.$$

(b) Plant of Fig. 12.3(a) and minimize $E\{Q_1 e_0^2(t) + R_1 u^2(t)\}$.

From (12.14):

$$YY^* = \Phi_0 + \frac{PRP^*}{AA^*} = \frac{DD^*}{AA^*}$$

and from (12.22):

$$M_0 = Y_1^{-1}\left\{(Y_1^*)^{-1}W^*Q_1\left(Y - \frac{PRP^*}{AA^*}\right)(Y^*)^{-1}\right\}_+ Y^{-1}$$

but

$Y = D/A$ thus:

$$M_0 = Y_1^{-1}\left\{(Y_1^*)^{-1}W^*Q_1 Y\right\}_+ Y^{-1} - Y_1^{-1}\left\{(Y_1^*)^{-1}W^*Q_1 \frac{PRP^*}{AD^*}\right\}_+ Y^{-1}.$$

If the measurement noise is assumed to be white (as in the remainder of the chapter) then $P \triangleq A$ and the last term above can be neglected. In this case the two expressions for M_0 and the two controllers become identical.

An alternative demonstration that the two controllers are identical for the case of white measurement noise is obtained by considering the respective cost functions. Let

$$J_1 \triangleq E\{Q_1 e_1^2(t)\} = E\{(e_0(t) - v(t))^2 Q_1\}$$
$$= E\{e_0^2(t)Q_1\} - 2E\{e_0(t)v(t)Q_1\} + E\{v^2(t)Q_1\}.$$

The last term of this expansion for J_1 does not affect the minimization problem, consequently it may be neglected. The plant has at least a unit delay, hence the cross-product term and the related term $E\{y(t)v(t)\}$ are both zero as (by assumption) is the term $E\{r(t)v(t)\}$. For this case, the optimal controller thus minimizes the remaining terms: $E\{Q_1 e_0^2(t) + R_1 u^2(t)\}$ so that the same optimal controller minimizes both J and J_1.

This establishes the important fact that for LQG optimal control laws and for systems with white measurement noise, the controller to minimize the tracking error is the same as the controller to minimize the feedback error, when the equivalent plant model of Fig. 12.3(b) is used. The same result may be demonstrated for the multivariable case.

Example 12.1. *Optimal LQG controller.* This example is based upon the non-minimum-phase system considered by Åström and Wittenmark [28] and by Clarke and Gawthrop [6]:

$$A(z^{-1})y(t) = B(z^{-1})u(t) + C(z^{-1})\xi(t)$$

where

$$A(z^{-1}) = (1 + a_1 z^{-1}), \qquad a_1 = -0.95$$
$$B(z^{-1}) = z^{-k}(1 + b_1 z^{-1}), \qquad b_1 = 2,\ k = 2$$
$$C(z^{-1}) = (1 + c_1 z^{-1}), \qquad c_1 = -0.7.$$

The performance criterion to be minimized is defined as:

$$J = E\{Q_1 e_0^2(t) + R_1 u^2(t)\}$$

where $Q_1 = 1$ and $R_1 = \rho > 0$. The white-noise covariances are defined as $Q = 1$ and $R = \mu > 0$. The reference $r \equiv 0$ and the plant is assumed known. The LQG optimal regulator for this system is calculated as follows.

The spectral factors are first obtained using equations (12.14) to (12.17):

$$DD^* = (1 + c_1 z^{-1})Q(1 + c_1 z) + \mu(1 + a_1 z^{-1})(1 + a_1 z)$$
$$= (Qc_1 + \mu a_1)(z^{-1} + m + z)$$

where

$$m \triangleq (Q(1 + c_1^2) + \mu(1 + a_1^2))/(Qc_1 + \mu a_1).$$

Hence,

$$DD^* = \sigma^2(\alpha + z^{-1})(\alpha + z)$$

where $\sigma^2 \triangleq (Qc_1 + \mu a_1)/\alpha$. Let α denote the solution $|\alpha| > 1$ of $\alpha = (m \pm \sqrt{m^2 - 4})/2$ then

$$D(z^{-1}) = \sigma(\alpha + z^{-1}), \quad \sigma > 0.$$

$$D_1^* D_1 = (1 + b_1 z)Q_1(1 + b_1 z^{-1}) + \rho(1 + a_1 z)(1 + a_1 z^{-1})$$

$$= (Q_1 b_1 + \rho a_1)(z^{-1} + m_1 + z)$$

where

$$m_1 \triangleq ((Q_1 + b_1^2) + \rho(1 + a_1^2))/(Q_1 b_1 + \rho a_1), \qquad \sigma_1^2 \triangleq (Q_1 b_1 + \rho a_1)/\alpha_1.$$

Let α_1 denote the solution of $\alpha_1 = (m_1 \pm \sqrt{m_1^2 - 4})/2$ for which $|\alpha_1| > 1$, then

$$D_1(z^{-1}) = \sigma_1(\alpha_1 + z^{-1}), \ \sigma_1 > 0.$$

The polynomials A and B are relatively prime and in this case a unique solution for the LQG optimal controller follows from (12.21) with

$$n_{g_0} \triangleq n_a - 1 = 0 \quad \text{and} \quad n_{h_0} = n_b - 1 = 2$$

$$AH_0 + BG_0 = D_1 D$$

$$(1 + a_1 z^{-1})(h_0 + h_1 z^{-1} + h_2 z^{-2}) + z^{-2}(1 + b_1 z^{-1})g_0 = \sigma\sigma_1(\alpha + z^{-1})(\alpha_1 + z^{-1}).$$

The solution of this equation follows as:

$$h_0 = \sigma\sigma_1\alpha\alpha_1$$

$$h_1 = -a_1 h_0 + \sigma\sigma_1(\alpha + \alpha_1)$$

$$h_2 = (-a_1 h_1 + \sigma\sigma_1)/(1 - a_1/b_1)$$

$$g_0 = -a_1 h_2/b_1$$

then

$$C_0(z^{-1}) = g_0/(h_0 + h_1 z^{-1} + h_2 z^{-2}).$$

If for example $\mu = \rho = 1$ then the controller becomes:

$$C_0(z^{-1}) = \frac{0.2143}{(3.753 + 1.187z^{-1} + 0.451z^{-2})}.$$

12.2.4. Observations weighted optimal control laws

Up to the present time it has not been possible to derive an implicit LQG self-tuning controller for general multivariable linear plant descriptions. For this reason a related controller which is termed an observations weighted controller was derived by Grimble [26], [59] which can be used in implicit self-tuning schemes. The problem in deriving an implicit LQG scheme is to obtain expressions from which H and G can be estimated and which ensure equation (12.20) is also satisfied. The observations weighted controller is defined to minimize a cost function which is a generalized minimum-variance cost-function (no explicit control weighting). However, by careful choice of this cost-function a very similar form to the usual LQG controller is derived. The details of self-tuning scheme are presented later but the controller is defined via the following theorem.

Theorem 12.2. ***Observations weighted optimal controller.*** *Consider the plant* (12.1b) *and feedback configuration shown in Fig.* 12.3(*a*). *The observations weighted optimal controller to minimize the performance criterion:*

$$J = E\{\phi^2(t+k)\} \tag{12.25a}$$

where $\phi(t+k) \triangleq D_1 e_1(t)/(A_0 B_2)$ *is defined as:*

$$C_0 = G_1/(H_1 B_1 A_0). \tag{12.25b}$$

Here, G_1, H_1 *represent the unique solution of the equation:*

$$A_1 H + B_2 G = D_1 D \tag{12.26}$$

and $n_{h_1} \triangleq n_2 - 1$, $A_1 \triangleq AA_0$. *The closed-loop characteristic polynomial:* $\rho_c(z^{-1}) \triangleq D_1 D$.

Proof. (The input to the optimal controller, neglecting the deterministic reference ($w = 0$) becomes:

$$e_1 = r - z = W_2\omega - W_0\xi - Wu - v.$$

This signal may be represented in an equivalent innovations form as:

$$e_1 = (D/A)\varepsilon - Wu$$

but from the Diophantine equation (12.26) obtain,

$$\frac{D}{A} = \frac{A_0 H_1}{D_1} + \frac{B_2 G_1}{A D_1}$$

so that the expressions for e_1 can be simplified to obtain:

$$e_1 = \frac{A_0 H_1 \varepsilon}{D_1} + \frac{B_2 G_1 e_1}{D_1 D} - \frac{H_1 B A_0 u}{D_1 D}.$$

The observations weighted signal $\phi(\cdot)$ now becomes

$$\begin{aligned}\phi(t+k) &= D_1 e_1(t)/(A_0 B_2)\\ &= \frac{H_1}{B_2}\varepsilon(t) + \frac{1}{DA_0}(G_1 e_1(t) - H_1 B_1 A_0 u(t)).\end{aligned} \tag{12.27}$$

The first term in (12.27) can be expressed in terms of a convergent series in powers of z. Assuming that the control is a stable signal the remaining terms in (12.27) can be expressed as a convergent series in powers of z^{-1}. These series involve future and past values of ε respectively, hence they are statistically independent. Let

$$\hat{\phi}(t+k \mid t) \triangleq \frac{1}{DA_0}(G_1 e_1(t) - H_1 B_1 A_0 u(t))$$

and

$$\tilde{\phi}(t+k \mid t) \triangleq \frac{H_1}{B_2}\varepsilon(t)$$

then

$$\begin{aligned}J &= E\{(\hat{\phi}(t+k \mid t) + \tilde{\phi}(t+k \mid t))^2\}\\ &= E\{\hat{\phi}(t+k \mid t)^2 + \tilde{\phi}(t+k \mid t)^2\}.\end{aligned}$$

The least-squares predictor for ϕ can be identified as $\hat{\phi}$ and the cost function J may be minimized by setting $\hat{\phi}(t+k \mid t) = 0$. The optimal control follows as:

$$u^0(t) = \frac{G_1}{H_1 B_1 A_0} e_1(t). \tag{12.28}$$

The assumption is made that the plant does not contain any unstable hidden modes and hence A_1, B_2 are relatively prime. It follows that the Diophantine equation has a unique solution.

The closed-loop characteristic polynomials for the LQG and observations weighted controllers are both $D_1 D$ but the zero polynomials can differ.

Special cases

If the control weighting $R_1 = 0$, the observations weighted and the LQG optimal controllers are identical (assuming the usual case of $n_{a_1} + n_2 - 1 > n_{d_1} + n_d$ applies). That is, the limiting case of the minimum-variance control law is identical to the observations weighted control law. This is true of controllers for both minimum- and non-minimum-phase plants.

When the plant has no minimum phase zeros $B_1(z^{-1}) = b_0$ and a comparison of (12.21) and (12.26) confirms that the observations weighted and LQG controllers are equivalent.

The observations weighted controller for minimum-phase plants is related to the controller derived by Hastings-James [7] which is called the generalized minimum variance controller. (Consider the problem described in Theorem 12.2 but assume that the plant is minimum phase and $B_2 = z^{-k}$.)

Corollary 11.4 ***Observations weighted controller (minimum phase plants).*** *The controller to minimize the cost (11.25a) may be identified as:*

$$C_0 = \frac{G_1}{H_1 A_0} \tag{12.29}$$

where G_1, H_1 are the unique solutions of the equation:

$$A_1 H + z^{-k} G = D_1 D \tag{12.30}$$

and

$$n_{h_1} \triangleq k - 1.$$

Remarks. Consider again the system described in Example 12.1 and let $A_0 = 1$. Note that $n_{h_1} = n_2 - 1 = 2$ which is the same as n_{h_0} in the previous example. Since the solution of the Diophantine equations (12.21) and (12.30) are unique when A_1, B are relatively prime and since $B = B_2$ for this problem, the solutions for the observations weighted and LQG controllers are identical.

Different performance characteristics may easily be achieved using the above control laws by judicious choice of dynamic Q_1 and R_1 matrices. The following example considers the observations weighted controller for the situation where the dynamic weighting term $A_0 \triangleq (1 - z^{-1})$.

Example 12.2. *Integral control.* From (12.15) and (12.17):

$$D_1^* D_1 = (1 + b_1 z) Q_1 (1 + b_1 z^{-1}) + \rho(1 + a_1 z)(1 - z)(1 + a_1 z^{-1})(1 - z^{-1})$$

and

$$D_1 = d_0 + d_1 z^{-1} + d_2 z^{-2}.$$

The controller is defined via the Diophantine equation:

$$A_1H_0 + BG_0 = D_1D$$

$$(1 - z^{-1})(1 + a_1z^{-1})(h_0 + h_1z^{-1} + h_2z^{-2}) + z^{-2}(1 + b_1z^{-1})(g_0 + g_1z^{-1})$$

$$= \sigma(d_0 + d_1z^{-1} + d_2z^{-2})(\alpha + z^{-1}).$$

The solution of this equation becomes:

$$h_0 = \sigma d_0\alpha$$

$$h_1 = -(a_1 - 1)h_0 + \sigma(d_1\alpha + d_0)$$

$$h_2 = -(a_1 - 1)h_1 + a_1h_0 - g_0 + \sigma(d_2\alpha + d_1)$$

$$g_0 = (\sigma d_2b_1 + a_1h_1b_1 + (1 - a_1)h_2b_1 - a_1h_2)/b_1^2$$

$$g_1 = a_1h_2/b_1$$

then

$$C_0 = \frac{(g_0 + g_1z^{-1})}{(1 - z^{-1})(h_0 + h_1z^{-1} + h_2z^{-2})}.$$

12.2.5. Single-stage optimal control laws

The single-stage optimal control laws are derived using a performance criterion of the form:

$$J_1 = E\{\phi_1(t + k)^2 \mid t\} \tag{12.31}$$

where

$$\phi_1(t + k) \triangleq -P_0e_1(t) - P_1w(t) + P_2u(t)$$

and

$$P_0 \triangleq D_1/(A_0B_2).$$

The generalized minimum-variance (GMV) controller is based on a single-stage criterion and is probably the most widely used controller in self-tuning systems. The controller may sometimes be used on non-minimum phase systems very successfully. However, if the plant is open-loop unstable and non-minimum phase it may be very difficult to select a control weighting parameter for which the closed-loop system is stable (the plant being imprecisely known in self-tuning problems). The weighted minimum-variance

(WMV) controller was introduced (Grimble [29]) to overcome this difficulty. This controller has the useful feature that for small control cost weightings, if the plant is known precisely, the control is guaranteed to be stabilizing, irrespective of whether the plant is non-minimum phase or open-loop unstable.

Consider the system described in equation (12.1b) where the plant is non-minimum phase.

Theorem 12.3. *Weighted minimum-variance* **(WMV)** *controller. The optimal control to minimize the performance criterion* (12.31) *is given as:*

$$u^0(t) = C_0 e_1(t) + C_1 w(t) \tag{12.32}$$

where

$$C_0 \triangleq \frac{G_1}{A_0(H_1B_1 + P_2D)}$$

$$C_1 = \frac{DP_1}{(H_1B_1 + P_2D)}$$

and G_1, H_1 *represent the unique solution of the equation:*

$$A_1H + B_2G = D_1D \tag{12.33}$$

and

$$n_{h_1} \triangleq n_2 - 1, \qquad n_g \triangleq \max(n_{a_1} - 1, n_{d_1} + n_d - n_2).$$

Proof. From equation (12.27) obtain:

$$\phi_1(t+k) = -\phi(t+k) - P_1w(t) + P_2u(t). \tag{12.34}$$

As in the previous section let

$$\hat{\phi}_1(t+k \mid t) = -\hat{\phi}(t+k \mid t) - P_1w(t) + P_2u(t)$$

$$\tilde{\phi}_1(t+k \mid t) \triangleq -\tilde{\phi}(t+k \mid t).$$

The signal $w(t)$ is a deterministic signal, hence

$$E\{w(t)\tilde{\phi}_1(t+k \mid t) \mid t\} = 0. \tag{12.35}$$

Thus, from similar arguments to those in Section 12.2.4 the optimal control satisfies $\hat{\phi}_1(t+k \mid t) = 0$, or

$$u^0(t) = \frac{G_1}{A_0(H_1B_1 + P_2D)} e_1(t) + \frac{DP_1}{(H_1B_1 + P_2D)} w(t). \qquad (12.36)$$

□

Special cases

If $P_1 = P_2 = 0$ the observations weighted and weighted minimum-variance controllers are identical. Similarly, if the control weighting is zero, the WMV controller is equivalent to the minimum-variance controller.

The generalized minimum-variance control law which is used by Clarke and Gawthrop [6] may also be derived from the above results. Consider the system described in Section 12.2 where the plant is assumed to be minimum phase.

Corollary 12.5. ***Generalized minimum-variance*** **(GMV)** ***control law.*** *The optimal control to minimize the criterion* (12.31) *is given by* (12.32) *where* G_1, H_1 *are the unique solution to the Diophantine equation:*

$$A_1H + z^{-k}G = D_1D \qquad (12.37)$$

where

$$n_{h_1} \triangleq k-1 \quad \textit{and} \quad B_2 = z^{-k}, \qquad B_1 = z^kB.$$

The main difference between the generalized and weighted minimum-variance control laws is now apparent. That is, as the control-weighting term goes to zero, the GMV controller reduces to a minimum-variance controller for a minimum phase system. If the plant is actually non-minimum phase the closed-loop system is then unstable. The WMV controller, however, allows for the presence of non-minimum phase zeros and produces a stabilizing control law in this situation.

It is important to have a control law which is stabilizing for non-minimum phase discrete systems because the sampled form of continuous-time minimum-phase plants is often of a non-minimum phase form. Consider, for example, a continuous-time system with delay $dy/dt = u(t-\tau)$ where $0 \le \tau < T$ and T is the sample interval. If the control $u(\cdot)$ is a piecewise constant signal from a hold device, then,

$$y(t+T) = y(t) + (T-\tau)u(t) + \tau u(t-T).$$

The pulse transfer function becomes:

$$H(z^{-1}) = \frac{((T-\tau)+\tau z^{-1})z^{-1}}{(1-z^{-1})}$$

and the discrete system has a zero at $z = -\tau/(T-\tau)$. This zero is non-minimum phase for $T/2 < \tau < T$. Thus, the minimum-phase continuous-time system can have either a minimum or a non-minimum phase sampled representation depending upon the magnitude of the intersample delay τ. A control strategy which is stabilizing in either situation is clearly desirable. It may be noted that the reverse situation to the above can also apply. That is if a continuous-time non-minimum-phase system is sampled slowly, the non-minimum-phase behaviour can be missed and the discrete model be of minimum-phase type.

12.3. PARAMETER ESTIMATION AND EXPLICIT SELF-TUNING CONTROL

The second component of a self-tuning scheme is the estimation or identification of the system or controller parameters depending on whether the explicit or implicit configuration is adopted. In practice, a least-squares algorithm is usually implemented and one purpose of this section will be to consider this technique and its niche in the theory of estimation.

Although several parameter estimation algorithms have been proposed for self-tuning applications, most have a similar structure, need approximately the same computer storage and execute similar numbers and types of arithmetic operations. The reader is referred to Soderström *et al.* [30] for an interesting survey of such methods. In the following, the parameter estimation problem is first considered and finally the least-squares identification algorithm is combined with a controller calculation to produce an explicit self-tuning regulator.

12.3.1. Desirable properties of estimators

A set of N input–output observation pairs is assumed and from these an estimate of a vector $\boldsymbol{\theta}$ is required whose entries comprise the unknown (or slowly varying) system parameters. The probability density function $p(\boldsymbol{\theta}, N)$ is usually unknown or too complicated for direct analysis, hence attention is directed at the most significant statistical characteristics of the estimator, its mean and covariance, and using them to prescribe various desirable properties.

(i) *Sufficiency.* The functional relationship $\hat{\boldsymbol{\theta}}$ based on a sample size N, retains or summarizes all the information in the sample regarding the parameter $\boldsymbol{\theta}$. In particular, there may be several sufficient statistics for estimating a parameter, for example

$$\frac{1}{n-1}\sum_{i=1}^{n}(x_i-\bar{x})^2 \quad \text{and} \quad \sum_{i=1}^{n}(x_i-\bar{x})^2$$

are both sufficient statistics for the variance of the normal distribution ($\bar{x} \triangleq E\{x\}$). However, to determine which is the better estimator further properties must be examined.

(ii) *Unbiasedness.* A statistic $\hat{\boldsymbol{\theta}}$ is an unbiased estimator of $\boldsymbol{\theta}$ if $E\{\hat{\boldsymbol{\theta}}\} = \boldsymbol{\theta}$.

(iii) *Squared error consistency.* If $\{\hat{\boldsymbol{\theta}}_i\}_{i=1}^{\infty}$ is a sequence of estimators for $\boldsymbol{\theta}$ (the i indexes the dependence on sample size) and

$$\lim_{i\to\infty} E\{\|\hat{\boldsymbol{\theta}}_i - \boldsymbol{\theta}\|^2\} = 0$$

then $\hat{\boldsymbol{\theta}}$ is a squared-error consistent estimator for $\boldsymbol{\theta}$. Expanding,

$$(\hat{\theta}_i(j) - \theta(j))^2 = [(\hat{\theta}_i(j) - E\{\hat{\theta}_i(j)\}) - (\theta(j) - E\{\hat{\theta}_i(j)\})]^2$$

obtains:

$$E\{(\hat{\theta}_i(j) - \theta(j))^2\} = E\{(\hat{\theta}_i(j) - E\{\hat{\theta}_i(j)\})^2\} + (\theta(j) - E\{\hat{\theta}_i(j)\})^2$$

$$= \text{variance } \hat{\theta}_i(j) + (\text{bias } \hat{\theta}_i(j))^2.$$

Thus if an estimator is squared-error consistent, then both the estimator variance and bias decrease to zero as the sample size is increased.

(iv) *Squared-error asymptotic efficiency.* An estimator $\{\hat{\theta}_i\}_{i=1}^{\infty}$ is squared error asymptotically efficient if there is no other squared-error consistent estimator $\hat{\theta}^*$ for which

$$E\{(\hat{\theta}_i - \theta)^2\} = E\{(\hat{\theta}_i^* - \theta)^2\}, \qquad i = 1, \ldots.$$

Minimum-variance unbiased estimators arise by seeking a square-error asymptotically efficient estimator in the class of unbiased estimators, in this case

$$E\{(\hat{\theta}_i - \theta)^2\} = E\{(\hat{\theta}_i - E\{\hat{\theta}_i\})^2\} = \text{var } \hat{\theta}_i \text{ and } \text{var}(\hat{\theta}_i) < \text{var}(\hat{\theta}_i^*), \qquad i = 1, 2, \ldots,$$

for all similar estimators $\hat{\theta}_i^*$.

It is clear that a consistent estimator is always unbiased. An estimator is unbiased if on average it yields the correct value of the parameter. A consistent sequence of estimators is one for which a sufficiently large sample is almost certain to produce an estimator close to the parameter value.

12.3.2. Identifiability

The convergence properties of least-squares identification algorithms when the control signal is persistently exciting (continuously varying) and independent of the noise have been established (see for example Åström and Eykhoff [31]). The special problem of feedback control is that the input u can be correlated with the noise. This may be the case in self-tuning systems where the open-loop plant parameters (or some equivalent set of parameters) are to be determined by tests on the closed-loop system. Thus the question of whether the open-loop characteristics of the system can be identified from input and output records, as the number of data points tends to infinity, must be considered. An overview of this identifiability problem for closed-loop processes has been presented by Gustavsson *et al.* [32].

The following example due to Åström and Wittenmark [5], illustrates the identifiability problem.

Example 12.3. *Identifiability.* Consider a first-order linear system:

$$(1 + az^{-1})y(t) = z^{-1}bu(t) + \xi(t)$$

or

$$y(t) = z^{-1}(-ay(t) + bu(t)) + \xi(t).$$

The minimum-variance regulator is clearly of the form $u(t) = (a/b)y(t)$. If the parameters are not known the constant feedback gain can be estimated using $k = \hat{a}/\hat{b}$ where $\hat{a}, \hat{b}$ are the least-squares estimates of a and b. The least-squares estimates are obtained (as described in the following section) so that the cost function:

$$V(a, b) = \sum_{s=1}^{t} (y(s) + ay(s-1) - bu(s-1))^2$$

is minimized with respect to a and b.

With the above feedback control the inputs and outputs are linearly related through:

$$\hat{a}y(t) - \hat{b}u(t) = 0.$$

Adding γ (an arbitrary real scalar) times this zero element to the cost function $V(a, b)$ gives:

$$\begin{aligned} V(a, b) &= \sum_{s=1}^{t} (y(s) + (a + \gamma\hat{a})y(s-1) - (b + \gamma\hat{b})u(s-1))^2 \\ &= V((a + \gamma\hat{a}), (b + \gamma\hat{b})). \end{aligned}$$

The cost function will thus have the same value for all estimates of a and b on a linear manifold (N.B. The translation of a subspace is referred to as a *linear manifold* or *linear variety*.) It follows that the parameters a and b, in the above model, are not identifiable when the above feedback is used. Note that the control

$$u(t) = \frac{(a + \gamma\hat{a})}{(b + \gamma\hat{b})}\, y(t)$$

is not optimal:

$$ay(t) - bu(t) = \frac{(a(b + \gamma\hat{b}) - b(a + \gamma\hat{a}))y(t)}{(b + \gamma\hat{b})}$$

$$= \gamma\,\frac{(a\hat{b} - b\hat{a})}{(b + \gamma\hat{b})}\, y(t) \neq 0$$

unless $a = \hat{a}$, $b = \hat{b}$ or $\gamma = 0$.

The example illustrates that it is, in general, not possible to estimate all of the parameters of the model when the input is generated by feedback. It is possible to estimate all of the parameters if the control law is changed to, for example, $u(t) = \alpha y(t-1)$. If the parameter estimates are incorrect then the control law based upon these estimates will, of course, become suboptimal.

Identifiability has been given several different definitions in the literature. The property is normally related to the consistency of the parameter estimate $\hat{\boldsymbol{\theta}}(t)$. The parameter $\hat{\boldsymbol{\theta}}$ is said to be identifiable if the sequence of estimates $\hat{\boldsymbol{\theta}}(t)$ converges to $\boldsymbol{\theta}$ in some stochastic sense.

The identifiability conditions which must be fulfilled to obtain consistent parameter estimates in closed loop (assuming only the disturbance input $\boldsymbol{\xi}$) have been established (Kurz and Isermann [33]) as:

(a) The order n and the time delay k of the process must be known *a priori*,

(b) $\max\{\mu; \nu + k\} - p \geq n$,

where ν and μ are the numerator and denominator degrees for the feedback controller and where p is the number of common poles and zeros of the disturbance transfer function:

$$W_c = D/(AC_d + BC_n).$$

If condition (b) is not fulfilled closed-loop parameter identifiability can be achieved by:

(i) Introducing an extra perturbation signal into the loop which is persistently exciting of order $\sigma \geq n$. The excitation signal must be introduced outside the points used for the identification algorithm measurements. Thus, if u and y are the measured variables, u may be formed from the feedback controller output plus the excitation signal.

(ii) Switching between two different sets of controller parameters.

If there are no pole-zero cancellations in the controller and the orders of the plant polynomials $A(z^{-1})$, $B(z^{-1})$ and $D(z^{-1})$ are denoted by n, and the feedback controller polynomials by n_{c0}, the following theorem can easily be established (Goodwin and Payne [49]):

Theorem 12.4. *A sufficient condition for identifiability of $A(z^{-1})$, $B(z^{-1})$ and $D(z^{-1})$ is that the order of the feedback law be greater than or equal to the order of the forward path ($n_{c0} \geq n$).*

Testing for identifiability

As noted above, some conditions involving the true order of the system must be evaluated before identifiability can be established (Gustavsson, Ljung and Soderström [48]). It is not possible to test after an experiment whether the system was really identifiable as the following example demonstrates.

Example 12.4. *Identifiability.* Consider the system:

$$\mathbf{A}(z^{-1})y(t) = z^{-k}\mathbf{B}(z^{-1})u(t) + \mathbf{C}(z^{-1})\xi(t)$$

with a feedback control law:

$$\mathbf{H}(z^{-1})u(t) = \mathbf{G}(z^{-1})y(t).$$

Clearly, the following system with arbitrary dynamics $\boldsymbol{\Sigma}(z^{-1})$,

$$(\mathbf{A}(z^{-1}) + \boldsymbol{\Sigma}(z^{-1})\mathbf{G}(z^{-1}))y(t) = (z^{-k}\mathbf{B}(z^{-1}) + \boldsymbol{\Sigma}(z^{-1})\mathbf{H}(z^{-1}))u(t) + \mathbf{C}(z^{-1})\xi(t)$$

will have the same input–output relation in closed-loop as the actual open-loop system. It follows that the true system cannot be distinguished by the use of input–output experiments and, in general, identifiability cannot be established *a posteriori*.

Identifiability plays a key role in the theory for the identification of linear systems. However, in most self-tuning examples the problem is successfully ignored! This may be partly explained by the fact that in self-tuning control the controller parameters are not fixed but initially are varying at each sample point. In addition, convergence of the controller parameters is of secondary importance in comparison with the stability of the closed-loop system. Note that stability refers to the boundedness of the control and output signals and convergence to the ability of the system to achieve a desired system performance asymptotically. For practical purposes it is not important if the converged plant parameter estimates are in error, providing the control achieved is almost the same as for the optimal case, based upon a known plant model.

12.3.3. Least-squares parameter estimation

To illustrate how the estimation algorithms are derived, a simplified problem will be considered where the delay $k = 1$ and the plant equation has the form:

$$A(z^{-1})y(t) = B(z^{-1})u(t) + \xi(t). \tag{12.38}$$

The model is referred to as a regression model since it is found using the least-squares method. Substituting for the system polynomials:

$$y(t) + a_1 y(t-1) + \cdots + a_n y(t-n) = b_0 u(t-1) + \cdots + b_{n-1} u(t-n) + \xi(t)$$

where $\{\xi(t), t = 1, 2 \ldots\}$ is a sequence of independent normal random variables with zero mean and variance Q. This is the special model structure considered by Wieslander and Wittenmark [34].

The model is linear in the parameters and has an input which is an uncorrelated zero-mean sequence. It may be written in a form which is similar to the output equation for a plant represented by state equations:

$$y(t) = \mathbf{X}(t-1)\boldsymbol{\theta} + \xi(t) \tag{12.39}$$

where

$$\mathbf{X}(t-1) \triangleq [-y(t-1), -y(t-2), \ldots, -y(t-n); u(t-1), u(t-2), \ldots, u(t-n)]$$

is the matrix of regressors, and

$$\boldsymbol{\theta} \triangleq [a_1, a_2, \ldots, a_n; b_0, b_1, \ldots, b_{n-1}]^T \tag{12.40}$$

is the vector of parameters.

The vector $\boldsymbol{\theta}$ may be a function of time if the plant parameters are slowly varying. Thus, $\boldsymbol{\theta}$ may be treated as the unknown state of the system and its time dependence may be represented by a the additional equations:

$$\boldsymbol{\theta}(t) = \mathbf{A}\boldsymbol{\theta}(t-1) + \boldsymbol{e}(t-1)$$

where $\boldsymbol{e}(t)$ is a sequence of independent normal random variables with zero-mean and covariance matrix $\mathbf{Q}_p$. In many applications no prior knowledge of the parameter noise is available and the parameter noise covariance matrix is assumed to be zero. In this case the equations can be assumed to be scaled, with variance $Q = 1$, without affecting the estimation of $\boldsymbol{\theta}$.

If the plant parameters are known to be constant, as is the usual case in self-tuning systems $\mathbf{e}(t) \equiv \mathbf{o}$ and $\mathbf{A} = \mathbf{I}$. Since the transition matrix is equal to the identity matrix, the above state equation represents a discrete integrator $(\boldsymbol{\theta}(t) = (z/(z-1))\boldsymbol{\theta}(t-1))$. Summarizing, the parameter estimation state equations can be represented as:

$$\boldsymbol{\theta}(t) = \boldsymbol{\theta}(t-1) + \boldsymbol{e}(t-1)$$

$$y(t) = \mathbf{X}(t-1)\boldsymbol{\theta}(t-1) + \xi(t)$$

where the noise sequences $\boldsymbol{e}$ and ξ are uncorrelated.

The system description is clearly in the usual form for a Kalman filtering problem and hence the state $\boldsymbol{\theta}$ may be estimated using the usual recursive equations:

$$\hat{\boldsymbol{\theta}}(t) = \hat{\boldsymbol{\theta}}(t-1) + \mathbf{K}_p(t)(y(t) - \mathbf{X}(t-1)\hat{\boldsymbol{\theta}}(t-1))$$

$$\mathbf{K}_p(t) = \mathbf{P}_p(t-1)\mathbf{X}^T(t-1)(Q + \mathbf{X}(t-1)\mathbf{P}_p(t-1)\mathbf{X}^T(t-1))^{-1} \tag{12.41}$$

$$\mathbf{P}_p(t) = \mathbf{P}_p(t-1) + \mathbf{Q}_p - \mathbf{K}_p(t)(Q + \mathbf{X}(t-1)\mathbf{P}_p(t-1)\mathbf{X}^T(t-1))\mathbf{K}_p^T(t).$$

The order in which the calculations must be performed is

$$\mathbf{P}_p(t-1) \rightarrow \mathbf{K}_p(t) \rightarrow \hat{\boldsymbol{\theta}}(t) \rightarrow \mathbf{P}_p(t).$$

If scaled equations are used ($\mathbf{Q}_p = \mathbf{O}$, $Q = 1$), the matrix $\mathbf{P}_p$ is only proportional to the parameter estimation error covariance. This method of estimating the parameters is equivalent to recursive least squares where the following cost function is minimized:

$$V(t) = \sum_{s=1}^{t} \| y(s) - \hat{y}(s) \|^2$$

$$= \sum_{s=1}^{t} \| y(s) - \mathbf{X}(s-1)\boldsymbol{\theta}(s) \|^2$$

with respect to $\boldsymbol{\theta}$. The advantage of the least-squares method is that it is simple but it suffers from the disadvantage that the estimates can be biased if $C(z^{-1}) \neq 1$, namely if $C(z^{-1})$ includes delay elements. To obtain unbiased estimates the noise term $C(z^{-1})\xi(t)$ in the output equation must be statistically independent of the elements in the matrix of regressors. Clearly, this condition may not be true if $C(z^{-1}) \neq 1$.

12.3.4. Explicit self-tuning control and numerical aspects of estimation

A problem which might occur upon implementing the recursive least squares algorithm is that of numerical ill-conditioning where the parameter estimates diverge and instability ensues. One solution is to replace the updating of $\mathbf{P}_p(t)$ by a square-root algorithm. Here, an upper triangular matrix $\mathbf{S}(t)$ is defined as the square root of $\mathbf{P}_p(t)$ via:

$$\mathbf{S}(t)\mathbf{S}^T(t) = \mathbf{P}_p(t).$$

Briefly, the square-root filtering algorithm has advantages (Anderson and Moore [35]) (a) the product $\mathbf{S}(t)\mathbf{S}^T(t) \geq \mathbf{O}$ and the calculation of $\mathbf{S}(t)$ cannot lead to a $\mathbf{P}_p(t)$ matrix which fails to be non-negative definite as a result of computational errors in the covariance update equations and (b) the numerical conditioning of $\mathbf{S}(t)$ is generally much better than that of $\mathbf{P}_p(t)$.

If $\mathbf{Q}_p$ is taken to be zero, the gain and $\mathbf{P}_p$ matrix can converge to null matrices. The estimates will not then be updated and the parameter estimates can diverge. The problem is more severe in systems with slowly varying parameters since the changing plant model will not be tracked by the identification algorithm, once the gain has reached zero. *Exponential forgetting* of past data is used in such situations so that the parameters depend most strongly on the recent observations. The use of an exponential-forgetting factor prevents the gain from tending to zero to ensure the parameters are continuously updated. The exponential weighting of past data is achieved by artificially inflating the *a priori* covariance $\mathbf{P}_0(t)$ at each instant. This matrix is divided by the exponential forgetting factor β where β is slightly less than unity (typically $\beta \epsilon [0.95, 0.995]$). The above algorithm for $\mathbf{Q}_p = \mathbf{O}$ then becomes:

$$\mathbf{K}_p(t) = \mathbf{P}_p(t-1)\mathbf{X}^T(t-1)(\beta + \mathbf{X}(t-1)\mathbf{P}_p(t-1)\mathbf{X}^T(t-1))^{-1}$$

$$\mathbf{P}_p(t) = [\mathbf{P}_p(t-1) - \mathbf{K}_p(t)(\beta + \mathbf{X}(t-1)\mathbf{P}_p(t-1)\mathbf{X}^T(t-1))\mathbf{K}_p^T(t)]/\beta$$

where $0.9 < \beta < 0.995$.

The algorithm may be written in an alternative form which is derived using the Householder rule. Assume $\mathbf{Q}_p = \mathbf{O}$, then

$$\mathbf{P}_p(t)^{-1} = \beta \mathbf{P}_p(t-1)^{-1} + \mathbf{X}^T(t-1)\mathbf{X}(t-1).$$

Since

$$\mathbf{P}_p(t)\mathbf{X}^T(t-1) = \mathbf{P}_p(t-1)\mathbf{X}^T(t-1)/(\beta + \mathbf{X}(t-1)\mathbf{P}_p(t-1)\mathbf{X}^T(t-1)),$$

the gain matrix may be expressed as:

$$\mathbf{K}_p(t) = \mathbf{P}_p(t)\mathbf{X}^T(t-1).$$

Let

$$\mathbf{S}(t-1) \triangleq \mathbf{P}_p(t)^{-1}$$

then the algorithm becomes:

$$\hat{\boldsymbol{\theta}}(t) = \hat{\boldsymbol{\theta}}(t-1) + \mathbf{S}(t-1)^{-1}\mathbf{X}^T(t-1)(y(t) - \mathbf{X}(t-1)\hat{\boldsymbol{\theta}}(t-1))$$

$$\mathbf{S}(t) = \beta \mathbf{S}(t-1) + \mathbf{X}^T(t)\mathbf{X}(t).$$

The order in which the calculations must be performed is

$$\mathbf{S}(t-1) \to \mathbf{S}(t-1)^{-1} \to \hat{\boldsymbol{\theta}}(t) \to \mathbf{S}(t).$$

The repeated calculation of a matrix inverse is the main disadvantage of this method.

If a plant parameter is to be fixed at an initial value, this can be achieved by choosing the initial covariance $\mathbf{P}_0 = \text{diag}\{p_{ii}(0)\}$ where the jth diagonal element corresponding to this parameter is set to zero ($p_{jj}(0) = 0$). It may easily be verified that the parameter estimate is not updated since the jth row and column in $\mathbf{P}_p(i)$ will be zero for all $i > 0$.

The use of a value of the exponential-forgetting factor $\beta < 1$ will ensure that the estimator $\mathbf{P}_p(t)$ matrix and gain $\mathbf{K}_p(t)$ will not tend to zero. The algorithm will therefore remain alert to track changing dynamics. Unfortunately, a small value of β will also result in greater sensitivity of the parameter estimates to the noise level. A small value of β therefore provides a faster tracking capability to parameter variations but the estimates will wander more with the noise signal.

Estimator windup

The ability of an adaptive controller to track variations in the process dynamics is the novel feature which fixed controllers cannot provide. This tracking action requires that old data be discounted by, for example, the use of exponential forgetting. A compromise must, however, be achieved (Wittenmark and Åström [51]). If fast discounting is used and the plant parameters are constant, the estimates will be uncertain. If the discounting is too slow, it will be impossible to track fast parameter variations.

Exponential forgetting works well when the process is continuously excited but problems can arise when the excitation changes or decays to zero. If there is no excitation for a long period the estimator will forget the actual parameter values and the uncertainty grows. The phenomenon is called *estimator windup* which can cause a burst in the control action. One method of avoiding windup and bursts is to provide an extra input excitation signal. To illustrate the mechanism of 'estimator windup' assume that $\mathbf{X}(\cdot) \to \mathbf{o}^T$, then if $\beta < 1$, $\mathbf{S}(\cdot) \to \mathbf{O}$ and $\mathbf{P}_p(\cdot)$ grows exponentially.

Explicit or indirect self-tuning control schemes

Both explicit and implicit schemes have their own particular merits but it is relatively straightforward to define an explicit self-tuning scheme (Fuchs [36]) depending as it does on separate parameter estimation and control routines.

Example 12.5. *Explicit LQG self-tuning regulator.* To illustrate the explicit self-tuning philosophy, consider the system described in Example 12.1 and the expression for the LQG controller $C_0(z^{-1})$. The explicit algorithm involves:

(a) Plant identification using an extended least-squares algorithm.
(b) Calculation of the controller $C_0(z^{-1})$.

The plant parameter estimates are shown in Fig. 12.5. These estimates have converged to the known values $a_1 = -0.95$, $b_1 = 2$ and $d_1 = -0.72$ after about 300 seconds (process noise $Q \triangleq 10$ and mesurement noise $R \triangleq 1$). The controller was used in a fixed spectrum mode, that is on-line spectral factorization was not used but instead the polynomial D_1 was chosen *a priori* to give good close-loop pole positions. For this example, D_1 was fixed at the spectral factor which resulted from weighting elements of

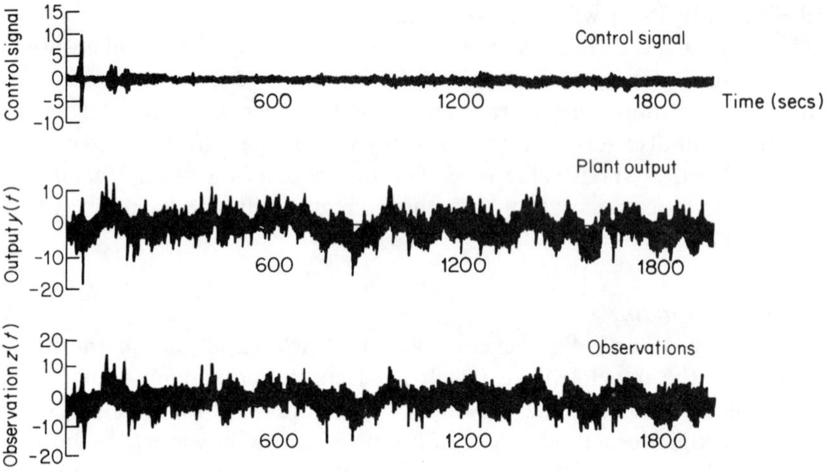

Fig. 12.4. Fixed spectrum explicit LQG self-tuning regulator

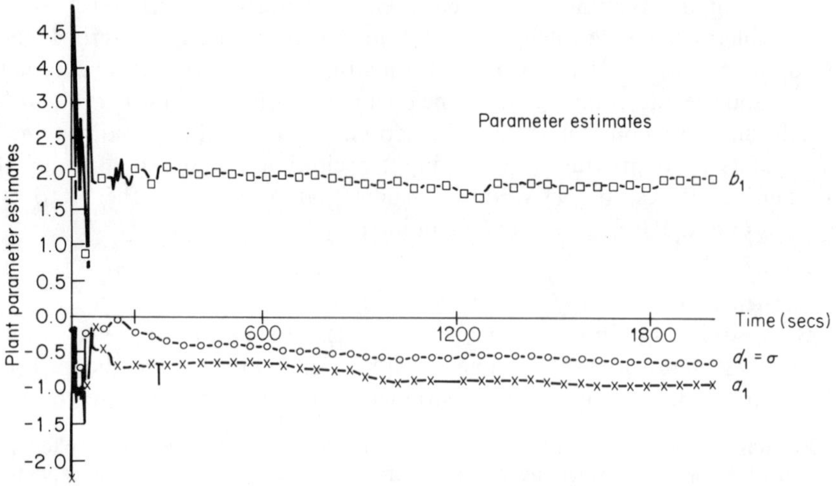

Fig. 12.5. Fixed spectrum explicit LQG self-tuning regulator

$Q_1 = 1$ and $R_1 = 0.01$. In practice, D_1 cannot be selected in this manner since the plant polynomials which determine D_1 in (12.17) are unknown at the start of the identification procedure. However, it is possible to use estimated values to calclate D_1. The resulting control (u), plant output (y) and observations (z) signals are shown in Figure 12.14.

The fixed spectrum or pole asignment self-tuning scheme can also be implemented using the generalized minimum variance controller as described by Allidina and Hughes [37].

Example 12.6. *Explicit WMV self-tuning regulator.* Consider again the system described in Example 12.1 but using the WMV controller (12.32). The performance criterion to be minimized is given by (12.31) where $P_0 = 1/B_2$ and $P_1 = P_2 = 0$. The process noise covariance $Q = 1$ and the measurement was set to zero. The system responses are shown in Fig. 12.6.

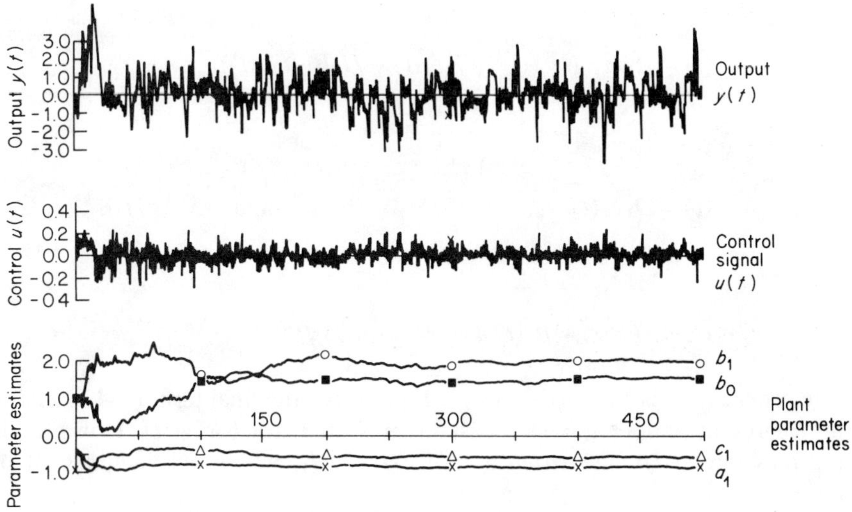

Fig. 12.6. Explicit weighted minimum variance regulator

The two types of controller will give similar results if the closed-loop poles are arranged to be the same. The differences are, in fact, more apparent in the multivariable case. For example, the LQG controller is complicated to implement in self-tuning form but it does have better stability characteristics.

The explicit identification approach involves separate estimation and control and thus further details are unnecessary. The implicit schemes enable the controller parameters to be estimated directly and there is therefore the possibility that less calculations will be required. The implicit or direct types of algorithm are considered next.

12.4. IMPLICIT SELF-TUNING CONTROL

The conceptually compact approach of an implicit self-tuning scheme creates its own special difficulties. It does not (at present) seem possible to derive an implicit self-tuner based on the LQG control laws, except for a special class of systems (Grimble [54]). The implicit self-tuner derived in this section is based upon the observations weighted optimal control law.

12.4.1. Observations weighted controller

To derive an implicit self-tuning controller an expression is required which involves the controller parameters to be estimated. This is normally in the form of a computable equivalent plant output which involves the sum of noise, control and output dependent terms. Such an expression is obtained below.

From the results in Section 12.2 and in particular equation (12.27), obtain:

$$D_1 e_1(t) = A_0 H_1 \varepsilon(t) + \frac{B_2}{D}(G_1 e_1(t) - H_1 B_1 A_0 u(t)).$$

Define

$$y_1(t) = D_1 e_1(t)/A_0, \qquad e_1'(t) = e_1(t)/A_0 \quad \text{and} \quad N_1 = H_l B_1,$$

thus

$$y_1(t) = H_1 \varepsilon(t) + B_2(G_1 e_1'(t) - N_1 u(t))/D. \tag{12.42}$$

The order of H_1 is $n_{h_1} = n_2 - 1 \geq k - 1$ and thus, the final term in (12.42) does not represent (unless $n_2 = 1$) a least-squares predictor for $y_1(t)$. However, let the polynomial H_1 be written as $H_1 = H_{11} + H_{12}$ where H_{11} includes all the terms in $z^{-1}, z^{-2}, \ldots, z^{-(k-1)}$:

$$\hat{y}_1(t \mid t-k) = (G_1 e_b(t) - N_1 u_b(t))/D + H_{12}\varepsilon(t)$$

where

$$u_b \triangleq B_2 u \quad \text{and} \quad e_b \triangleq B_2 e_1/A_0.$$

Noting that the optimal control sets the term $(\cdot)/D$ to zero and assuming without loss of generality that $D(0) = 1$ obtains,

$$\hat{y}_1(t \mid t-k) = G_1 e_b(t) - N_1 u_b(t) + H_{12}\varepsilon(t) \tag{12.43}$$

The signal $y_1(t) = D_1 e_1'(t)$ can be calculated assuming that D_1 is fixed *a priori* (corresponding to fixed closed-loop poles). The above equation for $\hat{y}_1(t \mid t-k)$ may now be used to identify the unknown controller polynomials

G_1 and N_1 by setting:

$$\hat{y}_1(t \mid t-k) = \mathbf{X}(t-k)\boldsymbol{\theta}$$

and by letting $\mathbf{X}(t)$ represent the matrix of regressors and $\boldsymbol{\theta}$ the vector of unknown controller parameters. To avoid complicating the present argument, assume that past values of ε and that the B polynomial, and hence B_1, B_2 are obtained via a separate identification test on the plant. Let

$$\mathbf{X}(t-k) = [\,e_b(t),\, e_b(t-1), \ldots;\ -u_b(t),$$

$$-u_b(t-1), \ldots;\ \varepsilon(t-k),\ \varepsilon(t-k-1), \ldots]$$

$$\boldsymbol{\theta} = [\,g_0, g_1, \ldots;\ n_0, n_1, \ldots;\ h_{1k}, h_{1k+1}, \ldots\,]^T. \tag{12.44}$$

If the matrix $\mathbf{X}(t-k)$ included only quantities which were known at time t and if the prediction error $\tilde{y}_1(t \mid t-k) = H_{11}\varepsilon(t)$ was uncorrelated with $\hat{y}_1(t \mid t-k)$ then the estimation of $\boldsymbol{\theta}$ could be via least-squares as described in the previous section. However, the plant parameters are unknown and thus the quantities in $\mathbf{X}(t-k)$ cannot be calculated exactly. These quantities are therefore calculated, by analogy with extended least-squares, using estimated values from the previous identification.

The optimal control signal can be found using the estimated polynomials $\hat{G}_1$ and $\hat{N}_1$, viz.:

$$u^0(t) = \hat{G}_1 e_1^{\prime}(t)/\hat{N}_1. \tag{12.45}$$

The second plant identification (to obtain the B polynomial) can be avoided using a bilinear identification technique developed by Åström [38], although this particular technique is still a topic of current research.

Example 12.7. *Observations weighted implicit self-tuner.* The system described in Example 12.1 is again considered with $R = Q = Q_1 = 1$ and $R_1 = 0.01$. The observations weighted controller can be calculated from the given plant parameters using:

$$C_0(z^{-1}) = G_1(z^{-1})/H_1(z^{-1})$$

where G_1 and H_1 satisfy $AH_1 + B_2G_1 = D_1D$ and $n_h \triangleq n_2 - 1$ and

$$n_g \triangleq \max(n_a - 1,\ n_{d_1} + n_d - n_2).$$

Then,

$$G_1(z^{-1}) = -0.217$$

$$H_1(z^{-1}) = -2.9 - 1.897z^{-1} - 0.456z^{-2}.$$

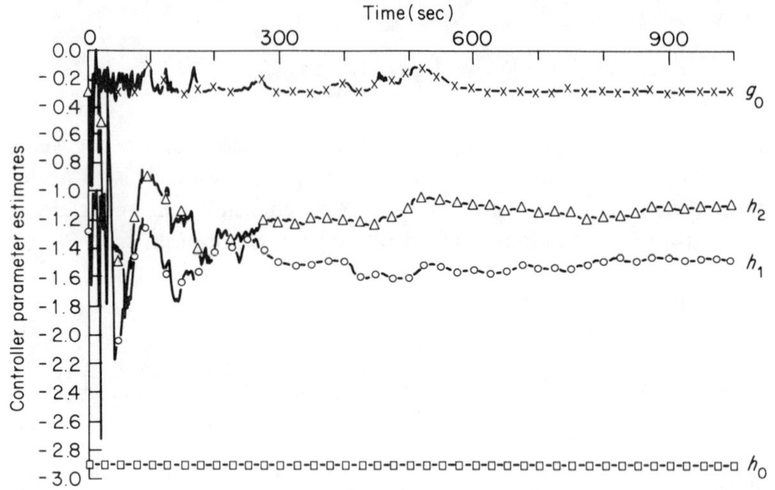

Fig. 12.7. Estimates of the controller parameters

The estimates of the controller parameters and the plant parameters are shown in Figs 12.7 and 12.8, respectively. The plant parameter $\alpha = 1.266$ and one of the controller parameters (in this case h_0) may be held constant without loss of generality. The estimates of the plant parameters, including $d_1 = \sigma = 1.1414$, are converging relatively slowly but the controller parameter estimates converge after 300 seconds. The system responses (shown in Fig. 12.9) are, however, good from almost time zero. This behaviour is typical of self-tuning systems.

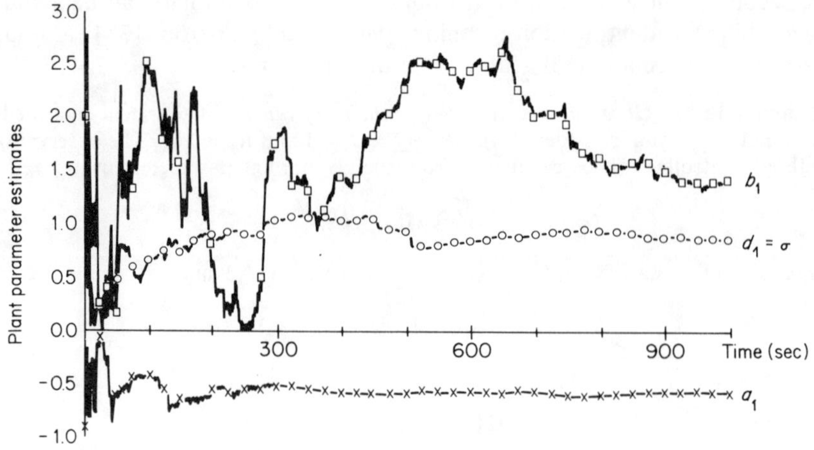

Fig. 12.8. Parameter estimates of the plant model

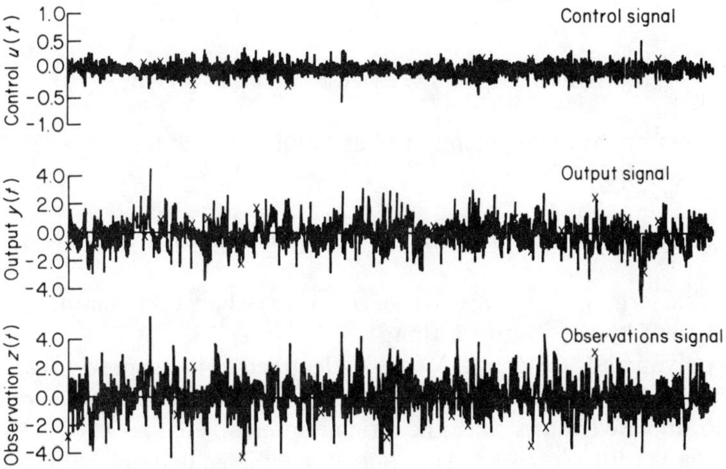

Fig. 12.9. Control, output and observations signal

12.4.2. Weighted minimum-variance controller

An implicit self-tuning controller based upon the weighted minimum-variance controller may be derived using the results in Section 12.2.5. From (12.34):

$$\phi_1(t+k) = -\frac{H_1}{B_2}\,\varepsilon(t) - \frac{1}{DA_0}\,(G_1 e_1(t) - H_1 B_1 A_0 u(t))$$
$$- P_1 w(t) + P_2 u(t). \tag{12.46}$$

Let

$$y_1(t) = B_2\phi_1(t+k) = -D_1 e_1'(t) - B_2(P_1 w(t) - P_2 u(t))$$

then

$$y_1(t) = -H_1\varepsilon(t) - \frac{B_2}{D}\,(G_1 e_1'(t) + DP_1 w(t)$$
$$-(H_1 B_1 + P_2 D)u(t))$$

and

$$\hat{y}_1(t \mid t-k) = -\frac{B_2}{D}\,(G_1 e_1'(t) + Dw_r(t) - Nu(t)) + H_{12}\varepsilon(t)$$

where

$$e_1'(t) = e_1(t)/A_0, \; w_r(t) = P_1 w(t) \quad \text{and} \quad N = H_1B_1 + P_2D.$$

Multiplying by the D polynomial and assuming without loss of generality that $D(0) = 1$ obtain:

$$\hat{y}_1(t \mid t-k) = -B_2(G_1 e_1'(t) + Dw_r(t) - Nu(t)) + DH_{12}\varepsilon(t) \quad (12.47)$$

where the term $(-\sum_{i=1}^{n_d} d_i \hat{y}_1 \; (t-i \mid t-k-i))$ has been omitted, since it is set to zero by the control action.

The polynomials G_1, D and N can be identified as in the previous section by using on-line parameter estimation techniques. The parameters should converge to unbiased values since the estimation error $\tilde{y}_1(t \mid t-k) = -H_{11}\varepsilon(t)$ is uncorrelated with $y(t \mid t-k)$. The control is obtained in terms of the estimated polynomials as:

$$u^0(t) = (\hat{G}_1 e_1' + \hat{D}w_r)/\hat{N}. \quad (12.48)$$

Further details of the weighted minimum-variance self-tuning controller are presented in Grimble [23].

12.4.3. PID self-tuning control

The proportional, integral, derivative (PID) self-tuning controller described below is based upon the philosophy proposed by Cameron and Seborg [52]. A PID controller has a digital realization:

$$u(t) = K_c\left(1 + \frac{T_s}{T_i}\frac{1}{1-z^{-1}} + \frac{T_d}{T_s}(1-z^{-1})\right)e(t)$$

or if

$$\nabla u(t) \triangleq u(t) - u(t-1),$$

the velocity form becomes:

$$\nabla u(t) = K_c(e(t) - e(t-1) + \frac{T_s}{T_i}e(t) + \frac{T_d}{T_s}(e(t) - 2e(t-1) + e(t-2)))$$

where T_s is the sampling interval, K_c the proportional gain, T_i the integral and T_d the derivative controller settings. The error signal:

$$e(t) \triangleq w(t) - y(t)$$

and $w(t)$ is a constant set point signal. Assume that the output $y(t)$ is filtered then the velocity form of the controller is given as:

$$\nabla u(t) = K_c\left[- y_f(t) + y_f(t-1) + \frac{T_s}{T_i}(w(t) - y_f(t)) - \frac{T_d}{T_s}(y_f(t) - 2y_f(t-1) + y_f(t-2))\right].$$

Rearranging gives:

$$\nabla u(t) = \frac{K_c T_s}{T_i} w(t) - K_c\left(1 + \frac{T_s}{T_i} + \frac{T_d}{T_s}\right) y_f(t) + K_c\left(1 + \frac{2T_d}{T_s}\right) y_f(t-1) - K_c \frac{T_d}{T_s} y_f(t-2).$$

Let the filtered output $y_f = y/A_0$ where A_0 is a polynomial defined from (12.31).

Consider now an application of the Generalized Minimum-Variance control law (Corollary 12.5) where

$$u(t) = \frac{DP_1 w(t) - G_1 y_f(t)}{(H_1 B + P_2 D)}.$$

Take $C = 1$ and $r \equiv 0$ (then $D = 1$) and define $E = H_1 B$. The optimal control signal is then given as:

$$u(t) = \frac{P_1 w(t) - G_1 y_f(t)}{(E + P_2)}.$$

To ensure that the Generalized Minimum-Variance (GMV) control law has a PID structure, let the degree of G_1 be $n_{g_1} = 2$. The output filter A_0 might be of degree 1 or 0. Assume that $n_{g_1} = n_{a_1} - 1$ in (12.37) then $n_{a_1} = 3$ and the process n_a is second- or third-order depending upon the filter choice (denote $A_0 = 1 + a_{01}z^{-1}$).

To introduce integral action let,

$$E + P_2 = (1 - z^{-1})/\nu$$

where ν is a design parameter. In the steady-state the output $y(t)$ should equal

$w(t)$, thus choose,

$$P_1 = P_{10} = \left(\frac{G_1}{A_0}\right)_{z=1} = \sum_{i=1}^{2} g_i/(1 + a_{01})$$

where $G_1 = g_0 + g_1 z^{-1} + g_2 z^{-2}$. The final value theorem of z-transforms then gives:

$$u(\infty) = \lim_{z\to 1} (1 - z^{-1}) \frac{(P_1 w(t) - G_1 y_f(t))}{E + P_2} = 0.$$

By this approach, offset is eliminated due to both set point and load changes.

The expression for the GMV controller, under these assumptions, becomes:

$$u(t) = \frac{(P_{10} w(t) - G_1 y(t)/(1 + a_{01} z^{-1}))\nu}{(1 - z^{-1})}$$

or

$$\nabla u(t) = \nu(P_{10} w(t) - (g_0 + g_1 z^{-1} + g_2 z^{-2}) y_f(t))$$

where $-1 < a_{01} \leq 0$. Comparison with the above expression for the velocity form of the PID controller yields:

$$K_c T_s / T_i = \nu(g_0 + g_1 + g_2)/(1 + a_{01})$$

$$K_c(1 + T_s/T_i + T_d/T_s) = \nu g_0$$

$$K_c(1 + 2T_d/T_s) = -\nu g_1$$

$$K_c T_d / T_s = \nu g_2.$$

Let $\alpha \triangleq 1 + a_{01}$ then

$$K_c = -\nu(g_1 + 2g_2)$$

$$T_i = \frac{T_s K_c \alpha}{\nu(g_0 + g_1 + g_2)}$$

$$T_d = T_s \nu g_2 / K_c$$

but

$$\nu(g_0 + g_1 + g_2) = \nu g_0 - K_c(1 + 2T_d/T_s) + K_c T_d / T_s$$

$$= \nu g_0 - K_c - K_c T_d / T_s$$

and

$$T_i = \frac{T_s K_c \alpha}{\nu g_0 - K_c - K_c T_d / T_s}.$$

The proportional gain and the integral and derivative controller setting are therefore completely determined by the GMV control law. Notice that the control weighting P_2 cannot be specified independently as in the usual GMV problem but it is constrained to satisfy:

$$P_2 = ((1 - z^{-1})/\nu) - H_1 B.$$

This polynomial need not be calculated separately in the self-tuning algorithm.

The scalar ν acts as a gain parameter. Large values of ν result in more vigorous control action and underdamped responses, where small values of ν give sluggish control and overdamped responses. The loop gain falls to zero as $\nu \to 0$.

The self-tuning algorithm is derived in the same manner as the previous WMV self-tuner. The only difference lies in the fact that the control-weighting term P_2 cannot be chosen freely. The value of ν which is selected can be based upon an assumed model for the plant and this determines P_2.

The system can be run as a PID self-tuning controller or the device can be used to find appropriate values of K_c, T_i, and T_d and these can then be fixed. Thus, the self-tuning system can be used occasionally to retune fixed PID controller settings or it can be used continuously to maintain an 'optimum' output. This subject is not pursued further here, however, some points which might be considered are as follows.

(i) The assumptions regarding the plant model are rather restrictive and most plants will not fall in this category, however, it may still be worth applying the PID strategy relying upon its inherent robustness.

(ii) The benefits of using a PID controller arise from its acceptance in industry rather than from its engineering advantages. Perhaps more effort should be devoted to popularizing the modern advanced control solutions.

(iii) The link between the GMV and PID control laws, established above, is rather artificial and the merits of the approach must be determined with applications experience.

12.5. MULTIVARIABLE SELF-TUNING CONTROL SYSTEMS

The multivariable self-tuning control problem is significantly more complicated than its scalar equivalent. For this reason, the simplest of the control strategies, the minimum-variance and the generalized minimum-variance control laws will be discussed first. Borisson [8] is generally credited with the first multivariable regulator. A brief description of this work opens this section on multivariable self-tuning.

A multivariable self-tuning controller is also defined based upon the generalized minimum-variance control law. The problem considered is an extension of that described by Koivo [9] who based his self-tuner almost directly on the Clarke and Gawthrop self-tuning controller for scalar systems. Finally, a multivariable form of the observations weighted self-tuning controller is proposed.

Some remarks on incorporating pole assignment into the self-tuning philosophy close the section.

12.5.1. Multivariable minimum-variance regulators

Consider the following process model which is assumed to be minimum phase:

$$\mathbf{A}(z^{-1})y(t) = \mathbf{B}(z^{-1})u(t-k) + \mathbf{C}(z^{-1})\xi(t). \qquad (12.49)$$

The polynomial matrices are all assumed to be m-square and det $\mathbf{B}(z^{-1})$ and det $\mathbf{C}(z^{-1})$ have all their zeros outside the unit disc. The delay k is the largest common delay term in the $\mathbf{B}(z^{-1})\ z^{-k}$ matrix coefficients. The disturbance $\{\xi(t)\}$ is a sequence of independent, equally distributed random vectors with zero-mean value and covariance $E\{\xi(t)\xi^T(t)\} = \mathbf{Q}$. The plant polynomials are assumed to be of the form:

$$\mathbf{A}(z^{-1}) = \mathbf{I}_m + \mathbf{A}_1 z^{-1} + \cdots + \mathbf{A}_{n_a} z^{-n_a}$$

$$\mathbf{B}(z^{-1}) = \mathbf{B}_0 + \mathbf{B}_1 z^{-1} + \cdots + \mathbf{B}_{n_b} z^{-n_b}$$

$$\mathbf{C}(z^{-1}) = \mathbf{I}_m + \mathbf{C}_1 z^{-1} + \cdots + C_{n_c} z^{-n_c}.$$

The assumption will now be made that $\mathbf{B}_0$ is non-singular. The polynomial matrices **A, B** and **A, C** comprising the matrix fraction descriptions $\mathbf{A}^{-1}\mathbf{B}$ and $\mathbf{A}^{-1}\mathbf{C}$ are assumed to be relatively left prime.

The following matrix Diophantine equation is required in the analysis for the minimum variance controller:

$$\mathbf{A}(z^{-1})\mathbf{H}(z^{-1}) + z^{-k}\mathbf{G}(z^{-1}) = \mathbf{C}(z^{-1}) \qquad (12.50)$$

where

$$\mathbf{H}(z^{-1}) = \mathbf{I}_m + \mathbf{H}_1 z^{-1} + \cdots + \mathbf{H}_{n_h} \ldots z^{-n_h}$$

$$\mathbf{G}(z^{-1}) = \mathbf{G}_0 + \mathbf{G}_1 z^{-1} + \cdots + \cdots + \mathbf{G}_{n_g} z^{-n_g}$$

and $n_h = k - 1$, $n_g = n_a - 1$. Since $\mathbf{A}(0)$ is non-singular the solutions for $\mathbf{H}$ and $\mathbf{G}$ are unique. This follows by equating terms of the same degree:

$$\begin{aligned} \mathbf{H}_1 + \mathbf{A}_1 &= \mathbf{C}_1 \\ \mathbf{H}_2 + \mathbf{A}_1\mathbf{H}_1 + \mathbf{A}_2 &= \mathbf{C}_2 \\ &\vdots \\ \mathbf{H}_{n_h} + \mathbf{A}_1\mathbf{H}_{n_h-1} + \cdots + \mathbf{A}_{n_h-1}\mathbf{H}_1 + \mathbf{A}_{n_h} &= \mathbf{C}_{n_h} \end{aligned}$$

and

$$\mathbf{G}_i = -\sum_{j=1}^{k-1} \mathbf{A}_{i+k-j}\mathbf{H}_j + \mathbf{C}_{i+k}$$

for $i = 0, 1, \ldots, n_g$. Whenever $\mathbf{A}_{n+1}$, $\mathbf{A}_{n+2} \ldots$ appear in the above, these are assumed to be zero. Now introduce the following relationship:

$$\tilde{\mathbf{H}}(z^{-1})\mathbf{G}(z^{-1}) = \tilde{\mathbf{G}}(z^{-1})\mathbf{H}(z^{-1}) \tag{12.51}$$

where

$$\det \tilde{\mathbf{H}}(z^{-1}) = \det \mathbf{H}(z^{-1}), \qquad \tilde{\mathbf{H}}(0) = \mathbf{I}_m \quad \text{and} \quad n_{\tilde{h}} = n_h,\ n_{\tilde{g}} = n_g.$$

The polynomial matrices $\tilde{\mathbf{H}}(z^{-1})$ and $\tilde{\mathbf{G}}(z^{-1})$ are not unique but always exist. Define also the polynomial matrix $\tilde{\mathbf{C}}$:

$$\tilde{\mathbf{C}}(z^{-1}) \triangleq \tilde{\mathbf{H}}(z^{-1})\mathbf{A}(z^{-1}) + z^{-k}\tilde{\mathbf{G}}(z). \tag{12.52}$$

From this equation and the relationships $\tilde{\mathbf{H}}\mathbf{G} = \tilde{\mathbf{G}}\mathbf{H}$ and $\mathbf{C} = \mathbf{A}\mathbf{H} + z^{-k}\mathbf{G}$ it follows that

$$\tilde{\mathbf{H}}(z^{-1})\mathbf{C}(z^{-1}) = \tilde{\mathbf{C}}(z^{-1})\mathbf{H}(z^{-1}) \tag{12.53}$$

and

$$\det \mathbf{C}(z^{-1}) = \det \tilde{\mathbf{C}}(z^{-1}).$$

From the process equation (12.49) obtain:

$$\tilde{\mathbf{H}}\mathbf{A}y(t+k) = \tilde{\mathbf{H}}\mathbf{B}u(t) + \tilde{\mathbf{H}}\mathbf{C}\xi(t+k)$$

or

$$\tilde{\mathbf{C}}y(t+k) - \tilde{\mathbf{G}}y(t) = \tilde{\mathbf{H}}\mathbf{B}u(t) + \tilde{\mathbf{C}}\mathbf{H}\xi(t+k)$$

hence

$$y(t+k) = \mathbf{H}\xi(t+k) + \tilde{\mathbf{C}}^{-1}(\tilde{\mathbf{G}}y(t) + \tilde{\mathbf{H}}\mathbf{B}u(t)).$$

Using arguments similar to those used in Section (12.4.1) let

$$\hat{y}(t+k \mid t) \triangleq \tilde{\mathbf{C}}^{-1}(\tilde{\mathbf{G}}y(t) + \tilde{\mathbf{H}}\mathbf{B}u(t))$$

and

$$\tilde{y}(t+k \mid t) \triangleq \mathbf{H}\xi(t+k) \tag{12.54}$$

and note that the control to minimize:

$$J = E\{y(t+k)^T\mathbf{Q}_1 y(t+k)\} \tag{12.55}$$

follows by setting $\hat{y}(t+k \mid t) = \mathbf{o}$. That is,

$$\tilde{\mathbf{H}}\mathbf{B}u^0(t) + \tilde{\mathbf{G}}y^0(t) = \mathbf{o}.$$

The optimal controller is independent of the noise covariance $\mathbf{Q}$ and the cost-function weighting matrix $\mathbf{Q}_1$. This is a consequence of the very simple control philosophy employed which results in a poor performance for some systems. Note that the control law is physically realizable (that is, it does not involve knowledge of future plant outputs) since $\tilde{\mathbf{G}}$ is a polynomial matrix in z^{-1} and $\tilde{\mathbf{H}}(0)$ and $\mathbf{B}(0)$ are both full rank. This latter condition ensures that there are no terms in positive powers of z in the matrix series expansions of $\tilde{\mathbf{H}}(z^{-1})^{-1}$ and $\mathbf{B}(z^{-1})^{-1}$ (see Problem 12.7.1).

12.5.2. Multivariable minimum-variance self-tuning regulators

The explicit self-tuning philosophy is readily extended to the multivariable case, thus in the following the implicit self-tuner is considered exclusively.

From the previous discussion $n_h = k-1$ and hence $\tilde{y}(t+k \mid t)$ is uncorrelated with $\hat{y}(t+k \mid t)$. Recall $\tilde{\mathbf{C}}(0) = \mathbf{I}_m$ and let $\tilde{\mathbf{N}} = \tilde{\mathbf{H}}\mathbf{B}$ then following the arguments of Section 12.4.2.

$$y(t+k) = \tilde{y}(t+k \mid t) + \hat{y}(t+k \mid t) \tag{12.56}$$

where

$$\hat{y}(t+k \mid t) \triangleq \tilde{\mathbf{C}}^{-1}(\tilde{\mathbf{G}}y(t) + \tilde{\mathbf{N}}u(t))$$

or

$$\hat{y}(t+k \mid t) = \tilde{\mathbf{G}}y(t) + \tilde{\mathbf{N}}u(t) + (\mathbf{I}_m - \tilde{\mathbf{C}})\hat{y}(t+k \mid t). \tag{12.57}$$

The final term may be neglected (due to the choice of u^0) and the matrix of regressors becomes:

$$\mathbf{X}(t) = [y^T(t), y^T(t-1), \ldots ; u^T(t), u^T(t-1), \ldots]. \tag{12.58}$$

The matrix of unknown controller parameters may be identified as:

$$\mathbf{\Theta} = [\mathbf{\theta}_1, \mathbf{\theta}_2, \ldots, \mathbf{\theta}_m] = [\tilde{\mathbf{G}}_0, \tilde{\mathbf{G}}_1, \ldots ; \tilde{\mathbf{N}}_0, \tilde{\mathbf{N}}_1, \ldots]^{\mathrm{T}}.$$

Each column of $\mathbf{\Theta}$ corresponds to a collection of the rows of $\tilde{\mathbf{G}}_i$ where $\tilde{\mathbf{G}}_i$ is the matrix coefficient of z^{-i} in the polynomial matrix $\tilde{\mathbf{G}}(z^{-1})$ and similarly for $\tilde{\mathbf{N}}(z^{-1})$, that is:

$$\mathbf{\Theta}_i = [g_{i1}^0, \ldots, g_{im}^0;\ g_{i1}^1, \ldots, g_{im}^1; \ldots ;\ n_{i1}^0, \ldots, n_{im}^0;\ n_{i1}^1, \ldots, n_{im}^1; \ldots]^T$$

Each row of (12.56) can be written in the form:

$$y_i(t) = \tilde{y}_i(t \mid t-k) + \mathbf{X}(t-k)\mathbf{\Theta}_i. \tag{12.59}$$

These equations are in the form of an implicit model with the controller parameters representing the unknown system coefficients. These coefficients may be identified and the control signal can be evaluated at each sample instant.

The controller parameters may be estimated using the recursive least squares algorithm:

$$\hat{\boldsymbol{\theta}}_i(t) = \hat{\boldsymbol{\theta}}_i(t-1) + \mathbf{K}_p(t)(y_i(t) - \mathbf{X}(t-k)\hat{\boldsymbol{\theta}}_i(t-1))$$

$$\mathbf{K}_p(t) = \mathbf{P}_p(t-1)\mathbf{X}^T(t-k)(1 + \mathbf{X}(t-k)\mathbf{P}_p(t-1)\mathbf{X}^T(t-k))^{-1}$$

$$\mathbf{P}_p(t) = \{\mathbf{P}_p(t-1) - \mathbf{K}_p(t)(1 + \mathbf{X}(t-k)\mathbf{P}_p(t-1)\mathbf{X}^T(t-k))\mathbf{K}_p^T(t)\}/\beta$$

for $i = 1, 2, \ldots, m$,. the approximation to the optimal control follows using:

$$\hat{\tilde{N}}u^0(t) + \hat{\tilde{G}}y^0(t) = \mathbf{o}. \tag{12.60}$$

The last term in (12.57) was neglected because the optimal control is chosen to set this term to zero. This reduces the number of parameters to be estimated and is a device which is employed on many of the algorithms discussed here.

However, if a non-optimal controller is to be used, this term cannot be neglected. This situation might arise when demonstrating that a self-tuning controller can replace an existing controller (say a fixed PID controller). The estimation algorithm would be used without employing the computed optimal controller within the loop. If the algorithm converged to what appeared to be a reasonable controller then this would give some confidence in the proposed change.

12.5.3. Multivariable generalized minimum variance controllers

The generalized minimum-variance controller is derived below using the performance criterion:

$$J = E\{\boldsymbol{\phi}_1^T(t+k)\boldsymbol{\phi}_1(t+k) \mid t\} \tag{12.61}$$

where the notation $E\{\cdot \mid t\}$ denotes the expectation conditioned upon observations and controls measured up to time t and

$$\boldsymbol{\phi}_1(t+k) \triangleq -\mathbf{P}_0\boldsymbol{e}_1(t) - \mathbf{P}_1\boldsymbol{w}(t) + \mathbf{P}_2\boldsymbol{u}(t)$$

Here, $\mathbf{P}_0$, $\mathbf{P}_1$ and $\mathbf{P}_2$ represent transfer-function matrices and

$$\mathbf{P}_0 \triangleq d_1\mathbf{A}_0^{-1}$$

where d_1 is a specified polynomial factor and $\mathbf{A}_0^{-1}$ is a non-singular stable polynomial matrix. The controller input $\boldsymbol{e}_1(t) \triangleq \boldsymbol{r}(t) - \boldsymbol{z}(t)$ and the system is that of previous sections but the minimum-phase restriction is removed, and $\mathbf{B}_0$ is not assumed non-singular.

Writing the system equation in an innovations form (Section 12.2.3):

$$\mathbf{A}\boldsymbol{e}_1(t+k) = \mathbf{D}\boldsymbol{\varepsilon}(t+k) - \mathbf{B}\boldsymbol{u}(t) \tag{12.62a}$$

where the matrix $\mathbf{D}$ follows from the spectral factorization

$$\mathbf{D}(z^{-1})\mathbf{D}^T(z) = \mathbf{E}(z^{-1})\mathbf{Q}_0\mathbf{E}^T(z) + \mathbf{C}(z^{-1})\mathbf{Q}\mathbf{C}^T(z) + \mathbf{A}(z^{-1})\mathbf{R}\mathbf{A}^T(z). \tag{12.62b}$$

The following matrix Diophantine equation is required:

$$\mathbf{A}_1\mathbf{H} + z^{-k}\mathbf{G} = d_1\mathbf{D} \tag{12.63}$$

where $\mathbf{A}_1 = \mathbf{A}\mathbf{A}_0$, $n_h = k-1$ and $n_g = \max(n_{a_1}-1, n_{d_1}+n_d-k)$. As in Section 12.5.1, define $\tilde{\mathbf{H}}$, $\tilde{\mathbf{G}}$ via $\tilde{\mathbf{H}}\mathbf{G} = \tilde{\mathbf{G}}\mathbf{H}$ where $\det(\tilde{\mathbf{H}}(z^{-1})) = \det(\mathbf{H}(z^{-1}))$ and

$\tilde{\mathbf{H}}(0) = \mathbf{I}_m$ and define $\tilde{\mathbf{D}}$ using:

$$\tilde{\mathbf{D}} = \tilde{\mathbf{H}}\mathbf{A}_1 + z^{-k}\tilde{\mathbf{G}}.$$

It then follows that $\tilde{\mathbf{D}}\mathbf{H} = \tilde{\mathbf{H}}d_1\mathbf{D}$. From the system equation (12.62):

$$d_1\tilde{\mathbf{H}}\mathbf{A}\boldsymbol{e}_1(t+\mathrm{k}) = \tilde{\mathbf{D}}\mathbf{H}\boldsymbol{\varepsilon}(t+k) - d_1\tilde{\mathbf{H}}\mathbf{B}\boldsymbol{u}(t)$$

but

$$\tilde{\mathbf{H}}\mathbf{A} = (\tilde{\mathbf{D}} - z^{-k}\tilde{\mathbf{G}})\mathbf{A}_0^{-1}$$

hence

$$\boldsymbol{\phi}(t+k) \triangleq d_1\mathbf{A}_0^{-1}\boldsymbol{e}_1(t+k)$$

becomes

$$\boldsymbol{\phi}(t+k) = \mathbf{H}\boldsymbol{\varepsilon}(t+k) + \tilde{\mathbf{D}}^{-1}d_1(\tilde{\mathbf{G}}\mathbf{A}_0^{-1}\boldsymbol{e}_1(t) - \tilde{\mathbf{H}}\mathbf{B}\boldsymbol{u}(t)).$$

The least-squares predictor and prediction error may be identified as:

$$\hat{\boldsymbol{\phi}}(t+k \mid t) = \tilde{\mathbf{D}}^{-1}d_1(\tilde{\mathbf{G}}\mathbf{A}_0^{-1}\boldsymbol{e}_1(t) - \tilde{\mathbf{H}}\mathbf{B}\boldsymbol{u}(t))$$

$$\tilde{\boldsymbol{\phi}}(t+k \mid t) = \mathbf{H}\boldsymbol{\varepsilon}(t+k)$$

and

$$\boldsymbol{\phi}_1(t+k) = -\boldsymbol{\phi}(t+k) - \mathbf{P}_1\boldsymbol{w}(t) + \mathbf{P}_2\boldsymbol{u}(t) \qquad (12.64)$$

thus

$$\hat{\boldsymbol{\phi}}_1(t+k \mid t) = -\hat{\boldsymbol{\phi}}(t+k \mid t) - \mathbf{P}_1\boldsymbol{w}(t) + \mathbf{P}_2\boldsymbol{u}(t)$$

and $$\tilde{\boldsymbol{\phi}}_1(t+k \mid t) = -\tilde{\boldsymbol{\phi}}(t+k \mid t).$$

The prediction error $\tilde{\boldsymbol{\phi}}_1$ is uncorrelated with $\hat{\boldsymbol{\phi}}_1$ and the cost-function may be written as:

$$J = E\{\hat{\boldsymbol{\phi}}_1^T(t+k \mid t)\hat{\boldsymbol{\phi}}_1(t+k \mid t) \mid t\}$$
$$+ E\{\tilde{\boldsymbol{\phi}}^T(t+k \mid t)\tilde{\boldsymbol{\phi}}(t+k \mid t) \mid t\}.$$

The cost-function is minimized when $\hat{\boldsymbol{\phi}}_1(t+k \mid t) = \mathbf{o}$ and hence the optimal control satisfies:

$$(d_1\tilde{\mathbf{H}}\mathbf{B} + \tilde{\mathbf{D}}\mathbf{P}_2)u^0(t) - (\tilde{\mathbf{G}}\boldsymbol{\phi}(t) + \tilde{\mathbf{D}}\mathbf{P}_1 w(t)) = \mathbf{o}. \tag{12.65}$$

The characteristic equation for the closed-loop system follows from:

$$\det(\mathbf{I}_m + (d_1\tilde{\mathbf{H}}\mathbf{B} + \tilde{\mathbf{D}}\mathbf{P}_2)^{-1}\tilde{\mathbf{G}} d_1 \mathbf{A}_1^{-1}\mathbf{B}\mathbf{z}^{-k}) = 0$$

or

$$\det((d_1\tilde{\mathbf{H}}\mathbf{B} + \tilde{\mathbf{D}}\mathbf{P}_2)^{-1}\tilde{\mathbf{D}}\mathbf{A}_1^{-1}(d_1\mathbf{B} + \mathbf{A}_1\mathbf{P}_2)) = 0.$$

If the control weighting is large ($\mathbf{P}_2 \to \infty$) the closed-loop poles of the system tend to the zeros of $\tilde{\mathbf{D}}$ and of $\mathbf{A}_1$, $\mathbf{P}_2$. Thus, if the plant is open-loop stable, the closed-loop system can also be guaranteed to be stable in this situation.

12.5.4. Multivariable generalized minimum-variance self-tuning controllers

The generalized system output was obtained as:

$$\boldsymbol{\phi}_1(t+k) = \tilde{\boldsymbol{\phi}}_1(t+k \mid t) + \hat{\boldsymbol{\phi}}_1(t+k \mid t) \tag{12.66}$$

where

$$\hat{\boldsymbol{\phi}}_1(t+k \mid t) = \tilde{\mathbf{D}}^{-1}\{(d_1\tilde{\mathbf{H}}\mathbf{B} + \tilde{\mathbf{D}}\mathbf{P}_2)u(t) - \tilde{\mathbf{G}}\boldsymbol{\phi}(t) - \tilde{\mathbf{D}}\mathbf{P}_1 w(t)\}.$$

Note that $\tilde{\mathbf{H}}(0) = \mathbf{I}_m$ and assuming $\mathbf{D}(0) = \mathbf{A}_0(0) = \mathbf{I}_m$, $d_1(0) = 1$, then

$$\mathbf{H}(0) = \mathbf{D}(0) = \mathbf{I}_m.$$

Thus, defining

$$\tilde{\mathbf{N}} = d_1\tilde{\mathbf{H}}\mathbf{B} + \tilde{\mathbf{D}}\mathbf{P}_2$$

and

$$\tilde{\mathbf{P}} = \tilde{\mathbf{D}}\mathbf{P}_1$$

then

$$\hat{\boldsymbol{\phi}}_1(t+k \mid t) = \tilde{\mathbf{N}}u(t) - \tilde{\mathbf{G}}\hat{\boldsymbol{\phi}}(t) - \tilde{\mathbf{P}}w(t) + (\mathbf{I}_m - \tilde{\mathbf{D}})\hat{\boldsymbol{\phi}}_1(t+k \mid t). \tag{12.67}$$

Once again the final term in the prediction equation may be neglected and the matrix of regressors becomes:

$$\mathbf{X}(t) = [-\boldsymbol{\phi}(t),\ -\boldsymbol{\phi}(t-1), \ldots;\ \boldsymbol{u}(t),\ \boldsymbol{u}(t-1), \ldots;$$

$$-\boldsymbol{w}(t),\ -\boldsymbol{w}(t-1), \ldots].$$

The unknown controller parameters are included in the matrix:

$$\boldsymbol{\Theta} = [\boldsymbol{\theta}_1, \boldsymbol{\theta}_2, \ldots, \boldsymbol{\theta}_m]$$

$$= [\tilde{\mathbf{G}}_0, \tilde{\mathbf{G}}_1, \ldots;\ \tilde{\mathbf{N}}_0, \tilde{\mathbf{N}}_1, \ldots;\ \tilde{\mathbf{P}}_0, \tilde{\mathbf{P}}_1, \ldots]^T.$$

The optimal control can be calculated form the estimated polynomial matrices $\hat{\tilde{G}}$, $\hat{\tilde{N}}$ and $\hat{\tilde{P}}$, that is using

$$\hat{\tilde{N}}u^0(t) - (\hat{\tilde{G}}\phi(t) + \hat{\tilde{P}}w(t)) = \mathbf{o}. \tag{12.68}$$

The assumption that $\mathbf{B}(0)$ is full rank is not necessary in this case. A sufficient condition for the control to be causal is that

$$\tilde{\mathbf{N}}(0) = (\tilde{\mathbf{H}}(0)\mathbf{B}(0) + \tilde{\mathbf{D}}(0)\mathbf{P}_2(0))$$

be full rank which can easily be arranged by choice of $\mathbf{P}_2$. This does, of course, imply a mild restriction on the specification of the performance criterion.

12.5.5. Multivariable observations weighted minimum-variance controllers

The multivariable observations weighted minimum-variance controller is of interest here. The system description follows that of Section 12.5.1 but with these differences:

(a) The delay term is included within the non-minimum phase spectral factor $\mathbf{B}_2$ where $\mathbf{B} = \mathbf{B}_2\mathbf{B}_1$ is defined below.
(b) The polynomial matrix $\mathbf{B}$ is assumed to have full normal rank. Note that this does not imply $\mathbf{B}$ is a regular matrix (a regular polynomial matrix is such that $\det \mathbf{B}_{2n} \neq 0$ where $\mathbf{B}_{2n}$ is the highest-order coefficient matrix in $\mathbf{B}_2$). Consequently, the spectral factors $\mathbf{B}_1$ and $\mathbf{B}_2$ are nonsingular.
(c) The constant term $\mathbf{B}_0$ in the expansion of $\mathbf{B}$ is not necessarily assumed to be full rank. This implies that different transport delay terms may exist in different loops of the multivariable system.

To perform the above factorization, recall from the Smith normal form that any m-square polynomial matrix $\mathbf{B}$ of full rank can be written in the form:

$$\mathbf{B} = \mathbf{E}_1 \operatorname{diag}\{b_1, b_2 \ldots, b_m\}\mathbf{E}_2.$$

The polynomials b_k are the invariant polynomials of $\mathbf{B}$ and satisfy

$$b_k \mid b_{k+1}, \; k = 1, 2, \ldots, \; m-1$$

(this notation means that b_k is a divisor of b_{k+1}). The matrices $\mathbf{E}_1$ and $\mathbf{E}_2$ are unimodular (their determinants are non-zero and independent of z). The $(m \times m)$ factors may be defined as:

$$\mathbf{B}_1 = \operatorname{diag}\{b_{11}, b_{12}, \ldots, b_{1m}\}\mathbf{E}_2$$

$$\mathbf{B}_2 = \mathbf{E}_1 \operatorname{diag}\{b_{21}, b_{22}, \ldots, b_{2m}\}$$

where $b_j = b_{2j}b_{1j}$; b_{1j} is minimum phase and $b_{1j}(0) = 1$ for $j = 1, 2, \ldots m$. It follows that $\mathbf{B}_1(0)$ is full rank.

The performance criterion to be minimized by analogy with the scalar case (Section 12.2.4) becomes:

$$J = E\{\boldsymbol{\phi}^T(t+k)\boldsymbol{\phi}(t+k)\} \tag{12.69}$$

where $\boldsymbol{\phi}(t+k) \triangleq d_1(z^{-1})(\mathbf{A}_0\tilde{\mathbf{B}}_2)^{-1}\boldsymbol{e}_1(t)$ and $\tilde{\mathbf{B}}_2$ is defined below. The following linear Diophantine equation is required in the analysis:

$$\mathbf{A}_1\mathbf{H} + \mathbf{B}_2\mathbf{G} = d_1\mathbf{D} \tag{12.70}$$

where

$$\mathbf{A}_1 = \mathbf{A}\mathbf{A}_0, \; n_h = n_2 - 1 \quad \text{and} \quad n_g = \max(n_{a1} - 1, \; n_{d1} + n_d - n_2).$$

The equation (12.70) has a solution if and only if the greatest common left divisor of $\mathbf{A}_1$, $\mathbf{B}_2$ is a left divisor of $d_1\mathbf{D}$ (Kučera [25]). From the definitions of $\mathbf{A}_1$ and $\mathbf{B}_2$, the greatest common left divisor of $\mathbf{A}_1$, $\mathbf{B}_2$ is unimodular and hence the equation (12.70) has a solution.

To investigate whether the solution to (12.70) is unique let $\mathbf{D}' = d_1\mathbf{D}$, then the Diophantine equation may be written in the form:

$$(\mathbf{I}_m + \mathbf{A}_{11}z^{-1} + \mathbf{A}_{12}z^{-2}\ldots)(\mathbf{I}_m + \mathbf{H}_1z^{-1} + \mathbf{H}_2z^{-2} + \cdots)$$
$$+ z^{-k}(\mathbf{B}_{20} + \mathbf{B}_{21}z^{-1} + \cdots)(\mathbf{G}_0 + \mathbf{G}_1z^{-1} + \cdots) = \mathbf{I}_m + \mathbf{D}_1'z^{-1} + \cdots.$$

This may be expressed as the following matrix equation (where $n_0 \triangleq n_2 - k$ and it is assumed that $n_2 + n_{a_1} - 1 \geq n_{d'}$):

$$
\begin{array}{l}
k-1\left\{\begin{array}{c}\\ \\ \\ \end{array}\right. \\
n_0+1\left\{\begin{array}{c}\\ \\ \\ \end{array}\right. \\
n_{a_1}-1\left\{\begin{array}{c}\\ \\ \\ \\ \end{array}\right.
\end{array}
\overbrace{}^{}
\left[\begin{array}{cccc|cccc}
\multicolumn{4}{c|}{\overbrace{\qquad\qquad}^{n_2-1}} & \multicolumn{4}{c}{\overbrace{\qquad\qquad}^{n_{a_1}}} \\
\mathbf{I} & \mathbf{O} & \cdots & \mathbf{O} & \mathbf{O} & \cdots & \cdots & \mathbf{O} \\
\mathbf{A}_{11} & \mathbf{I} & \cdot & \vdots & \mathbf{O} & \ddots & & \vdots \\
\vdots & & \ddots & \mathbf{O} & \mathbf{B}_{20} & & \ddots & \vdots \\
\vdots & \cdots & \cdots & \mathbf{I} & & \ddots & & \mathbf{O} \\
\hdashline
\vdots & & & \mathbf{A}_{11} & \mathbf{B}_{2n_0} & & \ddots & \mathbf{O} \\
\mathbf{A}n_{a_1} & & & & \mathbf{O} & \ddots & & \mathbf{B}_{20} \\
 & & & & \mathbf{O} & \mathbf{O} & \ddots & \vdots \\
\mathbf{O} & \ddots & & & \vdots & & \ddots & \vdots \\
\vdots & & \ddots & & \vdots & & & \vdots \\
\mathbf{O} & \cdots & \mathbf{O} & \mathbf{A}n_{a_1} & \mathbf{O} & \cdots & \mathbf{O} & \mathbf{B}_{2n_0}
\end{array}\right]
\left[\begin{array}{c}
\mathbf{H}_1 \\ \vdots \\ \vdots \\ \mathbf{H}_{n_h} \\ \hdashline \mathbf{G}_0 \\ \vdots \\ \vdots \\ \vdots \\ \vdots \\ \mathbf{G}_{n_g}
\end{array}\right]
=
\left[\begin{array}{c}
\mathbf{D}_1' \\ \vdots \\ \vdots \\ \vdots \\ \mathbf{D}_{n_d'}' \\ \mathbf{O} \\ \mathbf{O} \\ \vdots \\ \vdots \\ \mathbf{O}
\end{array}\right]
-
\left[\begin{array}{c}
\mathbf{A}_{11} \\ \mathbf{A}_{12} \\ \vdots \\ \vdots \\ \mathbf{A}_{n_a} \\ \mathbf{O} \\ \mathbf{O} \\ \vdots \\ \vdots \\ \mathbf{O}
\end{array}\right].
$$

The above matrix has $n_h + n_g = n_2 + n_{a_1} - 1$ columns and the same number of rows. For a unique solution to these equations to exist, the matrix on the left of this equation must be non-singular. During self-tuning this matrix will include the estimated parameters and due to the stochastic nature of the problem it is likely to be non-singular. However, numerical problems can, of course, be caused if the matrix is nearly singular. By considering the two diagonal blocks in this matrix it is clear that if $\mathbf{B}_{2n_0}$ is full rank (that is if $\mathbf{B}_2$ is regular) then the marix can be full rank and invertible. Thus, in this and other cases, unique solutions for the $\mathbf{H}$ and $\mathbf{G}$ polynomial matrices do exist. In practical terms, all that is required is for the self-tuning system to converge to an optimal controller, uniqueness has no practical premium.

The solution for the observations weighted controller is now considered. The following assignments may be made without loss of generality:

$$\mathbf{A}(0) = \mathbf{A}_0(0) = \mathbf{D}(0) = \mathbf{I}_m,\ d_1(0) = 1;\ \text{hence from (12.70)}\ \mathbf{H}(0) = \mathbf{I}_m.$$

From the system equation and (12.62b):

$$\mathbf{A}e_1 = \mathbf{D}\boldsymbol{\varepsilon} - \mathbf{B}u. \tag{12.71}$$

Dcfinc

$$e_1'(t) = \mathbf{A}_0^{-1} e_1(t) \tag{12.72}$$

and recall

$$\boldsymbol{\phi}(t+k) = d_1(\mathbf{A}_0\tilde{\mathbf{B}}_2)^{-1}\mathbf{e}_1(t)$$

then using (12.70):

$$\begin{aligned}\boldsymbol{\phi}(t+k) &= \tilde{\mathbf{B}}_2^{-1}(d_1\mathbf{A}_1^{-1}(\mathbf{D}\boldsymbol{\varepsilon}(t)-\mathbf{B}u(t)))\\ &= \tilde{\mathbf{B}}_2^{-1}\mathbf{A}_1^{-1}((\mathbf{A}_1\mathbf{H}+\mathbf{B}_2\mathbf{G})\boldsymbol{\varepsilon}(t)-d_1\mathbf{B}u(t))\\ &= \tilde{\mathbf{B}}_2^{-1}(\mathbf{H}\boldsymbol{\varepsilon}(t)+\mathbf{A}_1^{-1}\mathbf{B}_2\mathbf{G}\mathbf{D}^{-1}(\mathbf{A}_1e_1'(t)+\mathbf{B}u(t))\\ &\quad -\mathbf{A}_1^{-1}d_1\mathbf{B}u(t))\\ &= \tilde{\mathbf{B}}_2^{-1}(\mathbf{H}\boldsymbol{\varepsilon}(t)+\mathbf{A}_1^{-1}\mathbf{B}_2\mathbf{G}\mathbf{D}^{-1}\mathbf{A}_1e_1'(t)-\mathbf{H}\mathbf{D}^{-1}\mathbf{B}u(t)). \end{aligned} \tag{12.73}$$

The second term in this expression must be simplified, using (12.70), as follows:

$$\begin{aligned}\mathbf{A}_1^{-1}(d_1\mathbf{D}-\mathbf{A}_1\mathbf{H})\mathbf{D}^{-1}\mathbf{A}_1 = d_1\mathbf{I}-\mathbf{H}\mathbf{D}^{-1}\mathbf{A}_1 &= \mathbf{H}\mathbf{D}^{-1}(d_1\mathbf{D}-\mathbf{A}_1\mathbf{H})\mathbf{H}^{-1}\\ &= \mathbf{H}\mathbf{D}^{-1}\mathbf{B}_2\mathbf{G}\mathbf{H}^{-1}\end{aligned} \tag{12.74}$$

hence

$$\boldsymbol{\phi}(t+k) = \tilde{\mathbf{B}}_2^{-1}\mathbf{H}\boldsymbol{\varepsilon}(t)+\tilde{\mathbf{B}}_2^{-1}\mathbf{H}\mathbf{D}^{-1}\mathbf{B}_2(\mathbf{G}\mathbf{H}^{-1}e_1'(t)-\mathbf{B}_1u(t)). \tag{12.75}$$

The significance of the matrix $\mathbf{B}_2$ introduced in (12.69) may now be considered. It is first assumed that $\mathbf{D}^{-1}\mathbf{B}_2$ can be written in the form:

$$\mathbf{D}^{-1}\mathbf{B}_2 = \mathbf{B}_2^0\mathbf{D}_0^{-1} \tag{12.76}$$

where

$$\mathbf{B}_2^0(z^{-1}) = z^{-k}(\mathbf{B}_{20}^0+\mathbf{B}_{21}^0z^{-1}+\cdots+\mathbf{B}_{2n_0}^0z^{-n_0}).$$

Wolovich [39] has demonstrated the existence of non-unique right coprime polynomial matrices $\mathbf{D}_0$ and $\mathbf{B}_2^0$ which satisfy (12.76) where $\det(\mathbf{D}_0)=\det(\mathbf{D})$ and $\mathbf{D}_0(0)=\mathbf{I}_m$, and $n_{20}=n_2$, $n_{d_0}=n_d$. Now define $\tilde{\mathbf{B}}_2$ as:

$$\tilde{\mathbf{B}}_2 = \mathbf{H}\mathbf{B}_2^0$$

then

$$\tilde{\mathbf{B}}_2^{-1}\mathbf{H}\mathbf{D}^{-1}\mathbf{B}_2 = \mathbf{D}_0^{-1}$$

and

$$\tilde{\mathbf{B}}_2^{-1}\mathbf{H} = (\mathbf{B}_2^0)^{-1}. \tag{12.77}$$

Note that $\mathbf{D}_0^{-1}$ represents a stable causal transfer function but

$$(\mathbf{B}_2^0(z^{-1}))^{-1} = z^{n_2}(\mathbf{B}_{2n_0}^0 + \mathbf{B}_{2(n_0-1)}^0 z + \cdots + \mathbf{B}_{20}^0 z^{n_0})^{-1} \tag{12.78}$$

represents a non-causal transfer function, since $n_2 - n_0 = k \geq 1$.

The minimization of the cost function depends upon the above definition for $\tilde{\mathbf{B}}_2$ and may now proceed. Equation (12.75) may be written as:

$$\boldsymbol{\phi}(t+k) = (\mathbf{B}_2^0)^{-1}\boldsymbol{\varepsilon}(t) + \mathbf{D}_0^{-1}(\mathbf{G}\mathbf{H}^{-1}\boldsymbol{e}_1'(t) - \mathbf{B}_1\boldsymbol{u}(t)). \tag{12.79}$$

From the definition of the cost function and from Parseval's theorem it may be seen that for the cost to be finite the signal:

$$\boldsymbol{m}(t) \triangleq \mathbf{G}\mathbf{H}^{-1}\boldsymbol{e}_1'(t) - \mathbf{B}_1\boldsymbol{u}(t) \tag{12.80}$$

must be stable (the operator $\mathbf{D}_0^{-1}$ is stable). Note that the signals

$$\boldsymbol{n}(t) \triangleq (\mathbf{B}_2^0)^{-1}\boldsymbol{\varepsilon}(t) \tag{12.81}$$

and $\boldsymbol{m}(t)$ have poles outside or strictly within the unit-circle of the z-plane, respectively.

The cost function (12.69) may now be expressed in terms of $\boldsymbol{m}_0 \triangleq \mathbf{D}_0^{-1}\boldsymbol{m}$, as:

$$J = \frac{1}{2\pi j}\oint_{|z|=1} (\boldsymbol{n}^*\boldsymbol{n} + \boldsymbol{m}_0^*\boldsymbol{m}_0 + \boldsymbol{n}^*\boldsymbol{m}_0 + \boldsymbol{m}_0^*\boldsymbol{n})\,\frac{dz}{z} \tag{12.82}$$

where $\boldsymbol{n}^* \triangleq \boldsymbol{n}^T(z)$. The integral around the unit circle of the term $\boldsymbol{m}_0^*\boldsymbol{n}$, which is analytic within the unit-circle is zero. Similarly, for the term $\boldsymbol{n}^*\boldsymbol{m}_0$. The first term in (12.79) is independent of the control and hence the cost-function is minimized by setting $\boldsymbol{m}_0 = 0$. The optimal control follows as:

$$\boldsymbol{u}^0 = \mathbf{B}_1^{-1}\mathbf{G}\mathbf{H}^{-1}\boldsymbol{e}_1' \tag{12.83}$$

To show that the control law is admissible (physically realizable) note that $\mathbf{B}_1(0) = \mathbf{H}(0) = \mathbf{I}_m$. To investigate the stability of the closed-loop system consider the return difference matrix:

$$\begin{aligned}\mathbf{I}_m + \mathbf{A}^{-1}\mathbf{B}_2\mathbf{G}\mathbf{H}^{-1}\mathbf{A}_0^{-1} &= \mathbf{A}^{-1}(\mathbf{A}_1\mathbf{H} + \mathbf{B}_2\mathbf{G})\mathbf{H}^{-1}\mathbf{A}_0^{-1}\\ &= \mathbf{A}^{-1}d_1\mathbf{D}\mathbf{H}^{-1}\mathbf{A}_0^{-1}.\end{aligned}$$

Closed-loop stability is determined by the characteristic equation

$$(d_1)^m \det \mathbf{D} = 0$$

and is therefore assured (the controller is assumed to be implemented in its minimal form).

The optimal control signal may be expressed in an alternative form which is more convenient for the parameter estimation algorithm. Let $\mathbf{GH}^{-1}$ be written as:

$$\mathbf{GH}^{-1} = \mathbf{H}_1^{-1}\mathbf{G}_1$$

where $\det \mathbf{H}_1 = \det \mathbf{H}$, $n_{g_1} = n_g$, $n_{h_1} = n_h$ and $\mathbf{H}_1(0) = \mathbf{I}_m$. The optimal control can now be expressed as:

$$u^0 = \mathbf{N}_1^{-1}\mathbf{G}_1 e_1'$$

where $\mathbf{N}_1 \triangleq \mathbf{H}_1\mathbf{B}_1$. Define $m_1(t) = \mathbf{G}_1 e_1'(t) - \mathbf{N}_1 u(t)$, then the optimal control is chosen to set $m_1(t) = \mathbf{o}$.

Example 12.8. *Two-output two-input systems.* The following problem was considered by Koivo [9]. The plant is non-minimum phase and open-loop unstable. The plant equation becomes:

$$y(t) = -\mathbf{A}_1 y(t-1) + \mathbf{B}^1 u(t-1) + \mathbf{B}^2 u(t-2) + \xi(t) \tag{12.84}$$

where

$$\mathbf{A}_1 = \begin{bmatrix} -0.9 & 0.5 \\ 0.5 & -0.2 \end{bmatrix}, \qquad \mathbf{B}^1 = \begin{bmatrix} 0.2 & 1 \\ 0.25 & 0.2 \end{bmatrix}$$

$$\mathbf{B}^2 = \begin{bmatrix} 1 & 0 \\ 0 & 1 \end{bmatrix}, \qquad \mathbf{E}\{\xi(t)\xi^T(t)\} = \mathrm{diag}\{0.1, 0.1\}$$

or

$$\mathbf{A}y(t) = \mathbf{B}u(t) + \xi(t) \tag{12.85}$$

where

$$\mathbf{A}(z^{-1}) = \begin{bmatrix} 1 - 0.9z^{-1} & 0.5z^{-1} \\ 0.5z^{-1} & 1 - 0.2z^{-1} \end{bmatrix}$$

$$\mathbf{B}(z^{-1}) = z^{-1}\begin{bmatrix} 0.2 + z^{-1} & 1 \\ 0.25 & 0.2 + z^{-1} \end{bmatrix}.$$

Assume that the performance criterion to be minimized is given by (12.69) with $\mathbf{A}_0(z^{-1}) \triangleq \mathbf{I}_2$ and $d_1 \triangleq 1 - z^{-1}/2$. This choice of dynamic operators within the performance criterion is based upon the desired performance characteristics Grimble [40].

The solution for the observations weighted controller follows. Factorize $\mathbf{B} = \mathbf{B}_2\mathbf{B}_1$

where $\mathbf{B}_2$ includes the non-minimum phase and delay terms. The Smith form for $\mathbf{B}$ can be calculated using the algorithm in Kučera [25].

$$\begin{bmatrix}1 & 0\\0 & 1\end{bmatrix}\begin{bmatrix}0.2+z^{-1} & 1\\0.25 & 0.2+z^{-1}\end{bmatrix}\begin{bmatrix}0 & 1\\1 & 0\end{bmatrix}$$

$$\begin{bmatrix}1 & 0\\0 & 1\end{bmatrix}\begin{bmatrix}1 & 0.2+z^{-1}\\0.2+z^{-1} & 0.25\end{bmatrix}\begin{bmatrix}1 & -(0.2+z^{-1})\\0 & 1\end{bmatrix}$$

$$\begin{bmatrix}1 & 0\\-(0.2+z^{-1}) & 1\end{bmatrix}\begin{bmatrix}1 & 0\\0.2+z^{-1} & -(0.2+z^{-1})^2+0.25\end{bmatrix}\begin{bmatrix}1 & -(0.2+z^{-1})\\0 & 1\end{bmatrix}$$

$$\begin{bmatrix}1 & 0\\-(0.2+z^{-1}) & 1\end{bmatrix}\begin{bmatrix}1 & 0\\0 & 0.21-0.4z^{-1}-z^{-2}\end{bmatrix}\begin{bmatrix}1 & -(0.2+z^{-1})\\0 & 1\end{bmatrix}.$$

The Smith form $\mathbf{S} = \mathbf{UBV}$, where

$$\mathbf{S} = z^{-1}\begin{bmatrix}1 & 0\\0 & -(z^{-1}+0.7)(z^{-1}-0.3)\end{bmatrix} \tag{12.86}$$

and

$$\mathbf{U} = \begin{bmatrix}1 & 0\\-(0.2+z^{-1}) & 1\end{bmatrix}, \quad \mathbf{V} = \begin{bmatrix}0 & 1\\1 & -(0.2+z^{-1})\end{bmatrix}.$$

Both of the zeros in $\mathbf{S}$ lie outside the unit circle and hence $\mathbf{B}_2 = \mathbf{B}$, $\mathbf{B}_1 = \mathbf{I}_2$ for this system.

The Diophantine equation (12.70) may now be evaluated (recall $\mathbf{A}_0 = \mathbf{I}_2$ and $\mathbf{D} = \mathbf{I}_2$):

$$\mathbf{AH} + \mathbf{B}_2\mathbf{G} = d_1\mathbf{I}_2 \tag{12.87}$$

where $n_h = n_2 - 1$ and $n_g = n_a - 1$. Now $n_2 = 2$ and $n_a = 1$ hence $n_h = 1$ and $n_g = 0$ giving:

$$(\mathbf{I}_2 + \mathbf{A}_1 z^{-1})(\mathbf{I}_2 + \mathbf{H}_1 z^{-1}) + (\mathbf{B}_{21}z^{-1} + \mathbf{B}_{22}z^{-2})\mathbf{G}_0 = \mathbf{I}_2 - \tfrac{1}{2}\mathbf{I}_2 z^{-1}.$$

These equations may be written in the form:

$$\begin{bmatrix}\mathbf{I}_2 & \mathbf{B}_{21}\\\mathbf{A}_1 & \mathbf{B}_{22}\end{bmatrix}\begin{bmatrix}\mathbf{H}_1\\\mathbf{G}_0\end{bmatrix} = \begin{bmatrix}-\tfrac{1}{2}\mathbf{I}_2 - \mathbf{A}_1\\\mathbf{O}\end{bmatrix} \tag{12.88}$$

or

$$\begin{bmatrix}1 & 0 & 0.2 & 1\\0 & 1 & 0.25 & 0.2\\-0.9 & 0.5 & 1 & 0\\0.5 & -0.2 & 0 & 1\end{bmatrix}\begin{bmatrix}h_{11} & h_{12}\\h_{21} & h_{22}\\g_{11} & g_{12}\\g_{21} & g_{22}\end{bmatrix} = \begin{bmatrix}0.4 & -0.5\\-0.5 & -0.3\\0 & 0\\0 & 0\end{bmatrix}$$

hence

$$\begin{bmatrix} h_{11} & h_{12} \\ h_{21} & h_{22} \\ g_{11} & g_{12} \\ g_{21} & g_{22} \end{bmatrix} = \begin{bmatrix} 0.682 & -0.7 \\ -0.637 & -0.234 \\ 0.932 & -0.51 \\ -0.469 & 0.303 \end{bmatrix}. \tag{12.89}$$

The optimal controller may be identified as:

$$\mathbf{C}_0(z^{-1}) = \mathbf{B}_1^{-1}\mathbf{G}\mathbf{H}^{-1} = \mathbf{G}\mathbf{H}^{-1}$$

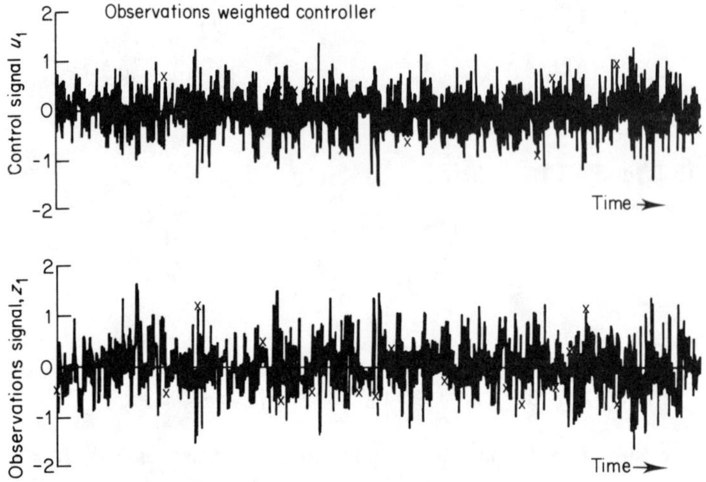

Fig. 12.10. Control and observations signals for Channel 1

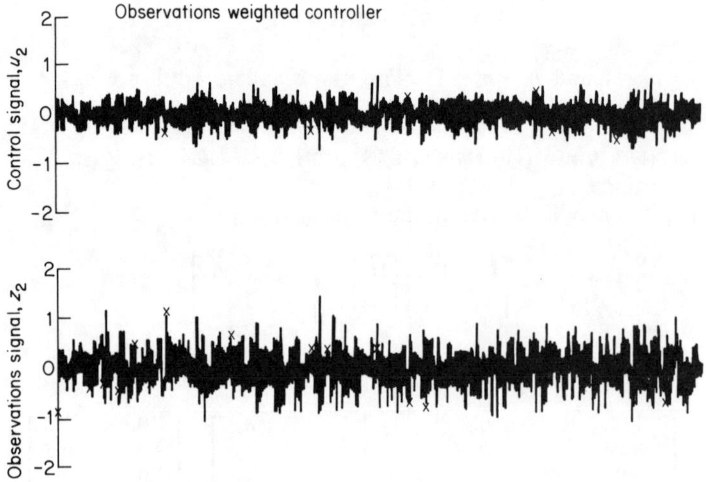

Fig. 12.11. Control and observations signal for Channel 2

where

$$\mathbf{G}(z^{-1}) = \begin{bmatrix} 0.932 & -0.51 \\ -0.469 & 0.303 \end{bmatrix}$$

$$\mathbf{H}(z^{-1}) = \begin{bmatrix} 1+0.682z^{-1} & -0.7z^{-1} \\ -0.637z^{-1} & 1-0.234z^{-1} \end{bmatrix}.$$

The operation of the observations weighted regulator for this system is illustrated in Figs. 12.10 and 12.11. Note that the controller is fixed in this example and is not being used in a self-tuning mode.

12.5.6. Multivariable observations weighted self-tuning controller

The estimation equations for the implicit observations weighted self-tuner can be derived as in Section 12.5.4. From the system equation (12.75) letting $y_1(t) \triangleq \tilde{\mathbf{B}}_2\boldsymbol{\phi}(t+k) = d_1\mathbf{A}_0^{-1}\boldsymbol{e}_1(t)$:

$$y_1(t) = \mathbf{H}\boldsymbol{\varepsilon}(t) + \mathbf{H}\mathbf{D}^{-1}\mathbf{B}_2\mathbf{H}_1^{-1}(\mathbf{G}_1\boldsymbol{e}_1'(t) - \mathbf{N}_1\boldsymbol{u}(t))$$

$$= \mathbf{H}\boldsymbol{\varepsilon}(t) + \mathbf{H}\mathbf{D}^{-1}\mathbf{B}_2\mathbf{H}_1^{-1}\boldsymbol{m}_1(t) \tag{12.90}$$

$$= \mathbf{H}\boldsymbol{\varepsilon}(t) + \mathbf{B}_2\boldsymbol{m}_1(t) + \sum_{i=1}^{\infty} \eta_i\boldsymbol{m}_1(t-i) \tag{12.91}$$

where the final term in equation (12.91) is obtained from the expansion of $(\mathbf{H}\mathbf{D}^{-1}\mathbf{B}_2\mathbf{H}_1^{-1} - \mathbf{B}_2)$, since $\mathbf{H}(0)\mathbf{D}(0)^{-1} = \mathbf{I}_m$ and $\mathbf{H}_1(0) = \mathbf{I}$. Let a k-steps ahead predictor be defined as:

$$\hat{y}_1(t \mid t-k) = \mathbf{B}_2\boldsymbol{m}_1(t) + \mathbf{H}_2\boldsymbol{\varepsilon}(t) + \sum_{i=1}^{\infty} \eta_i\boldsymbol{m}_1(t-i) \tag{12.92}$$

where $\mathbf{H} = \mathbf{H}_1 + \mathbf{H}_2$ and $\mathbf{H}_1$ includes all terms in $z^{-1}, z^{-2}, \ldots, z^{-(k-1)}$. In forming the estimation equation, as in the previous sections, the final term in (12.92) can be omitted because of the optimal control action.

The estimation equation follows as:

$$\hat{y}(t \mid t-k) = \mathbf{B}_2(\mathbf{G}_1\boldsymbol{e}_1'(t) - \mathbf{N}_1\boldsymbol{u}(t)) + \mathbf{H}_2\boldsymbol{\varepsilon}(t)$$

$$= \mathbf{X}(t-k)\boldsymbol{\theta} \tag{12.93}$$

This equation can form the basis of a bilinear identification algorithm from which $\hat{G}_1$ and $\hat{N}_1$ may be obtained (Åström [38]). The control signal then follows as:

$$\boldsymbol{u}^0(t) = \hat{N}_1^{-1}\hat{G}_1\boldsymbol{e}_1'(t). \tag{12.94}$$

Example 12.9. *Observations weighted self-tuning control.* To show how the estimation equation may be obtained in a suitable form consider the case of a square 2×2 system with

$$\mathbf{B}_2 = \begin{bmatrix} b_{11} & b_{12} \\ b_{21} & b_{22} \end{bmatrix}, \qquad \mathbf{G}_1 = \begin{bmatrix} g_{11} & g_{12} \\ g_{21} & g_{22} \end{bmatrix}, \qquad \mathbf{N}_1 = \begin{bmatrix} n_{11} & n_{12} \\ n_{21} & n_{22} \end{bmatrix}$$

hence

$$\mathbf{B}_2\mathbf{G}_1 e_1'(t) = \begin{bmatrix} b_{11}g_{11} + b_{12}g_{21} & b_{11}g_{12} + b_{12}g_{22} \\ b_{21}g_{11} + b_{22}g_{21} & b_{21}g_{12} + b_{22}g_{22} \end{bmatrix} \begin{bmatrix} e_1'(t) \\ e_2'(t) \end{bmatrix}. \tag{12.95}$$

Now define the signals:

$$v_{11} = b_{11}e_1', \qquad v_{12} = b_{11}e_2', \qquad v_{13} = b_{12}e_1', \qquad v_{14} = b_{12}e_2'$$

$$v_{21} = b_{21}e_1', \qquad v_{22} = b_{21}e_2', \qquad v_{23} = b_{22}e_1', \qquad v_{24} = b_{22}e_2'$$

hence

$$\mathbf{B}_2\mathbf{G}_1\mathbf{e}_1'(t) = \begin{bmatrix} v_{11} & v_{12} & v_{13} & v_{14} \\ v_{21} & v_{22} & v_{23} & v_{24} \end{bmatrix} \begin{bmatrix} g_{11} \\ g_{12} \\ g_{21} \\ g_{22} \end{bmatrix} = \mathbf{V}(t)\mathbf{g}.$$

The term $\mathbf{B}_2\mathbf{N}_1 u(t)$ can also be expressed in the similar form $\mathbf{U}(t)\mathbf{n}$ and hence

$$\mathbf{X}(t-k)\boldsymbol{\theta} = \left[\begin{array}{cccc|cccc|cccc} v_{11} & v_{12} & v_{13} & v_{14} & -u_{11} & -u_{12} & -u_{13} & -u_{14} & \varepsilon_1 & \varepsilon_2 & 0 & 0 \\ v_{21} & v_{22} & v_{23} & v_{24} & -u_{21} & -u_{22} & -u_{23} & -u_{24} & 0 & 0 & \varepsilon_1 & \varepsilon_2 \end{array}\right]\boldsymbol{\Theta}$$

$$\boldsymbol{\Theta} = [g_{11} \;\; g_{12} \;\; g_{21} \;\; g_{22} \;|\; n_{11} \;\; n_{12} \;\; n_{21} \;\; n_{22} \;|\; h_{11} \;\; h_{12} \;\; h_{21} \;\; h_{22}]^T. \tag{12.96}$$

If $g_{ij} = g_0^{ij} + g_1^{ij}z^{-1} + \cdots$ then $\boldsymbol{\Theta}$ may be written as:

$$\begin{aligned} \boldsymbol{\Theta} = [&g_0^{11}, g_1^{11}, \ldots;\; g_0^{12}, g_1^{12}, \ldots;\; g_0^{21}, g_1^{21}, \ldots;\; g_0^{22}, g_1^{22}, \ldots; \\ &n_0^{11}, n_1^{11}, \ldots;\; n_0^{12}, n_1^{12}, \ldots;\; n_0^{21}, n_1^{21}, \ldots;\; n_0^{22}, n_1^{22}, \ldots; \\ &h_0^{11}, h_1^{11}, \ldots;\; h_0^{12}, h_1^{12}, \ldots;\; h_0^{21}, \ldots h_1^{21}, \ldots;\; h_0^{22}, h_1^{22}, \ldots]^T. \end{aligned} \tag{12.97}$$

Special case $(\mathbf{B}_2 = b_2\mathbf{I}_m)$

The preceding results simplify considerably when the non-minimum phase spectral factor is a scalar, that is, $\mathbf{B}_2 = b_2\mathbf{I}_m$ where b_2 is a scalar. For this case equation (12.93) gives:

$$\hat{y}_1(t \mid t-k) = \mathbf{G}_1 \boldsymbol{e}_b(t) - \mathbf{N}_1 \boldsymbol{u}_b(t) + \mathbf{H}_2 \boldsymbol{\varepsilon}(t) = \mathbf{X}(t-k)\boldsymbol{\theta}$$

where $\boldsymbol{e}_b(t) = b_2\boldsymbol{e}_1'(t)$ and $\boldsymbol{u}_b(t) = b_2\boldsymbol{u}(t)$. The identification algorithm can be constructed as in Section 12.5.2. The factorization of $\mathbf{B} = b_2\mathbf{B}_1$ is more straightforward than in the general case. The optimal control follows from (12.94) and the algorithm becomes as follows.

Algorithm 12.1 $(\mathbf{B}_2 = b_2\mathbf{I}_m)$

1. $\boldsymbol{e}_1' = \mathbf{A}_0^{-1}\boldsymbol{e}_1$, $y_1 = d_1\boldsymbol{e}_1'$, $\boldsymbol{e}_b = b_2\boldsymbol{e}_1'$, $\boldsymbol{u}_b = b_2\boldsymbol{u}$.
2. Identify the plant polynomial matrix $\mathbf{B}$ and past values of $\boldsymbol{\varepsilon}$.
3. Spectrally factor $\mathbf{B} = b_2\mathbf{B}_1$.
4. Identify $\mathbf{G}_1$, $\mathbf{N}_1$ using the prediction model: $\hat{y}_1 = \mathbf{G}_1\boldsymbol{e}_b - \mathbf{N}_1\boldsymbol{u}_b + \mathbf{H}_2\boldsymbol{\varepsilon}$.
5. Calculate the optimal control as: $\boldsymbol{u}^0 = \hat{\boldsymbol{B}}_1^{-1}\hat{\mathcal{N}}_1^{-1}\boldsymbol{e}_1'(t)$. □

It may be possible to combine the two identification stages in this algorithm into a multivariable bilinear estimation algorithm.

Special case $(\mathbf{B}_2 = b_2\mathbf{I}_m$ and known $k)$
If the delay k is known then $(z^k\mathbf{B}_2)(0) = b_{20} \neq 0$ and the algorithm may be further simplified.

Algorithm 12.2 $(\mathbf{B}_2 = b_2\mathbf{I}_m$, known $k)$

1. $\boldsymbol{e}_1' = \mathbf{A}_0^{-1}\boldsymbol{e}_1$, $y_1 = d_1\boldsymbol{e}_1'$.
2. Identify the plant polynomial matrix $\mathbf{B}$ and past values of $\boldsymbol{\varepsilon}$.
3. Spectrally factor $\mathbf{B} = b_2\mathbf{B}_1$.
4. Identify $\mathbf{G}_1$, $\mathbf{N}_1$ using the prediction model:
   $\hat{y}_1(t \mid t-k) = \mathbf{G}_1\boldsymbol{e}_1'(t-k) - \mathbf{N}_1\boldsymbol{u}(t-k) + \mathbf{H}_2\boldsymbol{\varepsilon}(t)$.
5. Calculate $\mathbf{u}^0 = \hat{\boldsymbol{B}}_1^{-1}\hat{\mathcal{N}}_1^{-1}\boldsymbol{e}_1'(t)$. □

It is clear that similar arguments may be used to derive a multivariable version of the weighted minimum-variance self-tuning controller of Section 12.4.2. An explicit version of the WMV multivariable self-tuning controller has been described by Grimble and Fung [41] and an implicit WMV self-tuning system has been proposed by Grimble and Moir [50].

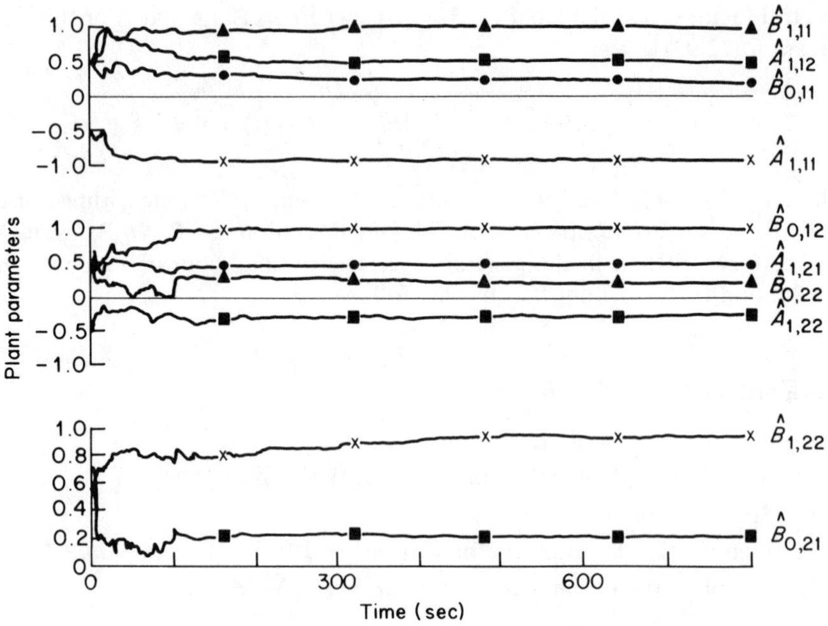

Fig. 12.12. Estimated plant parameters for $\mathbf{A}_1, \mathbf{B}^1$ and $\mathbf{B}^2$

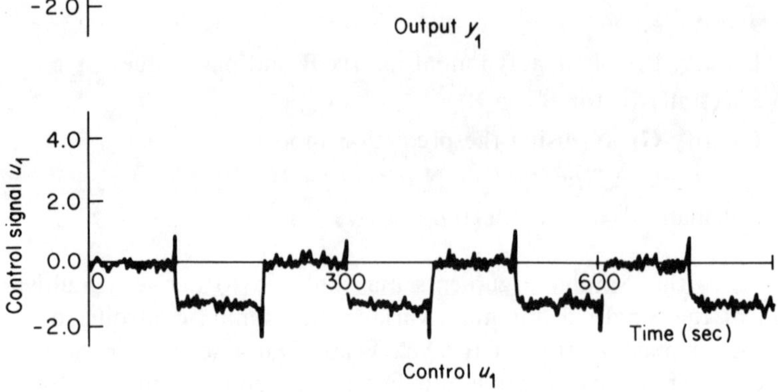

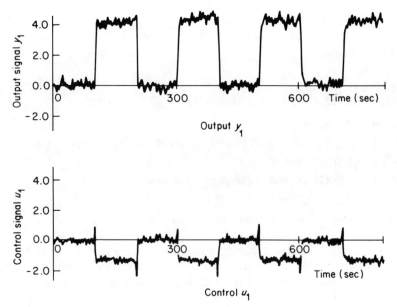

Fig. 12.13. Output and control signals (Channel 1)

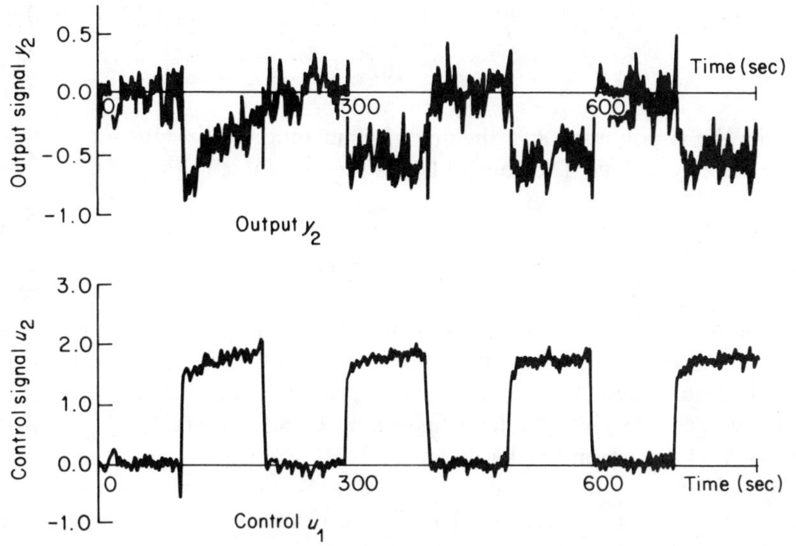

Fig. 12.14. Output and control signals (Channel 2)

Example 12.10. *Observations weighted self-tuner.* Consider the system of Example 12.8 where a square-wave disturbance is input to the system to demonstrate the tuning action. The plant parameter estimates are shown in Fig. 12.12. The observations weighted controller was calculated explicitly using the estimated plant parameters. The control and observations signals that result are shown in Figs. 12.13 and 12.14 for Channels 1 and 2, respectively.

12.5.7. Multivariable self-tuning regulators with pole assignment

Wellstead and co-workers (Wellstead *et al.* [42]) at UMIST, Manchester, have developed non-optimal self-tuning systems which accommodate pole assignment. The derivation of the control law which has the form $u = -\mathbf{G}\mathbf{F}^{-1}y$ is described briefly below.

Introduce the Diophantine equation:

$$\mathbf{AF} + \mathbf{BG} = \mathbf{CT}$$

where $n_t + n_c \leq n + n_b - 1$. The closed-loop system becomes:

$$y = \mathbf{FT}^{-1}\varepsilon.$$

Define $\tilde{\mathbf{F}}^{-1}\tilde{\mathbf{G}} = \mathbf{GF}^{-1}$ then it is convenient to calculate the control as:

$$u = (\mathbf{I} - \tilde{\mathbf{F}})u - \tilde{\mathbf{G}}y.$$

The explicit self-tuning algorithm is derived by using an on-line plant model:

$$\mathbf{A}y = \mathbf{B}u + \xi.$$

That is, the **C** polynomial in the actual plant model ($\mathbf{A}y = \mathbf{B}u + \mathbf{C}\xi$) is taken as the identity matrix. The control law is then evaluated using **F**, **G** calculated from:

$$\mathbf{AF} + \mathbf{BG} = \mathbf{T}.$$

It is the form of the control law which makes it possible for the $(\mathbf{C} - \mathbf{I})$ polynomial elements (Prager and Wellstead [43]) to be neglected during the identification stage. The closed-loop poles of the system, after the algorithm has converged, are given by det **T** (assuming **T** and **F** are relatively prime) and the matrix **T** is chosen by the designer.

The 'self-tuning principle' as introduced by Wellstead and co-workers states that if the system converges based upon an estimation equation with $(\mathbf{C} - \mathbf{I}) = \mathbf{O}$, it will converge to the desired closed-loop configuration even when this term is non-zero in the actual plant. The approach taken by Wellstead follows the type of analysis used by Åström and Wittenmark [28] for the optimal regulator. However, the pole-assignment philosophy is an important contribution because of its observed robustness in simulation trials.

12.6. CONCLUSIONS

Self-tuning control has provided an interesting area for academic research as some of the foregoing may have illustrated. There are, however, practical precautions to be taken when implementing these techniques and there follows a discussion of these engineering aspects in the next section. Subsequently, remarks which summarize the present position in the development of this very topical subject close the chapter.

12.6.1. Engineering and implementational aspects

Feedback tends to reduce the effect of parameter variations and if this is not sufficient, robust control or gain scheduling may be used. Since these approaches are simpler than adaptive control they are the natural first choice. The most useful practical feature of adaptive controllers is the ability to track variations in the process dynamics so that good control action over the total range of operation can be maintained.

Numerical problems

All least-squares estimation methods used in self-tuning systems are poorly conditioned if the plant models are overparameterized. Most numerical difficulties can, however, be reduced by the use of square-root algorithms, described in Section 12.3.4. Control design methods can also be poorly conditioned and this is often associated with a loss of controllability or observability in the process model. The cancellation of common factors in the plant transfer-function gives rise to this type of problem.

Unmodelled dynamics

The self-tuning systems are normally based upon a low-order approximation to the actual high-order real system. If the self-tuning system is to be robust the frequency range over which a good model is needed must be restricted. To reduce low-frequency sensitivity to modelling errors an integrator can be introduced as in classical design. To reduce the effect of unmodelled dynamics at high frequencies the gain must be kept low in this frequency range. This can be achieved by the use of presampling filters.

Start-up

During start-up of a self-tuning system the control and output signals can change violently if no action is taken to limit such variations. The problem, of course, arises from the initial error in the parameter estimates. This error can be avoided by initial identification experiments using any stabilizing controller. Alternatively, the control signal variations may be initially constrained in some way providing the stability of the system is not affected.

Integrator windup

One problem with the integral action, particularly in software implementations of fixed or adaptive controllers, is integral windup or saturation. This occurs when an error exists for a long time, relative to the integral time constant and the integrator holds the controller output at full scale for longer than desirable (Fensome [53]). If the error then reverses sign the integrator will take some time before its output decreases far enough to bring the controller output out of the saturation region. The effect is called controller paralysis which is undesirable since the system is not being controlled in this operating region.

There are standard methods to alleviate the worst effects of integrator windup. For example, the maximum value of the integrator output can be limited either numerically in software or by voltage level in an analogue controller. Alternatively, once full-scale controller output is reached (due to the sum of the proportional and integral terms) the integral term can be reduced. With this form of desaturation in operation, the controller has the advantage of coming off the limit as soon as the error starts to decrease. The paralysis period is therefore kept to a minimum. The integral action can of course be switched out completely when windup occurs and be reintroduced when the error falls. However, this can lead to rather unpredictable responses, although it is employed in the control of batch reactors.

Performance-related controller tuning inputs

A practical industrial self-tuning controller might be constructed which does not allow the operator to vary its performance and which does not require 'tuning' during commissioning. However, it is unlikely that such a 'knobless' device would be applicable to a wide range of applications. An advantage of the self-tuning approach is that any controller knobs can be performance related. That is, the operator can directly tune the cost or performance error. This should be much simpler than say tuning PID controller parameters to achieve a given performance.

12.6.2. Concluding remarks

Behind the popularity of the self-tuning philosophy has been the promise that self-tuners could supply solutions to control problems for which classical controllers could not be designed. Processes whose models contained unknown or slowly varying parameters (time constants or gains) and certain types of non-linear systems became suitable candidates. The gain of the non-linear elements can, for example, be treated as variable parameters. Self-tuners could be used to retune existing controllers to accommodate changing operating conditions. Over long periods of time, the maintenance of the optimum performance by the self-tuner would then obviate the need for frequent and possibly expensive on-line retuning. Despite the initial slow rate of real applications the technique still has much to offer and there is considerable interest from the chemical industry (Kershenbaum and Fortescue [44]).

A certain amount of skill and experience is needed when using self-tuning techniques and there are many situations where they are simply not appropriate. They should also be considered as a last resort after simple methods of control have been discarded. There are disadvantages, for example, in contrast to the extended Kalman filter, it is not always easy to take advantage of known or approximately known system parameters, they are often computed along with all the unknown variables. In addition, the need to use a simple control algorithm has often resulted in the use of controllers based upon single-stage cost-functions and these can result in unstable designs for some control weighting values, if the system is open-loop unstable or non-minimum phase, (Wittenmark and Rao [45]). Finally, the subject has developed sufficiently for several variants of optimal and non-optimal self-tuners to be available, hence the problem of selecting a self-tuner to match a particular application has become a more difficult problem. However, with recent work on expert systems, it may soon be easier to get an industrial self-tuning controller running than to go through the usual modelling and fixed controller design process.

Many of the results given in the sections on controller design are original and it is hoped that the approach goes some way towards unifying the possible optimal control strategies. There seems to be three basic approaches to controller design, namely Observations Weighted, Weighted Minimum Variance and LQG. The first two are closely related and differ only due to the presence of the P_1, P_2 terms in the cost function (12.31). The LQG and observations weighted controllers are related but in general the LQG multivariable designs are more complicated and cannot easily be incorporated into implicit self-tuning algorithms (Grimble [26], [57]). A comparison and review of WMV and LQG self-tuning controllers is given in Grimble *et al.* [46], [47]. the LQG design approach has the advantage that it may easily be extended to cope with more general system structures (Grimble [58]).

It is clear that a new control technology is emerging from which will evolve novel adaptive control products. Sophisticated algorithms will be used on particularly difficult control design problems and a spectacular improvement in the performance of some systems might be expected. With the ready availability of cheap computing power such techniques will be applied to the control of a wide range of industrial processes and domestic products.

12.7. PROBLEMS

1. A single-input/single-output system is represented in the form:

$$A(z^{-1})y(t) = z^{-k}B(z^{-1})u(t) + C(z^{-1})\xi(t)$$

where $zy(T) = y(t+1)$. The polynomials $A(z^{-1})$, $B(z^{-1})$ and $C(z^{-1})$ have the form:

$$A(z^{-1}) = 1 + a_1 z^{-1} + \cdots + a_n z^{-n}$$

$$B(z^{-1}) = b_0 + b_1 z^{-1} + \cdots + b_n z^{-n}.$$

$$C(z^{-1}) = 1 + c_1 z^{-1} + \cdots + c_n z^{-n}.$$

A closed-loop optimal control system is to be employed where the variance of the output is to be minimized. That is, the cost function to be minimized has the form:

$$J = E\{y^2(t)\}.$$

(a) Derive an expression for the minimum-variance controller using the Diophantine equation:

$$C(z^{-1}) = A(z^{-1})F(z^{-1}) + z^{-k}G(z^{-1})$$

where the orders of F and G are $n_f = k - 1$ and $n_g = n - 1$, respectively.

(b) Draw a block diagram showing how the above controller may be used in a self-tuning control system. State the purpose of such a system and briefly describe the operation of the self-tuner.

2. If $\mathbf{H}(z^{-1})$ represents a square polynomial matrix

$$\mathbf{H}(z^{-1}) = \mathbf{H}_0 + \mathbf{H}_1 z^{-1} + \cdots + \mathbf{H}_{n_h} z^{-n_h}$$

show that when $\mathbf{H}_0$ is full rank then $\mathbf{H}(z^{-1})^{-1}$ can be expressed as a causal series in powers of z^{-1}.

3. If measurement noise is white the controllers obtained by minimizing either output variance or observations are the same (Section 12.2.3). Show that this is also true for the following continuous-time systems.

(a) *Output regulator* (see Fig. 12.15). Let the power spectral densities for ξ and v be defined as $Q = 5$ and $R = 1$, respectively. The observations

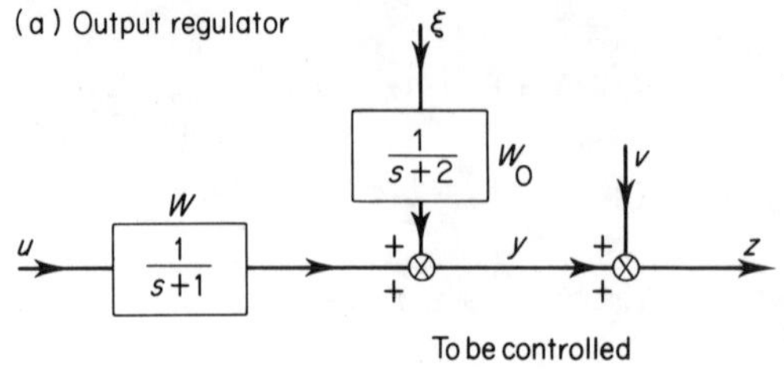

Fig. 12.15. Output regulator

spectrum follows as:

$$W_0(s)QW_0(-s)+R=\frac{Q}{-s^2+4}+R=\frac{(-s+3)(s+3)}{(-s+2)(s+2)}.$$

The optimal control cost-function weighting parameters are defined as $R_1 = 1$ and $Q_1 = 5$. The same noise spectrum and performance criterion is assumed for the following system.

(b) *Observations regulator.* (see Fig. 12.16). The noise source ε is assumed to have a unit spectral density so that the two plants generate the same total noise spectrums at the observations output z.

The transfer $M_0(s)$ and the closed-loop optimal regulators will be found to have the following forms:

$$M_0(s)=\frac{5(s+1)}{8(s+3)(s+6)}$$

$$C_0(s)=\frac{5(s+1)}{8s^2+72s+139}.$$

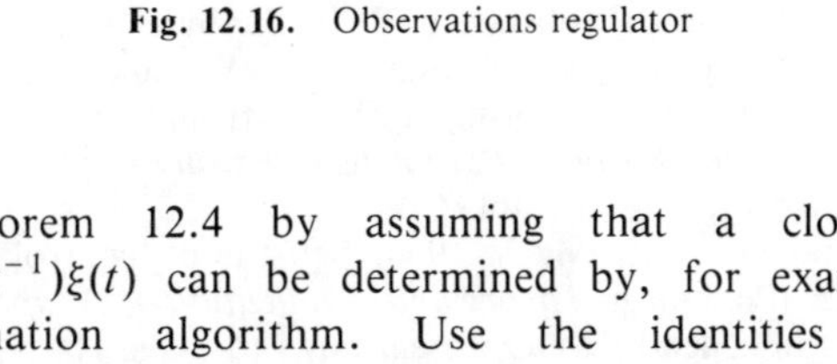

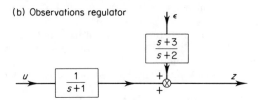

Fig. 12.16. Observations regulator

4. Prove Theorem 12.4 by assuming that a closed-loop model $P(z^{-1})y(t) = Q(z^{-1})\xi(t)$ can be determined by, for example, maximum likelihood estimation algorithm. Use the identities $Q = C_nD$ and $P = AC_d + BC_n$ and the fact that C_n, C_d are assumed to be coprime. Show that both a necessary and sufficient condition for the identifiability of $A(z^{-1})$ and $B(z^{-1})$ is that $n_{c0} \geq n$.

5. Self-tuning control schemes are now being introduced into more industrial applications. (i) Describe an explicit self-tuning control scheme and with

the aid of a block diagram detail the operation of a self-tuning system. (ii) List a typical self-tuning algorithm and name the types of identification algorithm employed. (iii) Describe the advantages of a self-tuning control approach and also note possible disadvantages. (iv). Assume that a fixed (non-self-tuning) controller is required for a system which is multivariable and includes stochastic disturbance and measures noise inputs. Draw a block diagram illustrating a stochastic optimal control solution to this problem and label the various subsystems.

12.8. REFERENCES

[1] Kalman, R. E. 1958. Design of a self-optimizing control system, *Trans. ASME*, **80**, 468–478.

[2] Peterka, V. 1970. Adaptive digital regulation of noisy systems, Second IFAC Symposium on Identification and Process Parameter Estimation, Prague.

[3] Åström, K. J. 1970. *Introduction to Stochastic Control Theory*, Academic Press, New York.

[4] Gawthrop, P. J. 1980. Hybrid self-tuning control, *IEE Proc.* **127**(5), Part D, 229–236.

[5] Åström, K. J. and Wittenmark, B. 1973. On self-tuning regulators, *Automatica*, **9**, 185–199.

[6] Clarke, D. W. and Gawthrop, P. J. 1975. Self-tuning controller, *Proc. IEE*, **122**(9), 929–934.

[7] Hastings-James, R. 1970. A linear stochastic controller for regulation of systems with pure time delay, Report No CN/70/3, Department of Engineering, Cambridge University.

[8] Borisson, U. 1975. Self-tuning regulators: Industrial application and multivariable theory, Report No 7513, Department of Automatic Control, Lund Institute of Technology, Sweden.

[9] Koivo, H. N. 1980. A multivariable self-tuning controller, *Automatica*, **16**, 351–366.

[10] Reeve, P. 1978. Self-tuning control of a reheat furnace, Colloquium Digest, University of Oxford, Published by IEE, Savoy Place, London.

[11] Gray, R., Macedo, F. X., Williams, A. W. and Williams, D. C. 1978. Identification of slab reheating furnace from normal operating records, *Identification and System Parameters Estimation* pp. 979–990, editor: Raybman; North-Holland Publ. Co.

[12] Fung, P. and Grimble, M. J. 1980. Self-tuning control of ship positioning systems, *IEE Workshop: Theory and Application of Adaptive Control*, Oxford University, Published by Peter Peregrinus Ltd., eds. Harris, C. J. and Billings, S. A.

[13] Harris, T. J., MacGregor, J. F. and Wright, J. D. 1980. Self-tuning and adaptive controllers: an application to catalytic reactor control, *Technometrics* **22**(2), 153–164, May.

[14] Allidina, A. Y. and Hughes, F. M. 1981. Self-tuning control of a nuclear reactor, *Proc. IEE. Conf. on Control and its Applications*, University of Warwick, Conf. Publ. No 194, IEE, London.

[15] Dahlquist, S. A. 1981. Control of a distillation column using self-tuning regulators, *Canadian J. Chem. Engineering*, **59**, 118–127.

[16] Jacobs, O. L. R. and Patchell, J. W. 1972. Caution and probing in stochastic control, *Int. Journal of Control*, **16**(1), 189–199.

[17] Clarke, D. W., Cope, S. N. and Gawthrop, P. J. 1975, Feasibility study of the application of microprocessors to self-tuning controllers, Report 1137/75; Department of Engineering Science, Oxford University.

[18] Sternby, J. 1977. Topics in dual control, Report LUTFD2/(TFRT-1012) pp. 1–135, Lund Institute of Technology, Lund, Sweden.

[19] Jacobs, O. L. R. 1974. *Introduction to Control Theory*, Clarendon Press, Oxford, UK.

[20] Grimble, M. J., Johnson, M. A. and Fung, P. T. K. 1981. Optimal self-tuning control systems: theory and application; Part 1; Introduction and controller design, *Trans. Inst. Meas. Control*, **2**(3), 115–120, July–Sept.

[21] Kurz, H., Iserman, R. and Schumann, R. 1980. Experimental comparison and application of various parameter adaptive control algorithms, *Automatica* **16**, 117–133.

[22] MacGregor, J. F. and Tidwell, P. W. 1977. Correspondence item: Discrete stochastic control with input constraints, *Proc. IEE*, **124**(8), 732–734.

[23] Grimble, M. J. 1982. Weighted minimum variance self-tuning control *Int. J. Control*, **36**(4) 597–609.

[24] Grimble, M. J. 1979. Solution of the discrete time stochastic optimal control problem in the z-domain, *Int. J. System Science*, **10**(12), 1369–1390.

[25] Kučera, V. 1979. *Discrete Linear Control: The Polynomial Equation Approach*, John Wiley & Son Ltd., London.

[26] Grimble, M. J. 1981. Observations weighted and LQG controllers for self-tuning systems, Research Report No EE/72/Jan 1981, Sheffield City Polytechnic, Sheffield, England.

[27] Åström, K. J. and Wittenmark, B. 1971. Problems of identification and control, *Journal of Mathematical Analysis and Applications*, **34**, 90–113.

[28] Åström, K. J. and Wittenmark, B. 1974. Analysis of a self-tuning regulator for non-minimum phase systems, IFAC Symposium on Stochastic Control, Budapest.

[29] Grimble, M. J. 1981. A control-weighted minimum-variance controller for non-minimum phase system, *Int. J. Control*, **33**(4), 751–762.

[30] Söderström *et al.* 1978. A theoretical analysis of recursive identification methods, *Automatica*, **14**, 231–244.

[31] Åström, K. J. and Eykhoff, P. 1977. System identification—A survey, *Automatica*, **7**, 123–162.

[32] Gustavsson *et al.* 1977. Identification of processes in closed loop identifiability and accuracy aspects, *Automatica*, **13**, 59–75.

[33] Kurz, H. and Isermann, R. 1975. Methods for on-line process identification in closed loop, 6th IFAC Congress, Boston, Mass., USA.

[34] Wieslander, J. and Wittenmark, B. 1971. An approach to adaptive control using real time identification, *Automatica*, **7**, 211–217.

[35] Anderson, B. D. O. and Moore, J. B. 1979. *Optimal Filtering*, Prentice-Hall Inc., Englewood Cliffs, New Jersey, USA.

[36] Fuchs, J. J. J. 1980. Explicit self-tuning methods, *IEE Proc.*, **127**(6), Part D, 259–264.

[37] Allidina, A. Y. and Hughes, F. M. 1980. Generalized self-tuning controller with pole assignment, *IEE Proc.*, **127**(1), Part D, 13–18.

[38] Åström, K. J. 1979. New adaptive pole-zero placement algorithms for non-minimum phase systems, Report No LUT FD2/TFRT-7172/1-012 (August), Dept of Automatic Control, Lund Institute of Technology, Sweden.

[39] Wolovich, W. A. 1974, *Linear Multivariable Systems*, Springer-Verlag, New York, USA.
[40] Grimble, M. J. 1980. The design of *s*-domain optimal controllers with integral action for output feedback control systems, *Int. J. Control*, **31**(5), 569–582.
[41] Grimble, M. J. and Fung, P. T. K. 1981, Explicit weighted minimum variance self-tuning controllers, IEEE Conference on Decision and Control, San Diego, California, USA.
[42] Wellstead, P. E., Prager, D. and Zanker, P. 1979. Pole assignment self-tuning regulator, *Proc. IEE*, **126**(8), 781–787.
[43] Prager, D. L. and Wellstead, P. E. 1980. Multivariable pole assignment self-tuning regulators, *IEE Proc*, **128**(1), Part D, Jan.
[44] Kershenbaum, L. S. and Fortescue, T. 1981. Implementation of on-line control in chemical process plants, *Automatica*, **17**(6), 777–788, Nov.
[45] Wittenmark, B. and Rao, K. 1979. Comments on single step versus multistep performance criteria for steady state SISO systems, *IEEE Trans. Automatic Control*, **AC-24**(1), 140–141, Feb.
[46] Grimble, M. J., Moir, T. J. and Fung, P. T. K. 1982, Comparison of WMV and LQG self-tuning controllers, presented at the IEEE Conf. on Applications of Adaptive and multivariable Control, Hull, July 19–21.
[47] Grimble, M. J. 1986, Self-tuners based on LQG theory, to be published in *Encyclopedia of Systems and Control*, Pergamon Press, ed. M. Singh.
[48] Gustavsson, I., Ljung, L. and Soderström. 1978. Identification of processes in closed loop—Identifiability and accuracy aspects, published in *Identification and System Parameter Estimation*, North Holland Publishing Company, ed. Rajbman.
[49] Goodwin, G. C. and Payne, R. L. 1977. *Dynamic System Identification: Experiment Design and Data Analysis*, Academic Press, London.
[50] Grimble, M. J. and Moir, T. J. 1983. Multivariable weighted minimum variance self-tuning controllers, presented at the IFAC workshop on Adaptive Systems in Control and Signal Proc., San Francisco, pp. 32–37; (extended version: *Int. J. of Control* **42**(6), 1283–1307, 1985).
[51] Wittenmark, B. and Åström, K. J. 1982. Implementation aspects of adaptive controllers and their influence on robustness, CDC Conf., Orlando, pp. 33–37.
[52] Cameron, F. and Seborg, D. E. 1983. A self-tuning controller with a PID structure, *Int. J. Control*, **38**(2), 401–417.
[53] Fensome, D. A. 1983. Understanding three-term control in industrial processes, *Electronics and Power*, pp. 647–650, September.
[54] Grimble, M. J. 1984. *Implicit and Explicit LQG Self-Tuning Controllers*, IFAC World Congress, Budapest, Hungary, July (published by Pergamon Press Ltd.); *Automatica*, **20**(5), 661–669.
[55] Söderström, T. and Stoica, P. G. 1983. *Instrumental Variable Methods for System Identification*, Springer-Verlag, Germany.
[56] Clarke, D. W. and Hastings-James, R. 1971. Design of digital controllers for randomly disturbed systems, *Proc. IEE*, **118**, 1503–1506.
[57] Grimble, M. J. 1984. LQG Multivariable controllers: minimum variance interpretation for use in self-tuning systems, *Int. J. Control*, **40**(4), 831–842.
[58] Grimble, M. J. 1986. Controllers for LQG self-tuning applications with coloured measurement noise and dynamic costing, *IEE. Proceedings*, **133**(1), Pt. D, January.
[59] Grimble, M. J. 1985. Observations weighted minimum variance control of linear and non-linear systems, *Int. J. Systems Sci.*, **16**(12), 1481–1492.

CHAPTER 13

Stochastic Industrial Control Systems

13.1. INTRODUCTION

Most industrial control problems involve noisy systems and are very often non-linear. The stochastic nature of such problems is frequently ignored, indeed, classical control owes much of its success and popularity to the fact that this approximation is adequate for many engineering solutions. If, however, the main design requirement is to accommodate a randomly varying disturbance or a noise signal then a stochastic formulation is usually required. Tacho-generator ripple in machine-control system is one example of a disturbance which does not normally fall into this category. As previously demonstrated in Chapter 2, deterministic design procedures may be used for this type of noise and a simple noise filter can be inserted in the tacho-feedback path. The industrial problems considered in this chapter must be treated as stochastic problems because of the dominant role played by the random signals.

Three applications of the stochastic approach to industrial problems are presented. The first section is concerned with the design of dynamic ship-positioning systems. These are employed in oil-rig drill ships or survey vessels to maintain a given position despite the various wind, wave and current disturbance which impinge on a vessel whilst at sea (Ball and Blumberg [1]). Kalman filters have been implemented in this type of system and most modern vessels carry a system designed using this technique (Balchen *et al.* [2], [3]). All the steps in the design process are discussed from model development through to simulation of the resulting Kalman filter. The success in accommodating the severe non-linearities which exist in the actual system is described.

The steel industry is the setting for the remaining two applications of this chapter: prediction in an ingot soaking pit and control of gauge in rolling mills. These are described in less detail than the ship-positioning problem and have been used to illustrate other application areas where an optimal stochastic approach is desirable.

13.2. DESIGN OF DYNAMIC SHIP-POSITIONING SYSTEMS

The stochastic formulation of a dynamic ship-positioning problem is presented. Starting with the specification in general terms describing what the system has to achieve, the design process is traced out until a Kalman filter solution is determined and successfully tested on a model simulation incorporating all the actual system non-linearities.

The ship-positioning problem is introduced in the following section, subsequently, the design requirements of the system and a simple classical control solution is described. In Sections 13.2.3 to 13.2.5 the non-linear thrust devices, position measurements and non-linear ship models are described. The optimal filter is designed using linearized equations for the model of the vessel these being found in Section 13.2.6.

13.2.1. Dynamic ship-positioning systems: An introduction

The objective of a dynamic-positioning system is to maintain the position and heading of a vessel at reference values by the use of thrust devices (Barton [4]). These are used to counteract the forces and moments imposed on the vessel by the incident wind, waves and sea currents. The design of the system involves a compromise between the accuracy of holding a position and the need to suppress excessive thruster response to wave motion (Brink *et al.* [5]). The latter is oscillatory in nature, and excessive thruster response involves the penalties of extra power demand and detrimental wear on the thruster mechanism. It is unnecessary to use the thruster to counteract first-order wave force and, in fact, they are not rated for this duty. Ideally, the thrusters should not respond to the oscillatory components of the wave motion and should be run at a steady power level to offset low-frequency disturbances, for example, sea-drift forces.

In a conventional dynamic ship-positioning system using PID controllers and notch filters, the wave-filters impose a phase lag on the position error signals. This phase lag restricts the allowable bandwidth which can be used for the controller, hence an inevitable conflict arises between controller bandwidth and filter attenuation. The more effective the wave-filter can be made in reducing thruster oscillations due to the waves, the less satisfactory is the position holding accuracy due to the increased restriction on the controller bandwidth. This type of problem has led to the development of a second generation of

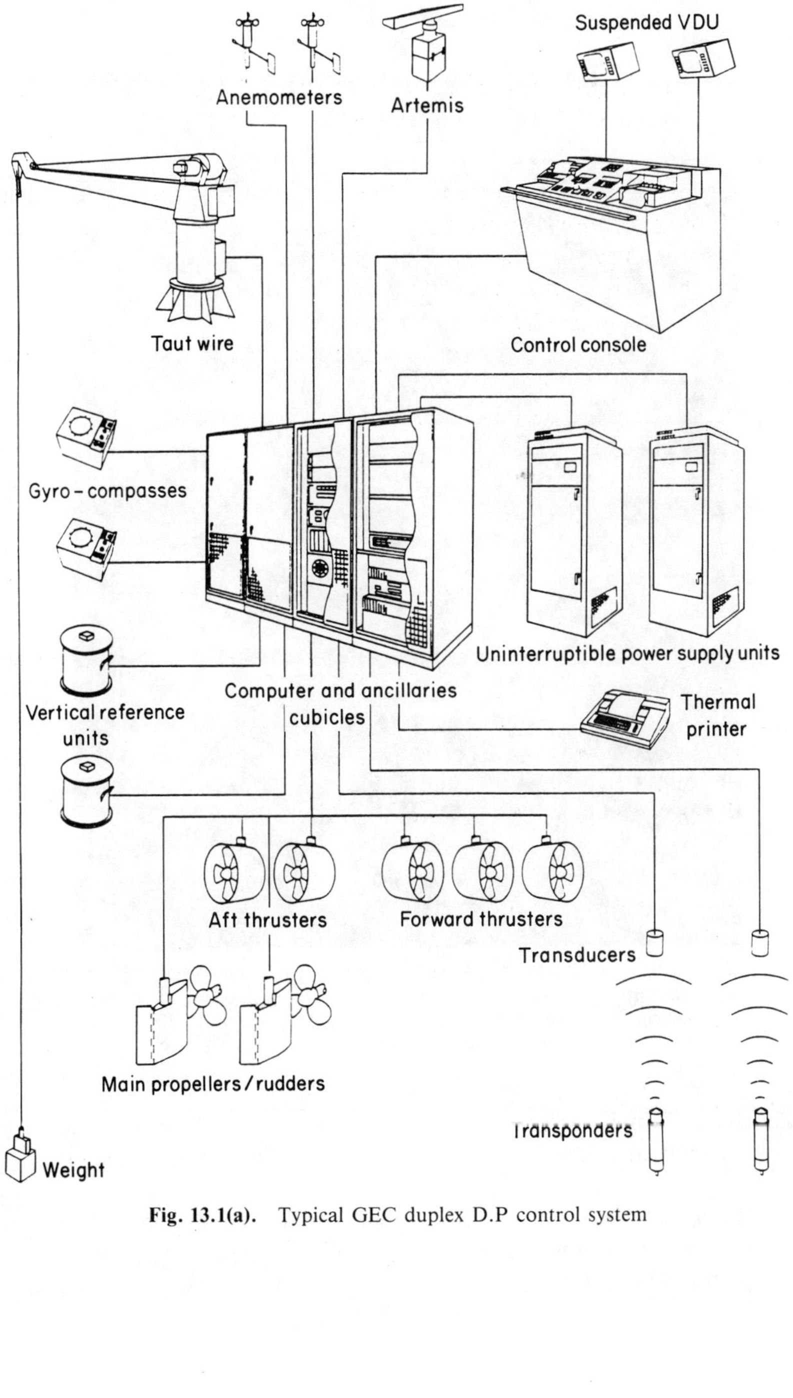

Fig. 13.1(a). Typical GEC duplex D.P control system

Fig. 13.1(b). STADRILL: A self-propelled semi-submersible drilling platform. The presence of a supply vessel alongside and a Sikorsky helicopter above demonstrate the type of duty for which a dynamic positioning control system is required

dynamic-positioning systems based on optimal stochastic control theory and employing Kalman filters (Balchen *et al.* [2], [3] and Grimble *et al.* [6], [7], [8], [9]).

The vessel motions are normally modelled under the assumption that they consist of the sum of low- (LF) and high- (HF) frequency components (Wise and English [10]). The LF motions are due to the thrusters, sea currents, wind

Fig. 13.1(c). STADIVE: A new multi-functional vessel capable of emergency fire-fighting, service and routine diving duties. This vessal operates in the Red Sector of the North Sea. Photograph by courtesy of Shell Petroleum Co. Ltd

and the second-order wave forces. The drift of the vessel is determined by the last three forces. The HF motions are assumed to be due to the first-order wave motions. These cause the oscillatory motion of the vessel at the wave frequencies 0.3–1.6 rads/sec. The LF motion components are to be controlled and the HF components which lead to thruster modulation are to be ignored.

The main elements of a dynamic ship-positioning system comprise (i) the thrusters (taken together with the main propellers in some cases), (ii) the position measurement system and (iii) the control computer. These are shown in Fig. 13.1(a) and are discussed further in the following sections. Two types of vessels which employ dynamic ship positioning systems are shown in the frontispiece photograph, and in Fig. 13.1(b) and (c).

13.2.2. Positioning system design requirements and the limitations of classical design solutions

A control system designer must arrive at a compromise between the desired quality of performance, the effort expended in the design process and the complexity of implementing the design solution. For dynamic ship-positioning systems, conflicting requirements arise between adequate attentuation of thruster modulation and position holding accuracy and the overall system response. In cases where this conflict cannot be resolved by the use of a fixed wave-filter, then adaptive wave-filters must be used. There is also a balance to be achieved between improved system response and accuracy and the amount of extra thruster activity and power consumption needed to bring about this improvement. Some vessel owners are willing to sacrifice accuracy for a more economical use of thruster power.

Normally vessel-heading is under the direct control of the operator. However, if required, the control system can be arranged so that the vessel-heading is altered until it reaches a point that requires minimum power from the thrusters.

Another feature of control system design is the incorporation of dead zones in the controller. This feature helps to effect further reductions in the thruster modulation than are possible by wave-filter action alone; these will not be considered further here.

The sway and yaw motions (Fig. 13.2) are coupled and the design of the sway and yaw position control system is therefore considered in the following. The surge motion is not dependent (in the linearized models) on the sway and yaw motions and will therefore be neglected here, since the single-input/single-output surge control system design is straightforward. The remaining motions

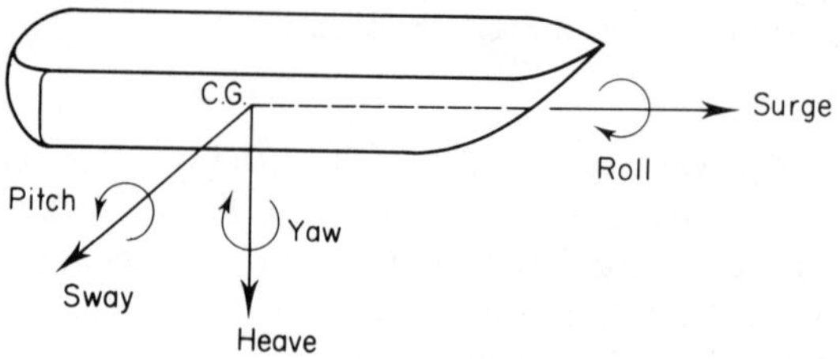

Fig. 13.2. Cartesian coordinate system

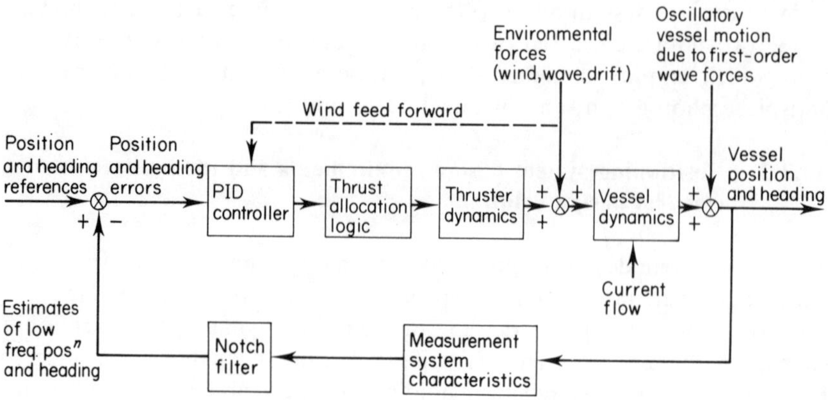

Fig. 13.3. Classical PID controller/notch filter scheme

of the vessel shown in Fig. 13.2 are not to be controlled since only the motions in the horizontal plane are of importance in survey or drilling vessels (Morgan [22]).

The first dynamic-positioning systems were designed using a classical approach and the usual feedback structure adopted is shown in Fig. 13.3. Three term (PID) controllers were employed together with notch wave filters.

Notch filter design

The oscillatory vessel motions due to the waves must be removed from the position and heading signals used for control. Failure to achieve this objective means that the controller will try to compensate for these motions and the resulting thruster oscillations (thruster modulation) will cause excessive wear on the thrusters and extra power consumption. In heavy weather, this thruster modulation can reduce the effectiveness of the thrusters in counteracting wind, wave-drift and current forces.

The notch filters produce estimates of the low-frequency part of the measured position and heading signals for use by the controller (namely, they remove the wave-motion components). The position signal is passed through a series of notch-filters, which are tuned to attenuate the wave-motion part of the signal, corresponding to a particular sea state.

The notch-filter normally consists of three cascaded second-order filters as follows:

$$H(s) = \prod_{j=1}^{3} \frac{(s^2 + 2\xi\omega_j s + \omega_j^2)}{(s^2 + 2\omega_j s + \omega_j^2)}$$

where $\xi = 0.1$ and ω_j is the notch centre-frequency of the jth second-order notch-filter. Using the bilinear transform:

$$s = (z - 1)/(z + 1)$$

where $z = e^{st}$, then

$$H(z) = \prod_{j=1}^{3} \frac{(1 + 2\xi\omega_j + \omega_j^2)z^2 + (2\omega_j^2 - 2)z + (1 - 2\xi\omega_j + \omega_j^2)}{(1 + 2\omega_j + \omega_j^2)z^2 + (2\omega_j^2 - 2)z + (1 - 2\omega_j + \omega_j^2)}.$$

In this type of marine problem, it is usual to adopt per-unit quantities thus the digital equivalent of $H(s)$ follows by replacing ω_j, by the per-unit frequency ω_j'. The centre frequencies for the notch-filters are typically $\omega_1 = 0.4$ rad/sec, $\omega_2 = 0.63$ rads/sec and $\omega_3 = 1.0$ rads/sec. The corresponding per-unit frequencies ($\omega_{\text{base}} = 1/T_{\text{base}} = 1/2.728$ rads/sec) become $\omega_1' = 1.091$, $\omega_2' = 1.72$ and $\omega_3' = 2.728$.

PID controller design

The PID controller operates on the low-frequency position and heading error signals, to command values of force and moment from the thrust devices, to reduce these errors to acceptable levels. The integral part of the control is required to offset the steady-state forces on the vessel due to wind, wave-drift and current.

The controller parameters are normally obtained from a frequency response analysis of the system and desired values of gain and phase margin are typically in the region of 12 dB and 50°. The bandwidth of the controller is limited by the thruster response time and the penalty of phase lags introduced by the notch filters, hence the inevitable conflict arises between attenuation of thruster modulation and control system response and accuracy.

The PID controller for the surge motions typically has the form:

$$C_s(s) = k_s(1 + \tau_d s + \tau_i/s)$$

where

$$k_s = 0.012, \ \tau_d = 50 \quad \text{and} \quad \tau_i = 0.005.$$

The differential term may be implemented in discrete equations as follows:

$$y(i) = (x(i) - x(i-1))/T$$

hence

$$y(z^{-1}) = \frac{1}{T}(1 - z^{-1})x(z^{-1}).$$

The integral term may be implemented using the following results:

$$y(i) = y(i-1) + \int_{(i-1)T}^{iT} x(t)\,dt$$

$$\simeq y(i-1) + \frac{T}{2}(x(i-1) + x(i))$$

giving

$$y(z^{-1}) = \frac{T}{2}\frac{(1 + z^{-1})}{(1 - z^{-1})}\,x(z^{-1}).$$

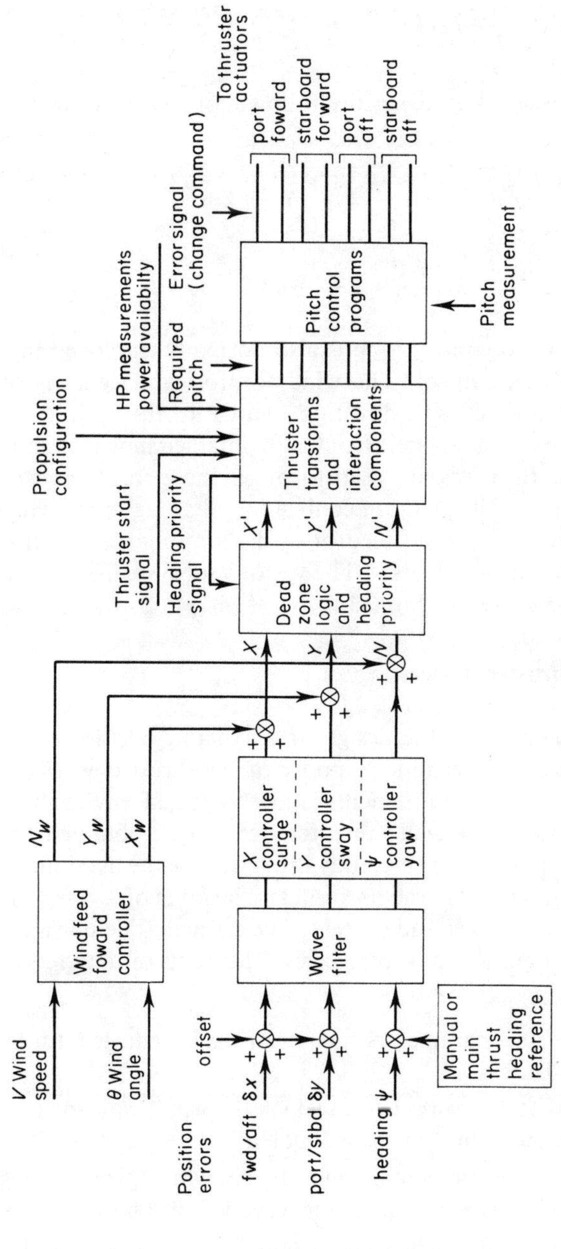

Fig. 13.4. Position control system

The PID controller therefore becomes:

$$C_s(z^{-1}) = k_s\left(1 + \frac{\tau_d}{T}(1 - z^{-1}) + \frac{\tau_i T}{2}\frac{(1 + z^{-1})}{(1 - z^{-1})}\right)$$

which may be implemented using the following recurrence relation:

$$y(i) = y(i-1) + k_s\left(1 + \frac{\tau_d}{T} + \frac{\tau_i T}{2}\right)x(i)$$

$$+ k_s\left(\frac{\tau_i T}{2} - 1 - \frac{2\tau_d}{T}\right)x(i-1) + k_s\,\frac{\tau_d}{T}\,x(i-2).$$

A wind feedforward-control loop is also normally employed in addition to the position feedback control. The wind feedforward signal is obtained by measuring the wind speed and direction relative to the vessel. The resulting forces and moments on the vessel (from data obtained in wind-tunnel tests) are computed and the thrusters are used to offset these forces and movements. Since the wave-filter in the position control loop has lag terms which limit the controller bandwidth, the wind feedforward action makes a useful contribution in accommodating wind gusts. The various subsystems of such a position control system are shown in Fig. 13.4.

13.2.3. The thruster devices

The first consideration in the design of a dynamic-positioning system is to ensure that the size and dynamic response of the thrust devices are adequate to meet the operational requirements under specified environmental conditions. Confirmation of adequate thruster size can be obtained from a 'capability study'. The purpose of this study is to estimate the maximum forces and moment required to maintain the position and heading of a vessel in a specified environment of wind, waves and current. The estimates include allowances for counteracting wind gusts and disturbances. The items of data required for the study include:

(a) The main vessel parameters such as displacement, length, breadth and operating draft.

(b) Coefficients for the vessel from which the wind, wave-drift and current forces and moments can be estimated.

(c) The limiting environmental conditions in which dynamic positioning is required (wind speed, significant wave height and current speed).

(d) Operational requirements (for example, details of vessel duties and whether the choice of heading is constrained or free).

The vessel on which the following discussions and parameters are based is the *Wimpey Sea-Lab*. Its thrusters are retractable a.c. motor driven with variable pitch propellers. The vessel has four rotatable thrusters, two in the bow and two in the stern, they are capable of 360° rotation and each produce about 12.5 tonnes of thrust.

Thrust allocation logic

The thruster dynamics are modelled in the following and the scheduling of the individual thrusters need not be considered. The thrust allocation logic takes the demanded forces and moments from the controller and sets the available thrust devices so that these demanded forces and moments are achieved as closely as possible. In the general case this is a static optimization problem, where minimum fuel consumption is the objective. Data required to design the logic includes the thruster configuration, together with thrust/pitch and power/pitch relationships, the limiting thrust available for each thrust device, and thruster/hull interaction relationships. Among many features of this logic are:

(a) ensuring that sufficient thrusters are available for dynamic positioning control,

(b) reallocation of thrusters, where this is possible, in the case of failure of one of the operational thrusters,

(c) provision of priority for heading control demands, over position control demands, when any thruster approaches its limiting thrust.

Thruster subsystems

The block diagram for a thruster is shown in Fig. 13.5. The thrusters are non-linear devices incorporating the following subystems.

Input servo

The input servo, is a bang-bang device (Fig. 13.6). An electrical input circuit compares a reference voltage against an electrical feedback from potentiometers measuring the movement of a ram. The error signal is applied to comparators which switch forward or reverse solenoid valves when the error exceeds a predetermined dead band. This dead band is set to stop the servo from hunting.

In practice the comparators are set when the position error is $\pm 2\%$ and $K_1 = 0.2$ so that the ram moves 100% in 5 seconds when a valve is fully open ($\int_0^5 0.2 \; dt = 1$).

The electrical comparators switch a voltage on to the solenoid of the electro-hydraulic valve which starts to open when the flux has built up sufficiently for the solenoid to exert a force greater than the spring and friction forces on the

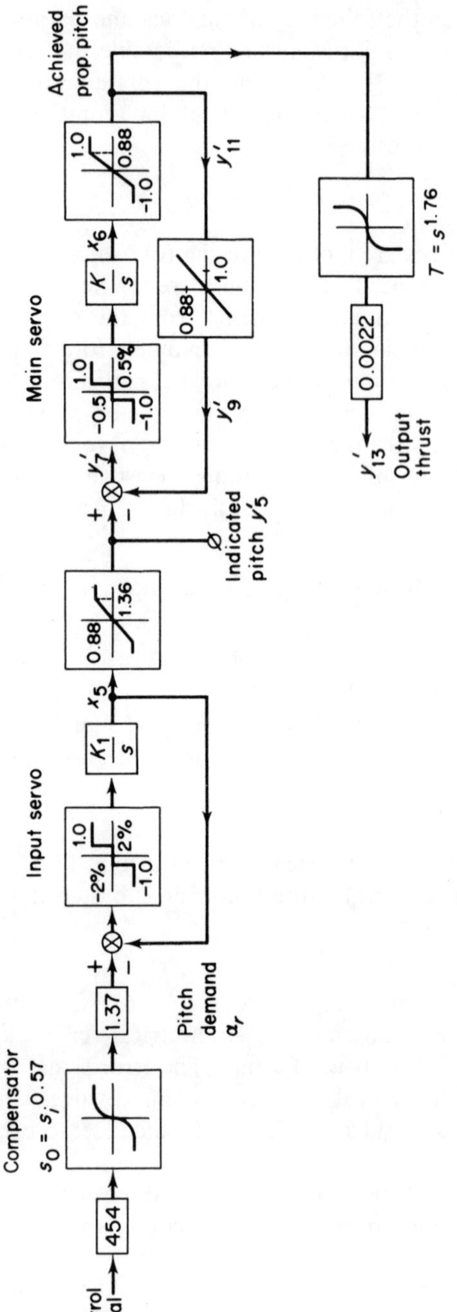

Fig. 13.5. Block diagram of the thrusters and compensator

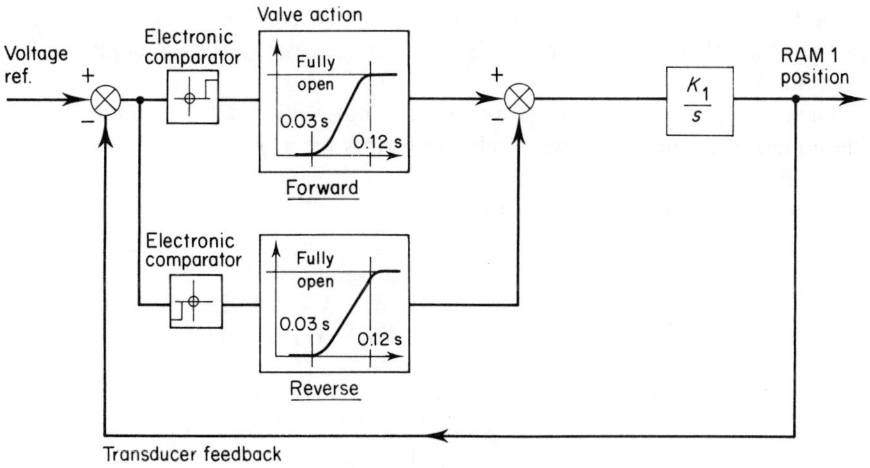

Fig. 13.6. Input servo

valve. The valve takes about 30 msec to start moving and about 125 msec to become fully open.

Opening the valve permits oil to flow to one side of the ram which then starts to move. When the valve is fully open the ram moves at maximum speed depending on how quickly the oil flow can fill the cylinder.

Spring box

This restricts the force exerted on the mechanical linkage between the two servos. The spring is sufficiently still to ensure that all stiction is overcome and the input to the second servo accurately follows the motion of ram 1.

Linkage

Force is transmitted to the main servo by a linkage between the spring box and the input to the main servo. A non-linearity arises from the use of finite radius arms but this is not great as the angular movement is small.

Main servo

The use of an underlapped valve ensures that the servo reacts when the input valve spool moves. The speed of ram 2 is limited by the oil flow into the cylinder. In practice the output servo is faster than the input servo so that K_2 should be greater than 0.2 to allow the ram 2 to move 100% in less than 5 seconds. The servo-control loop includes a small dead band.

Thrust v pitch law

The most severe non-linearity arises with the propeller pitch versus thrust characteristic. This relationship approximates $T = s^m$ where $m = 1.76$. A compensator $s_0 = s_i^{0.57}$ is normally added at the input to the thruster control circuits to make the input/output relationship approximately linear. The dead zones in the relay elements shown in Fig. 13.5 are necessary to avoid limit cycle oscillations.

The thruster models for the sway and yaw motions are defined by the following equations, where for sway (odd subscripts on y'):

$$\alpha_r = (454.5 \times 1.37)u_1^{0.57}$$

$$y_3 = \begin{cases} 1 \\ 0 \\ -1 \end{cases} \quad \text{if} \quad \begin{cases} (\alpha_r - x_5) > 0.02 \\ |\alpha_r - x_t| \le 0.02 \\ (\alpha_r - x_5) < -0.02 \end{cases}$$

$$\dot{x}_5 = K_1 y'_3$$

$$y'_5 = \begin{cases} 0.88 \\ 0.644x_7 \\ -0.88 \end{cases} \quad \text{if} \quad \begin{cases} x_5 > 1.364 \\ |x_5| \le 1.364 \\ x_5 < -1.364 \end{cases}$$

$$y'_7 = y'_5 - y'_9 \tag{13.1}$$

$$y'_{15} = \begin{cases} 1 \\ 0 \\ -1 \end{cases} \quad \text{if} \quad \begin{cases} y'_5 > 0.005 \\ |y'_5| \le 0.005 \\ y'_5 < -0.005 \end{cases}$$

$$\dot{x}_6 = k_2 y'_{15}$$

$$y'_{11} = \begin{cases} 1.0 \\ 1.1354x_5 \\ -1.0 \end{cases} \quad \text{if} \quad \begin{cases} y'_{11} > 1.0 \\ |y'_{11}| \le 1.0 \\ y'_{11} < 1.0 \end{cases}$$

$$y'_9 = 0.88y'_{11}$$

$$y'_{13} = 0.0022y'^{1.76}_{11}.$$

The input and output scaling factors of 454.5 and 0.0025 ($= 1/454.5$) are required to change from per-unit input and into per-unit output quantities respectively. Some care must be exercised when representing system non-linearities by linear models for the purposes of linear filter or control design. The thrusters can be approximated by either first-order or second-order models. The first-order models have the advantage of reducing the dimension of the Kalman filter. These have the form:

$$\begin{aligned} \dot{x}_5 &= -1.55x_5 + 1.55u_1 \\ \dot{x}_6 &= -1.55x_6 + 1.55u_2 \end{aligned} \tag{13.2}$$

where

$$x_5 = \text{sway thrust } (\equiv y_{13}) \text{ and } x_6 = \text{yaw thrust } (\equiv y_{14}).$$

The results of frequency-response tests on the non-linear thruster model suggested the use of a second-order linearization. The following time constants were estimated from Bode amplitude diagrams.

PEAK INPUT (pu)	τ_1	τ_2
0.0002	0.3981	0.3055
0.0005	0.7244	0.4266
0.001	1.059	0.6918
0.002	1.585	0.861

Since the maximum control input is 0.0022 pu, the average input of 0.001 pu can be used to select representative time constants, giving $1/\tau_1 + 1/\tau_2 = 2.3895$ and $1/(\tau_1\tau_2) = 1.3646$.

$$\left.\begin{aligned} \dot{x}_5 &= -2.3895x_5 - 1.3646x_6 + 1.3646u_1 \\ \dot{x}_6 &= x_5 \\ \dot{x}_7 &= -2.3895x_7 - 1.3646x_8 + 1.3646u_2 \\ \dot{x}_8 &= x_7 \end{aligned}\right\} \qquad (13.3)$$

where x_6, x_8 correspond to the thrust outputs in sway (y_{13}) and yaw (y_{14}), respectively.

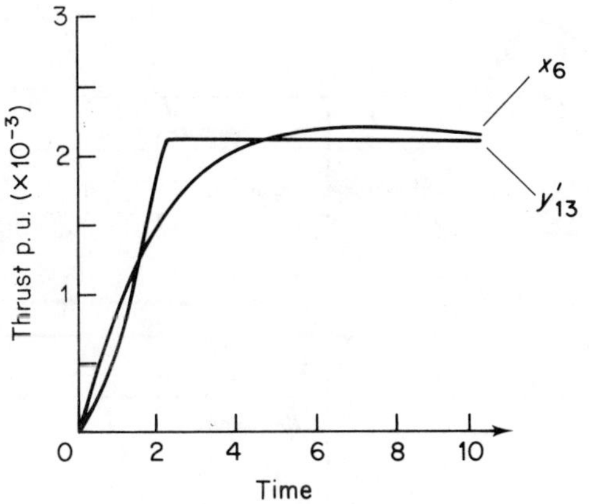

Fig. 13.7(a). Thruster step responses for linear and non-linear models (step input 0.0022 p.u.)

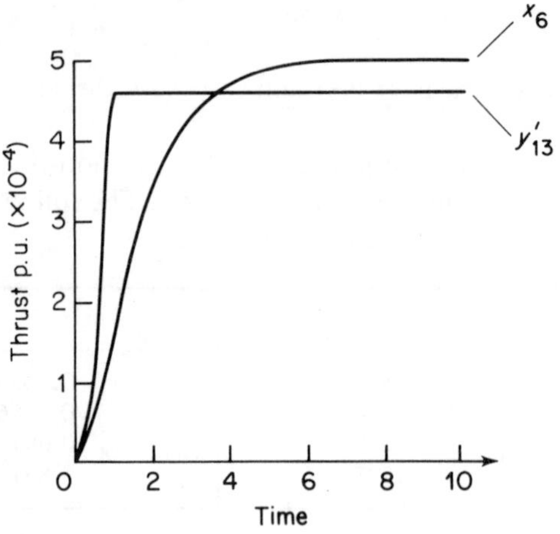

Fig. 13.7(b). Thruster step responses for linear and non-linear models (step input 0.0005 p.u.)

The linear and non-linear model step responses are compared in Figs. 13.7(a), (b) for step inputs of magnitude 0.0022 and 0.0005 per-unit respectively.

13.2.4. The position-measurement systems

Several types of position reference systems have been used in dynamic positioning control systems but the most commonly used involve hydroacoustic sensors (Van Calcar [23]). This system is supported by taut-wire angle sensors (Fig. 13.8). An acoustic system calculates the position of the vessel relative to

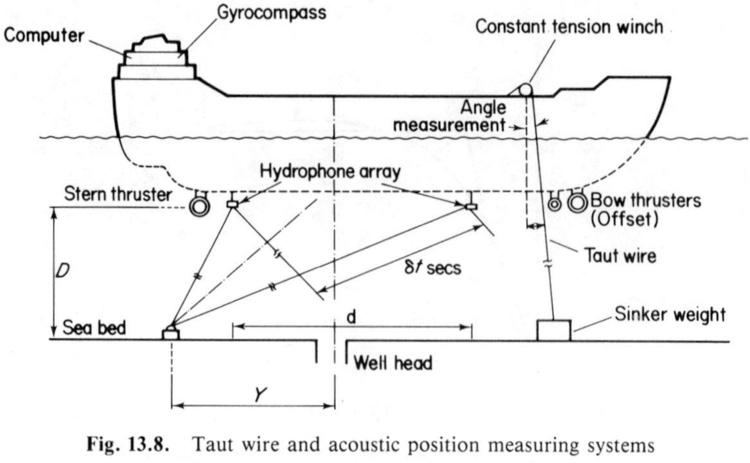

Fig. 13.8. Taut wire and acoustic position measuring systems

the ocean floor by measuring the time of arrival of pulses from beacon (life two months) or a transponder (life two years) on the seabed. These pulses are measured using a set of hydrophones mounted below the hull of the vessel. The noise level at the hydrophones is mainly determined by the noise of the main propellers and the thrusters, which produce a much higher level than the background sea noise and may even cause complete loss of the position reference for short periods.

Tamchiro *et al.* [11] gives the following requirements for a hydroacoustic reference system to maintain an accuracy of 1% of water depth in the measured position (assuming the vessel remains within a radius of 10% of the water depth relative to the beacon):

(i) The error of range estimation should be less than 1%.
(ii) The error in the sound velocity in water should be less than 0.05%.
(iii) The error of the time of arrival of pulses at the hydrophones should be less than 4 seconds.
(iv) The error of pitch and roll sensing should be less than 0.25 degrees.

The acoustic measurement signals often include spikes and these may be allowed for by increasing the measurement noise covariance matrix appropriately. The Kalman filter depends upon this covariance matrix and thus the best filter is automatically selected for this situation. The Kalman filter also performs well under a temporary loss of position measurements. In this case the estimates rely upon the model of the vessel rather than the observations input.

13.2.5. Equations of motion and disturbances

The first requirement in the design of a dynamic positioning system is to develop a composite mathematical model for the system. The subsystem models should:

(a) enable the wind, current and wave drift forces to be estimated,
(b) describe the vessel dynamics,
(c) describe the thrust device dynamics,
(d) quantify the interaction effects between the thrust devices, the vessel hull and the current flow.

These models are obtained from a series of tests carried out on models of the vessel in wind tunnels and towing tanks. Where such testing is not available then reliance must be placed on a combination of empirical and theoretical results, backed by tank and wind tunnel test experience obtained for similar vessels. To derive the dynamic equations the vessel is normally treated as a rigid body having freedom in surge, sway and yaw but restricted in heave, pitch and roll. Only motions in the horizontal plane are considered in ship-positioning problems.

The vessel dynamics are represented by non-linear differential equations. For control-system design, linearized equations are required. Unfortunately, the parameters of these linearized models are affected by the steady current flow and the transient force levels applied to the vessel. Selection of the appropriate parameters for the linearized model constitutes part of the art of control system design (English and Wise [12]).

The non-linear equations of motion, expressed in a formal style with white-noise inputs, have the form:

$$\begin{aligned}
\dot{x}_1 &= -2.4022\,|x_1|\,x_1 + 0.03696\,|x_3|\,x_3 - 0.5435\,|x_1|\,x_3 \\
&\quad + 0.5435\,y_{13} + 0.5435\,\xi_1 \\
\dot{x}_2 &= x_1 \\
\dot{x}_3 &= 2.5245\,|x_1|\,x_1 - 1.585\,|x_3|\,x_3 - 1.634\,y_{14} + 9.785\,\xi_2 \\
\dot{x}_4 &= x_3
\end{aligned} \tag{13.4}$$

where

x_1 = sway velocity

x_2 = sway position

x_3 = yaw velocity

x_4 = yaw angle

y_{13} = thrust 1

y_{14} = thrust 2

ξ_1, ξ_2 = process noises.

The design of the Kalman filter for the low frequency motions involves a linearized version of the above equations (see Wise and English [10], [12]) which can be obtained (under no current flow conditions) as:

$$\left.\begin{aligned} \dot{x}_1' &= -0.0546x_1' + 0.0016x_3' + 0.5435y_{13} + 0.5435\xi_1 \\ \dot{x}_2' &= x_1' \\ \dot{x}_3' &= 0.0573x_1' - 0.0695x_3' - 1.634y_{14} + 9.785\xi_2 \\ \dot{x}_4' &= x_3'. \end{aligned}\right\} \qquad (13.5)$$

The environmental forces acting on a vessel may be summarized as follows:

(a) *Wind.* The wind forces comprise a steady component, corresponding to the mean wind speed and a transient component, corresponding to the gusting of the wind about its mean value.

(b) *Current.* Current flow remains substantially constant over long periods of time and imposes steady forces and moments on a stationary vessel.

(c) *Waves.* Wave forces and moments on a vessel comprise two parts:

  (i) First-order wave forces which cause the vessel to oscillate in sympathy with the waves. These forces are normally much larger than those available from the thrust devices and hence no effective control is possible over these oscillatory motions.

  (ii) Second-order wave forces which impose relatively steady forces on the vessel known as wave drift forces. These forces are much smaller in magnitude than the first-order forces and must be countered by the thrust devices to prevent the vessel drifting off position.

It is generally believed that waves are created by the wind and for modelling purposes it is usual to assume that the wind has blown for a long time so that the sea is fully developed. The wind has a mean value which is a constant over a long period and the wind fluctuations can be treated as a Gaussian random variable. The ship will respond to wind gusts in the frequency range 0 to 0.25 radians/second.

Process noise depends upon the forces and moments imposed on the vessel by wind gusts. Estimates of the variances of the wind gust forces (surge and sway) and yaw moment for the vessel *Wimpey Sea-Lab* were obtained as:

Beaufort No.	Mean wind speed (m/sec)	σ^2_{surge} (pu)	σ^2_{sway} (pu)	σ^2_{yaw} (pu)
6	12.4	1.539×10^{-8}	8.625×10^{-8}	2.594×10^{-9}
7	15.4	3.661×10^{-8}	2.052×10^{-7}	6.170×10^{-9}
8	19.0	8.483×10^{-8}	4.755×10^{-7}	1.430×10^{-8}

The quantities are in per-unit form where the base quantities are:

base length = 94.5m, base force = 5.563×10^{-4} kN and

base moment = 5.257×10^{-6} kNm.

If allowance is made for a (non-optimal) wind feedforward control then these variances can be reduced to a quarter of the above values.

Manufacturers of the position measurement devices recommend the following measurement noise variances:

Heading: $\sigma = 0.8103^0$, $\sigma' = 1.414 \times 10^{-2}$ rads, $(\sigma')^2 = 2 \times 10^{-4}$
Position: $\sigma = 0.3274$m, $\sigma' = 3.464 \times 10^{-3}$ (per-unit) $(\sigma')^2 = 1.2 \times 10^{-5}$.

This completes the basic information needed to define the linearized ship model.

13.2.6. Linearized ship equations

The optimal filters which are to be defined are based upon the following linearized and discretized ship equations (Price and Bishop [13]). The thrusters and low-frequency ship dynamics can be represented by the LF subsystem S_l. This subsystem is stabilizable and detectable and is represented by the following discrete, time-invariant, state equations:

$$S_l: \begin{cases} \boldsymbol{x}_l(t+1) = \mathbf{A}_l \boldsymbol{x}_l(t) + \mathbf{B}_l \boldsymbol{u}(t) + \mathbf{D}_l \boldsymbol{\omega}(t) & (13.6) \\ y_l(t) = \mathbf{C}_l \boldsymbol{x}_l(t) \\ z_l(t) = y_l(t) + \boldsymbol{v}(t) & (13.7) \end{cases}$$

where

$$E\{\boldsymbol{\omega}(t)\} = \mathbf{o}, \qquad E\{\boldsymbol{\omega}(k)\boldsymbol{\omega}^T(m)\} = \mathbf{Q}\,\delta_{km}$$

$$E\{\boldsymbol{v}(t)\} = \mathbf{o}, \qquad E\{\boldsymbol{v}(k)\boldsymbol{v}^T(m)\} = \mathbf{R}\,\delta_{km} \tag{13.8}$$

and δ_{km} is the Kronecker delta function, $\boldsymbol{x}(t) \in \mathbb{R}^n$, $\boldsymbol{u}(t) \in \mathbb{R}^m$, $\boldsymbol{\omega}(t) \in \mathbb{R}^q$ and $y_l(t) \in \mathbb{R}^r$. The process noise $\boldsymbol{\omega}(t)$ is used to simulate the wind disturbance and $\boldsymbol{v}(t)$ represents a white-measurement-noise signal. The plant matrices $\mathbf{A}_l$, $\mathbf{B}_l$, $\mathbf{C}_l$ and $\mathbf{D}_l$ are assumed constant and known. The observed plant output includes the coloured noise (wave disturbance) signal $y_n(t)$ and is given as:

$$x(t) = z_l(t) + y_n(t). \tag{13.9}$$

The high-frequency wave disturbance (Pierson and Marks [14]) can be represented by the following multivariable autoregressive moving-average model:

$$S_h\colon \mathbf{A}_h(z^{-1})y_h(t) = \mathbf{C}_h(z^{-1})\xi(t) \tag{13.10}$$

which is assumed to be asymptotically stable and $y_h(t) \in \mathbb{R}^r$ and $\xi(t) \in \mathbb{R}^r$. Here $\xi(t)$ represents an independent, zero mean, random vector which is uncorrelated with $\omega(t)$ and $v(t)$ and has a diagonal covariance matrix $\mathbf{\Sigma}_\xi$. The polynomial matrices $\mathbf{A}_h(z^{-1})$ and $\mathbf{C}_h(z^{-1})$ are assumed to be square and of the form:

$$\mathbf{A}_h(z^{-1}) = \mathbf{I}_r + \mathbf{A}_1 z^{-1} + \mathbf{A}_2 z^{-2} + \cdots + \mathbf{A}_{n_a} z^{-n_a} \tag{13.11}$$

$$\mathbf{C}_h(z^{-1}) = \mathbf{C}_1 z^{-1} + \mathbf{C}_2 z^{-2} + \cdots + \mathbf{C}_{n_c} z^{-n_c} \tag{13.12}$$

where z^{-1} is the backward shift operator. The matrix $\mathbf{A}_h(z^{-1})$ can be taken to be regular (namely $\mathbf{A}_{n_a}$ is non-singular). The zeros of $\det(\mathbf{A}_h(\cdot))$ and $\det(\mathbf{C}_h(\cdot))$ are assumed to be strictly outside the unit circle. The orders of the polynomial matrices are known but the coefficient matrices $\{\mathbf{A}_i\}$ and $\{\mathbf{C}_j\}$, $i = 1, \ldots n_a$, $j = 1, \ldots n_c$, are treated as unknowns, since in practice the wave-disturbance spectrum varies slowly with weather conditions. It is also assumed that the disturbances in each observed channel are uncorrelated so that the matrices $\{\mathbf{A}_i\}$ and $\{\mathbf{C}_j\}$ have diagonal form.

The above two subsystems are illustrated in Fig. 13.9.

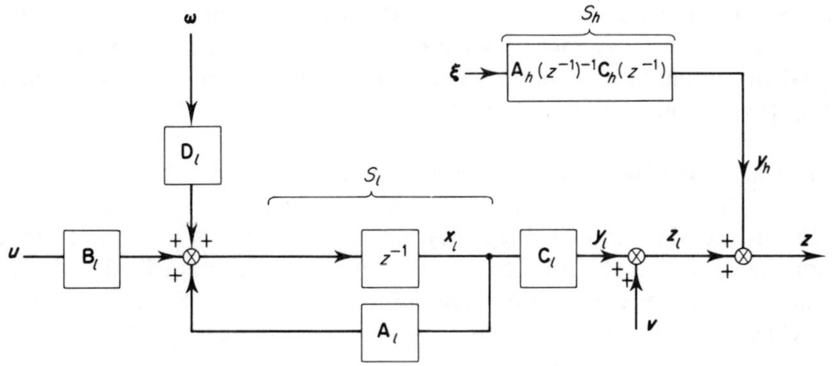

Fig. 13.9. LF and HF subsystems for a linearized ship model

13.2.7. Low-frequency motion estimator

If it is assumed that the coloured-noise signal y_n can be measured then the plant outputs z_l can be calculated (Fig. 13.9). With input z_l and assuming the ship equations and noise covariances are known, the plant states x_l can be estimated using a Kalman filter. The worst case process noise is assumed (this is taken as being stationary) and the LF subsystem is assumed to be stabilizable and detectable. Under these conditions, the Kalman gain is constant and may therefore be computed off-line.

The LF Kalman filter algorithm becomes:

$$\hat{x}_l(t \mid t-1) = \mathbf{A}_l \hat{x}_l(t-1 \mid \mathrm{t}-1) + \mathbf{B}_l u(t-1).$$

Predictor:
$$\hat{y}_l(t \mid t-1) = \mathbf{C}_l \hat{x}_l(t \mid t-1)$$

$$\mathbf{P}(t \mid t-1) = \mathbf{A}_l \mathbf{P}(t-1 \mid t-1)\mathbf{A}_l^T + \mathbf{D}_l \mathbf{Q} \mathbf{D}_l^T$$

$$\hat{x}_l(t \mid t) = \hat{x}_l(t \mid t-1) + \mathbf{K}_l(t)\boldsymbol{\varepsilon}_l(t). \tag{13.13}$$

Corrector:
$$\hat{y}_l(t \mid t) = \mathbf{C}_l \hat{x}_l(t \mid t)$$

$$\mathbf{P}(t \mid t) = \mathbf{P}(t \mid t-1) - \mathbf{K}(t)\mathbf{C}_l \mathbf{P}(t \mid t-1)$$

$$\mathbf{K}(t) = \mathbf{P}(t \mid t-1)\mathbf{C}^T[\mathbf{C}\mathbf{P}(t \mid t-1)\mathbf{C}^T + \mathbf{R}]^{-1}$$

where

$$\boldsymbol{\varepsilon}_l(t) \triangleq z(t) - \hat{y}_l(t \mid t-1) - y_h(t) \tag{13.14}$$

$$= z_l(t) - \hat{y}_l(t \mid t-1)$$

and $\mathbf{K}(\cdot)$ is the Kalman gain matrix, $\mathbf{P}(\cdot)$ is the error covariance matrix. Unfortunately $y_h(t)$ cannot be separated from $z(t)$ by measurement and the signal $\boldsymbol{\varepsilon}_l(t)$ cannot be calculated. This signal may, however, be approximated using (13.14) and the estimate $\hat{y}_h(t \mid t-1)$.

13.2.8. High-frequency motion estimator

In this section, the high-frequency motion estimator is constructed based upon the HF model (13.10). The assumption is made temporarily that the low-frequency motions can be estimated via the technique of Section 13.2.7.

Define a new variable $m_h(t)$ as:

$$m_h(t) = z(t) - \hat{y}_l(t \mid t-1) \tag{13.15}$$

and from (13.14):

$$\boldsymbol{m}_h(t) = \boldsymbol{\varepsilon}_l(t) + \boldsymbol{y}_h(t). \tag{13.16}$$

The innovations signal $\boldsymbol{\varepsilon}_l$ is white noise and $\boldsymbol{m}_h$ can be treated as the measured output of a plant S_h with measurement noise $\boldsymbol{\varepsilon}_l$. The covariance matrix for $\boldsymbol{\varepsilon}_l$ is denoted by $\boldsymbol{\Sigma}_{\boldsymbol{\varepsilon}_l}$. An innovations signal model may now be defined:

$$\mathbf{A}_h(z^{-1})\boldsymbol{m}_h(t) = \mathbf{D}_h(z^{-1})\boldsymbol{\varepsilon}(t) \tag{13.17}$$

where $\{\boldsymbol{\varepsilon}(t)\}$ is an independent random sequence with covariance matrix $\boldsymbol{\Sigma}_{\boldsymbol{\varepsilon}}$.

The matrix polynomial $\mathbf{D}_h(z^{-1})$ has the form:

$$\mathbf{D}_h(z^{-1}) = \mathbf{I}_r + \mathbf{D}_1 z^{-1} + \cdots + \mathbf{D}_{n_d} z^{-n_d}$$

where the zeros of $\det(\mathbf{D}_h(\cdot))$ lie strictly inside the unit circle. The parameters of $\mathbf{D}_h(z^{-1})$ are determined by the following spectral factorization:

$$\mathbf{D}_h(z^{-1})\boldsymbol{\Sigma}_{\boldsymbol{\varepsilon}}\mathbf{D}_h^T(z) = \mathbf{C}_h(z^{-1})\boldsymbol{\Sigma}_{\boldsymbol{\varepsilon}}\mathbf{C}_h^T(z) + \mathbf{A}_h(z^{-1})\boldsymbol{\Sigma}_{\boldsymbol{\varepsilon}_l}\mathbf{A}_h^T(z). \tag{13.18}$$

Note, that $n_d = n_a$ (since normally $n_a > n_c$) and that by multiplying both sides of equation (13.18) by z^{n_d} and taking the limit as $z \to 0$:

$$\mathbf{D}_{n_d}\boldsymbol{\Sigma}_{\boldsymbol{\varepsilon}} = \mathbf{A}_{n_a}\boldsymbol{\Sigma}_{\boldsymbol{\varepsilon}_l}. \tag{13.19}$$

Since $\mathbf{A}_h(z^{-1})$ is regular (that is $\mathbf{A}_{n_a}$ is non-singular) the following identity holds:

$$\mathbf{A}_{n_a}^{-1}\mathbf{D}_{n_d} = \boldsymbol{\Sigma}_{\boldsymbol{\varepsilon}_l}\boldsymbol{\Sigma}_{\boldsymbol{\varepsilon}}^{-1}. \tag{13.20}$$

The optimal estimate of $\boldsymbol{y}_h(t)$ can be calculated using:

$$\hat{\boldsymbol{y}}_h(t \mid t) = \boldsymbol{m}_h(t) - \boldsymbol{\Sigma}_{\boldsymbol{\varepsilon}_l}\boldsymbol{\Sigma}_{\boldsymbol{\varepsilon}}^{-1}\boldsymbol{\varepsilon}(t) \tag{13.21}$$

where

$$\boldsymbol{\varepsilon}(t) = \boldsymbol{m}_h(t) - \hat{\boldsymbol{y}}_h(t \mid t-1). \tag{13.22}$$

Using the identity in equation (13.20), $\hat{\boldsymbol{y}}_h(t \mid t)$ becomes:

$$\hat{\boldsymbol{y}}_h(t \mid t) = \boldsymbol{m}_h(t) - \mathbf{A}_{n_a}^{-1}\mathbf{D}_{n_d}\boldsymbol{\varepsilon}(t). \tag{13.23}$$

The wave-frequency model changes with environmental conditions and these variations are accounted for in (13.23) by estimating both $\mathbf{A}_{n_a}$, $\mathbf{D}_{n_d}$ and the innovations signal $\boldsymbol{\varepsilon}(t)$ (in the spirit of extended least-squares) as described below.

Approximation

The signal $\boldsymbol{y}_h(t)$ is not measurable and is replaced by the approximation $\hat{\boldsymbol{y}}_h(t \mid t)$ in the LF Kalman estimator. The calculated innovations then becomes:

$$\bar{\boldsymbol{\varepsilon}}(t) = \boldsymbol{z}(t) - \hat{\boldsymbol{y}}_l(t \mid t-1) - \hat{\boldsymbol{y}}_h(t \mid t). \tag{13.24}$$

This signal can be calculated using (13.15) and (13.22):

$$\bar{\boldsymbol{\varepsilon}}(t) = \mathbf{A}_{n_a}^{-1}\mathbf{D}_{n_d}\boldsymbol{\varepsilon}(t) \tag{13.25}$$

The signal $\boldsymbol{m}(t)$ is also modified and becomes:

$$\begin{aligned}\bar{\boldsymbol{m}}_h(t) &= \boldsymbol{m}_h(t) - \tilde{\boldsymbol{y}}_l(t \mid t-1)\\ &= \mathbf{A}_h(z^{-1})^{-1}\mathbf{D}_h(z^{-1})\boldsymbol{\varepsilon}(t) - \tilde{\boldsymbol{y}}_l(t \mid t-1).\end{aligned} \tag{13.26}$$

The quantity $\tilde{\boldsymbol{y}}_l$ is a slowly varying signal and can be treated as a constant over a short time interval. Let $\boldsymbol{s}(t) \triangleq \mathbf{A}_h(z^{-1})\tilde{\boldsymbol{y}}_l(t \mid t-1)$ then (13.26) becomes:

$$\mathbf{A}_h(z^{-1})\bar{\boldsymbol{m}}_h(t) = \mathbf{D}_h(z^{-1})\boldsymbol{\varepsilon}(t) - \boldsymbol{s}(t). \tag{13.27}$$

The innovations signal model can now be represented in the usual form for parameter estimation:

$$\bar{\boldsymbol{m}}_h(t) = \boldsymbol{\psi}(t)\boldsymbol{\theta} + \boldsymbol{\varepsilon}(t) \tag{13.28}$$

and the algorithm due to Panuska [50] can be employed to estimate the unknown parameters.

In the ship-positioning problem, the high-frequency disturbances can be assumed to be decoupled so that $\mathbf{A}_h(z^{-1})^{-1}\mathbf{D}_h(z^{-1})$ is a diagonal matrix and the parameters for each channel can be estimated separately (standard extended recursive least squares or maximum likelihood parameter identification algorithms may be used). For the ith channel:

$$\bar{m}_{h_i}(t) = \boldsymbol{\psi}_i(t)\boldsymbol{\theta}_i + \varepsilon_i(t) \tag{13.29}$$

where

$$\psi_i(t) \triangleq [-\bar{m}_{h_i}(t-1), \ldots, -\bar{m}_{h_i}(t-n_a); \; \varepsilon_i(t-1), \ldots, \varepsilon_i(t-n_d); \; -1] \tag{13.30}$$

$$\theta_i^T = [a_{i1}, \ldots, a_{in_a}; \; d_{i1}, \ldots, d_{in_d}; \; s_i]. \tag{13.31}$$

Past values of the innovations signal are approximated by:

$$\hat{\varepsilon}_i(t) = \bar{m}_{h_i}(t) - \hat{\psi}_i(t)\hat{\theta}_i \tag{13.32}$$

where $\hat{\psi}_i(t)$ is given by (13.30) with $\varepsilon_i(t-j)$ replaced by $\hat{\varepsilon}_i(t-j)$, $j = 1, 2, \ldots, n_d$ and $\hat{\theta}_i$ represents the estimated parameter vector.

13.2.9. Combined Kalman and self-tuning filter scheme

The LF Kalman estimator and the HF self-tuning filter are combined here in an algorithm which provides the low-frequency motion estimates needed for control purposes. The self-tuning filter gives automatic adaption to different sea states which was not a feature of the notch-filter schemes (Grimble [15]). The use of self-tuning filters in this application was first reported by Fung and Grimble [16]. An advantage of the self-tuning filtering scheme is that it only involves a small extension to existing fixed Kalman filtering systems. The structural simplicity of the scheme gives it an advantage over the use of a self-tuning controller, as discussed by Grimble *et al.* [17].

The recursive Kalman/self-tuning filter algorithm becomes:

Algorithm 13.1

1. Initialize $\hat{\theta}_i$. Initialize parameter covariance for each channel and assign the forgetting factor β. Initialize state estimates.
2. Generate the Kalman filter estimates $\hat{x}_l(t \mid t-1)$ and $\hat{y}_l(t \mid t-1)$ using (13.13).
3. Calculate $\bar{m}_{h_i}(t)$ using (13.26) and form $\hat{\psi}_i(t)$.
4. Parameter update:

$$\hat{\theta}_i(t) = \hat{\theta}_i(t-1) + \mathbf{K}_i^p(t)(\bar{m}_{h_i}(t) - \hat{\psi}_i(t)\hat{\theta}_i(t-1)).$$

5. Covariance and gain update:

$$\mathbf{P}_i^p(t) = \{\mathbf{P}_i^p(t-1) - \mathbf{K}_i^p(t)(\beta + \psi_i(t)\mathbf{P}_i^p(t-1)\psi_i^T(t))\mathbf{K}_i^p(t)^T\}/\beta$$

$$\mathbf{K}_i^p(t) = \mathbf{P}_i^p(t-1)\psi_i(t)(\beta + \psi_i(t)\mathbf{P}_i^p(t-1)\psi_i^T(t))^{-1}$$

where $0.95 < \beta < 1$.

6. Innovations update:

$$\hat{\boldsymbol{\varepsilon}}_i(t) = \bar{\boldsymbol{m}}_{h_i}(t) - \hat{\boldsymbol{\psi}}_i(t)\hat{\boldsymbol{\theta}}_i(t).$$

7. Calculate $\bar{\boldsymbol{\varepsilon}}_{li}(t)$ for channel i using (13.25)

$$\bar{\boldsymbol{\varepsilon}}_{l_i}(t) = \hat{a}_{n_a}^{-1}\hat{d}_{n_d}\hat{\boldsymbol{\varepsilon}}_i(t).$$

8. If $i < r$, number of channels go to step 3.
9. Generate the state $\hat{x}_l(t \mid t)$ (using equation (13.13)).
10. Calculate the estimated $\tilde{y}_l(t \mid t-1)$ as:

$$\hat{\hat{y}}_{li}(t \mid t-1) = \hat{s}_i(t)/\hat{A}_{h_i}(1). \tag{13.33}$$

11. Correct the position estimates using $\hat{\hat{y}}(t \mid t-1)$. Return to step 2. □

The structure of the Kalman/self-tuning filtering scheme for the dynamic positioning system is shown in Fig. 13.10. Note that to derive equation (13.33), the final value theorem for z-transforms was used together with an assumption that $\hat{s}_i$ can be treated as a constant.

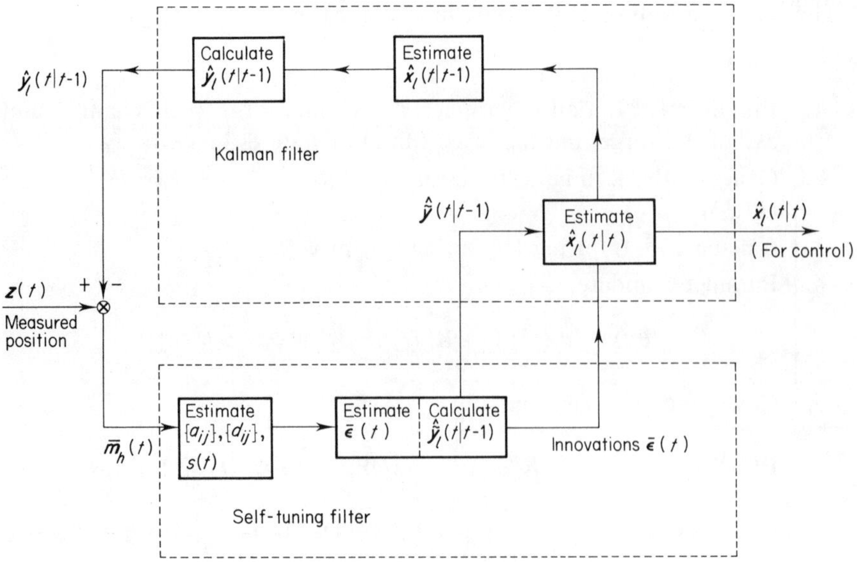

Fig. 13.10. Structure of the filtering scheme

13.2.10. Optimal controller design

The controller design is based on the separation principle of stochastic optimal control theory as described in Chapter 10. The controller with input z and output $\boldsymbol{u}$ is chosen to minimize the performance criterion:

$$J=\lim_{T\to\infty}\frac{1}{2T}E\left\{\int_{-T}^{T}\{\langle x_l-r_l,\mathbf{Q}_1(x_l-r_l)\rangle_{E_n}+\langle \boldsymbol{u},\mathbf{R}_1\boldsymbol{u}\rangle_{E_m}\}\,dt\right\}$$

where $\mathbf{Q}_1$ and $\mathbf{R}_1$ are positive definite weighting matrices. It may easily be shown that the control signal is generated from the LF Kalman filter cascaded with a control gain matrix $\mathbf{K}_c$:

$$\boldsymbol{u}(t)=-\mathbf{K}_c\hat{x}_l(t)$$

where the control-gain matrix is calculated from the steady-state matrix Riccati equation (Chapter 1).

Since low-frequency motions are to be controlled it seems intuitively reasonable that the feedback should only occur from the LF motion estimator. This can also be justified theoretically (Grimble [18]) by solving the control Riccati equation for the total HF + LF system. The control feedback gain from the HF subsystem is found to be the null matrix. This is perhaps an important role for theoretical results in that they can provide support for common sense engineering practice.

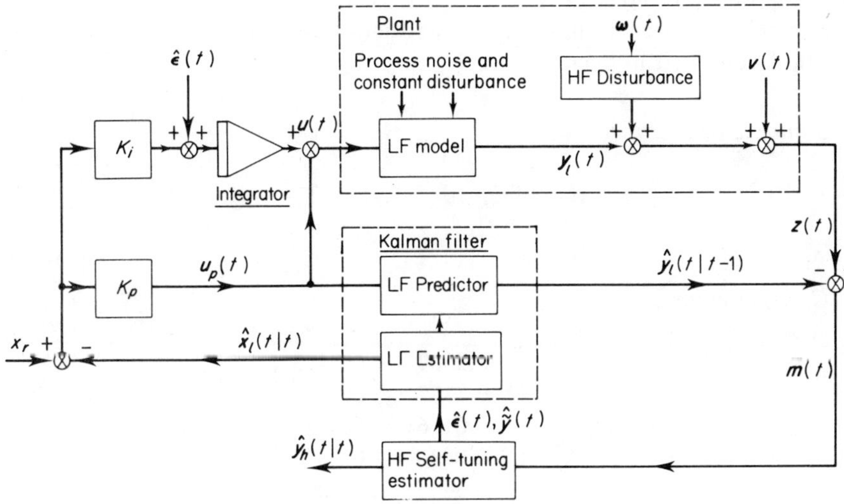

Fig. 13.11. Kalman/self-tuning filtering solution to the stochastic control problem

The vessel must not drift off position due to steady-state current forces or wind forces and thus integral control action must be used. This may be achieved by modifying the performance criterion as described by Grimble [18] and by solving a larger augmented state-space problem. The optimal control solution is then as shown in Fig. 13.11. Since the optimal controller and the Kalman filter depend only upon the low-frequency model equations they are both fixed and independent of variations in the HF or wave model. The controller and LF filter are therefore particularly easy to implement. Typical control and filter gains are given below for the relevant weighting or noise covariance matrices. The control weighting matrices were chosen using the asymptotic optimal multivariable root-loci results of Chapter 4 (Grimble [20]).

Control weightings

$$\mathbf{Q}_1 = \text{diag}\{50, 600, 50, 600, 10, 10\}$$

$$\mathbf{R}_1 = \text{diag}\{400, 4000\}.$$

Kalman filter noise covariances

$$\mathbf{Q} = \text{diag}\{4 \times 10^{-6}, 9 \times 10^{-8}\}$$

$$\mathbf{R} = \text{diag}\{5 \times 10^{-5}, 1.22 \times 10^{-4}\}.$$

The measurement noise covariance matrix is increased slightly above that for the actual measurement noise to allow for additional noise due to the approximation described in Section 13.2.7.

Simulated process noise

$$\mathbf{R}_0 = \text{diag}\{10^{-5}, 1.22 \cdot 10^{-5}\}.$$

Optimal gains

$$\mathbf{K}_c = \begin{bmatrix} 1.2907 & -0.0475 \\ 0.3873 & 0.0030 \\ 0.0116 & -0.8371 \\ 0.0030 & -0.3873 \\ 0.3815 & -0.0095 \\ -0.0095 & 0.6635 \end{bmatrix}$$

$$\mathbf{K}_f = \begin{bmatrix} 0.1263 & 0.0031 \\ 0.5023 & 0.0086 \\ 0.0170 & 0.2201 \\ 0.0210 & 0.6633 \\ 0.0 & 0.0 \\ 0.0 & 0.0 \end{bmatrix}$$

The above results are for a first-order thruster model.

The controller can be designed using frequency-response multivariable design methods but the resulting scheme has a similar structure and performance to the optimal system described above (Fotakis, Grimble and Kouvaritakis [24]).

13.2.11. Simulation of the dynamic ship-positioning system

After the design process has been completed it is usually necessary to simulate most industrial control systems of reasonable complexity. The simulation stage enables the effect of modelling errors to be observed and the performance of the controller to be fine-tuned. The effect of system non-linearities can also be investigated. Apart from the possible destabilizing influence of non-linearities they can also result in poor transient performance from a system designed using linear techniques.

Fortunately, as the following simulation results will show, the ship-positioning system can be designed quite satisfactorily using linear-control theory. This demonstrates that the LF subsystem linearization is reasonable and that the non-linear compensation in the thruster controls is also effective.

Two different techniques can be used to simulate the HF wave motions:

(1) *Sinusoidal generator*

The HF motion in one direction is represented by a summation of sinusoids:

$$y_h(t) = \sum_{i=1}^{m} y_i \sin(\omega_i t + \theta_i)$$

where θ_i is a random number which lies in the range $(\theta, 2\pi)$. Here y_i and ω_i are selected to fit the Pierson–Moskowitz sea spectrum (Neville [21]) and m is the number of equal energy sections into which the spectrum is divided (typically $m = 20$). This spectrum was obtained by the analysis of extensive wave data relating to fully developed sea conditions in the North Atlantic.

(2) *Spectrum generator*

In this case the HF motion is generated by passing white-noise sequences through a fourth-order filter. In state space notation:

$$\dot{x}_h = \mathbf{A}_h x_h + \mathbf{D}_h \xi$$

$$y_h = \mathbf{C}_h x_h$$

where

$$\mathbf{A}_h = \begin{bmatrix} \mathbf{A}_h^s & \mathbf{O} \\ \mathbf{O} & \mathbf{A}_h^y \end{bmatrix}, \qquad \mathbf{D}_h = \begin{bmatrix} \mathbf{D}_h^s & \mathbf{O} \\ \mathbf{O} & \mathbf{D}_h^y \end{bmatrix}$$

and the submatrices for the sway and yaw directions have the same form (here given for sway):

$$\mathbf{A}_h^s = \begin{bmatrix} 0 & 1 & 0 & 0 \\ 0 & 0 & 1 & 0 \\ 0 & 0 & 0 & 1 \\ -a_4^s & -a_3^s & -a_2^s & -a_1^s \end{bmatrix}, \qquad \mathbf{D}_h^s = \begin{bmatrix} 0 \\ 0 \\ 0 \\ k^s \end{bmatrix}$$

$$\mathbf{C}_h^s = [0 \quad 0 \quad 1 \quad 0].$$

The parameters of the system matrices are determined by minimizing an integral squared error criterion. For example:

$$J^s = \int_0^{\omega_m} (S(\omega) - S_0^s(\omega))^2 \, d\omega$$

where $(0, \omega_m)$ is the frequency-interval of interest. The power spectrum $S_0^s(\omega)$ follows as:

$$S_0^s(\omega) = | \mathbf{C}_h^s (j\omega \mathbf{I} - \mathbf{A}_h^s)^{-1} \mathbf{D}_h^s |^2 \sigma_\xi^s$$

where $S(\omega)$ represents the Pierson–Moskowitz sea-wave spectrum. The LF ship dynamics must be simulated in their full non-linear form, hence, the LF ship model is a direct implementation of the equations in Section 13.2.5.

Tests on the filters: Linear ship model

The simulation results presented in the following were obtained using a fourth-order wave spectrum model to generate the HF motions. The tests were based on weather conditions corresponding to Beaufort numbers 8 and 5 (wind speeds 19 m/sec and 9.3 m/sec, respectively) which are typical examples of rough and calm seas, respectively. The first set of results (Figs. 13.12 to 13.20) are for a sea state of Beaufort 8, without closed-loop control and for a linearized ship model. That is, this test is concerned with optimal filtering action. Control of the non-linear ship model is considered later in the section.

The total sway motion is shown in Fig. 13.12 and the esimated and modelled low-frequency sway motions are shown in Fig. 13.13. The estimate of the low-frequency motion is required for control purposes and it is clear the estimate

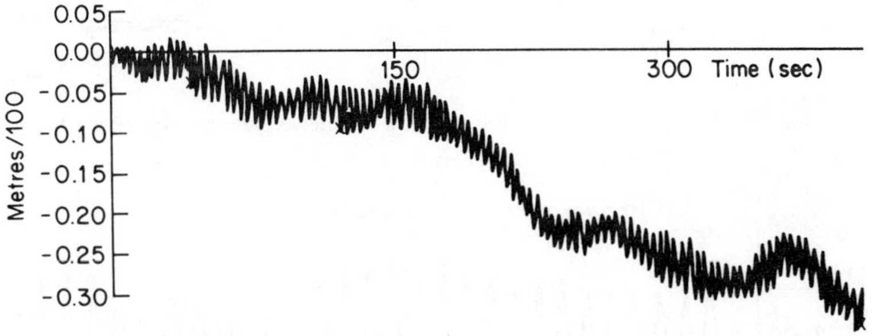

Fig. 13.12. Observed total sway motion (Beau. 8)

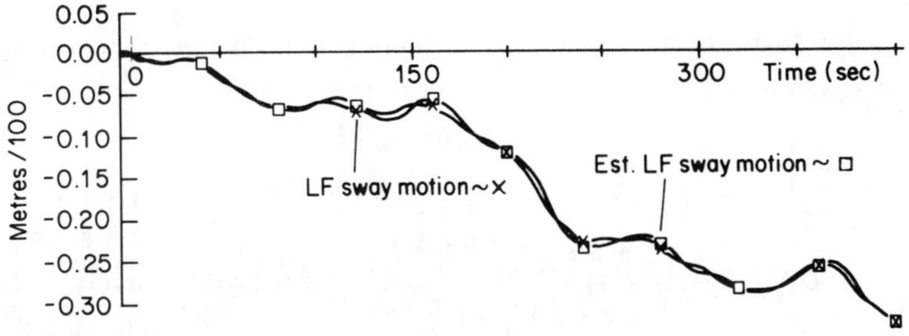

Fig. 13.13. Estimated and modelled low-frequency sway motion (Beau. 8)

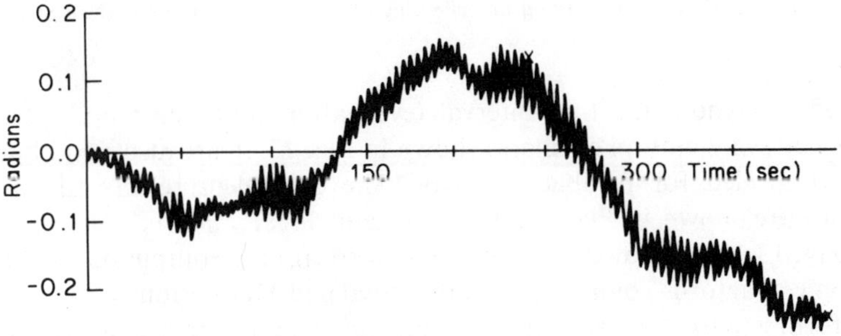

Fig. 13.14. Observed total yaw motion (Beau. 8)

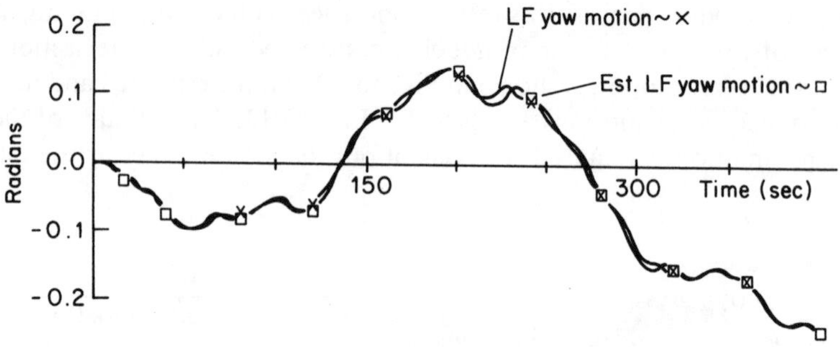

Fig. 13.15. Estimated and modelled low-frequency yaw motion (Beau.8)

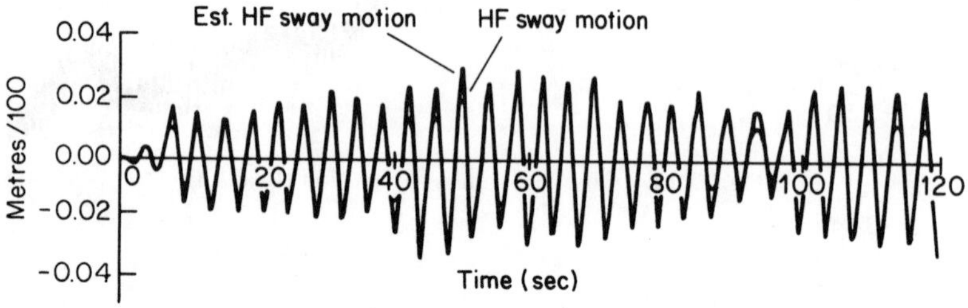

Fig. 13.16. Estimated and modelled high-frequency sway motion (Beau.8)

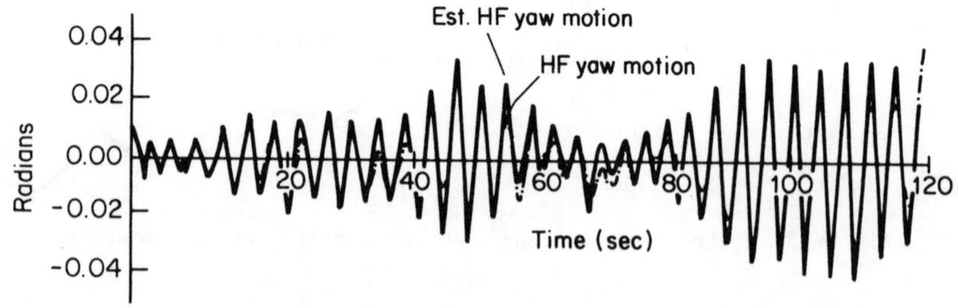

Fig. 13.17. Estimated and modelled high-frequency yaw motion (Beau.8)

is good thoughout the time-interval (even after initial start up). The high-frequency sway motion estimates shown in Fig. 13.16 are also good but these are not needed for feedback control. The equivalent results for the yaw motions are shown in Figs. 13.14, 13.15 and 13.17.

In Fig. 13.18, the power spectrum generated from the output of the HF sway motion estimator is compared with the simulated HF motions. A fast Fourier transform was used to calculate the spectrum. The spectra are similar as might be expected. The accumulative loss functions for the position errors in sway

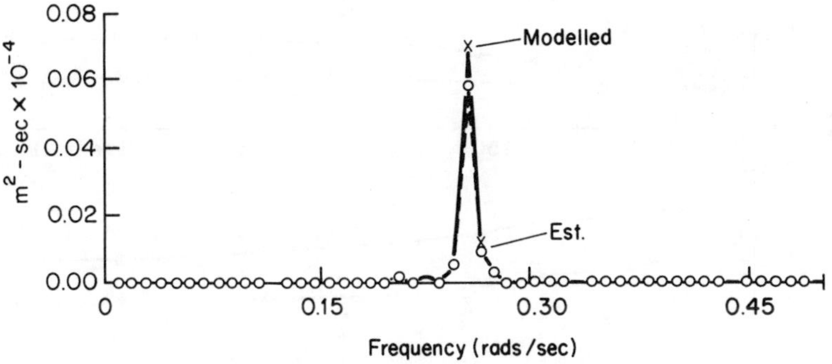

Fig. 13.18. Estimated and modelled HF sway motion (Beau. 8)

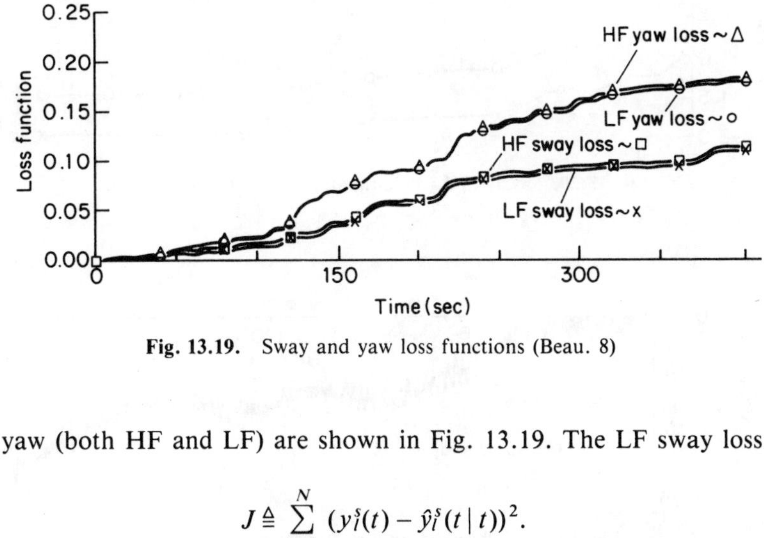

Fig. 13.19. Sway and yaw loss functions (Beau. 8)

and yaw (both HF and LF) are shown in Fig. 13.19. The LF sway loss:

$$J \triangleq \sum_{t=1}^{N} (y_l^s(t) - \hat{y}_l^s(t \mid t))^2.$$

In the optimal filter for a known plant the innovations signal is common to the HF and LF subsystems. Thus, the fact that the HF and LF loss functions are similar in Fig. 13.19 is an indication of optimal performance. The parameter estimates of the HF model are shown in Fig. 13.20.

The tests for a calm sea (Beaufort 5) condition are shown in Figs. 13.21 to 13.29 and these may be compared with the equivalent results for a rough sea shown in Figs. 13.12 to 13.20. The LF estimates are much closer to the true positions for the Beaufort 5 case. This is consistent with the operation of an optimal filter for a known plant and with the theory that shows that when the modelling errors are negligible, the term $\tilde{y}_l(t \mid t-1)$ is caused by the estimation

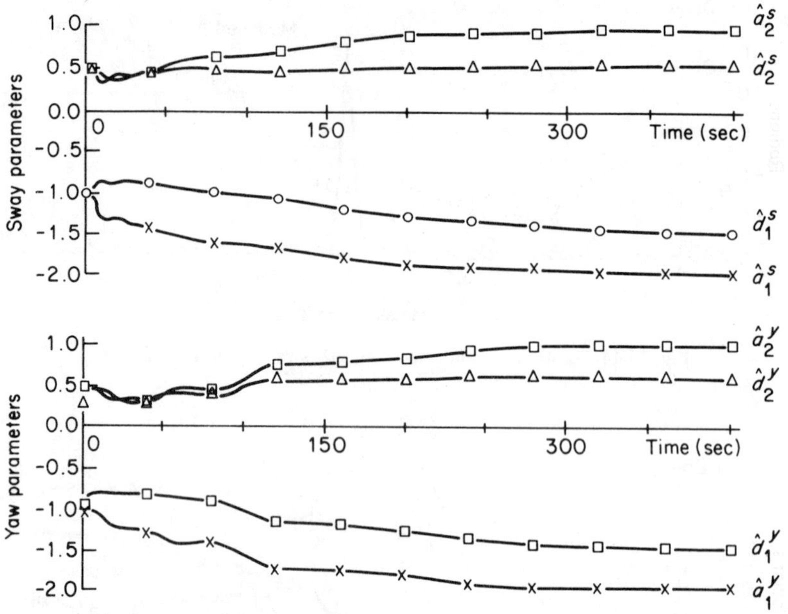

Fig. 13.20. Sway and yaw estimated parameters (Beau. 8)

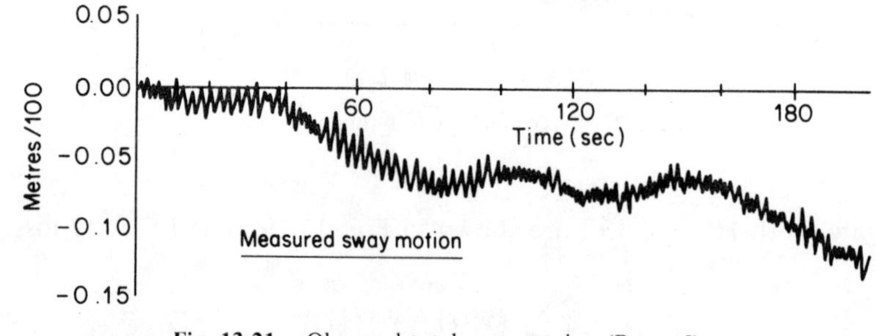

Fig. 13.21. Observed total sway motion (Beau. 5)

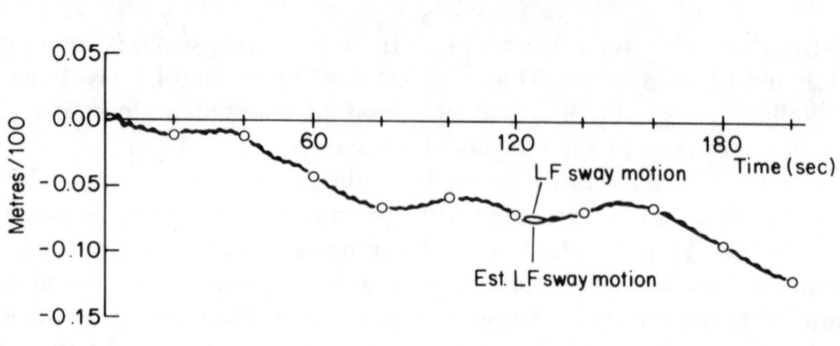

Fig. 13.22. Estimated and modelled low frequency sway motion (Beau.5)

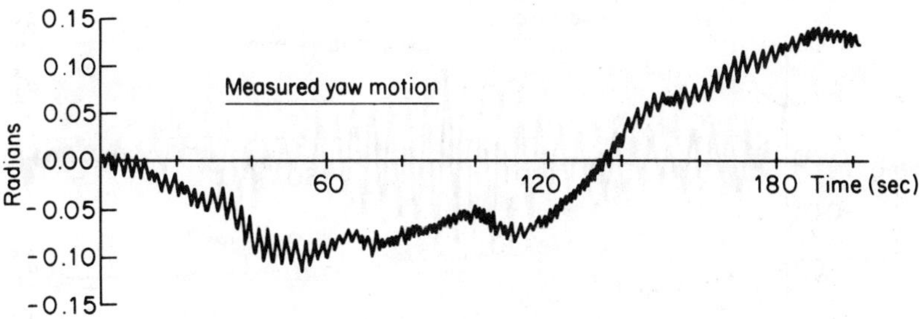

Fig. 13.23. Observed total yaw motion (Beau.5)

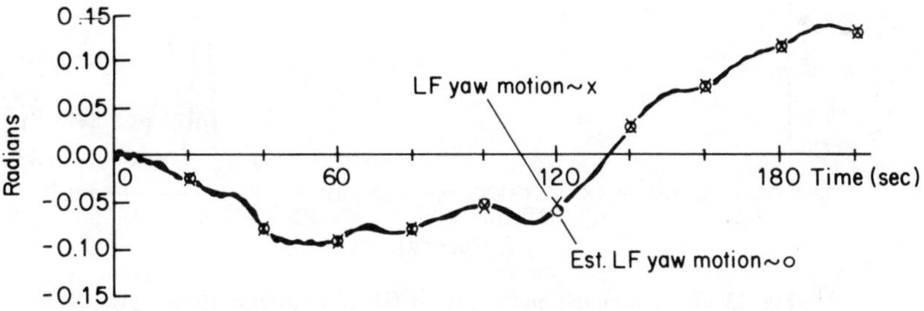

Fig. 13.24. Estimated and modelled low frequency yaw motion (Beau. 5)

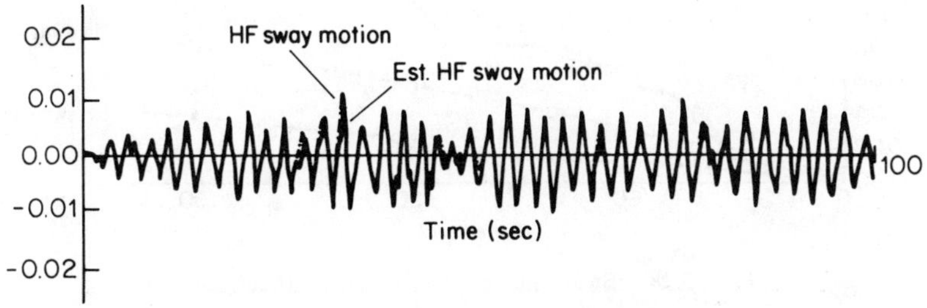

Fig. 13.25. Estimated and modelled high-frequency sway motion (Beau. 5)

error of the high-frequency motion. It follows that the estimation errors will be larger for the rough sea condition (compare the accumulative loss functions Figs. 13.19 and 13.28).

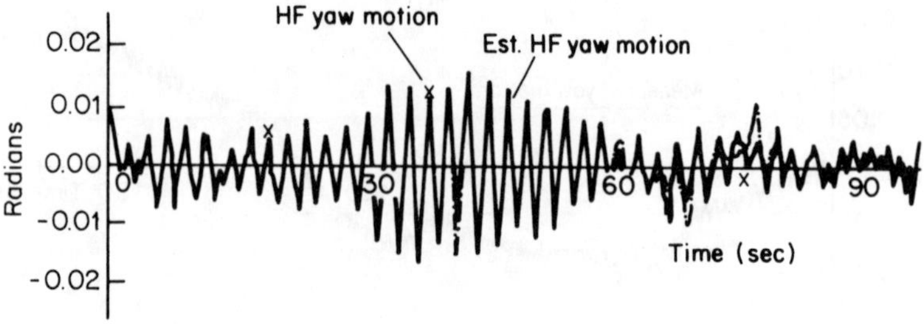

Fig. 13.26. Estimated and modelled high-frequency yaw motion (Beau.5)

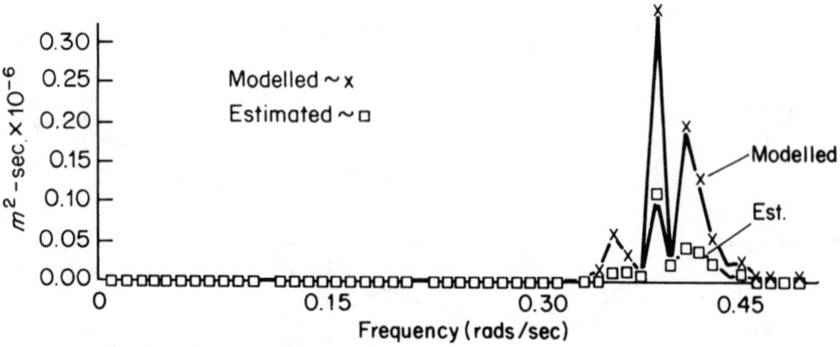

Fig. 13.27. Estimated and modelled HF sway motion (Beau. 5)

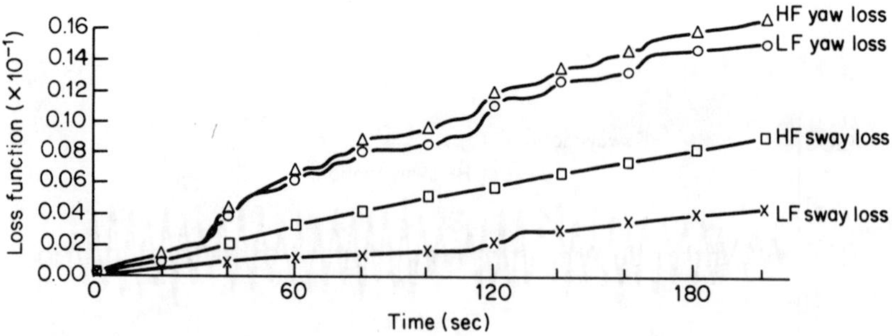

Fig. 13.28. Sway and yaw loss functions (Beau. 5)

Closed-loop control: Non-linear ship model

For a ship represented by a non-linear model, the results for closed-loop control (Beaufort 5) are shown in Figs. 13.30 to 13.37. The results are very similar to the Beaufort 8 case as shown in Fung, Chen and Grimble [19]. The low-frequency motions are, as expected, much reduced compared with the

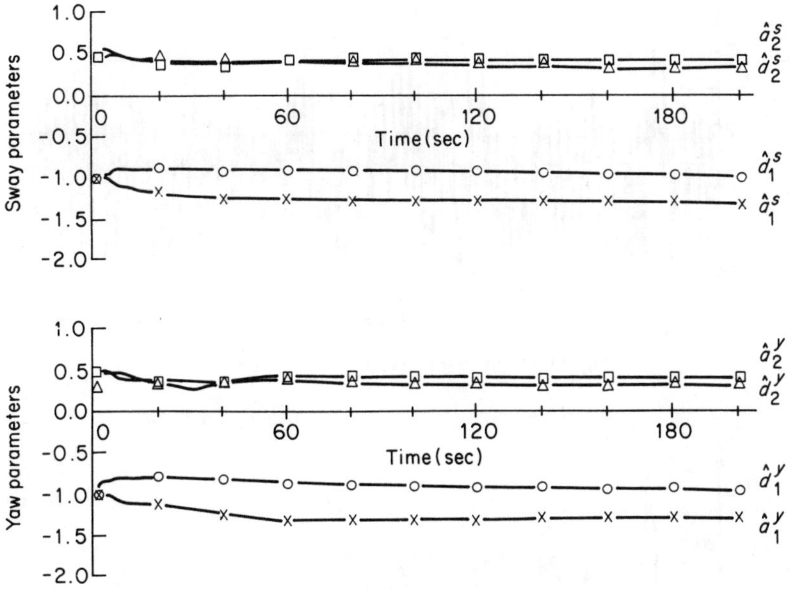

Fig. 13.29. Sway and yaw estimated parameters (Beau. 5)

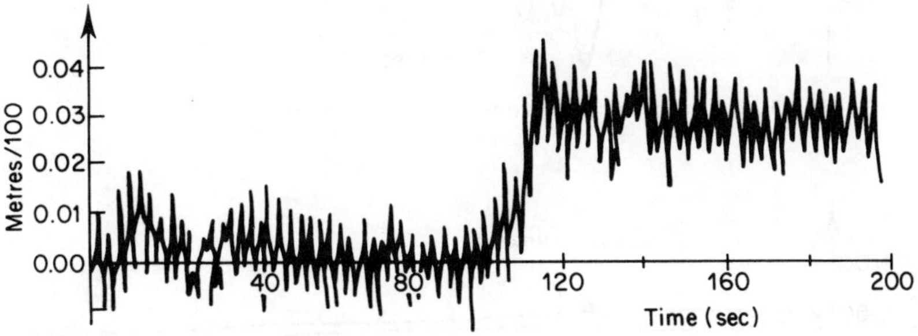

Fig. 13.30. Sway observations signal (step ref. 0.05 p.u. at 100 secs, Beau. 5)

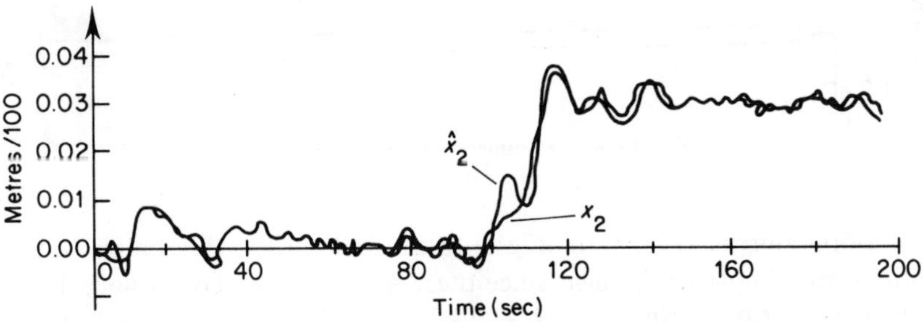

Fig. 13.31. Sway estimated and modelled low-frequency motions (Beau. 5)

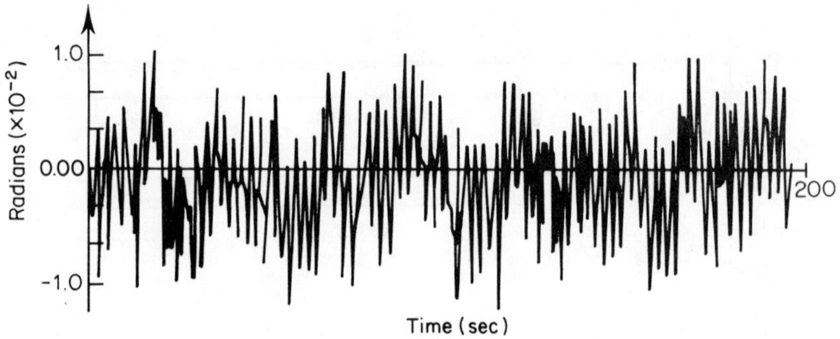

Fig. 13.32. Yaw observations signal (Beau. 5)

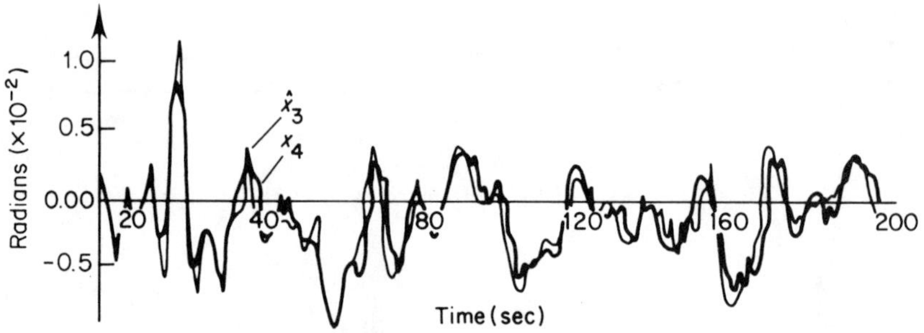

Fig. 13.33. Yaw estimated and modelled low-frequency motion (Beau. 5)

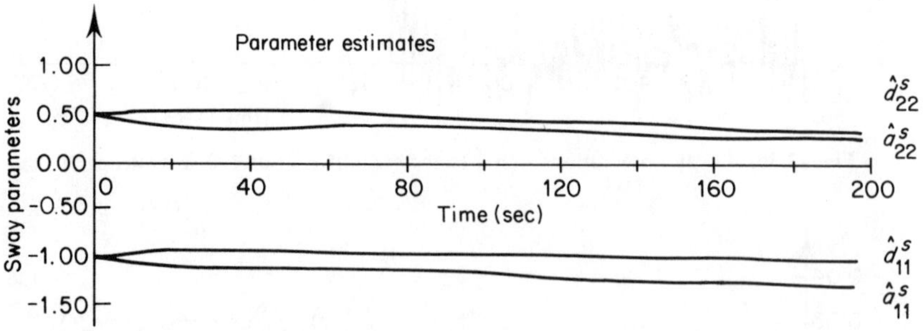

Fig. 13.34. Parameter estimates (Beau. 5)

open-loop control results of Figs. 13.21 and 13.23. The high-frequency motions are almost unchanged since these are not being controlled. The low-frequency motion estimates (Figs. 13.31 and 13.33) are smooth which is necessary if the control signals are to be relatively smooth. To allow the

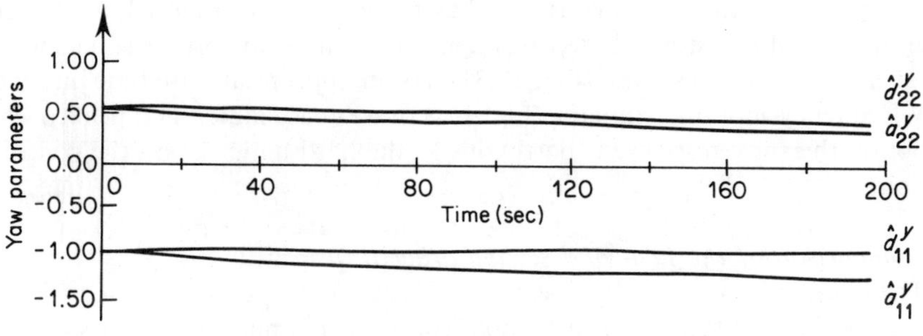

Fig. 13.35. Parameter estimates (Beau. 5)

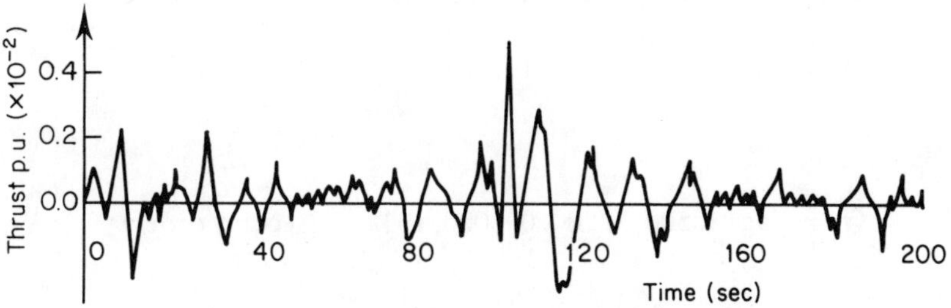

Fig. 13.36. Estimated and modelled thrust 1 (Beau. 5)

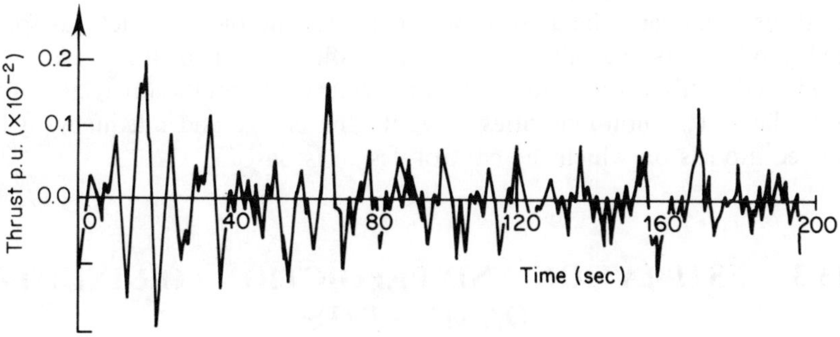

Fig. 13.37. Estimated and modelled thrust 2 (Beau. 5)

parameter estimates to converge, the step response of the sysem is measured from time 100 seconds. The second-order linearized models for the thrusters were used within the LF Kalman filter to provide the best results in this non-linear system case. The thruster responses are shown in Figs. 13.36 and 13.37.

The variations in the estimated thrust, based upon the linear model, are larger than the actual thrust which depends upon the non-linear thrusters. The low-frequency transient response (Fig. 13.31) has an appropriate rise time and the low-frequency motion estimates (Fig. 13.33) are reasonable. The peak overshoot on the sway response is partly due to the hard limits set on the control action.

Comparison of the first- and second-order thrusters

The first- and second-order linear models for the thrusters were used within the LF Kalman estimator in a comparison of performance. The first-order model could not be used together with a non-linear model for the ship, since the optimal controller did not stabilize the system. The poor performance of the first-order model was not evident when a linearized ship simulation was used. The optimal control-weighting matrices for these tests were defined as:

$$\mathbf{Q}_1 = \text{diag}\{3, 500, 3, 500, 10, 10\} \qquad \text{(first order)}$$

or

$$\mathbf{Q}_1 = \text{diag}\{3, 500, 3, 500, 0, 10, 0, 10\} \qquad \text{(second order)}$$

and

$$\mathbf{R}_1 = \text{diag}\{500, 200\}.$$

It is possible that different choices for $\mathbf{Q}_1$ and $\mathbf{R}_1$ would have enabled the first-order thruster model to be used, however, the second-order model was found to give good results over a wide range of conditions and for this reason is to be preferred. This demonstrates the importance of simulation trials which include the system non-linearities to verify the choice and adequacy of the linearized models on which the control design is based.

13.3. ESTIMATION AND PREDICTION FOR INGOT-SOAKING PITS

Like many processes in the steel industry, ingot-soaking pits are both deceptively simple in concept and extremely energy intensive. Usually these processes employ only a rudimentary control scheme and virtually no process optimization. Consequently, any small improvement in efficient operation can bring substantial savings in energy usage and increased throughput of products with satisfactory uniformity. Major design problems with this type of unsophisticated technology are a dearth of measurements and the related difficulty of determining a reliable process model upon which to base an estimation, control or prediction scheme.

Ingot-state estimation and prediction are two desirable features of any control scheme for soaking pits. Kalman filtering has been suggested as one possible solution to the problem (Cheek, Nobbs and Munro [25], Wick [26], Lumelsky [27]. On-line implementation of the proposed Kalman filtering schemes is favoured, thus it is essential the model and associated filter be as simple as possible. However, this simplicity must be achieved whilst maintaining a sufficient degree of accuracy in the representation of the physical process, a difficult task!

Although there will be some discussion of the filtering aspects of the problem most of the following section is devoted to the preliminary step, namely the development of a process model. Two different routes to a lumped parameter model for a soaking pit are developed and the influence of the formulation on the filtering scheme assessed.

13.3.1. The teeming to rolling cycle and the soaking pit problem

The complete cycle for the processing of an ingot is shown in Fig. 13.38. It comprises five distinct phases in a cycle designed to permit the ingot to solidify and to be reheated to accomplish a rolling operation. A key requirement of the reheating process is the achievement of a small temperature differential within the soaked ingot to permit treatment by a rolling process.

The physical separation between the teeming shop and the stripping bay usually requires the mould to be mounted on a bogie. In the teeming shop a number of moulds may be filled in one session. Although not drawn, a lid may be placed on the filled mould. The moulds themselves may be preheated and the molten steel is at a temperature of about 1600° C.

Onc. poured, the train of moulds and ingots are shunted out to stand cooling in air. This rate of cooling will differ according to the daily weather and the siting of the steel works.

After several hours, the train is then moved in to the stripping bay where the mould (and lid) are removed. This may be performed immediately, or several hours may pass before a decision to strip an ingot is taken. Once stripped, the ingot stands cooling, prior to charging the soaking pit. The length of time passing before charging the soaking pit will determine whether the ingots comprise a hot charge with possibly a molten core, or a cold charge when the ingot has been permitted to cool completely.

The operation of soaking requires the ingots to be charged to a top loaded chamber (roughly rectangular and refactory lined) and soaked in heat energy. The charge of seven, eight or more ingots may be stacked according to the work's practice, ingot size, or even the crane operator's convenience. The

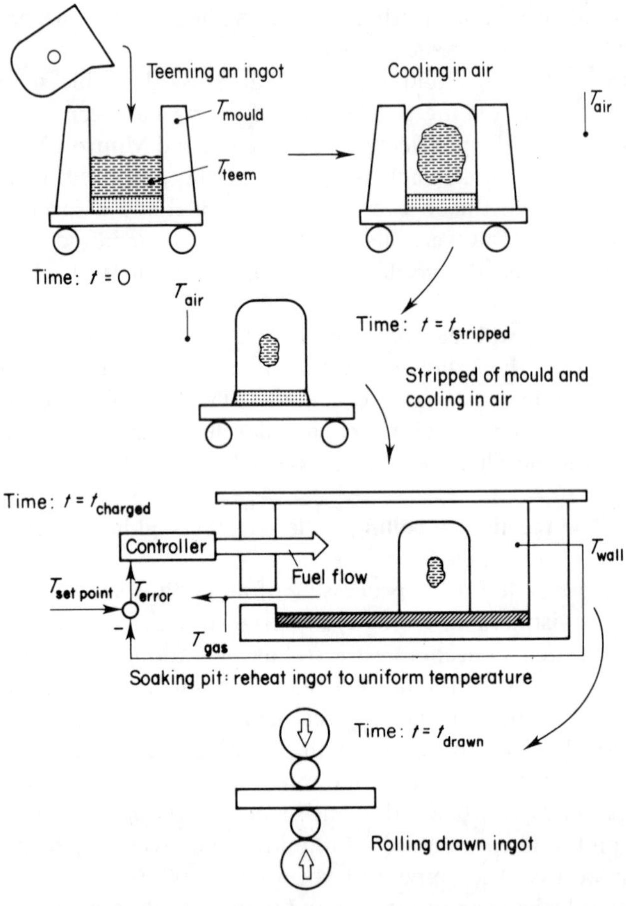

Fig. 13.38. Processing of an ingot; from teeming to rolling

charge is usually all cold or all hot ingots although they may be of mixed sizes depending on product demand. The soaking pit operates in two regimes, first a maximum fuel flow until the wall temperature reaches a setpoint value and then an in-control period. The in-control period signifies that the controller (usually proportional plus integral controller), driven by the error between wall and set-point temperature, now regulates the fuel flow. When deemed appropriate by the soaking-pit chargehand, the ingots are drawn for processing in the rolling mill.

If an ingots centre and surface temperature trajectories are superimposed on the measured variables (Fig. 13.39), the result is as shown in Fig. 13.40. The

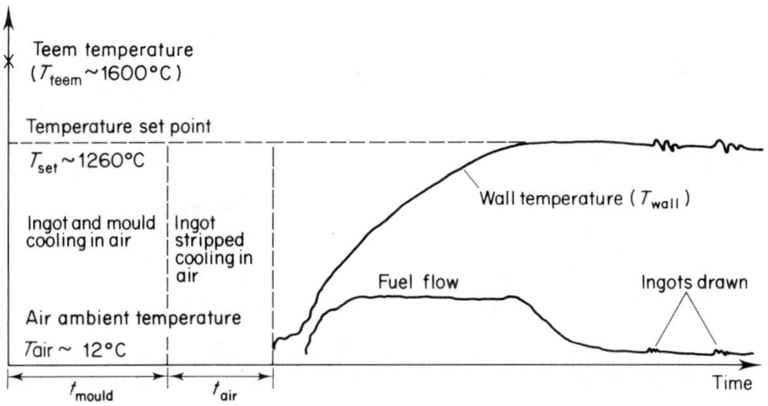

Fig. 13.39. Some measured variables, soaking pit operation (T ~ temperature values t ~ time values)

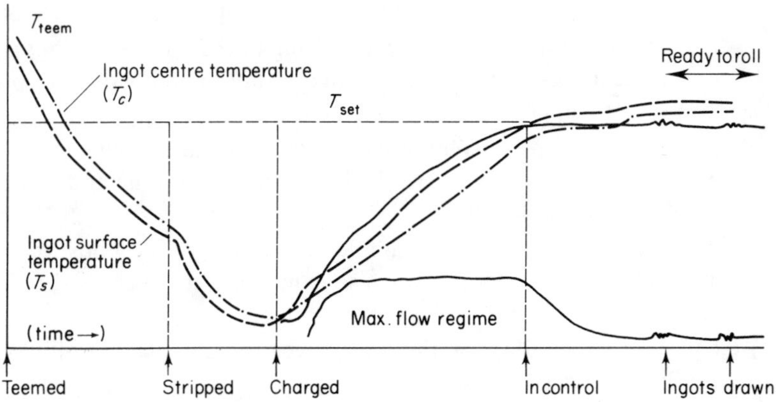

Fig. 13.40. Ingot centre and surface temperatures superimposed on the measured variables

soaking-pit problem is thus to provide values which

(i) predict when the 'ready-to-roll' stage *will* be reached and
(ii) to indicate when it actually *has been* reached.

A technique to effect this will facilitate the efficient scheduling of the battery of soaking pits and should eliminate unnecessary fuel expenditure caused by oversoaking. Other potentially dangerous situations like the processing of ingots still having liquid cores should be eliminated by ingot temperature estimation.

Clearly any solution for this problem must be extremely robust, since the variability within the process is considerable. A comprehensive list of relevant factors and parameters follows:

Ingot prehistory factors and parameters

(i) Teem times vary between ingots in any one teeming operation.
(ii) Moulds may be preheated.
(iii) During the liquid stages, cores may be stirred.
(iv) The stand-in-air time for mould and ingot varies.
(v) The inclement weather conditions affect cooling rates.
(vi) The ingot's stripped time varies from ingot to ingot.
(vii) The collection of any of these times for a batch of ingots is only likely in a macroscopic manner, ignoring the considerable variance of actual times within a batch.

Soaking-pit factors and parameters

(i) The charge may be cold or hot depending on the ingot's prehistory.
(ii) The stacking practice will vary according to the number of ingots, the size of the ingots, the state of soaking-pit floor and possibly according to the crane operators daily preference.
(iii) The pit-ageing process may effect soaking times appreciably. The refactory lining of the pit slowly deteriorates with time and will eventually require replacing.
(iv) Poor pit maintenance, for example, inaccurately fitting lids, or defective thermocouple loops can also contribute to increased soaking times.
(v) The gradual buildup of the scale layer on the soaking-pit floor is another aspect of pit ageing.
(vi) The fuel type, variations in its calorific value and the efficiency of gas recouperation loops can all affect soaking times.

The solutions for the soaking-pit prediction/estimation problem range from look-up tables for operator guidance to proposals to install microprocessor or minicomputer based Kalman filtering schemes. It is predominately the latter which are described here.

13.3.2. Mathematical models for the soaking pit process

A process as well established as the soaking pit has generated an extensive literature describing its operation and control. An appreciation of the techniques applied previously is frequently useful to assess the viability of new proposals. A historical view of these techniques is also remarkably indicative of the technological development which has occurred to solve the problem. In the case of soaking pits, the move from off-line computed lookup tables to the use of an on-line Kalman filter is occasioned by the rise of the minicomputer and microprocessor. (Since the ready-to-roll time is often determined by the experience of the soaking-pit chargehand on the basis of visual inspection (Kung *et al.* [28]) it would be interesting to know why the development of a portable optical device was never pursued with any enthusiasm in the last decade. One of the few such schemes incorporating such a device is that of Wick [39].)

The basis of most mathematical models developed since 1954 is the conduction equation. If the temperature at any point in an isotropic media is given the representation: $T(\cdot, t)$, then (Carslaw and Jaeger [29]) the unsteady conduction equation is:

$$\nabla \cdot (K(T) \nabla T) = \rho(T)c(T) \frac{\partial T}{\partial t} \tag{13.34}$$

where $K(T)$ is the thermal conductivity, $\rho(T)$ the density and $c(T)$ specific heat, all as functions of temperature $T \triangleq T(\cdot, t)$. Temperature $T(\cdot, t)$, itself is a function of spatial position and time.

Expanding (13.34) to give:

$$K(T)\nabla^2 T + \frac{\partial K}{\partial T} (\nabla T)^2 = \rho(T)c(T) \frac{\partial T}{\partial t}. \tag{13.35}$$

Assuming the thermal conductivity $K(\cdot)$ to be independent of temperature would simplify (13.35) via $\partial K/\partial T = 0$. Although this is not the case, Massey and Sheridan [30] demonstrated that for air cooling and the reheating of ingots, it is an adequate approximation. For example, 0.23% carbon mild steel has the following thermal conductivities.

Table 13.1 Thermal conductivities 0.23% carbon mild steel

TEMP °C	$K(T)$	$\partial K/\partial T$
300	0.106	0.4×10^{-2}
600	0.085	0.8×10^{-4}
700	0.075	0.16×10^{-3}
800	0.062	0
1150	0.07	-0.2×10^{-4}

Hence

$$\frac{\partial K}{\partial T} (\nabla T)^2 \ll K\nabla^2 T$$

so that (13.35) can be approximated by:

$$\nabla^2 T = D(T)\,\frac{\partial T}{\partial t} \tag{13.36}$$

where

$$D(T) \triangleq \frac{\rho(T)c(T)}{K(T)}$$

is the thermal diffusivity of the medium.

If an appropriate coordinate system is assumed for equation (13.36) and suitable boundary conditions chosen, finite difference schemes can be utilized to follow all various stages in the history of the ingot from teeming to soaking.

All the studies based on off-line computations used equations (13.35) or (13.36) and a finite difference scheme of some description, (Sarjant and Slack [31], Corlis *et al.* [32], Kung *et al.* [28], Massey and Sheridan [30], den Hartog *et al.* [33], Cummings [34]).

The first attempt to use sophisticated on-line control techniques for the control of soaking pits was reported by Hinami *et al.* [35]. That particular study recognized the value of systematically collecting data from the ingot's prehistory period and transmitting it to the soaking-pit chargehand. This was intimated to remove much of the uncertainty from the soaking process. A one-dimensional heat equation model for an ingot (using cylindrical coordinates) was used in a real time mode to estimate and predict the progress of the soaking operation.

The use of this model was validated against an instrumented ingot trial (a trial with temperature measurements taken within and on the ingot in a production pit). Although an innovative approach, the paper records: 'Effort will be continued for a further speeding up of the computation and improvement of the prediction of the computational model for better representation of the actual conditions.'

A second attempt to devise a scheme for the on-line control of soaking pits was that due to Maeda *et al.* [36], [37]. The methodology of these researchers was quite novel since they replaced the finite difference solutions of the conduction equation by solutions obtained using the integral profile method (Goodman [38]). Thus, the physical basis of the problem was used to obtain a non-linear equation for the soaking pit and two lumped parameter models for the ingot and the soaking-pit walls. The second innovation, perhaps not quite so successful, was to adopt a spherical geometry for the ingot and its mould. The replacement was made on the basis of equivalent volumes for the rectangular ingot/mould configuration. As with the trial by Hinami *et al.*, an instrumented ingot was used for validation purposes.

Subsequent to this research, proposals to use a Kalman filtering scheme implemented on minicomputers or microprocessors appeared in the literature.

However, there is a need for a robust model on which to base a filtering scheme. The next sections describe two approaches to achieve this aim and reflect on the implications for Kalman filtering applications.

13.3.3. Models for soaking pit filters I: Least-square fits

The measured pit variables are few (pit wall temperature, fuel flow, waste gas temperature and air flow) and from these it is desired to estimate the current ingot centre and surface temperatures. The prediction of future ingot surface and centre temperatures will be an important extension of the filtering scheme if an optimized scheduling of the soaking pit battery is to be achieved.

The trajectories of these measured variables, as shown previously in Fig. 13.39, exhibit two distinct phases: the maximum fuel flow and the in-control phase. Thus, Cheek *et al.* [25] approximated this non-linear distributed parameter process by two low-order lumped parameter models and a switch mechanism. Lumelsky [27] rejected this dichotomous approach and used just one lumped pit and ingot model for the whole trajectory. It was intimated by Lumelsky that this unification would simplify the computational burden associated with the Kalman filtering algorithm.

The models chosen by Cheek *et al.* [25] are pursued here because they illustrate the ideas of model structures. The basic control-loop to be modelled was given as shown in Fig. 13.41. For the pit and the ingots, the linear constant parameter discrete model was chosen as:

$$\begin{bmatrix} T_s(k+1) \\ T_c(k+1) \\ T_{\text{wall}}(k+1) \end{bmatrix} = \begin{bmatrix} a_{11} & a_{12} & a_{13} \\ a_{21} & a_{22} & 0 \\ a_{31} & 0 & a_{33} \end{bmatrix} \begin{bmatrix} T_s(k) \\ T_c(k) \\ T_{\text{wall}}(k) \end{bmatrix} + \begin{bmatrix} b_{11} \\ 0 \\ b_{31} \end{bmatrix} f(k) \qquad (13.37)$$

where

$$T_s = \text{ingot surface temperature}$$

$$T_c = \text{ingot centre temperature}$$

$$T_{\text{wall}} = \text{pit wall temperature}$$

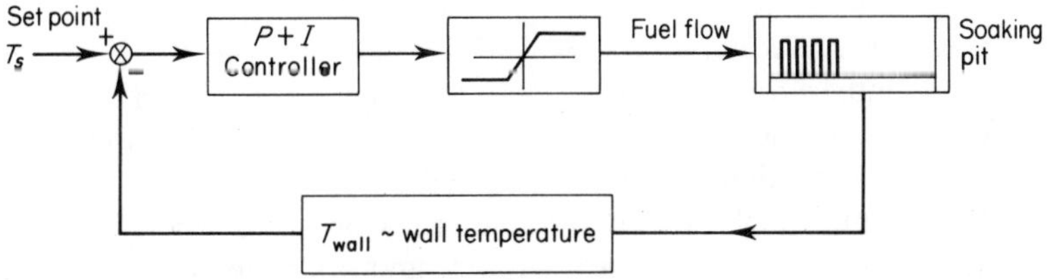

Fig. 13.41. Soaking pit control scheme (after Cheek *et al.* [25])

and

$$f = \text{fuel flow}.$$

This was considered to be a macroscopic model for all the ingots in the pit. For the pilot scale pit considered in Cheek *et al.*, the hot gas dynamics were omitted, as was a model for the recouperator dynamics. The null entries (positions a_{23}, a_{32} and b_{21}) were inserted to reflect the exclusive coupling of the ingot centre temperature to the ingot surface temperature. The pit outputs were modelled as:

$$\begin{bmatrix} T_{\text{wall}}(k) \\ T_{\text{gas}}(k) \\ f(k) \end{bmatrix} = \begin{bmatrix} 0 & 0 & 1 \\ c_{21} & 0 & c_{23} \\ 0 & 0 & 0 \end{bmatrix} \begin{bmatrix} T_s(k) \\ T_c(k) \\ T_{\text{wall}}(k) \end{bmatrix} + \begin{bmatrix} 0 \\ d_{21} \\ 1 \end{bmatrix} f(k) \qquad (13.38)$$

where T_{gas} = waste gas temperature. The estimation procedure should be constrained since there are no self-arising energies within the pit- and heat-transfer coefficient cannot be negative, hence $0 \leq a_{ij} \leq 1$ for all i, j.

For the in-control phase, the pit and ingot model is adjoined by additional equations representing the $P + I$ controller.

Pit and ingot model

$$\begin{bmatrix} T_s(k+1) \\ T_c(k+1) \\ T_{\text{wall}}(k+1) \\ x_4(k+1) \end{bmatrix} = \begin{bmatrix} a_{11}^1 & a_{12}^1 & a_{13}^1 & a_{14}^1 \\ a_{21}^1 & a_{22}^1 & 0 & 0 \\ a_{31}^1 & 0 & a_{33}^1 & a_{34}^1 \\ 0 & 0 & -\Delta t & 1 \end{bmatrix} \begin{bmatrix} T_s(k) \\ T_c(k) \\ T_{\text{wall}}(k) \\ x_4(k) \end{bmatrix} + \begin{bmatrix} b_{11}^1 \\ 0 \\ b_{31}^1 \\ \Delta t \end{bmatrix} T_{\text{set}}(k). \qquad (13.39)$$

Output equation

$$\begin{bmatrix} T_{\text{wall}}(k) \\ T_{\text{gas}}(k) \\ f(k) \end{bmatrix} = \begin{bmatrix} 0 & 0 & 1 & 0 \\ c_{21}^1 & 0 & c_{23}^1 & c_{24}^1 \\ 0 & 0 & c_{33}^1 & c_{34}^1 \end{bmatrix} \begin{bmatrix} T_s(k) \\ T_c(k) \\ T_{\text{wall}}(k) \\ x_4(k) \end{bmatrix} + \begin{bmatrix} 0 \\ d_{21}^1 \\ d_{31}^1 \end{bmatrix} T_{\text{set}}(k) \qquad (13.40)$$

where x_4 is an integrator state. Once again the equations exhibit a sparse structure dictated by physical considerations. The pair of equations (13.37) and (13.38) model the maximum flow mode, whilst equations (13.39) and (13.40) are for the in-control phase. The switchover is obtained when T_{wall} (model value) exceeds the wall temperature setpoint and once this occurs there is no switching back to the maximum flow model.

The outstanding problem is then to obtain values for the model parameters, viz., model identification. Data sources for soaking-pit models arise in three ways.

(i) *Measured plant data.* This can be recorded (that is fuel flow, pit wall temperature and so forth) and the missing ingot surface and centre temperatures can be obtained using one of the comprehensive finite difference models, like that devised by Massey and Sheridan [30]. This approach has the advantages of flexibility and economy and its appropriate for simulation purposes and software development. Another significant advantage is that hot charges (namely ingots with a short cooling time) may be studied. The major disadvantage is that the finite difference model itself will be a function of a number of parameters, for example, view factors, cooling coefficients and ambient temperatures whose choice will determine the response of the model.

(ii) *Pilot scale soaking-pit trials.* Normally used for burner design studies, a pilot scale soaking pit can readily be adapted for soaking-pit modelling activities. The model of Cheek *et al.* [25] was devised in the first instance for this purpose. An instrumented ingot (that is an ingot fitted with internal and external thermocouples) was used to obtain ingot surface and centre temperature data points. Whilst having all the advantages of an on-line trial and useful for testing prediction schemes, only cold or warm charges may be simulated.

(iii) *Production instrumented ingot trial.* Most expensive of all is an instrumented ingot trial in a production soaking pit. Very few have been performed since such an exercise requires extensive data logging which can disrupt production schedules. Apart from difficulties like inaccurately positioned thermocouples or the failure of thermocouples during the trial, a major drawback is the restriction to cold charges.

Given that data has been acquired, by possibly one of the above means, it is necessary to estimate the model coefficients and assess the accuracy of the representation. Essentially a data-processing exercise, this comprises two parts:

(i) ensure that the basic model representation is accurate (selection of regression variables), and
(ii) assess the sparseness assumed, that is the model structure.

The reassessment of the model structures serves to confirm (or otherwise) the physical reasoning used to introduce the null elements of equations (13.37) to (13.40) and may uncover some relationships not already suspected by the design engineer. The introduction of sparseness into the model parameter matrices also has the bonus of requiring less computer storage and may lead to a structural simplification of the filter equations.

Several model structures were applied to the data of Cheeks *et al.* [25] and model structural assessments performed. A sample assessment is given below. Assume that the following model is to be investigated:

In-control mode; pit and ingot model

$$\begin{bmatrix} T_s(k+1) \\ T_c(k+1) \\ T_{\text{gas}}(k+1) \\ \Sigma_f(k+1) \end{bmatrix} = \begin{bmatrix} \alpha_{11} & \alpha_{12} & \alpha_{13} & \alpha_{14} \\ \alpha_{21} & \alpha_{22} & \alpha_{23} & \alpha_{24} \\ \alpha_{31} & \alpha_{32} & \alpha_{33} & \alpha_{34} \\ \alpha_{41} & \alpha_{42} & \alpha_{43} & \alpha_{44} \end{bmatrix} \begin{bmatrix} T_s(k) \\ T_c(k) \\ T_{\text{gas}}(k) \\ \Sigma_f(k) \end{bmatrix} + \begin{bmatrix} \beta_1 \\ \beta_2 \\ \beta_3 \\ \beta_4 \end{bmatrix} (T_{\text{set}} - T_{\text{wall}}(k)). \tag{13.41}$$

where Σ_f = integral of fuel flow. The dynamic model for T_{gas}, the waste gas temperature, is now given the following structural assessment. First calculate the least-square equation subject to the constraints $\alpha_{3j} = 0$ $j = 1, \ldots, 4$; $\beta_3 = 0$ for various combinations of α_{3j} and β_3. For the data of Cheeks *et al.* the following table is obtained:

Table 13.2. Structural assessment for T_{gas} dynamic model

T_s	T_c	T_{gas}	Σ_f	e_ω	RUN 1	RUN 2	RUN 3	
α_{31}	α_{32}	α_{33}	α_{34}	β_3	23.66	63.94	51.42	$\hat{J}$(least square fit)
α_{31}	0	α_{33}	α_{34}	β_3	3.19	5.16	0.4	
α_{31}	α_{32}	α_{33}	0	β_3	0.78	0.47	0.15	
α_{31}	α_{32}	α_{33}	α_{34}	0	48.2	25.7	48.04	Values of
α_{31}	0	α_{33}	0	β_3	6.6	5.43	0.45	$\dfrac{\hat{J}(\text{struct}) - J(\text{full})}{J(\text{full})}$
α_{31}	0	α_{33}	α_{34}	0	63.11	38.15	49.1	
α_{31}	α_{32}	α_{33}	0	0	48.25	32.2	48.6	
α_{31}	0	α_{33}	0	0	67.41	39.29	50.41	

The diagnostic on the right-hand side monitors the degradation in the regression accuracy of a particular null structure choice for the model coefficients. It has the form:

$$\frac{\hat{J}(\text{structure}) - \hat{J}(\text{full coefficient set}),}{\hat{J}(\text{full coefficient set})}$$

where $\hat{J}(\cdot)$ indicates the optimal cost value of the least-squares fit algorithm. The table reveals a satisfactory dynamic model for T_{gas} is obtained if $\alpha_{32} = 0$ and $\alpha_{34} = 0$. It also indicates that further sparseness, for example, $\alpha_{32} = 0$, $\alpha_{34} = 0$, $\beta_3 = 0$ yields an unacceptable representation. When tests of this nature were performed for the complete linear model (13.41), the following sparse structure was obtained:

$$\begin{bmatrix} T_s(k+1) \\ T_c(k+1) \\ T_{gas}(k+1) \\ \Sigma_f(k+1) \end{bmatrix} = \begin{bmatrix} \alpha_{11} & \alpha_{12} & \alpha_{13} & 0 \\ \alpha_{21} & \alpha_{22} & 0 & 0 \\ \alpha_{31} & 0 & \alpha_{33} & 0 \\ \alpha_{41} & 0 & \alpha_{43} & \alpha_{44} \end{bmatrix} \begin{bmatrix} T_s(k) \\ T_c(k) \\ T_{gas}(k) \\ \Sigma_f(k) \end{bmatrix} + \begin{bmatrix} 0 \\ 0 \\ \beta_3 \\ \beta_4 \end{bmatrix} (T_{set} - T_{wall}(k))$$

A thorough analysis of plant data can reveal a satisfactory external structure for the linear model, in terms of the variables whose evolution is to be described by the model, and a useful sparse internal structure. The introduction of sparseness indicates the uncorrelated relationship the physical variables of process. This may provide a reassuring confidence in the designer's intuitive view of the process physics. Other benefits of sparseness may accrue, for example, a structured Kalman filtering algorithm may follow or computation burdens may be reduced.

The design of a Kalman filter requires the selection of covariance matrices. For the soaking pit problem this was rather difficult since there was little physical information available to determine or guide their selection. In the event, simple forms were adopted for these so that they could be used as design parameters. This is a typical problem in industrial filtering situations and is usually treated in this empirical manner. One particular problem the filter has to overcome is the variability of pit parameters from charge to charge. The available data was therefore divided into two sets, one for tuning purposes and one for testing purposes. A diagonal covariance matrix was used and as more experience of Kalman filter simulations was accumulated, this was modified to obtain an improved filter response. Care was exercised to avoid overtuning with one data set, otherwise dramatically poor results were occasionally obtained on other data sets. Yet another indication of the wide variability which exists in the soaking pit process. Root loci could be drawn for the filter poles as functions of the few parameters of the chosen covariance matrices, but little theory is available for the use of such loci. *In toto*, the process of filter design reduced to obtaining suitable responses for a sufficient number of data records over a wide range of operating conditions. Once a suitable degree of robustness was achieved, the approach was then ready for production pit pilot trials.

13.3.4. Models for soaking pit filters II: A phenomenological approach

The method of deriving linear models from simulation or plant data via a thorough data analysis exercise is to be compared and contrasted with the phenomenological approach to soaking pit models reported by Maeda *et al.*

[36], [37]. This work is presented here with the modification that a cylindrical instead of a spherical ingot is assumed.

The main model comprises a heat balance applied to the soaking pit. This is driven by the fuel flow and produces a mean gas temperature for the pit plus two average flux values one for the ingots and the other for the wall. Two other models are then utilized to calculate the evolution of wall and ingot temperatures. The connection between model parameters and the physical configuration will be less tenuous in this approach. The model will also have a modular structure, a common feature of the phenomenological approach.

Soaking-pit heat balance

The prime assumption of the heat balance equation is that heat transfers to the ingots within the pit, to the pit walls, and to the recouperator arise from the mean gas temperature of the pit interior. The important transfers are given in Fig. 13.42 where

Q = heat flow rate from burner
Q_g = heat flow rate leaving pit to the recouperator
η = efficiency of the recouperator
Q_w = heat flow rate to the pit wall
Q_I = heat flow rate to the ingot

and the temperatures are

T_g = temperature of waste gas
T_G = mean gas temperature
T_w = interior wall temperature.

The waste gas temperature, T_g and the mean gas temperature are assumed to be linearly related (Kung *et al.* [28]):

$$T_g = aT_G + b$$

Similarly the heat flows Q and Q_g are also assumed to be linearly related:

$$\frac{Q_g}{Q} = \frac{T_g}{T_F} = \frac{aT_G + b}{T_F}$$

to yield

$$Q_g = (aT_G + b)Q/T_F \tag{13.42}$$

where $T_F \triangleq$ temperature of the flame. The constants a and b are determined empirically and T_F may be acquired from a suitable fuel data book. In some

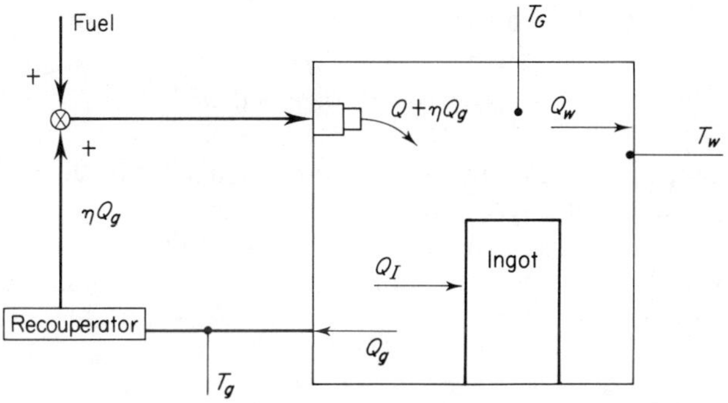

Fig. 13.42. Heat flows in a soaking pit (after Maeda *et al.* [36])

cases, T_g the waste gas temperature may be available as an online measurement, thus eliminating the above step for the analysis.

The heat balance for the soaking pit is

$$Q + \eta Q_g = Q_g + Q_I + Q_w.$$

Substituting from (13.42) for Q_g obtains

$$\left(1 + \frac{(\eta - 1)(aT_G + b)}{T_F}\right) Q = Q_I + Q_w.$$

The heat flow Q is related to the fuel flow; and parameters η, a, b, and T_F are assumed known, thus the two unknowns are Q_I and Q_w.

The constituent heat-transfer mechanisms are shown in Fig. 13.43, these comprise radiation flows (superscript R), and convective flows (superscript C) between gas and wall (subscript GW), gas and ingot (subscript GI) and wall and ingot (subscript WI). Thus heat flow to the ingot is given by:

$$Q_I = q^R_{GI} + q^C_{GI} + q^C_{GW}$$

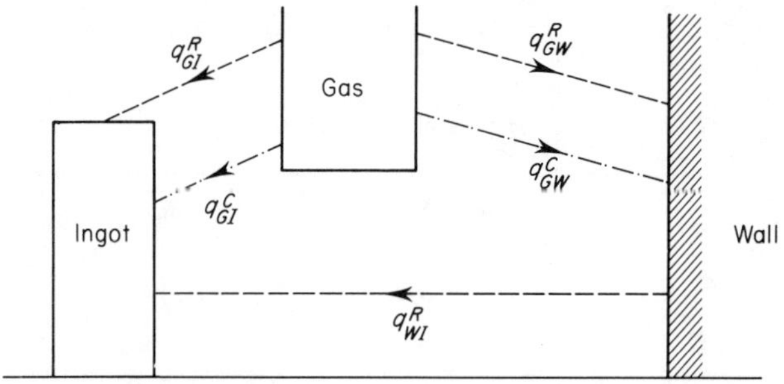

Fig. 13.43. Heat transfers within the soaking pit (after Maeda *et al.* [36])

and the heat flow to the wall is:

$$Q_w = q_{GW}^R + q_{GW}^C - q_{GI}^R.$$

when evaluated, these heat flows are used to calculate the average heat fluxes for driving the wall and ingot models.

Average wall flux

$$f_w \triangleq Q_w/S_w$$

Average ingot flux

$$f_I \triangleq Q_I/S_I$$

where S_w and S_I are the total internal pit wall area, and the total exposed ingot surface area, charged to the pit, respectively.

The actual derivation of the various radiation and conduction flows can be found in the literature (Maeda *et al.* [36], [37]) and is not therefore repeated here. The formula are collated in the following soaking-pit algorithm.

Algorithm 13.2. *Soaking-pit heat balance*

1. *Input*

$$\begin{aligned} t &= \text{time} \\ Q(t) &= \text{heat input} \\ T_w(t) &= \text{interior wall temperature} \\ T_s(t) &= \text{ingot surface temperature.} \end{aligned}$$

2. *Solve simultaneously for T_G, q_{GW}^R, q_{GI}^R, q_{WI}^R, q_{GI}^C, q_{GW}^C, B_w, B_I*

$$\left[1 + \frac{(\eta - 1)aT_G + b)}{T_F}\right] Q = Q_I + Q_w = q_{GW}^R + q_{GI}^R + q_{GI}^C + q_{GW}^C$$

$$q_{WI}^R = S_w F_{WI} \tau_G (B_w - B_I)$$

$$q_{GW}^R = S_w F_{WG} \varepsilon_G (\sigma T_G^4 - B_w)$$

$$q_{GI}^R = S_I F_{IG} \varepsilon_G (\sigma T_G^4 - B_I)$$

$$q_{GW}^R - q_{WI}^R = \frac{\varepsilon_w S_w}{(1 - \varepsilon_w)} (\sigma T_w^4(k) - B_w)$$

$$q_{GI}^{R} + q_{WI}^{R} = \frac{\varepsilon_I S_I}{(1 - \varepsilon_I)} (B_I - \sigma T_s(k)^4)$$

$$q_{GI}^{C} = h_{GW} S_w (T_g - T_w(k))$$

$$q_{GI}^{C} = h_{GI} S_I (T_G - T_s(k)).$$

3. *Complete flux values*

$$f_w = Q_w/S_w = (q_{GW}^{R} - q_{WI}^{R} + q_{GW}^{C})/S_w$$

$$f_I = Q_I/S_I = (q_{GI}^{R} + q_{WI}^{R} + q_{GI}^{C})/S_I.$$

4. *Update wall and ingot temperatures*

(a) Use f_w in wall module to obtain $T_w(t+1)$.
(b) Use f_I in ingot module to obtain $T_s(t+1)$.

5. *Update time increment*

$$t := t + 1 \text{ and goto 1.}$$ □

The simultaneous equations contain various heat transfer quantities: B = radiosities, ε = emissivities, t = transmissivities, S = surface areas, F = view factors, h = conductances and σ = Stefan–Boltzmann constant, all of which have to be prespecified. The simultaneous equations simplify to a fourth-order polynomial in mean gas temperature $T_G(k)$. This is readily solved by a Newton rule using the previous solution $T_G(k-1)$ as an initial starting value. To complete the soaking-pit heat balance, the details of the ingot and pit wall models are now required.

A lumped parameter model for an ingot

The ingot model of Maeda *et al.* [37] was based on an equivalent volume spheroid for the rectangular prismatic ingot. Here, an equivalent surface area cylinder is assumed. The analysis is for the temperature evolution at midheight so that heat transfer in the vertical direction is neglected.

In cylindrical coordinates, the conduction equation is

$$\frac{\partial(rT)}{\partial t} = D_I \frac{\partial}{\partial_r}\left(r\frac{\partial T}{\partial r}\right), \qquad 0 \le r \le R$$

where $T = T(r, t)$ is the temperature as a function of time t and radial distance r. Coefficient D_I is the ingot diffusivity where $D_I \triangleq K_I/\rho_I C_I$. The boundary conditions are:

(i) $T_c = T(0, t)$, ingot centre temperature

(ii) $T_s = T(R, t)$, ingot surface temperature

(iii) $\dfrac{\partial T}{\partial r} = 0$ at $r = 0$, (symmetry at the ingot centre)

(iv) $-K_I \left.\dfrac{\partial T}{\partial r}\right|_{r=R} = f_I$, surface flux, f_I. (13.43)

The lumped parameter model is based on two features, (i) the assumption of a parametric form for $T(r, t)$ and (ii) the use of the integral balance technique to produce time derivatives of the parameters. Assume the temperature profile may be modelled by the parametric form:

$$T(r,t) = a_0(t) + a_1(t)\frac{r}{R} + a_2(t)\frac{r^2}{R^2} + a_3(t)\frac{r^3}{R^3}.$$

Applying the boundary conditions (13.43) yields:

$$T(0,t) = a_0(t) = T_c(t)$$

$$T(R,t) = a_0(t) + a_1(t) + a_2(t) + a_3(t) = T_s(t)$$

$$\frac{\partial T(r,t)}{\partial r} = \frac{a_1(t)}{R} + 2a_2(t)\frac{r}{R^2} + 3a_3(t)\frac{r^2}{R^3}$$

at

$$r = 0, \frac{\partial T}{\partial r}(0,t) = 0 \quad \text{implies } a_1(t) = 0 \quad \text{for all } t \geq 0.$$

To use the heat balance technique requires a set of weighting functions to be determined. These are usually chosen empirically and for this example, the following set were selected:

$$w_0(r) = 1; \qquad w_1(r) = r; \qquad w_2(r) = r^2, \qquad 0 \leq r \leq R.$$

Some of the following integral heat balances were used to obtain the lumped parameter model:

$$\int_0^R w_i(r)\frac{\partial T}{\partial t}\,dr = \int_0^R D_I w_i(r)\left[\frac{1}{r}\frac{\partial}{\partial r}\left(\frac{r\partial T}{\partial r}\right)\right]dr, \qquad i = 0, 1, 2. \quad (13.44)$$

Evaluating equation (13.44) yields the following equation for $\mathbf{a}^T = [a_0, a_2, a_3]$:

$$\begin{bmatrix} 1 & \frac{1}{3} & \frac{1}{4} \\ \frac{1}{2} & \frac{1}{4} & \frac{1}{5} \\ \frac{1}{3} & \frac{1}{5} & \frac{1}{6} \end{bmatrix} \begin{bmatrix} \dot{a}_0(t) \\ \dot{a}_2(t) \\ \dot{a}_3(t) \end{bmatrix} = \frac{D_I}{R^2} \begin{bmatrix} 0 & 4 & \frac{9}{2} \\ 0 & 0 & 0 \\ 0 & \frac{4}{3} & \frac{9}{4} \end{bmatrix} \begin{bmatrix} a_0(t) \\ a_2(t) \\ a_3(t) \end{bmatrix} + \begin{bmatrix} 0 \\ -1 \\ 0 \end{bmatrix} \frac{D_I f_I}{K_i R}. \tag{13.45}$$

As it stands, the model is given in terms of the time evolution of the parameters of $\mathbf{a}(\cdot) \in \mathbb{R}^3$. This can be translated into a system appropriate to the ingot temperatures as follows. Define a midpoint temperature:

$$T_M(t) \triangleq T(\tfrac{1}{2}R, t) = a_0(t) + \tfrac{1}{4} a_2(t) + \tfrac{1}{8} a_3(t)$$

then collating,

$$\begin{bmatrix} T_c(t) \\ T_M(t) \\ T_s(t) \end{bmatrix} = \begin{bmatrix} 1 & 0 & 0 \\ 1 & \frac{1}{4} & \frac{1}{8} \\ 1 & 1 & 1 \end{bmatrix} \begin{bmatrix} a_0(t) \\ a_2(t) \\ a_3(t) \end{bmatrix}$$

and this may be used in conjunction with (13.45) to yield

$$\begin{bmatrix} \dot{T}_c(t) \\ \dot{T}_M(t) \\ \dot{T}_s(t) \end{bmatrix} = \frac{D_I}{R^2} \begin{bmatrix} 56.5 & -134.0 & 77.5 \\ -61.0 & 116.0 & -55.0 \\ 181.5 & -354.0 & 172.5 \end{bmatrix} \begin{bmatrix} T_c(t) \\ T_M(t) \\ T_s(t) \end{bmatrix} + \begin{bmatrix} 20.0 \\ -15.0 \\ 40 \end{bmatrix} \frac{D_I f_I}{K_I R}. \tag{13.46}$$

Define the ingot state vector as $\mathbf{x}_I^T(t) = [T_c(t), T_M(t), T_s(t)]$, then in the usual state-space notation (13.46) becomes

$$\dot{\mathbf{x}}_I(t) = \mathbf{A}_I \mathbf{x}_I(t) + \mathbf{b}_I f_I. \tag{13.47}$$

Equation (13.47) is thus a continuous time-lumped parameter model for centre, midpoint and surface-ingot temperatures. Its initial condition is just the $(t = 0)$ values of these three ingot temperatures.

A lumped parameter model for the pit wall

The derivation of this model corresponds to that given by Maeda *et al.* [36] consequently only the important features are given. A one dimensional con duction equation is utilized:

$$\frac{\partial T}{\partial t} = D_w \frac{\partial^2 T}{\partial x^2}, \qquad 0 \le x \le b \tag{13.48}$$

where D_w is the wall diffusivity and b the wall thickness. The boundary conditions are:

(i) Internal pit wall flux condition $-K_w \left.\dfrac{\partial T}{\partial x}\right|_{x=0} = f_w,$ (13.49)

(ii) External pit wall flux condition $-K_w \left.\dfrac{\partial T}{\partial x}\right|_{x=b} = f_b.$ (13.50)

To facilitate the use of the integral balance technique, three weighted temperature profiles were introduced, each related to a weight function, namely

$$\Theta_0 = \int_0^b T(x,t)\,dx, \qquad \omega_0(x) = 1, 0 \le x \le b, \tag{13.51}$$

$$\Theta_1 = \int_0^b (b-x)T(x,t)\,dx, \qquad \omega_1(x) = (b-x), 0 \le x \le b, \tag{13.52}$$

$$\Theta_2 = \int_0^b (b-x)^2 T(x,t)\,dx, \qquad \omega_2(x) = (b-x)^2, 0 \le x \le b. \tag{13.53}$$

The integral balance is applied as

$$\int_0^b \omega_i(x)\frac{\partial T}{\partial t}\,dx = \int_0^b \omega_i(x) D_w \frac{\partial^2 T}{\partial x^2}\,dx \qquad i = 0, 1, 2, \tag{13.54}$$

to yield three ordinary differential equations for Θ_0, Θ_1 and Θ_2, namely:

$$\begin{aligned} \dot{\Theta}_0 &= (D_w/K_w)(f_w - f_b) \\ \dot{\Theta}_1 &= (bD_w f_w)/K_w + (T_b(t) - T_w(t))D_w \\ \dot{\Theta}_2 &= D_w(b^2 f_w/K_w - 2bT_w(t) + 2\Theta_1). \end{aligned} \tag{13.55}$$

Three lumped-parameter temperatures are introduced:
Internal wall temperature

$$T_w(t) \triangleq T(0,t),$$

Mean wall temperature

$$\bar{T}_w(t) = \frac{1}{b}\int_0^b T(x,t)\,dt, \tag{13.56}$$

External wall temperature

$$T_b(t) = T(b, t),$$

and the wall profile is assumed to be:

$$T(x, t) = c_0(t) + c_1(t)\,\frac{x}{b} + c_2(t)\,\frac{x^2}{b^2}. \tag{13.57}$$

Using (13.48) to (13.50), (13.51) to (13.54), (13.55), (13.56) and (13.57), three eliminations are performed to obtain a lumped parameter model for the temperatures $T_w(t)$, $\bar{T}_w(t)$, $T_b(t)$, viz.:

1. The linear relationship between $\mathbf{c}^T = [c_0(t), c_1(t), c_2(t)]$ and $\mathbf{T}_{\text{wall}}^T \triangleq [T_w(t), \bar{T}_w(t), T_b(t)]$ is established.

2. The linear relationship between $\mathbf{\Theta}^T \triangleq [\Theta_0(t), \Theta_1(t), \Theta_2(t)]$ and $\mathbf{c}^T = [c_0(t), c_1(t), c_2(t)]$ is used to define the linear relationship between $\mathbf{\Theta}(t) \in \mathbb{R}^3$ and $\mathbf{T}_{\text{wall}}(t) \in \mathbb{R}^3$.

3. Using equation (13.55), $\dot{\mathbf{\Theta}}(t)$ and $\mathbf{\Theta}(t)$ are eliminated to yield a lumped parameter pit wall model as:

$$\begin{bmatrix} \dot{T}_w(t) \\ \dot{\bar{T}}_w(t) \\ \dot{T}_b(t) \end{bmatrix} = \frac{D_w}{b^2} \begin{bmatrix} -36 & 60 & -24 \\ 0 & 0 & 0 \\ -24 & 60 & -36 \end{bmatrix} \begin{bmatrix} T_w(t) \\ \bar{T}_w(t) \\ T_b(t) \end{bmatrix} + \frac{b}{k_w} \begin{bmatrix} 9 & -3 \\ 1 & -1 \\ 3 & -9 \end{bmatrix} \begin{bmatrix} f_w \\ f_b \end{bmatrix} \tag{13.58}$$

where the model is driven by f_w = internal wall flux and f_b = external wall flux. Define the pit wall state vector as

$$\mathbf{x}_w^T(t) = [T_w(t), \bar{T}_w(t), T_b(t)]$$

and the flux vector as

$$\mathbf{f}_{\text{wall}}^T(t) = [f_w(t), f_b(t)],$$

then equation (13.58) has state space form:

$$\dot{\mathbf{x}}_w(t) = \mathbf{A}_w\mathbf{x}_w(t) + \mathbf{B}_w\mathbf{f}_{\text{wall}}. \tag{13.59}$$

Clearly, this approach to the description of the soaking-pit process produces a more detailed and complex model. An overview of all the modules and their interaction may be seen in Fig. 13.44. An increase in complexity may give a more accurate representation of the process under investigation but there are some disadvantages. There are quite a number of physical parameters to be

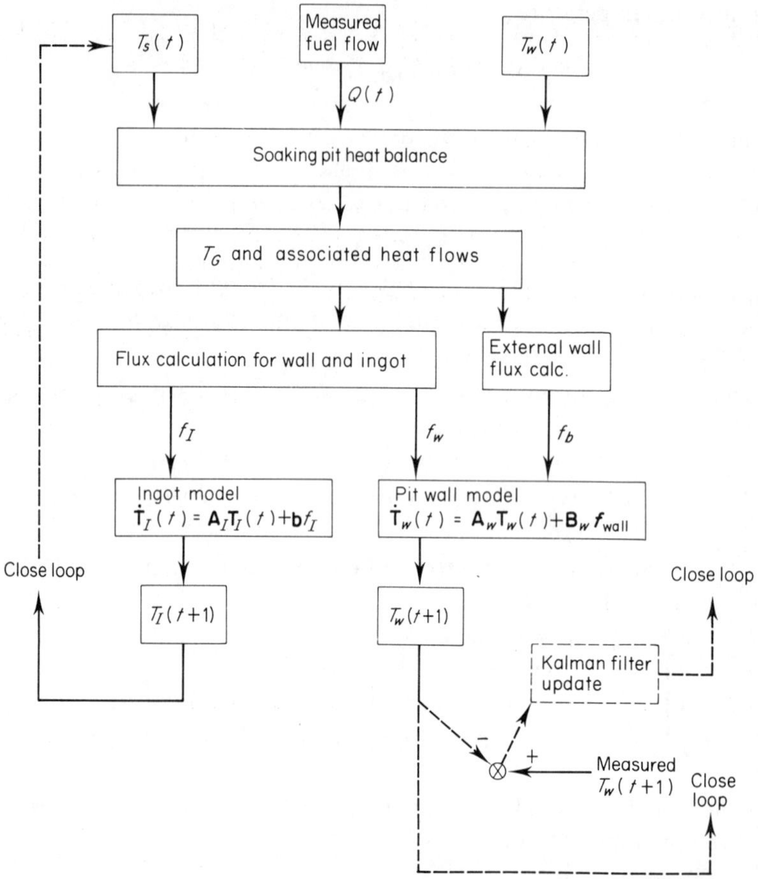

Fig. 13.44. Overview of a phenomenological approach to soaking pit modelling

determined with this model and parts of the computation require the solution of non-linear equations. Both of these may not permit the use of the model for on-line purposes. However, there are other interesting possibilities for investigation. One is to drive the model using measured values of the fuel flow and pit wall temperature and omit the pit wall model altogether. The measured waste gas temperature could be brought into the non-linear calculation for the heat fluxes; this may increase the accuracy of the model. Finally, a Kalman filter could be used with the pit wall model and the measured pit wall temperature to provide a filtered value of pit wall temperature. This filtered value could be input direct to the non-linear calculation for pit gas temperature T_g and the fluxes. As can be seen, the exploitation of this type of model is rather more involved than the method of using least-square fitting.

13.4. GAUGE CONTROL AND THE BACK-UP ROLL ECCENTRICITY PROBLEM

Rolling mills are an essential component of the present-day technology for processing steel. Once ingots or slabs have been through a reheating process, the next stage is often a rolling process of some type. Steel being rolled into strip will meet the hot-rolling mill and subsequently the cold-rolling mill. The problem of back-up roll eccentricity arises with both types of mills whether or not they have roll screw-down controlled by hydraulic actuators or electric motors. It has proven to be very difficult to eliminate the eccentricity problem using conventional control schemes and in fact these can exacerbate the condition. This causes an important quality control problem given the throughput of a typical rolling mill.

In this section, the details of the back-up roll eccentricity problem are elaborated and the conventional control system described. To overcome the shortcomings of this approach, a Kalman filter solution is prescribed and its implementation discussed.

13.4.1. Rolling-mill gauge-control systems

The production of off-gauge material from a rolling mill can have numerous causes, but these can be divided into two main groups (Mornas and Planté [40]):

Inherent strip deficiences

In this first group, the reasons for off-gauge strip lies with the profile or mechanical properties of the incoming strip, for example:

- ★ Incoming strip already off-gauge.
- ★ Yield strength variations within incoming strip.
- ★ Friction factor variations with lubrication and rolling speed.

Mill variability

A second grouping arises from variations in the mill setup, for example:

- ★ Eccentricity of the backup roll (BUR).
- ★ Gap in the oil-bearing film varying with rolling speed.
- ★ Roll hardness, roll expansion, roll wear and inaccurate roll profile.

The first group act on the separating force between the work rolls, and hence, on the mill spring or stretch. Consequently, the lower the value of the stiffness ratio between the mill and product strip, the greater the influence of these parameters on the off-gauge material. An *increasing trend* in *output gauge* is associated with an *increase in roll force* due to these factors.

The second group of reasons for off-gauge material include spurious roll gap variations. For these factors, the higher the mill stiffness ratio, the greater is the influence of these effects on the off-gauge strip. Consequently, an *increasing output gauge* is *associated with a decrease in the roll force.*

The fact that these two types of disturbances act in opposite senses produces a difficult classical control problem but a stochastic optimal control approach provides a formal procedure for dealing with such cases. This proposal is discussed later in the section; first, the conventional gauge-control philosophy, based on the BISRA–Davy gauge-meter principle is presented.

13.4.2. The BISRA–Davy gauge-meter principle

Many gauge-control systems are based on the British Iron and Steel Research Association–Davy gauge-meter principle. In this scheme, the strip gauge is controlled indirectly by manipulating the roll force. A common problem with such methods is the adverse effect of roll eccentricity on the performance of the gauge control. This eccentricity causes a periodic component in the roll-force signal which results in a similar variation in the output gauge.

The main cause of the eccentricity component in the roll-force signal in a four high mill (Fig. 13.46) is the eccentricity of the *back-up rolls.* This eccentricity can arise during the roll-grinding process, when the rolls are being refurbished or during the operation of the mill when a crash (disruption caused by strip snapping) occurs. Eccentricity may also be caused by thermal changes or wear on the bearing sleeves. Originally the back-up rolls may have been inherently non-cylindrical, a condition the grinding process has failed to remove. Alternatively, the rolls may have been ground with an off-axis mounting (that is, the axis does not correspond with the axis of the roll when mounted in normal bearing chocks). The main cause of the eccentricity variations in the strip thickness is due to eccentricity defects in the back up rolls (Waltz and Reed [41]).

In contrast, *work-roll* eccentricity does not contribute significantly to the eccentricity signal since the work rolls are able to move vertically in their bearings. When the roll gap is reduced, the increase in roll force tends to push the work rolls apart. The freedom of movement vertically permits them to move apart by an amount determined by the bending and flattening of the backup rolls. Consequently, work-roll eccentricity tends to be absorbed by this mechanism.

To appreciate the gauge-meter principle, it is necessary to discuss some of the basics of rolling theory, in particular, the mill-stretch and roll-force characteristics.

Mill-stretch relationship

The mill stand is an elastic body which deforms in accordance with Hooke's law. The rolls also bend and flatten under load. The load-extension relationship for a rolling stand may be obtained by removing the strip and measuring the load-screwdown relationship, as shown in Fig. 13.45(a). (If the mill has hydraulic actuators, then this relationship involves the hydraulic capsule position.) Several features of this characteristic are important:

(i) Over a significant range of roll-force values, the relationship is linear. The error at zero-load caused by using the linear characteristic is denoted by Δ.

(ii) The form of this characteristic can be assumed to be independent of the initial screw setting s_i.

(iii) The mill-stretch modulus M_m is defined in terms of this characteristic:

$$M_m = \frac{1}{k_m(f)} \quad \text{where } k_m(f) \triangleq \frac{\partial}{\partial f}\phi_m(f) > 0. \tag{13.60}$$

In the operating range where the characteristic $\phi_m(\cdot)$ is linear, one value of M_m may be used.

If several characteristics for a range of initial roll-gap settings are plotted (they will be required in subsequent discussions) then Fig. 13.45(b) is obtained where $s_3 < s_2 < s_1$. This family of characteristics is a set of shifted inverse relationships $\phi_m^{-1}(\cdot)$ so that a given rolling load f_0 will generate several roll gap thicknesses depending on the initial strip thickness from which screwdown began, namely from the appropriate s_i, $i = 1, 2, 3$.

The relationship between the output gauge, roll-gap setting and the exten-

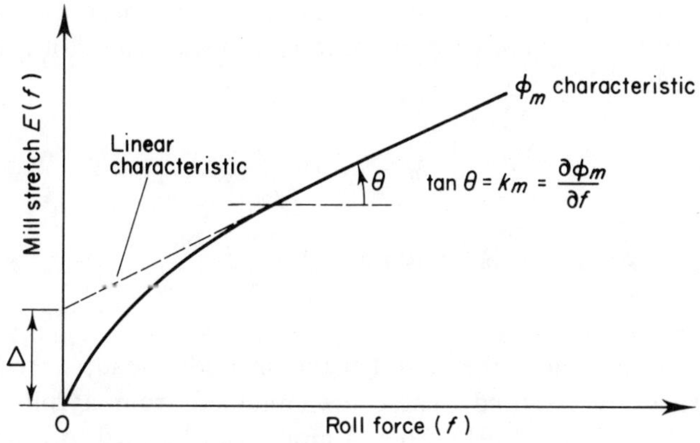

Fig. 13.45(a). Mill stretch-roll force characteristic ϕ_m

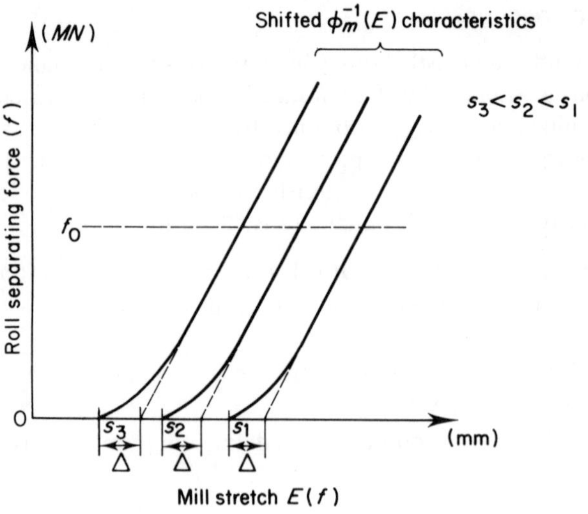

Fig. 13.45(b). Mill stretch characteristics for several initial screwdown settings

sion due to mill stretch may be written:

$$h = s_i + E = s_i + \phi_m(f) \tag{13.61}$$

where h = output gauge, s_i = rollgap setting and E = mill extension or stretch. For first-order changes the following linearized form is appropriate:

$$\delta h = \delta s_i + \frac{\partial \phi_m}{\partial f}\,\delta f.$$

However, the linearity of the $\phi_m(\cdot)$ characteristic, over a wide range of operating roll-force values, yields a linear relationship, valid for changes larger than first order, viz.:

$$\delta h = \delta s_i + \frac{\delta f}{M_m} \tag{13.62}$$

where M_m is assumed to be constant.

Roll-force characteristic

To completely define the mill behaviour under load, a second set of characteristics are required. These are obtained from a roll-force model (Golten [48], Grimble [49]). Such a model can be used to generate a roll-force-output gauge characteristic for different sets of rolling parameters (such

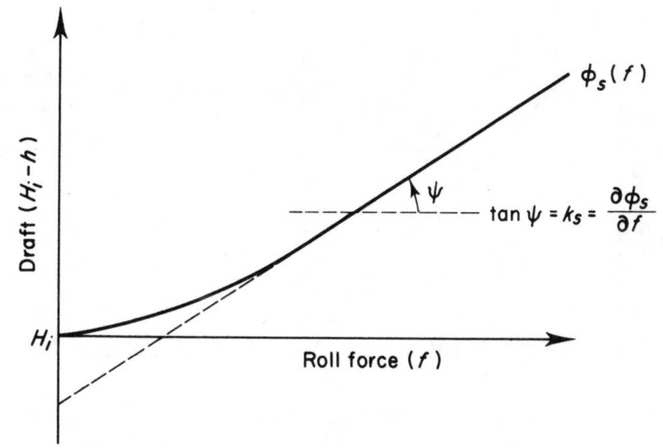

Fig. 13.45(c). Roll force characteristic for input gauge, H_i

as strip tensions, yield stress and friction coefficients) and a given input gauge. A typical characteristic is given in Fig. 13.45(c) for which the following features are important.

(i) This characteristic is produced from a highly non-linear model for which a dependence on the input gauge H_i exists. However, for present purposes such a dependence is considered to be small, and is therefore ignored.

(ii) The slope of the characteristic is used to define the strip modulus, M_s as:

$$M_s = \frac{1}{k_s(f)} \quad \text{where } k_s(f) = \frac{\partial \phi_s}{\partial f} > 0. \tag{13.63}$$

(iii) Although non-linear, the characteristic is well behaved (continuous, and strictly monotonic increasing) so that

(a) $f_1 > f_2$ implies $\phi_s(f_1) > \phi_s(f_2)$

and (13.64)

(b) $\phi_s(f_1) > \phi_s(f_2)$ implies $f_1 > f_2$.

In the sequel, these characteristics are transposed so that the situation for several input gauges is shown as in Fig. 13.45(d). These are the shifted inverse characteristics $\phi_s^{-1}(\cdot)$, and as can be seen, to achieve a given output gauge, the thicker input gauges require higher rolling forces. The relationship between

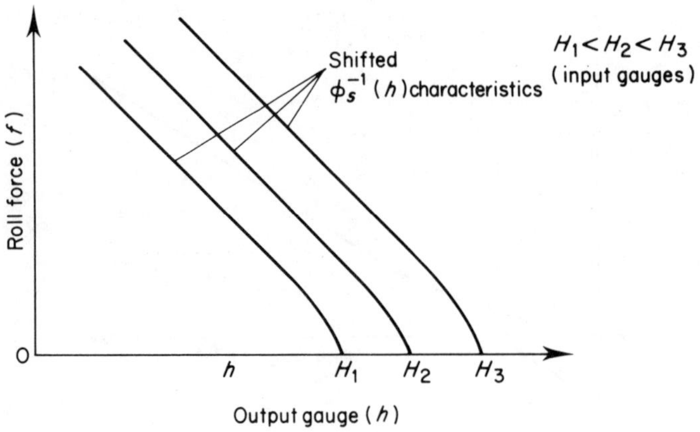

Fig. 13.45(d). Roll force characteristics for several input gauges $H_1 < H_2 < H_3$

input and output gauge may be written:

$$h = H_i - \phi_s(f) \tag{13.65}$$

where h = output gauge, H_i the input gauge and $\phi_s(f)$ the reduction achieved by roll force f.

Mill modulus

To measure the mill modulus (M_m), relating the roll-separating force to the stand stretch, the load is measured with no strip in the mill. The screw movement with the work rolls in contact reveals the total stand stretch. This curve of separating force against screw movement is shown in Fig. 13.45(a). If required, the roll-force-strip thickness relationship can be calculated from reasonably accurate models (Grimble, Fuller and Bryant [42]). A typical curve is shown in Fig. 13.45(*c*).

Gauge-meter principle

The gauge-meter principle, for an electric screwdown roll-gap control scheme, is illustrated in Fig. 13.46. Material of thickness H_1 is assumed to be entering the stand and producing a roll force f_1 which in turn causes a mill stretch, given by the mill-stretch curve. Thus, the operating point may be written as the tuple (s_1, H_1, h_1, f_1). If the strip hardness increases or the input gauge increases, this disturbance may be written as a new operating point (s_1, H_2, h_2, f_2). For such changes in the linear operating region, the equation (13.62) gives:

$$\delta h = \delta s_i + \delta f / M_m.$$

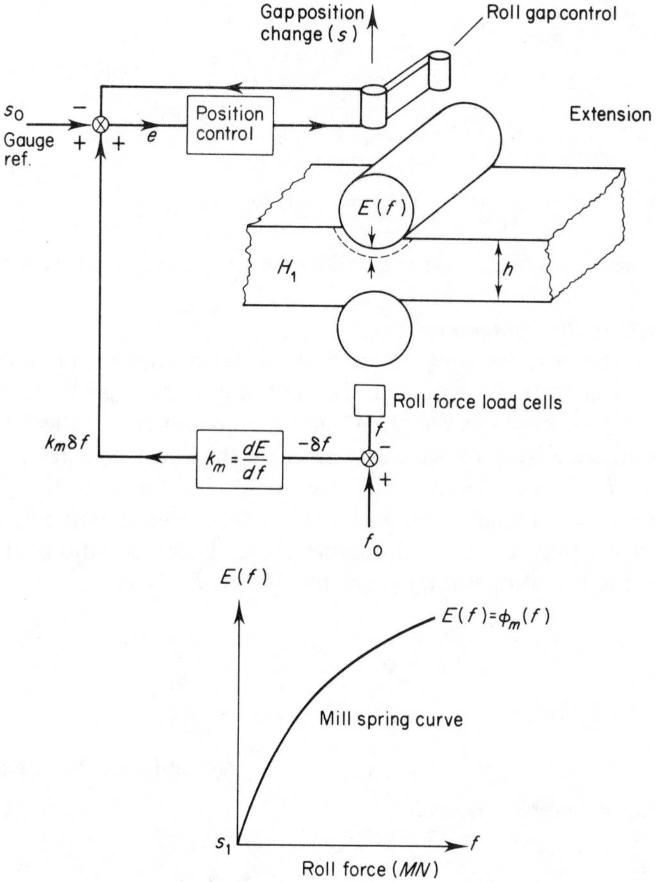

Fig. 13.46. Basic principle of the BISRA–Davy gaugemeter (roll load increase leads to extension and screwdown act

Thus, in order to restore the output thickness to its original value, the unloaded roll gap must be altered by the gauge-meter to compensate for the mill stretch. Since the change in roll force can be measured and a value assigned to the mill-stretch modulus (which in reality is roll-force dependent), the calculated gauge error (relative to the initial gauge setting) becomes:

$$e = -\delta h = -\delta s_i - \delta f / M_m \tag{13.66}$$

Note that $\delta s > 0$ implies an increase in the unloaded roll-gap size. The gauge-meter decreases the roll gap compensating for mill stretch due to the roll-force change δf until the error becomes zero. The controller continuously tries to achieve $\delta h = 0$, viz.:

$$\delta s_i = -\delta f / M_m$$

giving

$$s - s_i = -\delta f / M_m$$

and hence $s = s_i - \delta f / M_m$ and the roll-gap setting is reduced as required.

Operation of the gauge-meter

The characteristics for mill stretch and roll force enables the operation for the gauge-meter to be investigated. The initial operating position is the tuple (s_1, H_1, h_1, f_1) as shown in Fig. 13.47 and it is assumed that there is an input gauge disturbance (due to strip hardness or thickness changes). Hence the input gauge H_1 is considered to increase to H_2 and the operating conditions change. No screw changes are considered at this stage the operating point is free to move along the $\phi_m^{-1}(\cdot)$ characteristic. Using equations (13.61) and (13.65), at the initial operating point (s_1, H_1, h_1, f_1), gives:

$$h_1 = s_1 + \phi_m(f_1) = s_1 + \Delta + \frac{f_1}{M_m} \tag{13.67}$$

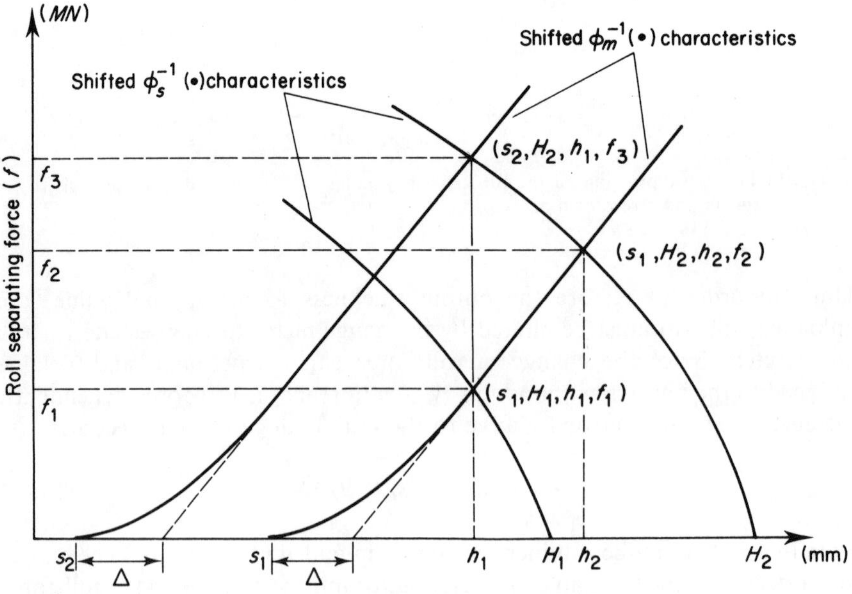

Fig. 13.47. Operation of the BISRA–Davy gaugemeter.

and

$$h_1 = H_1 - \phi_s(f_1) \tag{13.68}$$

whilst at a new operating condition (s_1, H_2, h, f)

$$h = s_1 + \phi_m(f) = s_1 + \Delta + \frac{f}{M_m} \tag{13.69}$$

and

$$h = H_2 - \phi_s(f). \tag{13.70}$$

Eliminating h_1 between equations (13.67) and (13.68) yields (13.72) and similarly eliminating h between (13.69) and (13.70) yields (13.71):

$$s_1 + \Delta + \frac{f}{M_m} = H_2 - \phi_s(f) \tag{13.71}$$

$$s_1 + \Delta + \frac{f_1}{M_m} = H_1 - \phi_s(f_1). \tag{13.72}$$

Subtracting (13.72) from (13.71) obtains

$$H_2 - H_1 = \frac{f - f_1}{M_m} + \phi_s(f) - \phi_s(f_1) \tag{13.73}$$

and the assumption of an increase in input gauge coupled to the property (13.64) for characteristic $\phi_s(\cdot)$ ensures an increase in roll-force occurs. Subtracting equation (13.68) from (13.67) yields:

$$\delta h = h - h_1 = \frac{f - f_1}{M_m} > 0$$

and an increase in output gauge also follows.

Equation (13.73) may be rearranged as:

$$\frac{f}{M_m} + \phi_s(f) = C \tag{13.74}$$

where $C = H_2 - H_1 + (f_1/M_m) + \phi_s(f_1) > 0$ and the property $\phi_s(0) = 0$, with (13.64), ensures the existance of a unique solution $f = f_2$ to equation (13.74).

A new operating condition thus occurs at (s_1, H_2, h_2, f_2) and this is shown on Fig. 13.47.

The next step in the gauge-meter operation can be considered to be a screw change which reduces the output gauge back to h_1. This operating point is denoted by the tuple (s, H_2, h_1, f), obtained by moving along the roll-force characteristic (in practice, these separate steps do not occur under continuous gauge-meter control). The equations which apply are at the initial position (s_1, H_1, h_1, f_1):

$$h_1 = s_1 + \Delta + f_1/M_m \tag{13.67}$$

$$h_1 = H_1 - \phi_s(f_1). \tag{13.68}$$

Intermediate operating position (s_1, H_2, h_2, f_2):

$$h_2 = s_1 + \Delta + f_2/M_m \tag{13.75}$$

$$h_2 = H_2 - \phi_s(f_2) \tag{13.76}$$

and a final operating condition (s, H_2, h_1, f):

$$h_1 = s + \Delta + f/M_m \tag{13.77}$$

$$h_1 = H_2 - \phi_s(f). \tag{13.78}$$

From equations (13.76) and (13.78)

$$h_2 - h_1 = \phi_s(f) - \phi_s(f_2) > 0$$

which, using property (13.64), yields $f > f_2 > f_1$ and the new operating position corresponds with an increase in roll force. From equations (13.68) and (13.78) obtains:

$$\phi_s(f) = C_1 \tag{13.79}$$

where $C_1 = H_2 - H_1 + \phi_s(f_1) > 0$. Property (13.64) and $\phi_s(0) = 0$ ensures the existence of a unique $f = f_3$ to satisfy equation (13.79), thus the new operating position may be written (s_2, H_2, h_1, f_3) as shown on Fig. 13.47.

Using equation (13.67) and (13.77) the new screw setting $(s_2 = h_1 - \Delta - f_3/M_m)$ may be derived as:

$$s_2 = s_1 - \frac{(f_3 - f_1)}{M_m}$$

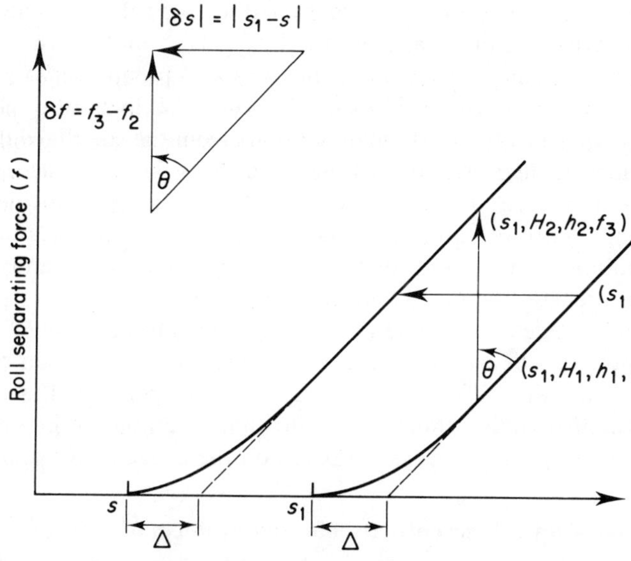

Fig. 13.48. Geometric justification for gaugemeter operation

or in terms of screw and roll-force changes as:

$$\delta s = \frac{\delta f}{M_m}.$$

Thus, complete gauge correction can be achieved given exact knowledge of M_m and the roll force change δf.

The geometric justification for this result is quite direct depending only on the constancy of the $\phi_m(\cdot)$ and the property of parallel characteristics as shown in Fig. 13.48. Thus, for a roll-force increase δf, as screw change of $|\delta s| = \delta f \tan\theta$ namely,

$$s = s_1 - \frac{\delta f}{M_m},$$

is required. Clearly, whatever the magnitude of the δf change, due to an input gauge variation, the change δs is sufficient to return the output gauge to its original value.

Hydraulic gap control

Although the mill diagram (Fig. 13.46) above, referred to electric screw-down position control, the eccentricity problem is also of importance in hydraulic position controlled mills. These rely upon the accurate control of hydraulic cylinders for the adjustment and control of the roll gap. The main components of a hydraulic mill stand are the hydraulic cylinders which vary the rolling load, a hydraulic power pack (control valve + pump), servo valves, position and pressure transducers and load cells. A typical hydraulic cylinder is of 3000 tonnes and has a stroke of 50 mm. Apart from the simplified design and lower capital cost, their main attraction is the claim that more accurately rolled strip can be produced to give a higher yield than from screw-down mills. Hydraulic roll-gap control is sufficiently fast that complete compensation for eccentricity variations is theoretically feasible. (Typical position and force loop bandwidths are 15–20 and 2–3 Hz respectively.)

Since the gauge-meter enables the same output gauge to be maintained in the presence of changes in the incoming gauge and hardness, its operation can be considered to increase the stiffness of the stand. As an operational measure, the stiffness of the first rolling stand in a multistand hydraulic rolling mill is often increased by using a control system based on the gauge-meter principle.

13.4.3. The back-up roll eccentricity filtering problem

The eccentricity in the back-up rolls causes the actual roll gap and rolling load to vary. Neglecting harmonics, the period of the eccentricity signal is equal to the period of revolution of the back-up roll. This eccentricity will cause a gauge deviation but, more debilitating, a signal is passed to the gap position control in such a direction so as to exaggerate this eccentricity deviation. For example, consider when the eccentricity is causing a reduction in the roll gap; this causes an increase in roll force which reduces the roll gap, while the gauge-meter loop uses the roll-force increase to further reduce the roll gap. Thus, not only is the natural eccentricity deviation imprinted on the strip by way of a small gauge variation, but this is exacerbated by the action of the gauge-meter loop.

Over long periods of operation, the top and bottom back-up rolls often wear to slightly different diameters and may both cause different eccentricity components. In this case, the total effect on the roll-force signal is to cause an amplitude modulated variation, the magnitude of which can be as significant as 8% of the total load in hot-rolling mill applications. The degree of eccentricity can also vary across the barrel of the rolls. There is the possibility of accommodating this variation when compensating for eccentricity since there are independent load measurements on both sides of the rolling stand.

The fundamental component of the eccentricity signal (ω_0) changes with the line speed and if this frequency is low enough the roll-gap actuators can respond to the oscillations. Even when electric screw motors do not respond, the eccentricity signal has a detrimental effect on the control system degrading performance and causing maintenance problems. Variations in the screw-motor current causes increased heating, brush and commutator wear and gear train wear (lubricants are rubbed away by small rapid movements). The problem is more acute in hydraulic mills which can respond to the eccentricity signal and can cause a gauge deviation which is worse than the natural eccentricity variations.

The basic layout of the classical solutions using a turnable notch filter network to reduce the eccentricity effect is shown in Fig. 13.49. In this figure, the physical model is based on the linearized version of equation (13.61) and the non-linear equation (13.65):

$$h = s + \Delta + \frac{f}{M_m} \tag{13.80}$$

$$h = H - \phi_s(f). \tag{13.81}$$

Note that equation (13.80) is rearranged and utilized in the model as:

$$f = M_m(h - s - \Delta). \tag{13.82}$$

Other approaches to reducing the eccentricity effect includes reducing the position loop gain K or introducing a dead band to limit the gain at the eccentricity frequency. The performance of the roll-force control loop is degraded by all of these solutions. The eccentricity signals contain significant distortion and higher-order harmonics hence, it is difficult to design conventional frequency-domain filters which are both effective and do not degrade the controller regulating action.

The gauge-meter scheme shown in Fig. 13.49 has a simplified small signal form as shown in Fig. 13.50. The small signal version involves linearizing the characteristic $\phi_s(f)$ about some nominal equilibrium value f^*, thus, introducing the coefficient $k_s \triangleq M_s^{-1}$ into the analysis.

The transfer-function relating roll-force variations to input gauge variations becomes:

$$\delta f = \frac{M_m}{[1 + M_m/M_s]\{1 - [M_m P(s)/\hat{M}_m(1 + M_m/M_s)]\}}\,\delta H$$

$$= \frac{M_m\,\delta H}{[1 + (M_m/M_s) - (M_m P(s)/\hat{M}_m)]}. \tag{13.83}$$

If the gauge-meter loop is opened:

$$\delta f = \frac{M_m}{[1 + (M_m/M_s)]}\,\delta H. \tag{13.84}$$

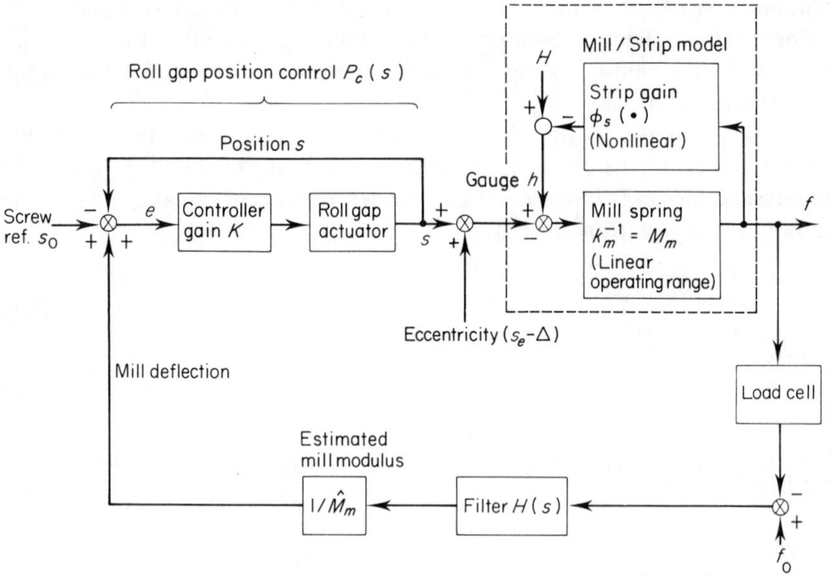

Fig. 13.49. BISRA gaugemeter without X-ray gauge feedback

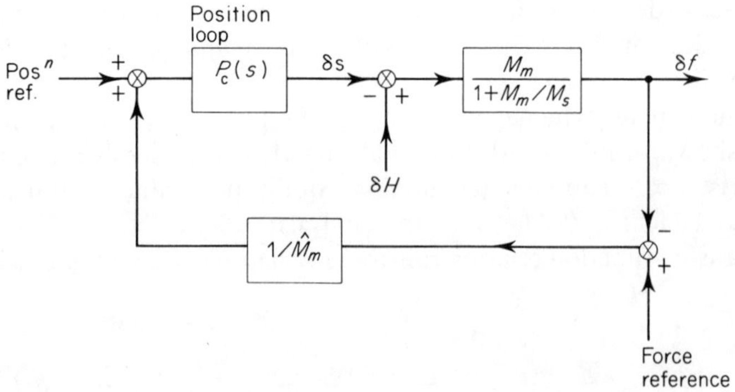

Fig. 13.50. Simplified small signal block diagram of gaugemeter scheme

Example 13.1. Let $M_m/M_s = 1$, filter $H(s) = 1$, $\hat{M}_m = M_m$ and $P(s) = a/(s + a)$. In the steady state, with the gauge-meter loop open, $\delta f = M_m\, \delta H/2$, $\delta h = \delta H/2$ but with the gauge-meter loop closed $\delta f = M_m\, \delta H$, $\delta s = -\delta H$ and $\delta h = 0$ (from 13.62). The modulus of steel strip is usually much greater than that of the mill and $M_m/M_s \ll 1$, typically $M_m/M_s = 1/10$. However, for aluminimum $M_m/M_s \gg 1$ and typically $M_m/M_s = 10$.

Stability

Consider the characteristic equation for the system

$$\left(1+\frac{M_m}{M_s}-\frac{M_m}{\hat{M}_m}P(s)\right)=0$$

where typically $P(s)=a/(s+a)$ and $a>0$, hence

$$s+a\left[1-\frac{M_m}{\hat{M}_m}\frac{1}{(1+M_m/M_s)}\right]=0. \tag{13.85}$$

Clearly, if $|M_m/\hat{M}_m|<1$ then since $|1+M_m/M_s|>1$

$$[1-(M_m/\hat{M}_m)/(1+M_m/M_s)]>0 \Rightarrow a[\cdot]>0 \quad \text{for all } M_m/M_s>0$$

and hence, the system is stable for all such cases.

Performance

The mill modulus $\hat{M}_m \simeq k_m^{-1}=1/(\partial\phi_m/\partial f)$ used in the gauge-meter calculations should ideally match the actual stand modulus $M_m=k_m^{-1}$ during rolling. If $M_m<\hat{M}_m$ (or $k_m>\hat{k}_m$) then the system is degraded in performance. Consider, for example, the steady-state situation. From (13.68) and (13.83):

$$\delta f=\frac{M_m}{(1+(M_m/M_s)-(M_m/\hat{M}_m))}\,\delta H$$

$$\delta s=-\delta f/\hat{M}_m$$

and

$$\delta h=(-1/\hat{M}_m+1/M_m)\,\delta f=(1-M_m/\hat{M}_m)\,\delta f/M_m \tag{13.86}$$

$$\neq 0.$$

If the mill modulus $M_m>\hat{M}_m(k_m<\hat{k}_m)$ then from the characteristic equation (13.85), roll-gap overcorrection can occur causing an AGC runaway. This is a type of positive feedback situation which must be avoided. Consequently, the modulus used in the mill-stretch calculation is set greater than the maximum stand modulus ($\hat{M}_m>\max M_m$ or $\hat{k}_m<\min k_m$). By this means runaway is avoided, but unfortunately the performance of the gauge-meter deteriorates. If the position-loop feedback is broken, the system is, of course, unstable because of the positive feedback in the force-control loop.

The mill partly compensates for input gauge and eccentricity variations naturally, since from (13.84), without the gauge-meter action

$\delta f = M_s\, \delta H/(1 + M_s/M_m)$. Thus, force changes caused by δH are attenuated by the factor of $1/(1 + M_s/M_m)$. This attenuation factor tends to unity when the mill is infinitely stiff or when soft material is being rolled. For steel strip of say $M_s/M_m = 10$, the attenuation factor is 1/11, and this reduces the effect of the input gauge and eccentricity disturbances.

Other disturbances which degrade gauge-meter performance originate within the stand itself (Wood and Ivacheff [43]). For example, from roll-diameter changes, caused by thermal gradients or from variations in the back-up roll bearing oil film thickness, caused by speed changes. These can also cause incorrect gap control movements and must be compensated for using thermal models and tachometer signals, respectively.

X-ray gauge feedback

A feedback signal from a downstream thickness gauge is often required (Edwards [44]) to compensate for initial condition errors and drift in the gauge-meter thickness estimate. Such a scheme is illustrated in Fig. 13.51. The original force-control loop is effectively an open-loop gauge-control scheme and has the advantage of being as fast as the roll positioning system dictates. The X-ray gauge feedback loop is intended to compensate for slowly varying disturbances. The speed of this loop is limited by the thickness gauge response and the inherent transport delay time between the roll gap and the gauge measurement device. The controller must usually have speed dependent gains to cater for the large change in the transport delay with line speed. The fact that the basic gauge-meter loop is not limited by a transport delay time is a considerable advantage when the gauge disturbances are varying rapidly.

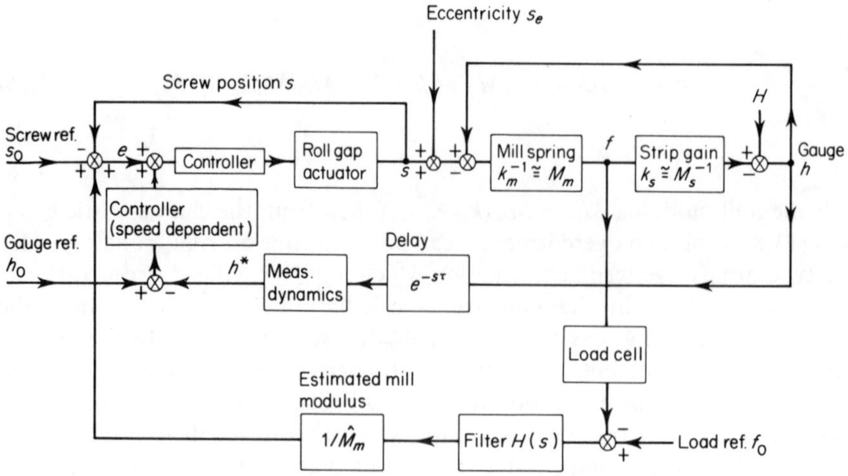

Fig. 13.51. BISRA–Davy gaugemeter with X-ray gauge feedback

Advantages of BUR eccentricity filters

The removal of the BUR eccentricity signal reduces variations in the roll-gap actuators and hence, reduces maintenance problems in electric screwdown systems. The gauge variations in the product strip should also be reduced. Equally important, the tolerance in the roll-grinding process can be increased which will decrease the periodic roll-grinding costs.

13.4.4. Control scheme philosophy

There are two basic ways in which back-up roll eccentricity can be mitigated:

(i) Estimate the low-frequency force variations and use feedback control from these estimates to maintain the reference force level (tunable notch-filter schemes). The inherent variation in gauge due to the back-up roll eccentricity will not be affected by this technique but ideally the control loop should not increase these variations.

(ii) Estimate the low-frequency force variations and the variations due to the back-up roll eccentricity. The eccentricity estimates are used to drive the screws (or hydraulics) in the opposite direction to the inherent roll eccentricity variations. By this means, the roll force and hence gauge can be maintained constant. However, the control system and actuators must be very fast acting and this makes the approach more suitable for hydraulic mills.

Disadvantages of the gaugemeter philosophy

There seems to be a fundamental control theoretic weakness in the gauge-meter principle. Consider, for example, the eccentricity which causes an increase in the roll load as the roll gap decreases. In a negative feedback situation the gap would be increased to compensate for the eccentricity (the same applies to the oil film and thermal camber disturbances). However, the gauge-meter loop employs positive feedback, of admittedly low gain.

The positive feedback occurs because the gauge-meter is designed to regulate against disturbances which arise from changes in input gauge or input strip hardness. In this case, load increases caused by mill stretch result in the gap actuators being moved down and the rolling load increasing still further (positive feedback in terms of load change).

The roll-force change is not, therefore, a reliable indicator of gauge changes since an increase in load may demand opposite screw-position changes, depending upon the cause of the disturbance. This fundamental problem would be resolved if the disturbances were directly measurable which they are not. However, they might be estimated using a Kalman filter since the disturbances have different frequency-response characteristics. The eccentricity signal is periodic with a dominant high-frequency sinusoidal characteristic. The output-gauge disturbances, due to hardness changes and input-gauge variations, are of low frequency and might be modelled as constants.

Given estimates of the disturbances, a more consistent control philosophy can be developed using PID or optimal controllers. If a Kalman filter is used to provide these estimates, a simple solution is to cascade the Kalman filter with a constant feedback gain matrix derived from LQG optimal control theory.

13.4.5. Kalman filtering solution for the control of gauge

Consider the basic plant model for a system without X-ray gauge feedback shown in Fig. 13.49. The roll-gap position control loop can be represented by a state equation subsystem S_p and the eccentricity and input gauge/hardness variations can be modelled by subsystems S_e and S_h, respectively. The gap control subsystem S_p can be taken to be linear which is a good approximation in the presence of the feedback of gap position. Thus, the plant shown in Fig. 13.49 can be represented in state equation form, as shown in Fig. 13.52. The dynamics of the load measurement device can be neglected for the present and note that the subsystems representing k_m and k_s are also non-dynamical. The gauge is not assumed measurable in this case but the force can be measured with measurement noise v.

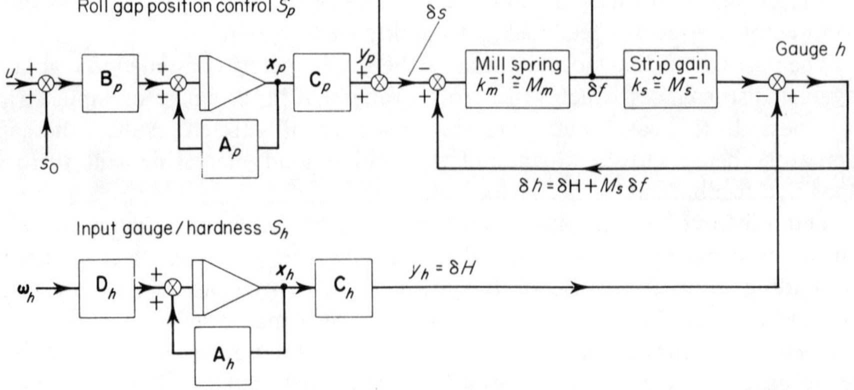

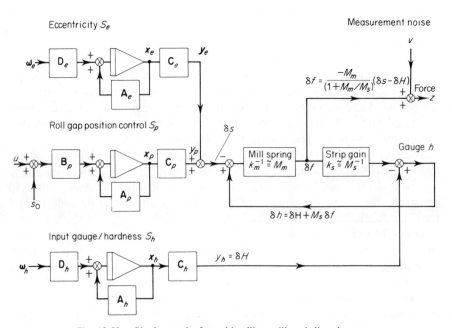

Fig. 13.52. Single-stand of a cold-rolling mill and disturbances

The state equations may be written using the system of Fig. 13.52 as:

$$\begin{bmatrix} \dot{x}_p \\ \dot{x}_e \\ \dot{x}_h \end{bmatrix} = \begin{bmatrix} \mathbf{A}_p & \mathbf{O} & \mathbf{O} \\ \mathbf{O} & \mathbf{A}_e & \mathbf{O} \\ \mathbf{O} & \mathbf{O} & \mathbf{A}_h \end{bmatrix} \begin{bmatrix} x_p \\ x_e \\ x_h \end{bmatrix} + \begin{bmatrix} \mathbf{O} \\ \mathbf{D}_e\omega_e \\ \mathbf{D}_h\omega_h \end{bmatrix} + \begin{bmatrix} \mathbf{B}_p(u+s_0) \\ \mathbf{O} \\ \mathbf{O} \end{bmatrix}$$

$$z = v + \phi_m(y_e + y_p; y_h). \tag{13.87}$$

The controlled gauge becomes:

$$h = \phi_s(y_e + y_p; y_h). \tag{13.88}$$

If the extended Kalman filter were to be used for this application the above non-linear equations would form part of the filter model. The difficulty with the extended filter is that the computational load is normally excessive although a constant gain extended Kalman filter (see Safonov and Athans [45]) might be utilized here. However, for this discussion assume that a linear Kalman filter is to be determined for a number of operating points. The gain of such a filter will have to be switched for different material schedules but this is not a problem.

By linearizing the functions ϕ_m and ϕ_s about the operating points determined by the gap setting s_0, gains (C_{11}^m, C_{12}^m) and (C_{11}^s, C_{12}^s) are obtained, respectively. Now assume that the above state variables and outputs represent changes about the nominal operating point. The linearized state equations becomes:

$$\begin{bmatrix} \dot{x}_p \\ \dot{x}_e \\ \dot{x}_h \end{bmatrix} = \begin{bmatrix} \mathbf{A}_p & \mathbf{O} & \mathbf{O} \\ \mathbf{O} & \mathbf{A}_e & \mathbf{O} \\ \mathbf{O} & \mathbf{O} & \mathbf{A}_h \end{bmatrix} \begin{bmatrix} x_p \\ x_e \\ x_h \end{bmatrix} + \begin{bmatrix} \mathbf{B}_p u \\ \mathbf{O} \\ \mathbf{O} \end{bmatrix} + \begin{bmatrix} \mathbf{O} \\ \mathbf{D}_e\omega_e \\ \mathbf{D}_h\omega_h \end{bmatrix} \tag{13.89}$$

$$z = [C_{11}^m\mathbf{C}_p \quad C_{11}^m\mathbf{C}_e \quad C_{12}^m\mathbf{C}_h]x + v \tag{13.90}$$

$$h = [C_{11}^s\mathbf{C}_p \quad C_{11}^s\mathbf{C}_e \quad C_{12}^s\mathbf{C}_h]x. \tag{13.91}$$

These equations are in the standard form:

State equation: $\dot{x} = \mathbf{A}x + \mathbf{B}u + \mathbf{D}\omega$. (13.92)

Observations: $z = \mathbf{C}x + v$. (13.93)

Controlled output: $h = \mathbf{H}x$. (13.94)

Once the linearized system equations have been defined and the noise covariances for the signals $\boldsymbol{\omega}$ and v have been specified, the solution of the above control problem follows from LQG regulator theory. There is still a remaining design stage, since although the Kalman filter follows directly from the system description, the controller is determined by the optimal control cost function. As seen previously in Chapter 4, the cost-function specification requires particular consideration.

The eccentricity signal varies with line speed and thus part of the system (subsystem S_e) is speed dependent. This can be accommodated by measuring the line or roll speed and by changing the controller and filter gains (and filter subsystem S_e) at different speed ranges. The speed measurement need not be very accurate as required in some eccentricity filtering schemes. An alternative approach would involve the use of a self-tuning or extended Kalman filter (Panuska [50]), which would estimate the speed dependent parameters. The resulting control scheme would be similar to that described for the adaptive ship-positioning problem. However, the added complexity may not justify this approach, unless sufficient computing power is readily available.

If X-ray feedback of gauge is also available then the system model in Fig. 13.52 has an additional delayed and noisy output. A second observations equation becomes:

$$z_h(t) = h(t - \tau) + v_h(t).$$

The stochastic optimal control solution to this delayed output problem involves a Smith predictor (Marshall [51]) type of structure (Grimble [52]) which is well known, having found many applications in the chemical industry.

13.4.6. Concluding comments

A gauge-control scheme without X-ray gauge feedback and using switched gains with line speed and schedule may be based upon a standard LQG Kalman filtering solution. This solution might be further simplified by using a reduced-order filter (Grimble [46], [47]) given the measurable states of the position control subsystem S_p. The use of combined state and state estimate feedback in this way improves both the robustness and tolerance of the scheme to actuator nonlinearity.

A more elegant solution to the problem would involve the use of self-tuning or extended Kalman filters which could estimate speed and schedule dependent parameters. The additional cost being increased technical and computational complexity. If X-ray gauge feedback is available, this also increases the complexity of the solution although this may be justified from the obvious advantages of direct feedback. The increase in complexity can also be limited by careful modelling and implementation of the Smith-predictor-type of controller.

13.5. REFERENCES

[1] Ball, A. E. and Blumberg, J. M. 1975. Development of a dynamic ship positioning system, *GEC. Journal of Science and Technology,* **42**(1), 29–36.

[2] Balchen, J. G., Jenssen, J. A. and Saelid, S. 1976. Dynamic positioning using Kalman filtering and optimal control theory. *Automation in Offshore Oil Field Operation, Proceedings of IFAC/IFIP Symposium on Automation in Offshore Oil Field Operation,* pp. 183–188, North Holland Publishing Co. Ltd.

[3] Balchen, J. G., Jenssen, N. A., Mathisen, E. and Saelid, S. (1980). A dynamic positioning system based on Kalman filtering and optimal control, *Modelling, Identification and Control,* **1**(3), 135–163.

[4] Barton, P. H. (1978). Dynamic positioning systems, *GEC Journal for Industry.* **2**(3), 119–125, Oct.

[5] Brink, A. W., Van Den Brug, J. B., Ton, C., Wahab, R. and Van. Wijk, W. R. 1972. Automatic position and heading control of a drilling vessel, Inst. TNO for Mechanical Constructions, The Netherlands.

[6] Grimble, M. J. 1976. The application of Kalman filters to dynamic ship positioning control, GEC Engineering Memorandum, No. EM188.

[7] Grimble, M. J., Patton, R. J. and Wise, D. A. 1980. Use of Kalman filtering techniques in dynamic ship positioning systems, Oceanology International Conf., Brighton, March, 1978, *IEE Proc.* **127**(3), Pt. D, 93–102.

[8] Grimble, M. J., Patton, R. J. and Wise, D. A. 1979. The design of dynamic ship positioning control systems using extended Kalman filtering techniques, Oceans 1979 Conf. (IEEE), San Diego, California.

[9] Grimble, M. J., Patton, R. J. and Wise, D. A. 1980. The design of dynamic ship positioning control systems using stochastic optimal control theory, *Optimal Control Applic. and Methods,* **1**, 167–202.

[10] Wise, D. A. and English J. W. 1975. Tank and wind tunnel tests for a drill ship with dynamic position control, Offshore Technology Conference, Dallas, Paper No. OTC 2345.

[11] Tamchiro, M., Akasaka, N., Kasui, H. and Miwa, E. 1977. On dynamic positioning system design in particular reference to the postitional signal filtering technique, *ISNA. Japan,* **142**, 173–188, Dec.

[12] English, J. W. and Wise, D. A. 1975. Hydrodynamic aspects of dynamic positioning, *Transactions, North East Coast Inst. of Engineers and Shipbuilders,* **92**(3), 53–72.

[13] Price, W. G. and Bishop, R. E. D. 1974. *Probabilistic Theory of Ship Dynamics*, p. 159, Chapman and Hall.

[14] Pierson, W. G. and Marks, W. 1952. The power spectrum analysis of ocean wave record, *Trans. American Geophysical Union,* **33**(6), 834–844.

[15] Grimble, M. J. 1978. Relationship between Kalman and notch filters used in dynamic ship positioning systems, *Electronics Letters,* Pt. 13, **14**, 399–400.

[16] Fung, P. T. K. and Grimble, M. J. 1981. Self-tuning control of ship positioning systems, *IEE Workshop on the Theory and Application of Adaptive and Self-Tuning Control,* Oxford University, March. Also published by Peter Peregrinus Ltd, edited by C. J. Harris and S. A. Billings.

[17] Grimble, M. J., Fung, P. T. K. and Johnson, M. A. 1982. Optimal self-tuning control systems: theory and applications, Part 2: Identification and self-tuning, *Trans. Inst. MC,* **4**(1), 25–36, Jan–March.

[18] Grimble, M. J. 1979. Design of optimal stochastic regulating systems including integral action, *Proc. IEE,* **126**(9), 841–848.

[19] Fung, P. T. K., Chen, Y. L. and Grimble, M. J. 1982. Dynamic ship positioning control systems design including non-linear thrusters and dynamics, presented at the NATO Advanced Study Institute on Non-linear Stochastic Problems 16818, Algarve, Portugal.

[20] Grimble, M. J. 1981. Design of optimal output regulators using multivariable root loci, *Proc. IEE,* Part D, **128**(2) 41–49.

[21] Neville, E. J. 1971. Standard wave spectra, National Physical Laboratory, Ship T.M. 301.

[22] Morgan, M. J. (1978). *Dynamic Positioning of Offshore Vessels,* Petroleum Publishing Company.

[23] Van Calcar H: (1969). Acoustic position reference methods for offshore drilling operations, Offshore Technology Conf., Paper No. OTC 1141, (567–582).

[24] Fotakis, J., Grimble, M. J. and Kouvartakis, B. 1982. A comparison of characteristic locus and optimal designs for dynamic ship positioning systems, *IEEE Trans. Auto Control*, Vol **AC-27**, No 6.

[25] Cheek, C. N., Knobs, K. and Munro, N. 1978. Estimation and prediction of ingot temperatures on a pilot scale soaking pit, *IFAC Proceedings,* Helsinki.

[26] Wick, H. J. 1981. On-line sequential estimation of ingot centre temperatures in a soaking pit, Rolling-Mill Conference, Cincinnati, May.

[27] Lumelsky, V. J. 1983. Estimation and prediction of unmeasurable variables in the steel mill soaking pit control system. *IEEE Trans. Automatic Control,* **AC-28**(3), March.

[28] Kung, E. Y., Dahn, J. R. and Delancy, G. B. 1967. A mathematical model of soaking pits, *ISA Transactions,* **6**, 162–168.

[29] Carslaw, H. S. and Jaeger, J. C. 1959. *Conduction of Heat in Solids,* Oxford University Press, Oxford.

[30] Massey, I. D. and Sheridan, A. T. 1971. Theoretical predictions of earliest rolling times and solidification times of ingots, *Journal of the Iron and Steel Institute,* **209**, 391–395, May.

[31] Sarjant, R. J. and Slack, M. R. 1954. Internal temperature distribution in the cooling and reheating of steel ingots, *Journal of Iron and Steel Institute,* **177**, 428–444, August.

[32] Corlis, R. G., Brookes, C. H. P., Dyer, D. and Jones, N. 1966. Optimization of the soaking pit process, *The BHP Technical Bulletin,* **29**, 11–20.

[33] den Hartog *et al.,* 1975. Application of a mathematical model in the study of the ingot solidification process, *Iron making and Steel making (quarterly),* **2**, 134–144.

[34] Cummings, S. D. G. 1977. Thermal model for ingot solidification and reheating, *J. Australian Institute of Metals,* **22**, 11–16, Pt 1.

[35] Hinami, M., Konoeda, S., Oiwa, T. and Inui, A. 1975. Development of a computerised system for predicting the progress of soaking in the soaking pit, *The Sumitomo Search,* **13**, 1–7, May.

[36] Maeda *et al.* 1977. Mathematical models for ingot processing and control Preprints Joint Automatic Control Conference, Paper FR28-4:30, pp. 1713–1720.

[37] Maeda, T., Nachtigal, C. L. and Cook, J. R. 1977. A mathematical model of an ingot soaking pit and its application to time optimum control of the heating cycle, Preprints Joint Automatic Control Conference, Paper FP28-4150, pp. 1721–1727.

[38] Goodman, T. R. 1958. The heat balance integral and its application to problems involving a change of phase. *Transactions of the ASME,* **80**, 335–342.

[39] Wick, H. J. 1981. Estimation of ingot temperatures in a soaking pit using an extended Kalman filter, Paper 88.4, Vol XVIII, IFAC 8th Triennial Conference, Kyoto, Japan.

[40] Mornas, J. P. and Planté, J. 1971. Gauge regulation for flat cold rolled products, *Trans. Iron and Steel Inst. Japan,* **11**(4), 785–786.

[41] Waltz, M. D. and Reed, L. E. 1971. Eccentricity filter for rolling mills, *Instrumentation Metals Industry,* **21**, 1–9.

[42] Grimble, M. J., Fuller, M. A. and Bryant, G. F. 1978. A non-circular arc roll force model for cold rolling, *Int. J. for Numerical Meth. in Eng.* **12**, 643–663.

[43] Wood, G. E. and Ivacheff, D. P. 1977. Mill modulus variation and hysteresis—their effect on hot strip mill AGC, *Iron and Steel Eng.,* **54**(1), 65–71, January.

[44] Edwards, W. J. 1978. Design of entry strip thickness controls for Tandem cold mills, *Automatica,* **14**, 429–441.

[45] Safanov, M. G. and Athans, M. 1978. Robustness and computational aspects of non-linear stochastic estimators and regulators, *IEEE Trans. on Aut. Contr.* **AC-23**(4), 717–725.

[46] Grimble, M. J. 1980. Reduced-order optimal controller for discrete-time stochastic systems, *IEE Proc.* **127**, Part D, No 2, 55–63.

[47] Grimble, M. J. 1980. A combined state and state estimate feedback solution to the ship positioning control problem, *Optimal Contr. Applics. and Methods,* **1**, 55–67.

[48] Golten, J. W. 1969. Analysis of cold rolling with particular reference to roll deformations, University of Wales, Swansea, Doctorial Dissertation, April.

[49] Grimble, M. J. 1976. A roll-force model for tinplate rolling, *GEC Journal of Science and Technology,* **43**(1), 3–12.

[50] Panuska, V. 1980. A new form of the extended Kalman Filter for parameter estimation in linear systems with correlated noise, *IEEE Trans. Autom. Control,* **AC-25**(2), 229–235, April.

[51] Marshall, J. E. 1979. *Control of Time—Delay Systems,* Peter Peregrinus Ltd. Stevenage.

[52] Grimble, M. J. 1979. Solution of the stochastic optimal control problem in the *s*-domain for systems with time delay, *Proc. IEE,* **126**(7), 697–704.

Index

Page numbers in bold face indicate a definitive source reference within the book for particular topics